Conservation of Tropical Plant Species

M.N. Normah • H.F. Chin
Barbara M. Reed
Editors

Conservation of Tropical Plant Species

Springer

Editors
M.N. Normah
Institute of Systems Biology
Universiti Kebangsaan Malaysia
Selangor, Malaysia

H.F. Chin
Department of Crop Science
Universiti Putra Malaysia
Serdang, Selangor
Malaysia

Barbara M. Reed
National Clonal Germplasm Repository
USDA-ARS, Corvallis
OR, USA

ISBN 978-1-4899-8724-2   ISBN 978-1-4614-3776-5 (eBook)
DOI 10.1007/978-1-4614-3776-5
Springer New York Heidelberg Dordrecht London

Printed on acid-free paper

Springer is part of Springer Science+Business Media (www.springer.com)

Preface

The richness of tropical plant diversity is widely recognized, and the full range of this diversity is often not well appreciated. This wide diversity includes wild relatives of existing crops or landraces and also many neglected or underutilized species having potential as food sources and medicinal or ornamental uses. The rapid loss of these plant species is also very much recognized as they are under threat from a rapidly increasing population pressure and from natural and man-made disasters. The importance of germplasm conservation is being increasingly realized with expanded effort to conserve tropical plant species by a wide range of countries and international agencies.

In situ and ex situ approaches are both needed for optimal conservation. Research on utilization, characterization of the germplasm, and development of conservation techniques is being carried out in order to fully safeguard the diversity and to obtain the best storage available for the collections. Information on these approaches is scattered in journals, book chapters, and technical reports from scientists in all corners of the world, but not readily available to all interested parties. Hence, we feel that it is timely to present these various conservation efforts all in one place. To this end, we approached researchers involved in conservation of tropical plant species with diverse viewpoints and from various locations to contribute to this book.

We hope to provide a review of the methods and current status of conservation of a range of tropical plant species. Plants included in this volume are from the major crops, fruit, oil palm, coconut, and forestry species. In addition, ornamentals, with a focus on orchids, and spices and medicinal plants are represented. This book also provides information on the richness of tropical plant diversity, the need to conserve, and the potential utilization of these genetic resources. Future perspectives of conservation of tropical species are discussed. Besides being useful to researchers and graduate students in the field, we hope to create a reference for a much wider audience interested in the conservation of tropical plant diversity.

M.N. Normah
H.F. Chin
Barbara M. Reed

Introduction

The range of plant diversity is overwhelming. There are over a quarter million known species of plants; many are not yet discovered, identified, or classified. Among these, over 30,000 are edible species. This biological diversity has important economic, technological, and social implications for everyone. Of particular significance are the economic benefits of plants for food security and environmental stability. This rich biological heritage is of great value for recreational, educational, and scientific use as well. Hence, conservation is vital, as many countries have undergone rapid development, often at the expense of losing forests and wetlands. This development often results in the loss or degradation of biodiversity and also endangers species, habitats, and ecosystems.

A recent estimation and calculation by Hammer (2006) of the world's plant genetic resources (PGR) is around 100,000 species. These are plants commonly used in association with agrobiodiversity under the well-known term of "biodiversity," the big umbrella classification of all living organisms. Spray and McGlothlin (2003) estimated there are 75–100 million species, while Hammer (2006) includes 1.75 million known species of animals, plants, fungi, and bacteria. Many species from resource-poor or Global Biodiversity Hotspots (GBHs) and island countries are on the brink of extinction due to the rapid loss of habitat. Often, these species are endemic, and only a few populations or individuals remain in the wild (Reed et al. 2011).

Many countries realize the loss of biodiversity, the pressing conservation issues, and the need for saving tropical biodiversity. A few prominent global assessments, such as the Millennium Ecosystem Assessment (MEA) and Global Biodiversity Outlook (GBO), emphasized that tropical biodiversity is continuously being assaulted by a number of threats and climate change further impacts our tropical biodiversity. Red List assessments are now available, but many countries involved have failed to reach the 2010 Biodiversity Target set up by the Conference of Parties of the Convention on Biological Diversity.

The term "biodiversity" was first coined by Norse and McManus (1980), while agrobiodiversity was first mentioned in 1994 and first published in 1999; both terms are widely used today. In the past, these terms were used mainly in scientific meetings and conferences; now people are more familiar with them and recognize their importance in their daily lives. The term "genetic resources" was first used at the International Biological Programs Conference and published in the conference proceedings in 1970. This was followed by the establishment of the International Board for Plant Genetic Resources (IBPGR), renamed Institute of Plant Genetic Resources (IPGRI), and now Bioversity International, based in Rome, Italy.

Plant genetic resources are conserved for many various reasons. Due to the increase in world population, it is necessary to find more crops from the wild species that may have potential as new food crops. The wild species can also serve as germplasm for the improvement of productivity of our present agricultural crops through conventional breeding or modern biotechnologies. Thus, we can sustain biodiversity and prevent the extinction of plant species. At the same time, we must attempt to improve the plants that we all depend on for our food, shelter, fiber, and medicine. Conservation of plant genetic resources is of utmost importance and is the key to our survival and that of future generations.

The Food and Agriculture Organization (2010) in their "Report of the State of Ex Situ Conservation" on an overview of genebanks reported that there are now 1,750 individual genebanks worldwide and 130 of them possess over 10,000 accessions each. In addition, there are 2,500 botanical gardens around the world that maintain numerous additional accessions. In the past 35 years, international genebanks under the auspices of the Consultative Group on International Agricultural Research (CGIAR) are held in trust for the world community. It is estimated that about 7.4 million accessions are currently maintained globally. Of these, a total of 1.9–2.2 million accessions are distinct, with the remainder being duplicates.

Conservation of seed was the first method of biodiversity preservation. Seed storage facilities are well established for orthodox seeds. Most of these facilities involve storage at low humidity and low temperature (–20 °C). The latest and, perhaps, also the oldest seed storage method is using natural permafrost; the Svalbard Global Seed Vault is located 130-m deep in a mountainside near the North Pole. Norway built this facility as a service to humanity; it presently holds 412,000 accessions and is the world's largest seed bank to preserve crop genetic diversity for future generations. However, there are many plant species not amenable to seed storage for conservation of their germplasm.

Many of the tropical underutilized species produce recalcitrant seeds that cannot be dried or cooled; hence, they cannot be stored in the normal seed banks like orthodox seeds that are resistant to drying and to cold temperatures. Some of these species are sterile or do not produce seeds annually as they do not flower regularly. Therefore, seed banks are not the answer for their conservation. Methods such as (a) field genebanks: arboreta, botanical gardens, and crop museums; (b) slow-growing seedlings; (c) pollen storage; and (d) in vitro genebanking (tissue culture and cryopreservation) of shoot tips, embryos, and other vegetative materials are being used for these species.

The plant species referred to in this volume are grouped as crops; they are important as food, fruits, fibers, medicinal plants, industrial crops, timber species, and ornamentals. The tropical regions where they are found are well endowed with a rich heritage of rainforest species, and a number of the countries are considered as centers of mega diversity. Information on the storage of tropical crops is widely scattered throughout the scientific literature, in research station reports, and in abstracts that are not easily available to the public. Consolidating this important information into one volume should make it more accessible both to scientists and to the public.

This book is designed to provide a review of the methods and current status of conservation of many tropical plant species. Future perspectives of conservation of tropical species will also be discussed. The section on "Conservation Methods" covers the range of conservation techniques, in situ; seed banking, in vitro; and cryopreservation. Chapters on collection and biomarkers are included as they are important aspects of conservation. The section on "Current Status" provides comprehensive information on various conservation efforts on tropical fruits, orchids, oil palm and coconut, legumes, root and tuber crops, cereals, forestry, vegetables, spices, and medicinal plants. In the section on "Future Perspectives," global and major research challenges and directions are discussed. We hope the volume will be valuable to researchers and graduate students in the field, and it will be a reference to a much wider audience who are interested in conservation of tropical plant diversity.

M.N. Normah
H.F. Chin
Barbara M. Reed

References

FAO (2010) Report on the state of the world's plant genetic resources for food and agriculture. FAO, Rome, pp 55–88

Hammer K (2006) Biodiversity, agrobiodiversity and plant genetic resources. In: Proceedings of the APEC genebank management workshop, Suwon, pp 13–20

Norse EA, McManus RE (1980) Ecology and living resources biological diversity. In: Environmental quality 1980: the eleventh annual report of the Council on Environmental Quality, pp 31–90

Reed B, Sarasan V, Kane M, Bunn E, Pence V (2011) Biodiversity conservation and conservation biotechnology tools. In Vitro Cell Dev Biol Plant 47(1):1–4

Spray SL, McGlothlin KL (eds) (2003) Loss of biodiversity. Rowman and Littlefield, Lanham

Book Notes

Conservation of Tropical Plant Species

Contents

Section I
Conservation Methods

Chapter 1
Conservation of Tropical Plant Genetic Resources: In Situ Approach

Ramanatha Rao and Bhuwon Sthapit

1.1 Introduction

We are all aware that the plant genetic resources (PGR) collections are assemblies of genotypes or populations representative of cultigens (landraces as well as advanced cultivars), genetic stocks, and related wild and weedy species and these may be conserved in the form of plants, seeds, tissue cultures etc. (Frankel and Soulé 1981). Landraces may contain coadapted gene complexes that have evolved over decades (Harlan 1992) and are the most important of the plant genetic resources. Advanced cultivars, genetic stocks and wild relatives of crop plants play an important role in crop improvement and therefore need to be preserved (Frankel 1990). Over the years there has been much loss of genetic diversity, but the remaining genetic diversity in the genepools still retains vast potential for present and future uses. Generally speaking, plant genetic resources are non-renewable resources and it is essential that these important resources are conserved and used, be it at species level, genepool level or at the ecosystem level. Countries that still have a significant amount of genetic diversity and species diversity have a responsibility to themselves as well as to the world at large to conserve it and make it available to for use.

Ramanatha Rao (✉)
Bioversity International, Rome, Italy

Bioversity International, Office for South Asia, New Delhi, India

ATREE, Bangalore, India
e-mail: vramanatharao@gmail.com

B. Sthapit
National Agricultural Science Centre, DPS Marg,
Pusa Campus, New Delhi 110012, India
e-mail: B.STHAPIT@cgiar.org

M.N. Normah et al. (eds.), *Conservation of Tropical Plant Species*,
DOI 10.1007/978-1-4614-3776-5_1, © Springer Science+Business Media New York 2013

To date, the available national and international plant genetic resources related agreements recognize the sovereignty of countries where these resources occur within their borders. However, the onus to conserve (using both ex situ and in situ approaches) and use them rests with countries and current agreements stress the importance of equable sharing of these resources and technologies related to their utilization. This chapter focuses on the in situ conservation approach.

The Convention of Biological Diversity (CBD) requires, under the Article 8, that the countries develop guidelines for selecting areas for in situ conservation, establishing protected areas, regulating the use of resources for sustainable use and protecting ecosystems and natural habitats. Countries also require promoting environmentally sound development, rehabilitating degraded lands and ecosystems, and controlling or eradicating exotic species that threaten the native species, ecosystems, and habitats. Countries also should ensure compatibility between conservation of biodiversity and sustainable use. They should also respect and preserve the knowledge, innovations and practices of indigenous and local communities. Additionally, countries are required to provide regulatory mechanisms for the protection of threatened species and populations. They need to regulate and manage relevant processes and categories of activities and provide financial and other support for in situ conservation. In the case of crop genetic resources, in situ conservation and on-farm conservation involve the maintenance of traditional crop cultivars (landraces) or farming systems by farmers within traditional agricultural systems (Frankel et al. 1995; Altieri and Merrick 1987; Brush 1991). This approach to conservation has been gaining importance in recent years, though farmers have been using it for centuries.

1.2 Definition of and Conceptual Basis for In situ and On-Farm Conservation

In situ conservation is concerned with maintaining species' populations in the habitats where they naturally occur, whether as uncultivated plant communities or in farmers' fields as parts of existing agro-ecosystems (Brush 1995; Bellon et al. 1997). It involves the conservation on-farm of local crop landraces with the active participation of farmers. Its goal is to encourage farmers to select and maintain local crop diversity for the benefit of humankind in general and for livelihood needs in particular.

In situ conservation means the conservation of ecosystems and natural habitats and the maintenance and recovery of viable populations of species in their natural surroundings and, in the case of domesticated or cultivated species, in the surroundings where they have developed their distinctive properties (CBD Article 2). It is the sustainable management of genetic diversity of locally developed crop varieties (landraces) with associated wild and weedy species or forms by farmers within traditional agricultural, horticultural or agri-silvicultural systems (Maxted et al. 1997). In other words, on-farm conservation of agricultural biodiversity refers to the maintenance of traditional crop varieties (landraces) or cropping systems by farmers within in the natural habitats where they occur-in farmers' fields and uncultivated plant communities (Altieri and Merrick 1987; Brush 1995).

In situ conservation is dynamic in contrast to the semi-static nature of ex situ conservation, and these approaches complement each other help to maintain much more genetic diversity than it would be possible otherwise. Due to recent awareness on biodiversity conservation, in situ *conservation* has been generally given higher priority over ex situ. This is mainly because of its ability to maintain the evolutionary potential of species and populations (Hodgkin 1993; IPGRI 1994, 1996; Sthapit and Joshi 1996; Jarvis 1999; Sthapit and Jarvis 1999a; Jarvis et al. 2000b, 2011), and because it helps increase the access to and control of local communities over their genetic resources. However, given the fact that human activities can cause habitat destruction and loss of biodiversity in some cases, it will be necessary to complement it with ex situ conservation.

1.3 Role of In Situ Conservation

Despite the implementation of various ex situ and in situ agrobiodiversity conservation projects, the efforts to improve farmers' access to germplasm and associated information within communities have been limited. Interventions such as seed/biodiversity fairs could improve access to information and germplasm within and between communities (Grum et al. 2003; Adhikari et al. 2005). Community seed bank can help to improve the access to traditional crop varieties by communities (Shrestha et al. 2006). Small but significant efforts such as distribution of small quantities of germplasm seeds can lead the concept of informal research and development (diversity kits) and can be effective in both remote and accessible areas in terms of adoption of varieties (Joshi et al. 1997). Lessons learned from various informal research activities is that there must be hundreds of unique and useful local crop diversity that can be assessed, multiplied and distributed to farmers and communities and provide direct benefits to the farming community. Many roles of the community seed bank include characterizing the germplasm, multiplying healthy seed, and selling to communities, variety selection, and plant breeding (Shrestha et al. 2006). However, these methodologies have been evolving over time. This kind of plant breeding that farmers were doing in the past could be done by grassroots institutions (Sthapit and Ramanatha Rao 2009). All these efforts can improve the access to materials by farmers as well as improve the germplasm on-farm. On-farm conservation can also play a role in other aspects of the ecosystem (such as ecosystem health, services and functions) and in socioeconomics of communities that are involved in such conservation efforts.

We know now that the in situ or on-farm conservation of agrobiodiversity helps not only to conserve the genetic diversity in target plant species, but also continues the evolutionary processes that selectively improve genetic diversity and the ecosystems that host the genetic diversity. In addition, on-farm conservation can play a role in other aspects of the ecosystem (such as ecosystem health, services and functions) and in socioeconomics of communities that are involved in such conservation efforts. Some areas in which on-farm conservation has a role to play include (Ramanatha Rao et al. 2000):

1. Conservation of the processes of evolution and adaptation
2. Conservation of diversity at all levels (ecosystems, species, intra-specific)
3. Integrating farmers/communities into national plant genetic resources conservation systems
4. Contribution to ecosystem services and ecosystem health
5. Maintaining the process of local crop development by strengthening capacity of farming communities in landrace assessment and the selection and exchange of crop germplasm
6. Improving the livelihoods and quality of life of farmers
7. Empowering farmers and communities on their crop genetic resources and improving access to them
8. Providing information for national seed policy decisions regarding importance traditional seed supply system
9. Involving a component of complementary conservation strategy-linking farmers to genebank

Conceptually, a well organized in situ conservation programme should be able to accomplish several things, including:

- Nurturing responsibility and improving one's ability to respond to adverse conditions.
- Promoting leadership qualities among the active participants
- Empowering those at the bottom of the pyramid
- Promoting the ability to work as a team and capacity to manage teams
- Encouraging participants to be aware of their social responsibilities and Trusteeship
- Consolidating community role in community based management of agricultural biodiversity as a proxy method to realize on-farm conservation.

1.4 Factors that Shape Crop Genetic Diversity

Before an in situ conservation programme is devised, the type of genetic diversity that is being conserved must be understood. Crop plant genetic diversity in agricultural systems, in addition to being affected by population dynamics (e.g., mutation rates, migration, population size, isolation, breeding systems and genetic drift) and natural selection arising from conditions in the surrounding environment (e.g., soil type, climate, disease, competition), is affected by human selection and management. Often, plant genetic resources are passed on from generation to generation in

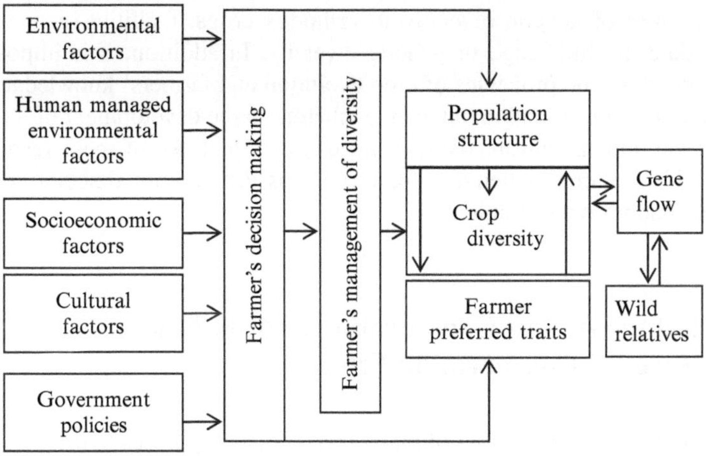

Fig. 1.1 A conceptual model of factors influencing farm's management of crop diversity (Source: Bellon et al. 1997)

families and are subject to different natural pressures and human pressures. Environmental, biological, cultural and socio-economic factors influence a farmer's decision whether to select or maintain local crop diversity at any given time (Jarvis et al. 1998). In the process of planting, managing, selecting, rouging, harvesting and processing, farmers make decisions on their crops that affect the genetic diversity of the crop populations. Strengthening farmer seed system is most likely the best way to support on-farm conservation of agricultural biodiversity. Over time, farmers may alter the genetic structure of a crop population by continuously selecting plants with preferred agro-morphological or quality characteristics. Thus, most crop plant landraces are products of farmer selection as well as farmer breeding (Riley 1995).

Farmers may also influence the survival of certain local varieties by choosing a particular farm management practice or by planting a crop population in a site with a particular microenvironment. Farmers make decisions on the size of the population of each crop variety to plant each year, the percentage of seed or planting materials to be saved from their own stock, and the percentage to be bought or exchanged from other sources. Each of these decisions, which can affect the genetic diversity of cultivars over time, is linked to a complex set of environmental and socio-economic influences on the farmer (Fig. 1.1).

There are growing pressures on farmers who maintain significant crop genetic diversity in the form of local cultivars. Such pressures, such as increased human population, poverty, land degradation, environmental change, government subsidies to plantation crops and the introduction of modern crop varieties, have contributed to the erosion of crop genetic resources.

In recent decades, agricultural scientists have responded to the threat of genetic erosion by developing a worldwide network of genebanks and botanical gardens for conserving available genetic resources ex situ. While this has been the main strategy against the loss of genetic diversity in orthodox crops, facilities are unlikely to accommodate the full range of genetic diversity. In addition, it is impossible to conserve the dynamic processes of crop evolution and farmers' knowledge of crop selection, management and maintenance inherent in the development of local cultivars. Nor can anyone ensure the continued access and use of these resources by farmers. This is more challenging for farmers as the traditional social network is weakening and growing smaller.

1.5 Need for In Situ Conservation and Its Role in Overall Conservation Efforts

The potential threat that the loss of plant genetic diversity poses to the world's food supply has already been recognised for many years and has led to the ex situ storage of genetic materials in genebanks (Bommer 1991). Whilst this form of conservation remains no doubt a useful method, it has major drawbacks in terms of effectiveness and extensiveness. Firstly, as noted earlier, ex situ genebank freezes the natural evolutionary process (Altieri and Merrick 1988). Secondly, ex situ collections are more vulnerable to mismanagement and the transmission of seed-borne pathogens (Wood 1993). The major advantage of in situ conservation over ex situ approach is its capacity to store large numbers of alleles and genotypes in crop populations conserved on farm with continuous inputs of farmers and communities.

The added advantage of in situ conservation is the process of adaptation that is consequent upon continuing evolution of the material conserved on-farm. It allows for the evolution of ecosystems in a landscape responding to climatic changes and the dynamics of biotic and abiotic stresses. It is important to understand why we want to conserve genetic diversity on-farm, as in situ conservation has the potential to:

1. Conserve the processes of local adaptation of crops to their environments
2. Conserve diversity at all levels – within the ecosystem, between and among the species
3. Improve the livelihood of farmers
4. Maintain or increase farmer control of and access to genetic resources
5. Integrate farmers into the national PGR system and involve farmers directly in the value-adding process
6. Link farming communities to the genebank for conservation and utilisation

The importance of conservation of agrobiodiversity for future global food security lies in its potential to supply the required germplasm to crop breeders and other users. Lastly, on farm conservation is a powerful instrument to allow the implementation of benefit sharing. It has been recommended by the CBD that recognized the continued maintenance of traditional varieties on-farm as an essential component of sustainable agricultural development. Effective in situ/on-farm conservation contributes significantly to this main goal of agrobiodiversity conservation.

1.6 Local and Global Relevance of In Situ Conservation

The soaring food prices and global warming have brought food security and climate change concerns to the top of the international agenda (FAO 2010). At the local level local crops and farmers' varieties are the main source of household level food security in food insecure areas (Baniya et al. 2003). For example, there are 100 million farms in India alone (Swaminathan 1998), out of which only 15–20 % use seeds from the regular seed trade. The remaining 80 million farms depend on seed supply from other farmers through the informal sector. These figures are similar in most developing countries (Cromwell 2000; Tripp 2001; Jarvis et al. 2004) and these farms with an informal seed supply constitute the sector where contribution to on-farm management on local crop diversity is significant for ensuring food security and sustainable livelihoods. There is the need to focus on the farmers, who are the main custodians of many neglected and under-researched crops. Supporting capacity of farmers and rural institutions in strengthening technical components of farmer's seed system (germplasm base, production, quality, knowledge and regulatory regime) is what could be achieved through in situ conservation programme. This is very relevant to develop community resilience in the context of food security and climate change impacts in developing countries (Sthapit and Ramanatha Rao 2009; Sthapit et al. 2010).

In situ conservation is an important component of the conservation and management of genetic resources. It supplements ex situ conservation efforts of local, national, and international agencies and provides some important advantages. Most agencies dealing with plant genetic resources conservation have been facing the dilemma of how to implement in practical terms in situ conservation of agricultural biodiversity. Sthapit et al. (2010) pointed out the major challenges that include:

1. Lack of a clear understanding of the scientific basis of in situ conservation of agricultural biodiversity and how it can be practically implemented on the ground,
2. Difficulty in changing the mindset of current PGR institutional set up to work closely with farmers and communities,
3. Rationale of identifying the least cost conservation areas and dealing with inspiration of local communities,
4. Difficulties in identifying sustainable incentive mechanisms to support on farm conservation,
5. Obstacles faced when trying to canvass policy support to empower the communities in diversity rich areas for community based management of agricultural biodiversity.

What makes these challenges particularly complex is the fact that they are highly interlinked and dependent upon a mix of socio-cultural, economic and political factors. On-farm conservation is not a purely technical intervention (as is the case in ex situ conservation) but a much more complex social and collective-action type of endeavour.

At the local level, on-farm conservation is the by-product of on-farm production activities that ensure household level food and nutrition security. This means that on-farm conservation efforts must be carried out within the framework of farmers' livelihood needs. For those reasons, the mobilization of support to on farm conservation needs to be conceived and designed within the broader objective of creating a more enabling environment for agricultural development in its various aspects (Sthapit et al 2008). Individual farmers search, select and keep their own locally adapted seeds and breeding stocks as a traditional practice of on-farm management of agricultural biodiversity (Sthapit and Jarvis 1999b). But such sociality embedded good practice of exchange such as gifts, barter and sales that spread biodiversity across communities are weakening fast. In the context of climate change, communities with strong social seed networks are better equipped to cope with the effect of climate change compared to communities with weak and disturbed social seed networks (Subedi et al. 2003; Poudel et al. 2008).

These practices are increasingly facing challenges because of commercialisation supported by a strong policy support and regulatory barriers. Also, the management of ecosystems as a strategy for in situ conservation of fruit and fodder trees is underappreciated.

1.7 Impact of Climate Change on In Situ Conservation

1.7.1 Climate Change, Agriculture and Agricultural Biodiversity

The implications of climate change for agriculture are still a bit vague and mostly based on modelling and estimations. The actual impact is not very clear and is expected to vary from area to area and region to region. However, even what little that we know indicates that there will be major reduction in food supply (Rosenzweig and Parry 1994; Hijmans 2003; Hijmans and Graham 2006; Jones and Thornton 2003). This reduction is the main reason for a discussion on the future role of agricultural biodiversity in agriculture (Kotschi 2007) under climate change. Further, the impact of climate change on conservation and use of agricultural biodiversity is still vague and anything we can say on in situ conservation is only a scientific guess at this stage.

There is evidence that climate change is already affecting biodiversity and will continue to do so. The Millennium Ecosystem Assessment ranks climate change among the main drivers affecting ecosystems and biodiversity. Consequences of climate change on the species component of biodiversity include:

- Changes in distribution
- Increased extinction rates
- Changes in reproduction timings, and
- Changes in length of growing seasons for plants
- Changes in plant community composition
- Changes in ecosystems

These changes will result in significant changes in current farming practices and genetic resources. The available evidence is still being debated; however, most researchers working in the area of climate change agree that there will be drastic changes in available water supply in different regions of the globe. This will have major effect on agricultural systems as we know as well as on total productivity. Current information available indicates that sub-tropical regions received less precipitation and were subjected to more frequent droughts, while the northern hemisphere received higher rainfall in recent past. Nevertheless, recent research suggests this trend is less predictable but the degree of variation will be more pronounced (IPCC 2001, 2002). All of these will have serious consequences on agricultural practices, crop improvement and agricultural biodiversity. However, we will focus here on the impact of climate change on in situ conservation of agrobiodiversity.

Enhancing resilience to the effects of climate change is important for all these systems and functional diversity, particularly in tolerance traits for abiotic and biotic stress, is one of the most effective targets for improved sustainability (Newton et al. 2011). Newton and his colleagues also report that the effects of disease on crops may be large but climate change will also have direct effects, complicating analyses. Diversity to give more options and build spatial and temporal heterogeneity into the cropping system will enhance resilience to both abiotic and biotic stress challenges (Newton et al. 2009, 2011). Thus, in the context of climate change induced variation in increased pest damage, diversity of agricultural crops in ecosystem maintained through on-farm conservation has much relevance.

1.7.2 Climate Change and In Situ *Conservation of Agrobiodiversity*

Earlier we have seen that in situ conservation of agricultural biodiversity is defined as the management of a diverse set of crop populations by the farmers in the ecosystem where the crop evolved. It allows the maintenance of the processes of evolution and adaptation of crops to their environment. We also use this term for managing useful plants and crop wild relatives in the wild. Various climate change predictions made it is clear that many regions around the globe are going to change in various ways. Thus, a good question to ask is how these various changes will affect different in situ conservation efforts of landraces and wild species. Although ecosystems have adapted to changing conditions in the past, current changes are occurring at rates not seen historically. In general, the faster the climate changes, the greater the impact on people and ecosystems. Reductions in greenhouse gas emissions can lessen these pressures, giving these systems more time to adapt (CBD 2007). In addition to mitigation, there is an urgent need to develop and implement climate change adaptation plans. There is a significant research gap in understanding the genetic capacity to adapt to climate change. An examination of available literature indicates that while a broad range of studies examine the generic impacts of climate change on crop productivity, few studies examine varietal level changes in adaptation (Jarvis et al. 2008a). An attempt was made by one of the authors to elucidate the impact of climate change on in situ/on farm conservation (Ramanatha Rao 2009) and much of the following discussion is drawn from it.

1.7.2.1 Changes in Range and Size of Species Distribution

Climate is one of the major factors governing the distribution of wild plant species and cultivation of crops. It impacts physiological and reproductive processes and influences ecological factors such as competition for resources (Shao and Halpin 1995). There have been many recorded cases of climatic changes having significant impacts on the distribution, abundance, phenology and physiology of a wide range of species, It is now possible to apply species distribution models and predict range shifts and assess extinction risks due to climate change (Walther et al. 2002; Parmesan and Yohe 2003; Root et al. 2003; Parmesan 2006; Thomas et al. 2004; IPCC 2007; Araújo and Rahbek 2006; Hijmans and Graham 2006; Howden et al. 2007; Lawler et al. 2006) that can lead to better understanding if migration rates are known for particular plant species (Menendez et al. 2006; Midgley et al. 2006).

Jarvis and his colleagues (2008b) used current and projected future climate data for ~2055, and a climate envelope species distribution model to predict the impact of climate change on the wild relatives of groundnut (*Arachis*), potato (*Solanum*) and cowpea (*Vigna*). They report that wild groundnut were the most affected group, with 24 to 31 (depending on the migration scenario) of 51 species projected to go extinct and their distribution area on average reduced by 85–94 %, depending on the migration scenario over the next 50 years. In terms of species extinction, *Vigna* was the least affected of the three groups crops studied. Their results suggest that there is an urgent need to identify and effectively conserve crop wild relatives that are at risk from climate change. While increased habitat conservation will be important to conserve most species, those that are predicted to undergo strong range size reductions should be a priority for collecting and inclusion in genebanks (Jarvis et al. 2008b). We need similar studies of crop wild relatives in other countries. An additional factor that may have to be used for most such studies is to taking into consideration the capacity to adapt to changed conditions. Can the species that are shown to be at risk adapt fast to changing climatic conditions or have they really run out of their time? This is one of the right questions to ask even through the answer may not be within our grasp for a while. Diversity conserved at the in situ areas will be interesting to monitor as rich biodiversity is expected to buffer much better against unpredictable temperature and precipitation change than do the areas with increased uniform farming system (Ramanatha Rao 2009).

1.7.2.2 Protected Areas (PA)

While considering the in situ conservation of useful wild plants and crop wild relatives it is important to consider the effects of climate change on protected areas. Although, as noted earlier, there is little empirical data, it can safely be assumed that significant amount of species and genetic diversity of related agricultural biodiversity occurs in protected areas (need survey and determination of distribution). Thus, the mitigation of negative effects of climate change on protected areas will help in conserving valuable agricultural biodiversity that is present in them. Hannah et al. (2007)

studied the range shifts due to climate change and species range dynamics that reduce the relevance of current fixed protected areas in future conservation strategies. They applied species distribution modelling and conservation planning tools in three regions (Mexico, the Cape Floristic Region of South Africa, and Western Europe) to examine the need for additional protected areas in light of anticipated species range shifts caused by climate change. Their findings indicate that protected areas can be an important conservation strategy in such a scenario and that early action may be both more effective and less costly than inaction or delayed action. According to their projections, costs may vary among regions and none of the three areas studied will fully meet all conservation targets, even under a moderate climate change scenario. This suggests that limiting climate change is an essential complement to adding protected areas for conservation of biodiversity. We need more studies on these lines for making appropriate conservation decisions.

1.7.2.3 More Work Is Needed

At the same time, it is important to note that key risks associated with projected climate trends for the twenty-first century include the prospects of future climate states unlike the current states (novel states) and the disappearance of some extant climates. Williams et al. (2007) conclude that there is a close correspondence between regions with globally disappearing climates and previously identified biodiversity hotspots. For these regions, standard conservation solutions (e.g., assisted migration and networked reserves) may be insufficient to preserve biodiversity. By extrapolation, we can assume that this applies to agricultural biodiversity found in areas affected by climate change. This further strengthens earlier statement that there is as large gap in research to make correct conservation decisions.

1.8 Roles and Responsibilities of Different Stakeholders in In Situ Conservation

The Convention of Biological Diversity (CBD 1992) assigned the responsibility of in situ conservation to nations. in situ conservation has been recognized for the conservation and sustainable use of agricultural biodiversity in several international conventions and agreements, including the CBD, the Global Plan of Action (GPA) of the FAO. Each of these instruments not only recognizes the countries' responsibilities to conserve and use their PGRFA, but recommends the importance of equitable sharing of the benefits derived from the use of resources and technologies. The international community has also recognized the critical role of local institutions of genetic resources whether they are identified as farmers, indigenous and local communities, notably in the preamble to the CBD, which has been ratified by 181 countries. The contribution of the CGIAR is mainly in identifying biodiversity hotspots and in developing methods and best practices for in situ conservation. In many

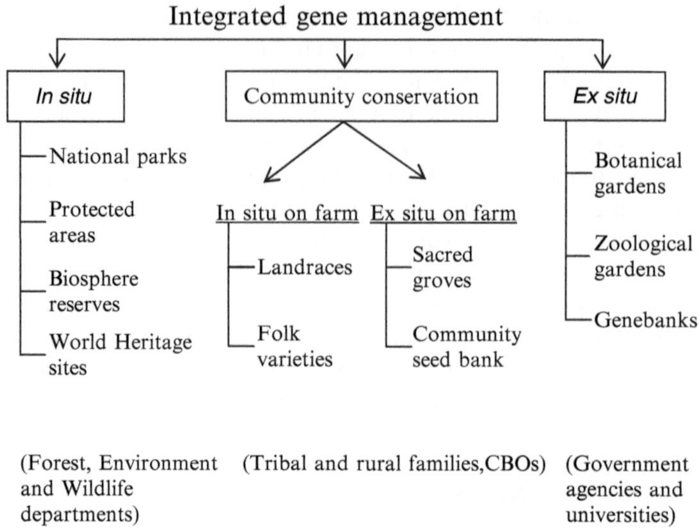

Fig. 1.2 Role of community conservation in integrated gene management

countries NPGR centres who is responsible for plant genetic resource is confused with their role on management of on-farm conservation (community conservation) with the traditional in situ conservation of natural ecosystems or ex situ conservation (gene banks and botanical gardens) (Fig. 1.2). It is essential to change the mind-set of institutions to consolidate the role of local communities of diversity rich areas as innovators, users, custodians and conservers. This requires PGR institutions to cultivate and innovate partnerships with farmers, local institutions and other service providers as a precondition to empower farmers and develop incentive mechanisms so that farmer and their institution can continue community-based management of local biodiversity for community benefits.

1.9 Feasibility and Context

The last decade has seen an increase in the use of participatory approaches and multistakeholder teams implementing on farm conservation projects (FAO 2010). A survey carried out by FAO in recent years illustrates that the role of in situ approach is increasingly appreciated as complementing ex situ conservation. Hence, it is important to understand how farmers value local crop diversity and how much they are willing to pay for such genetic resources. Agricultural biodiversity provides many products and services of environmental, economic, social and cultural importance; these environmental products and services contribute to sustainable economy in a number of ways.

On-farm conservation will be the least costly in sites that are most highly ranked in terms of public benefits (richness of genetic diversity) and where the private benefits those farmers obtain from growing genetically diverse varieties is greatest. The economic concept that farmers' varieties embody both (1) 'private' value in the harvest the farmer enjoys, either directly as food or feed, or indirectly through the cash obtained by selling the seed/grain and purchasing other items, and (2) 'public' value in its contribution to the genetic diversity from which future generations of farmers and consumers will also benefit (Smale et al. 2004). Those crop genetic resources, which have both low farmer utility (current private value) and public value, will be difficult to conserve on-farm unless public interventions are made for additional benefits. One of the often-cited disadvantages of on-farm conservation is the difficulty of accessing the material conserved. This is mainly because the on-farm conservation efforts to date have not been mainstreamed and are not linked to national PGR efforts.

1.10 Benefits to Community

Cultivation of a particular crop or crop cultivar by farmers solely depends on the benefits they derive from it. The benefit may be in the form of subsistence or livelihood, cash income, cultural uses, pride, adaptation to particular abiotic or biotic stresses, etc. At the same time, needs and preferences of farmers may change over time, leading to cessation of cultivation of particular crop species or a cultivar. So, if we aim to conserve a particular crop or cultivar, it is necessary to enhance the benefits for farmers from local crop diversity. This essentially means that the farmers would continue to cultivate the same crop or cultivar and at the same time benefit from it in terms of economics, socio-cultural or ecological value. Enhancing benefits could be achieved through several means, some of which are listed below (Ramanatha Rao et al. 2000).

1. Increasing crop genetic diversity's competitiveness for farmers – i.e. plant genetic diversity available with farmers should be able to compete with other commercial cultivars or provide better options.
2. Improving the crop itself through participatory plant breeding (Sthapit and Jarvis 1999b), strengthening farmers' seed management (Gyawali et al. 2010) and improving the agroecosystem health.
3. Improved farmers' access to genetic materials. Attempts are underway to do this in some crops through maintaining Community Biodiversity Registers (CBR), seed exchange networks, linking farmers' seed/seedling supply systems to the formal sector, incorporating local crop resources into agricultural extension packages and by organizing diversity fairs.
4. Increasing consumer demand. Demand can be increased by adding value through processing, diversifying the product base, creating respect for the environment and organic farming, increasing public awareness, changing policy, linking to eco-tourism and agro-tourism, etc.

1.11 Implementing Community Driven In Situ Conservation Programme

It has been argued that farmers have been conserving plant genetic resources over centuries. However, as the need exists for integrating rural areas into national and global development efforts, interventions are required to promote on-farm conservation practices by farmers to ensure that they benefit from it. Several suggestions as to how this can be achieved have already been made (Altieri and Merrick 1987; Brush 1993, 2000; Bellon et al. 1997; Sthapit and Joshi 1996; Sthapit et al. 2008). On-going research at Bioversity International has shown that it is possible to conserve much of the genetic diversity on farm by assisting the farmers and communities in doing so while ensuring that they benefit from it. In situ (on-farm) conservation can play a significant role in empowering the local, indigenous, and rural communities to treasure and control their PGR, thereby putting the FAO's concept of Farmers' Rights in practice. Bioversity International has developed a guide for establishing an in situ/on-farm conservation research programme (Jarvis et al. 2000b, 2011). However, much of the discussion therein will be useful for establishing actual in situ conservation programmes as well.

To establish an in situ conservation programme for agricultural crops we will need to consider the following broad requirements:

1.11.1 Institutional Framework

Although farmers have been practising in situ conservation informally for a long time, a focused programme that can bring it within the framework of national plant genetic resources conservation programme would require institutionalising the effort at local levels. Relevant organizations that are committed to in situ conservation need to be identified and guidelines for their joint participation should be developed and implemented. Roles of local and national institutions are central to enhance the local management decisions for agro biodiversity conservation and utilization. The role and responsibilities of individual organizations need to be clearly identified from the very beginning (though these could change during the period of the programme) and appropriate linkages for working together will be developed.

Implementation of conservation on-farm requires multi-institutional, multidisciplinary frameworks at central and local levels. Thus, before even the selection of sites, multidisciplinary, multi-institutional and multi-level linkages must first be developed and established by the national partners at the central and local levels. National personnel are then trained to carry out the work at those levels, and the teams are required to promote equity at all project levels, from farmer participation to research to project management and decision making. In addition, as the approach is largely community based, much time is devoted to building or creating rapport with the farmers in whose field much of the work is being undertaken, and whose experiences is an integral part of the project.

Once understanding between institutions, collaborators and farming communities has been reached, the actual effort of on-farm conservation can begin. Preparation, site selection, sampling and participatory approaches to on-farm management of agrobiodiversity are the initial actions needed. Prior to site selection, the existing data such as descriptor information, databases of ex situ collections, herbarium collections, published literature in the natural and social sciences and other unpublished information should be collected and used for eliminating inappropriate sites. Personal knowledge of experts, including personnel from NGOs, CBOs, and others existing local institutions would be most valuable.

1.11.2 Information Needs for Implementing In Situ *Conservation*

As in any conservation effort, information is a major key for success of the efforts. Information on the following four main research questions will provide a scientific basis for designing and planning effective on-farm conservation:

- What is the extent and distribution of the genetic diversity maintained by farmers over space and over time?
- What are the processes used to maintain the genetic diversity on-farm?
- Who maintains genetic diversity within farming communities (men, women, young, old, rich, poor, certain ethnic groups) and how?
- What factors influence farmer decisions on maintaining traditional varieties: market, non-market, social, and environmental?

Answering these four questions together with strengthening formal and informal capacity and links will enable the national programme to support farmers in the conservation and use of crop genetic resources. It will enhance the social, economic, ecological and genetic benefits from local crop resources to farmers and other stakeholders.

So as to fully develop the context in which in situ conservation would be successful, one should try to understand why, when and where farmers grow landraces and how they maintain and use them. In order to be an effective in situ conservation programme in a system wide perspective, the above key questions as well as the gaps need to be thoroughly reviewed. These questions could be examined during the group discussion session with farmers/communities and find out what activities could be carried out. It is important to note that this in situ approach can only succeed in some contexts and not in others.

1.11.3 Criteria for Selecting In Situ *Conservation Sites*

A fundamental problem faced by any in situ conservation effort is locating crop populations to focus on and identifying locally driven methods to support maintenance of local crop diversity. It is essential to consider some generalized criteria for selection of sites: ecosystems, intra-specific diversity of target species, species adaptation, genetic erosion, diverse use values, and interests of farming community, partners and government agencies and logistics for monitoring (Jarvis et al. 2000b). Depending upon the available resources and government commitment, selection of on-farm conservation sites should consider two broad guidelines:

1. Identification of the least cost conservation site.
2. Potentiality of "win-win" situation in terms of livelihood gains and ecological costs for the site.

The criteria for site and farmer selection have to be well defined. Broadly speaking, the criteria would be based on the genetic diversity, accessibility and interest of the farmers to continue to grow the varieties that are being targeted and these will have to be evaluated through a survey (Ramanatha Rao and Sthapit 2002). Some generalized criteria that could be used for developing an on-farm conservation programme could include:

1. Ecosystems: It will be important to select sites in diverse agroecosystems preferably with different ecotypes. This will increase the chances of conserving genetic diversity, as this may be associated with agroecosystem diversity.
2. Intra-specific diversity within target species: It is important that the areas selected are grown to different landraces.
3. Specific adaptations: Efforts should be made while selecting different agroecosystems (see 1 above) such as sites with extreme environmental conditions (high soil salinity, cold temperatures, etc.) and variation in pests. This will help to include types with specific adaptations.
4. Genetic erosion: It is better to select sites with less threat of genetic erosion to increase the life of conservation efforts.
5. Diverse use values: It is possible to ensure conservation of hidden genetic diversity by selecting sites with diverse use values of crops for food and other uses. It is important to note that for many farming communities, a crop is not just a matter of food production but also an investment and is important in maintaining social relations and religious rituals.
6. Farmers and communities: Farmer's interest and willingness to participate are keys in site selection. This may require preliminary work in community sensitization on the benefits to farmers of conserving crop varieties. Site selection should also include sites with: socio-cultural and economic diversity; diversity of livelihoods; cultural or economic important target crops for various ways of life; farmers' knowledge and skills in seed selection and exchange; and market opportunities
7. Partners: Partners with interest in the community and who have experience in conservation interventions will be beneficial to the programme. Partners with distinct community participation expertise will have comparative advantage in dealing with community.

8. Logistics: These would include mainly the accessibility of the site throughout the year (in situ conservation monitoring is essential) and availability of resources.

The existing data should be combined with an exploratory survey, using a Rapid Rural Appraisal (RRA), Participatory Rural Appraisal (PRA), or a similar approach. The community needs to be sensitised to issues on hand and for this use of participatory approach is recommended (Ramanatha Rao et al. 2000; Friis-Hansen and Sthapit 2000; Sthapit et al. 2006).

1.11.4 On-Farm Diversity Assessment

Genetic diversity is central to conservation and utilization of plant genetic resources. To do so effectively, it is important to have some understanding of the distribution of genetic diversity and its use value. Presently little information is available on the status of genetic diversity on farms. In this context, crop history (origin etc.) can help to some extent. However, another source of information on genetic variation and uses of crop plants is the farmers' knowledge/traditional knowledge on propagation and exchange of germplasm within and between communities. If time and resources permit, other tools of genetic diversity determination can be made, but we feel that it is not particularly needed as the goal of genetic diversity assessment in the context of in situ conservation needs to focus on genotypic diversity and not just allelic diversity. In addition, it has also needs to integrate the adaptive variation that can be better assessed from the farmers' experiences.

Conservation sites can first be established with the criteria mentioned above together with traditional knowledge of crop diversity and then genetic diversity within the sites can be determined. Information on genetic diversity can be used to rationalize the number of sites. It should, however, be noted that currently only morphometric methods are used to locate and characterize diversity. It is now possible to monitor and estimate genetic diversity using molecular markers, but actual use of these in countries where in situ conservation may be planned may take some more time. This may be further complicated as only limited resources are available for such work and thus it would be imperative to do the work in collaboration with other resource rich countries. We recommend that project leaders begin to immediately systematically document traditional knowledge on crop diversity. This information as well as other historical information can help guide them to sites where they wish to invest resources in molecular measurements of genetic diversity. Experiences of IPGRI's global projects in Nepal and Vietnam suggest that community based organisations could be mobilized to locate and monitor crop plant diversity through participatory methods such as diversity fairs and biodiversity register (Subedi et al. 2008). Based upon these information molecular markers can be used to estimate genetic diversity from diversity rich regions.

1.11.5 Sensitizing and Strengthening Local Community

Identification of grassroot institutions with objectives that are related to on-farm conservation is essential. The capacity of local organisations to implement the programme should also be strengthened. This is a fundamental requirement as the strategy for on-farm conservation will only succeed if indigenous communities and grassroot organisations are involved at different stages and their needs and problems are understood and addressed. This requires better understanding of what they do, how they do it and why they do it. Scientists working in this field often tend to ignore this important step while establishing framework for collaboration and partnership (Sthapit and Jarvis 1999a), leading to failure of the efforts. Firstly, it is essential to develop an appropriate understanding of the extent and distribution of diversity in a system and how it is maintained through local institutions and practices. Secondly, the analysis is likely to lead to the identification of a number of complementary supporting actions. Thirdly, the success of any actions will depend centrally on local knowledge, the strength of local institutions and the leadership of farmers and communities.

To achieve this community sensitization is essential to understanding farmer management and developing local strategies for the conservation and sustainable utilization of biodiversity. It raises awareness among farming communities who learn about the value of local crop diversity. It strengthens community-based organizations' capacity and shifts behaviour towards conservation and diversity. Farmers' sense of pride in their cultural heritage is fostered as well. Various tools can be used, including the biodiversity and seed fair, the teej geet (folk song) competition, the rural poetry journey, the traditional food fair and rural roadside drama. These tools are people friendly and effective in giving rural people access to the required information. The choice of tools depends upon the cultural context of the community.

1.11.6 *Collaboration and Sharing Benefits from* In Situ *Conservation Efforts*

Several different organizations and people with different backgrounds will need to come together and work for successful in situ conservation. There is a need for multidisciplinary and multi-institutional teams as diverse expertise is required to carryout of-farm conservation programme. Such teams need to be well coordinated through effective communication, networking and participatory approaches. Any hurdles to collaboration need to identify ahead of time remedial measures that need to be taken, so that the work goes on smoothly.

In order to have institutional understanding for the people from different organizations and different background to work together in teams, agreements such as Memorandum of Understanding (MOU) may have to be formalised. Close rapport between teams and rural people has to be built up to enhance collaboration with farmers and farming communities. Building the rapport and making collaborative arrangements with farming communities is essential for the success of in situ conservation programme and could be done by developing a framework for collaboration. The participation of farmers is crucial in setting the goal as they know their needs better in their local varieties. There can be four different categories of farmer participation: contractual, consultative, collaborative and collegial (Biggs 1988). As the decision-making capacity of local institutions improves, the quality of participation is enhanced from collaboration to collegial participation. Linking on-farm conservation of PGR with various market outlets and incentives is essential so that the farmers see the value of conservation.

On-farm conservation initiatives should also promote equity at all project levels, from farmer participation to research to project management and decision-making. Equitable gender treatment and ethnic and other minority involvement is the key for ensuring representative partnerships and benefit sharing. They should be included as members of research and management teams. Increased women, minority, and farmer participation in decision-making is essential to ensure that diverse perspectives are incorporated into project objectives and that all stakeholders feel ownership in the project (Ramanatha Rao and Sthapit 2002).

1.12 Conclusions

There are a few critical questions that need some discussion in the context of each partner country. There is also a need to analyse the pros and cons of methods and options: Who should be involved? What is the role of the local community? What are the mechanisms for collaboration and resource allocation between partners? What are the long-term and short-term threats to the in situ conservation? How do benefits of on-farm conservation address the needs of the local community? How can we link on-farm conservation to value-adding activities? How can a farming community be integrated into the national PGR system?

A challenge is to develop the framework of knowledge to determine where, when and how in situ conservation will be effective, and to develop broad guidelines for research and practice in situ conservation for national programmes that address the needs of the farming community as well. Another major challenge is to integrate the conservation of plant genetic resources with agricultural development, and in particular to conserve as much diversity as possible and nourish the processes that give birth to it.

In situ conservation or on-farm conservation of crop genetic diversity is an essential component of an integrated approach to conservation of germplasm. Efforts to collect, conserve and use crop genetic diversity are in progress in many countries. As noted in the FAO's survey, currently ex situ approach is the main focus of many national programmes (Kar-Ling Tao and Murthi Anishetty 2001; FAO 2010). However, the role of in situ approach is being appreciated more and more, as together the two approaches – in situ and ex situ – would help us effectively conserve and access maximum crop genetic diversity. To do this successfully, we need to understand how farmers value crop diversity, how they select and exchange the materials within and between communities so that sustainable crop genetic resources conservation in situ can be achieved. Additionally, on-farm conservation helps linking farmers and communities with national ex situ collections and provides additional security for preserving locally adapted crop cultivars. On-farm conservation will assist the researchers to work towards the development goals of the conservation efforts, i.e. food security and the well being of our main partners, the farmers. At the same time, it is important to realize that, unlike ex situ conservation in situ conservation is highly contextual and can be successful only when genetic diversity is important to its immediate users, i.e. farmers.

Efforts are underway in different places to understand the genetic basis of on-farm conservation. There is a need to make some assumptions, as the genetic basis for in situ conservation is not fully understood. However, it can be noted that the efforts in this direction would only lead to a win-w*in situ*ation, i.e. conserving and using crop genetic diversity for the benefit of those who depend on it. In addition, it can help us to contribute to environmental health through its contribution to ecosystem functions in general.

Sustainable on-farm conservation is possible only when farmers, communities, and national institutions perceive benefits in terms of social, economic, and environmental services. Once we understand that the farmer management of local crop diversity is integral part of production and the primarily livelihood option for rural community, and then cost of on-farm conservation is much cheaper than ex situ. In the process of farming, farmers derive social, economic and environmental benefits from local genetic resources and maintain the evolutionary potential of such genetic resources. It is also important to note that on-farm conservation per se is not a panacea on its own as it is neither recommended as a universal practice nor a feasible method in all circumstances. On-farm conservation has a place and time as on-farm conservation can be transient and subject to change over time. It provides a major link with ex situ conservation and both approaches complement each other. It is important to see that conservation is a kind of spectrum extending from strictly in situ to completely static ex situ and that it is possible to have various degrees of ex situ and in situ in our efforts to conserve genetic resources.

References

Adhikari A, Rana RB, Sthapit BR, Subedi A, Shrestha PK, Upadhayay MP, Baral KP, Rijal DK, Gyawali S (2005) Effectiveness of diversity fair in raising awareness on agrobiodiversity management. In: Sthapit BR, Upadhyay MP, Shrestha PK, Jarvis D (eds) On-farm conservation of agricultural biodiversity in Nepal, vol II. Managing diversity and promoting its benefits. In: Proceedings of the second national workshop, 25–27 Aug 2004, Nagarkot, IPGRI, Rome, pp 236–253

Altieri MA, Merrick LC (1987) *In situ* conservation of crop genetic resources through maintenance of traditional farming systems. Econ Bot 41:86–96

Altieri MA, Merrick LC (1988) Agroecology and in-situ conservation of native crop diversity in the third world. In: Wilson E, Peter F (eds) Biodiversity. National Academy of Sciences, Washington, DC, pp 361–369
Araújo MB, Rahbek C (2006) How does climate change affect biodiversity? Science 313:1396–1397
Baniya BK, Subedi A, Rana R, Tiwari RK, Chaudhary P, Shrestha S, Tiwari P, Yadav RV, Gauchan D, Sthapit B (2003) What are the processes used to maintain genetic diversity on-farm? In: Gauchan D, Sthapit BR, Jarvis DI (eds) Agrobiodiversity conservation on-farm: Nepal's contribution to a scientific basis for policy recommendations. IPGRI, Rome, pp 20–23
Bellon MR, Pham JL, Jackson MT (1997) Genetic conservation: a role for rice farmers. In: Maxted N, Ford-Lloyd BV, Hawkes JG (eds) Plant genetic conservation. Chapman and Hall, London, pp 263–289
Biggs SD (1988) Resource poor farmer participation in research: a synthesis of experiences in nine national research systems. On-farm client oriented research, Comparative study paper no. 3, ISNAR, The Hague
Bommer DFR (1991) The historical development of international collaboration in plant genetic resources. In: van Hintum Th JL, Frese L, Perret PM (eds) Searching for new concepts for collaborative genetic resources management. Papers of the EUCARPIA/IBPGR symposium, Wageningen, 3–6 Dec 1999. International Crop Networks series no. 4, IPGRI, Rome, pp 3–12
Brush SB (1991) A farmer-based approach to conserving crop germplasm. J Econ Bot 45(2): 153–165
Brush SB (1993) *In situ* conservation of landraces in centres of crop diversity. In: Symposium on global implications of germplasm conservation and utilization at the 85th annual meetings of the American Society of Agronomy, Cincinnati
Brush SB (1995) *In situ* conservation of landraces in centres of crop diversity. Crop Sci 35:346–354
Brush SB (2000) Genes in the field: on-farm conservation of crop diversity. Lewis Publishers/ IDRC/IPGRI, Boca Raton/Ottawa/Rome
CBD (1992) Convention on biological diversity. United Nations. (http://www.cbd.int/doc/legal/ cbd-en.pdf). 22/4/2012
CBD (2007) Biodiversity and climate change. Booklet produced on the occasion of International Day for Biological Diversity. (www.cbdint/doc/bioday/2007/booklet 01-en.pdf). 25/04/2012
Cromwell E (2000) Local-seed activities: opportunities and challenges for regulatory frameworks. In: Tripp R (ed) New seed and old laws: regulatory reform and the diversification of national seed systems. ODI/Intermediate Technology, London, pp 214–231
FAO (2010) The second report on the state of the World's plant genetic resources for food and agriculture. FAO, Rome, 370 pp
Frankel OH (1990) Germplasm conservation and utilization in horticulture. In: Wiley-LissFrankel OH, Soulé ME (eds) Horticultural biotechnology, conservation and evolution. Cambridge University Press, Cambridge, pp 5–17
Frankel OH, Soulé ME (1981) Conservation and evolution. Cambridge University Press, Cambridge, 327 pp
Frankel O, Brown ADH, Burdon JJ (1995) The conservation of plant biodiversity. Cambridge University Press, Cambridge, UK, 299 pp
Friis-Hansen E, Sthapit B (2000) Participatory approaches to the conservation and use of plant genetic resources. IPGRI/CDR, Rome
Grum M, Gyasi EA, Osei C, Kranjac-Berisavljevic G (2003) Evaluation of best practices for landrace conservation: farmer evaluation. Unpublished donor report
Gyawali S, Sthapit BR, Bhandari B, Bajracharya J, Shrestha PK, Upadhyay MP, Jarvis D (2010) Participatory crop improvement and formal release of Jethobudho rice landrace in Nepal. Euphytica 176(1):59–78. doi:10.1007/s10681-010-213-0
Hannah L, Midgley G, Andelman S, Araújo M, Hughes G, Martinez-Meyer E, Pearson R, Williams P (2007) Protected area needs in a changing climate. Front Ecol Environ 5(3):131–138
Harlan JR (1992) Crops and man. American Society of Agronomy and Crop Science Society of America, Madison

Hijmans RJ (2003) The effect of climate change on global potato production. Am J Potato Res 80:271–280
Hijmans RJ, Graham CH (2006) The ability of climate envelope models to predict the effect of climate change on species distributions. Glob Change Biol 12:2272–2281
Hodgkin T (1993) Managing the populations-some general considerations. A workshop organized by the Council of Europe in cooperation with Swiss Authorities and Neuchâtel University – conservation of the wild relatives of European cultivated plants: developing integrated strategies, Neuchâtel University, Switzerland, 14–17 Oct 1993
Howden SM, Soussana J-F, Tubiello FN, Chhetri N, Dunlop M, Meinke H (2007) Climate change and food security special feature: adapting agriculture to climate change. Proc Natl Acad Sci USA 104:19691–19696
IPCC (Intergovernmental Panel on Climate Change) (2001) Climate change: the scientific basis. Cambridge University Press, Cambridge
IPCC (Intergovernmental Panel on Climate Change) (2002) Climate change and biodiversity. Technical paper V. IPCC, Geneva
IPCC (Intergovernmental Panel on Climate Change) (2007) IPCC fourth assessment report: climate change 2007 (AR4). IPCC, Geneva
IPGRI (1994) *In situ* conservation of crop and agroforestry species. Prepared for the CGIAR Mid-term meeting, 23–26 May 1994, New Delhi
IPGRI (1996) Strengthening the scientific basis of *in situ* conservation of agricultural biodiversity – a global collaborative project prepared by the International Plant Genetic Resources Institute, Project summary. IPGRI, Rome
Jarvis DI (1999) Strengthening the scientific basis of *in situ* conservation of agricultural biodiversity on farm. Botanica Lithuanica Suppl 2:79–90
Jarvis D, Hodgkin T, Eyzaguirre P, Ayad G, Sthapit B, Guarino L (1998) Farmer selection, natural selection and crop genetic diversity: the need for a basic data set. In: Jarvis D, Hodgkin T (eds) Strengthening the scientific basis of *in situ* conservation of agricultural biodiversity on farm. Options for data collecting and analysis. In: Proceeding of a workshop to develop tool and procedures for *in situ* conservation on-farm, 25–29 Aug 1997. International Plant Genetic Resources Institute, Rome, pp 5–19
Jarvis D, Sthapit B, Sears L (2000a) Conserving agricultural biodiversity *in situ*: a scientific basis for sustainable agriculture. In: Proceedings of a workshop, 5–12 Jul 1999, Pokhara. IPGRI, Rome
Jarvis DI, Myer L, Klemick H, Guarino L, Smale M, Brown AHD, Sadiki M, Sthapit B, Hodgkin T (2000b) A training guide for *in situ* conservation on-farm. International Plant Genetic Resources Institute, Rome, 161 pp
Jarvis D, Sevilla-Panizo R, Chavez-Srvia JS, Hodgkin T (2004) Seed systems and crop genetic diversity on-farm. In: Proceedings of a workshop, 16–20 Sept 2003, Pucallpa. IPGRI, Rome
Jarvis A, Lane J, Hijmans RJ (2008a) The effect of climate change on crop wild relatives. Agric Ecosyst Environ 126:13–23
Jarvis A, Upadhyaya H, Gowda CL, Aggerwal PK, Fujisaka S (2008b) Climate change and its effect on conservation and use of plant genetic resources for food and agriculture and associated biodiversity for food security. Thematic study for the SoW Report on PGRFA. FAO, Rome
Jarvis D, Hodgkin T, Sthapit BR, Fadda C, Lopez-Noriega I (2011) A heuristic framework for identifying multiple ways of supporting the conservation and use of traditional crop varieties within the agricultural production system. Crit Rev Plant Sci 30(12):1–49
Jones PG, Thornton PK (2003) The potential impacts of climate change on maize production in Africa and Latin America in 2055. Glob Environ Change 13:51–59
Joshi KD, Subedi M, Rana R, Kadayat KB, Sthapit BR (1997) Enhancing on-farm varietal diversity through participatory varietal selection: a case study for Chaite rice in Nepal. Exp Agric 33:335–344
Kar-Ling Tao, Murthi Anishetty N (2001) The implementation of the global plan of action in the APO region. APO Newsletter no 35, New Delhi, India

Kotschi J (2007) Agricultural biodiversity is essential for adapting to climate change. GAIA 16(2):98–101

Lawler J, Whit D, Nelson R, Blaustein AR (2006) Predicting climate-induced range shifts: model differences and model reliability. Glob Chang Biol 12:1568–1584

Maxted N, Ford-Lloyd BV, Hawkes JG (1997) Complementary conservation strategies. In: Maxted N, Ford-Lloyd BV, Hawkes JG (eds) Plant genetic conservation: the *in situ* approach. Chapman and Hall, London, pp 15–39

Menendez R, Gonzalez A, Hill JK, Braschler B, Willis S, CollinghanY FR, Roy D, Thomas CD (2006) Species richness changes lag behind climate change. Proc R Soc Lond B Biol Sci 273(1593):1465–1470

Midgley GF, Hughes GO, Thuiller W, Rebelo AG (2006) Migration rate limitations on climate change-induced range shifts in Cape Proteaceae. Divers Distribut 12:555–562

Newton AC, Begg GS, Swanston JS (2009) Deployment of diversity for enhanced crop function. Ann Appl Biol 154:309–322

Newton AC, Johnson SN, Gregory PJ (2011) Implications of climate change for diseases, crop yields and food security. Euphytica 179:3–18

Parmesan C (2006) Ecological and evolutionary responses to recent climate change. Annu Rev Ecol Syst 37:637–669

Parmesan C, Yohe G (2003) A globally coherent fingerprint of climate change impacts across natural systems. Nature 421(6918):37–42

Poudel D, Shrestha P, Basnet A, Shrestha P, Sthapit B, Subedi A (2008) Dynamics of farmers' seed networks in rice seed flow systems: implications for on-farm conservation. In: Sthapit BR, Gauchan D, Subedi A, Jarvis D (eds) On-farm management of agricultural biodiversity in Nepal: lessons learned. Proceedings of the national symposium, 18–19 Jul 2006, Kathmandu, pp 88–96

Ramanatha Rao V (2009) *In situ*/on-farm conservation of crop biodiversity. Indian J Genet Plant Breed 69(4):284–293

Ramanatha Rao V, Sthapit B (2002) Towards *in situ* conservation of plant genetic resources. In: Zhang Z, Zhou M, Ramanatha Rao V (eds) Plant genetic resources network in East Asia. Proceedings of the meeting for the regional network for conservation and use of plant genetic resources in East Asia, 13–16 Aug 2001, Ulaanbaatar. IPGRI Office for East Asia. Beijing, pp 100–1007

Ramanatha Rao V, Jarvis D, Sthapit B (2000) Towards *in situ* conservation of coconut genetic resources. Paper presented at the 9th COGENT Steering Committee meeting, 20–12 Jul 2000, Chennai

Riley KW (1995) *In situ* conservation and on-farm conservation. East Asia coordinators meeting on plant genetic resources, 23–25 Sept 1994, CAAS, Beijing, IPGRI Office for East Asia

Root TL, Price JT, Hall KR, Schneider SH, Rosenzweig C, Pounds JA (2003) Fingerprints of global warming on wild animals and plants. Nature 421(6918):57–60

Rosenzweig C, Parry ML (1994) Potential impact of climate change on world food supply. Nature 367:133–138

Shao G, Halpin PN (1995) Climatic controls of eastern North American coastal tree and shrub distributions. J Biogeogr 22:1083–1089

Shrestha P, Sthapit BR, Subedi A, Paudel D, Shrestha PK, Upadhyay MP, Joshi BK (2006) Community seed bank: a good practice for on-farm conservation of agricultural biodiversity. Paper presented at the incentives for supporting on-farm conservation, and augmentation of agro-biodiversity through farmers' innovations and community participation: an international policy consultation for learning from grassroots initiatives and institutional interventions. Indian Institute of Management, Ahmadabad, 27–29 May 2006,

Smale M, Bellon MR, Jarvis D, Sthapit B (2004) Economic concepts for designing policies to conserve crop genetic resources on farms. Genet Res Crop Evol 51(20):121–135

Sthapit BR (2008) On-farm conservation of agricultural biodiversity: concepts and practices. In: Sthapit BR, Gauchan D, Subedi A, Jarvis D (eds). On-farm management of agricultural biodiversity on-farm in Nepal: lessons learned. Proceedings of a national workshop, 18–19 Jul 2006, Kathmandu, pp 1–18

Sthapit BR, Jarvis D (1999a) On-farm conservation of crop genetic resources through use. In: Mal B, Mathur PN, Ramanatha Rao V (eds) South Asia Network on Plant Genetic Resources (SANPGR), Proceedings of fourth meeting, Kathmandu, 1–3 Sept 1998. IPGRI South Asia Office, New Delhi, pp 151–166

Sthapit BR, Jarvis D (1999b) PPB for on-farm conservation. ILEA 15 no ¾

Sthapit BR, Joshi KD (1996) *In situ* conservation possibilities through participatory plant breeding and selection approach: a case study from Nepal. Paper presented in expert meeting on agricultural biodiversity. Chinese Academy of Sciences, Chengdu, 11–12 Jun 1990

Sthapit BR, Ramanatha Rao V (2009) Consolidating community's role in local crop development by promoting farmer innovation to maximise the use of local crop diversity for the well-being of people. Acta Hort 806:669–676, http://www.actahort.org/books/806/80683.htm

Sthapit BR, Shrestha PK, Upadhyay MP (2006) On-farm management of agricultural biodiversity: good practices. NARC/LI-BIRD/IPGRI, Nepal, pp 1–20

Sthapit BR, Gauchan D, Subedi A, Jarvis D (2008) On-farm management of agricultural biodiversity on-farm in Nepal: lessons learned. In: Proceedings of a national workshop. Kathmandu, 18–19 Jul 2006

Sthapit BR, Padulosi S, Mal B (2010) Role of *in situ* conservation and underutilized crops in the wake of climate change. Indian J Plant Genet Res 23(2):145–156

Subedi A, Chaudhary P, Baniya B, Rana R, Tiwari R, Rijal D, Jarvis D, Sthapit B (2003) Who maintains crop genetic diversity and how: implications for on-farm conservation and utilization. Cult Agric 25(2):41–50

Subedi A, Sthapit B, Shrestha P, Gauchan D, Upadhyay MP, Shrestha PK (2008) Community biodiversity register: consolidating the community role in the management of agricultural biodiversity. In: Sthapit BR, Gauchan D, Subedi A, Jarvis D (eds) On-farm management of agricultural biodiversity on-farm in Nepal: lessons learned. Proceedings of a national workshop, 18–19 Jul 2006, Kathmandu, pp 107–111

Swaminathan MS (1998) The speech cited by Tryge Berg's key note speech, the role of *in situ* genetic resource management: a summary of conclusions of the workshop. The keynote speech presented at the international workshop: towards a synthesis between crop conservation and development. Baarlo, 30 Jun–2 Jul 1997

Thomas CD, Cameron A, Green RE, Bakkenes M, Beaumont LJ, Collingham YC, Erasmus BFN, Ferreira De Siqeira M, Grainger A, Hannah L, Hughes L, Huntley B, Van Jaarsveld AS, Midgley GF, Miles L, Ortega-Huertas MA, Peterson AT, Phillips OL, Williams SE (2004) Extinction risk from climate change. Nature 427:145–148

Tripp R (2001) Seed provision and agricultural development: the institutions of rural change. Overseas Development Institute/James Currey/Heinemann, London/Oxford/Portsmouth, 174

Walther GR, Post E, Convey P, Menzel A, Parmesan C, Beebee TJC, Jean-Fromentin M, Hoegh-Guldbergand O, Bairlein F (2002) Ecological responses to recent climate change. Nature 416:389–395

Williams JW, Jackson ST, Kutzbach JE (2007) Projected distributions of novel and disappearing climates by 2100 AD. Proc Natl Acad Sci USA 104(14):5738–5742

Wood D (1993) Agrobiodiversity in global conservation policy, vol 11, Biopolicy Int. ACTS Press, Nairobi

Chapter 2
Collecting the Wild Relatives of Crops in the Tropics

Duncan Vaughan, Yasuko Yoshida, Masaru Takeya, and Norihiko Tomooka

2.1 Introduction

The wild relatives of crops, particularly those most closely related to crops, are very often plants of disturbed habitats. Such habitats are road sides and areas where there is much human disturbance. The adaptation to disturbance predisposes these wild species to the agricultural environment and hence this is one reason they were domesticated. These same habitats are ones that are vulnerable to urban expansion. Many populations of the wild relatives of rice that used to exist along the roads out of Bangkok, Thailand, have gone (Morishima et al. 1996). Whereas genetic erosion of crops is generally due to replacement by new varieties or new crops, the wild relatives of crops are most vulnerable to either habitat loss or habitat fragmentation (Hammer and Teklu 2008).

Wild relatives of crops have furnished important genes of crop improvement and help increase crop productivity. Two important genes from wild rice, resistance to grassy stunt virus not found in the cultivated rice genepool and a gene for male sterility that lead the way for hybrid rice development in China, are examples of the use of crop wild relative (Vaughan and Sitch 1991). Economic analysis of rice and other crops suggests genebank genetic resources including wild relatives of crops

D. Vaughan (✉) • M. Takeya • N. Tomooka
National Institute of Agrobiological Sciences,
Kannondai 2-1-2, Tsukuba, Ibaraki 305-8602, Japan
e-mail: Duncan@affrc.go.jp

Y. Yoshida
University of Tsukuba, 1-1-1 Tennondai, Tsukuba, Ibaraki 305-8572, Japan

M.N. Normah et al. (eds.), *Conservation of Tropical Plant Species*,
DOI 10.1007/978-1-4614-3776-5_2, © Springer Science+Business Media New York 2013

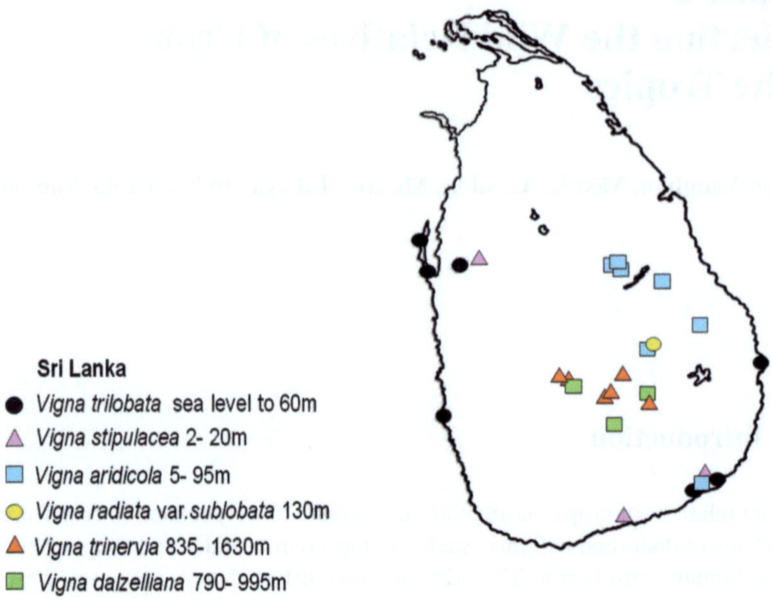

Fig. 2.1 *Vigna* species found in Sri Lanka during collecting missions (2000–2003). Prior to these missions' only two *Vigna* species were reported from Sri Lanka in the Revised Flora of Ceylon (From Tomooka et al. 2011 with permission)

directly and unambiguously have led to high economic return through increased crop production (Day Rubenstein et al. 2005; Evenson and Gollin 1997)

Scientist have been studying and collecting plant genetic resources for hundreds of years. Particularly with regards well known genera, therefore, surely we know all the species and these have now been collected. This is far from true, there remains much work to be done to collect and understand the wild species related to crops. *Vigna* is a genus that includes some of the world's most important legume crops such as cowpea and mungbean but much is still not known about the species in this genus. Prior to a comprehensive species specific collecting for wild *Vigna* in Sri Lanka only two *Vigna* species were recorded in the Revised Flora of Ceylon. After collecting six *Vigna* species are now known to grow in the country including a new *Vigna* species, *Vigna aridicola* (Fig. 2.1) (Tomooka et al. 2011). *Oryza*, the genus of rice, has not been surveyed in some areas of Latin America (Ecuador) and Asia (parts of Irian Jaya, Indonesia). Since there is still much to be learned about major crop genera for less well studied crops the gaps in our knowledge of their wild relative's remains even more incomplete.

Wild relatives of crops are still being described. A consequence is that inventory and survey is only possible for species that are known to science. Collecting is still necessary to uncover variation as yet unknown.

There are three main reasons for collecting the wild relatives of crops (von Bothmer and Seberg 1995):

(a) To provide germplasm for taxonomic, phylogenetic and biosystematic research;
(b) To conserve germplasm so that it may be characterized and evaluated;
(c) To furnish germplasm for plant breeders.

Collecting germplasm is only of value when there is a repository that can handle effectively the germplasm that is collected such as a genebank or botanic garden. Even a genebank or botanic garden may not be able to handle all germplasm if there is no specialist that understands the requirements for growing and conserving specific germplasm. In planning a collecting mission it is necessary to know where the collected materials will be maintained effectively. This chapter provides a guide to collecting wild plant genetic resources. For a much more comprehensive coverage of the collecting plant genetic resources subject readers are directed to Guarino et al. 1995 and Smith et al. 2003.

2.2 International and Legal Considerations

The "Global Plan of Action for the conservation and sustainable use of Plant Genetic Resources for Food and Agriculture (PGRFA)" (GPA) was adopted by the global community in 1996. The GPA has become a road map for monitoring PGRFA conservation and use. The first of the 20 priority areas in the Global Plan of Action (and its updated version FAO 2011) is "Survey and inventorying PGRFA" and the seventh priority area is "Supporting planned and targeted collecting of PGRFA".

The Convention on Biological Diversity (CBD), an international treaty, came into force in 1993 and now has more than 185 contracting parties. The CBD is a watershed agreement because it recognizes the sovereign rights of States over their natural resources, while having provisions for access and benefit sharing. The CBD and more recent treaty the International Treaty of Plant Genetic Resources for Food and Agriculture (ITPGRFA 2009) require that those involved with seed conservation and collecting are fully aware of the implications of these Treaties on their work.

The CBD requires that collectors of germplasm follow the correct, up-to-date, legal procedures. A set of rules for collecting germplasm has been produced to help guide collectors (FAO 1993). Among the guiding principles is the need for collectors to obtain "prior informed consent". A second principal is clear understanding that after collecting, the germplasm is properly documented to ensure that any benefits from the use of the collected germplasm are fairly and equitably shared with the donor of that germplasm. Both of these principals reflect the CBD that countries have sovereign rights over the genetic resources within their borders.

With respect to international partnership on collecting and conserving germplasm, agreements between the parties involved are needed. These usually need to be prepared well in advance, based on discussions among the parties involved. A helpful guide to developing access and benefit sharing agreements can be found in Cheyne (2003). Collecting conducted within a country often requires a collecting permit(s) and they need to be obtained in advance from the appropriate authorities

2.3 Preparation Phase

The main activity in the preparatory phase is making sure "all the homework is done". This includes having very clear collecting objectives. Legal aspects of collecting are one of the preparatory activities of collecting PGRFA (see above). Another major preparatory activity is obtaining as much information as possible about the target germplasm and area where the collecting is to take place. It may be necessary to undertake a comprehensive eco-geographic survey (Maxted and Guarino 2003). Much information may be available via the internet but it is often also very worthwhile visiting herbaria with material of the target species to accumulate data that might not be available on the internet. While there may well be much information on target taxa it is important to realize that, particularly in tropical regions, there is still much that is unknown and undocumented.

Contact with local officials in the target area in advance may facilitate the collecting mission. If given advanced warning local people knowledgeable of the target area can assist during the collecting mission and provide contacts and seek local permission for collecting.

However much preparation is done, it is necessary to expect the unexpected and be as prepared as possible for potential "good" and "bad" situations. For example, finding a new species will require that herbarium specimens are prepared to enable the new species to be described; the herbarium specimens are essential for naming a new species and must be deposited in reputable herbaria. If the vehicle you are using breaks down and you need to stay out at night have the necessary preparations been taken including taking the medicine needed to prevent malaria infection? A comprehensive check list of various types of equipment that may be needed during a collecting trip is provided (Appendix 1)

2.3.1 In the Field

The objective of collecting for ex situ conservation is to efficiently gather and document the genetic diversity of the target species in a particular area. Hence the job of collecting in the field has two distinct components (a) sampling the plant populations and (b) documenting as thoroughly as possible the sampled population and its environment.

2.3.1.1 Sampling

Sampling in the field needs to be as comprehensive as possible and this may involve, not just collecting the seeds or vegetative parts, but also herbarium specimens, DNA samples and associated symbiotic micro-organism (ENSCONET 2009).

Specific recommendations to four questions for a basic sampling strategy have been proposed:

(a) How many sites should be collected? About 50 populations in an eco-geographic area or during a mission;
(b) How many plants per population should be collected? Sample about 50 plants per population;
(c) How should samples be selected in a population? Sample individuals randomly within and among plants at each site but with attention to microenvironments if the habitat is heterogeneous;
(d) What quantity of material should be collected per plant? Sample sufficient good quality seeds or vegetative material per plant to assure that each plant is represented in duplicates of original collections (modified from Brown and Marshall 1995).

These recommendations are guidelines and highlight some important issues. For example, when collecting wild species ensure that attention is paid to varying habitats at a site. During a collecting mission for wild *Vigna* in northern Thailand at one site four species were found in different parts of the site that varied in disturbance and shade cover (Fig. 2.2). At the end of a collecting mission samples collected will be divided for safe duplication hence it is necessary to have a sufficiently large sample of good quality material to permit duplication.

While a basic strategy about sampling is a useful guide perhaps as important is the good advice that follows:

> There is no substitute for common sense, based on biological knowledge, to guide the collection of seeds of wildland plants. (Young and Young 1986)

The collector's biological knowledge of germplasm being collected depends on how well the plants being collected have previously been studied. However, generally it is known in advance if the material is inbreeding, outcrossing or vegetatively reproduces (including parthenocarpy). This information is core to determining sampling strategy because the genetic system determines the apportionment of genetic variation among and within populations. The approach to sampling will be affected by what plant parts are being collected – seeds, vegetative parts, or whole plants. In addition, there is the "common sense" elements that includes not causing irreparable damage to the population being sampled and using time wisely.

Collected germplasm needs to be as healthy and in the case of seeds fully mature on collecting. It is one thing to sample according to an appropriate design for the species of interest, however, it is also necessary to ensure that what is collected is of "good quality" and that the quality is retained until planted for rejuvenation. Therefore, timing of collecting is crucial to ensure that it is the best time for collecting

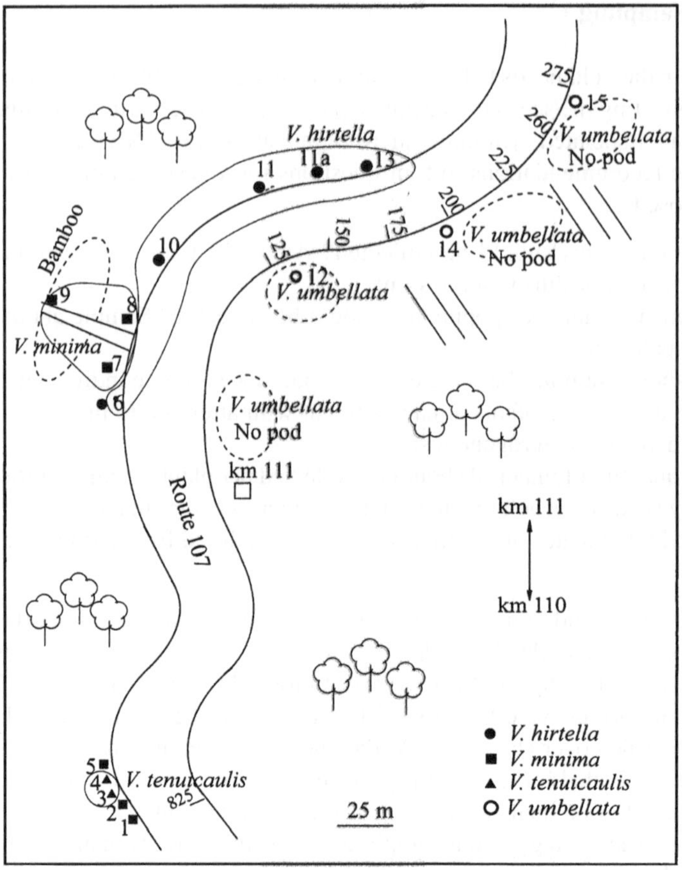

Fig. 2.2 Diversity of *Vigna* species found at one site in northern Thailand on the road between Chiang Mai and Fang. The different species were isolated by ecological conditions such as in forest (*V. minima*) or out of the forest (*V. hirtella, V. tenuicaulis* and *V. umbellata*) and time of flowering (*V. umbellata* flowers later than the other species at this site)

germplasm of excellent quality. The long term value of the collection will be impacted by the quality of the germplasm that is collected and the speed with which it is processed for temporary storage or planting. Issues associated with collecting seeds at the appropriate time can be found in Hay and Smith (2003).

One of the greatest constraints for sampling in the field is time. Initially it is necessary to canvass an area to find target plants and then determine their distribution in a locality and any obvious ecological or phenotypic variation in the sampling area. There is the need to balance the need for surveying a location, taking samples, including herbarium specimens, and documentation of these samples within and between populations. The approach taken will depend on multiple factors but the nature of the germplasm being collected will be the main determinate. For example, sampling large vegetative propagating plants may be more time consuming than collecting seeds of herbaceous species. For examples of sampling different types of species see Guarino et al. (1995) and for palms Dransfield (1986) and Martin (2004).

Techniques for collecting herbarium specimens have been described by various authors (e.g. Martin 2004; Zippel et al. 2010). Herbarium specimens are particularly important for the less well known or indeed unknown wild species that are being collected.

There are now simple methods available for collecting DNA in the field such as using Whatman FTA® paper (Vitha and Yoder 2005). If a collector on return to base has access to molecular biology laboratory that can analyze DNA this may enable a collection to be rapidly characterized. Such molecular characterization can be helpful in guiding future collecting (Fay 2003).

When collecting some species, it is necessary also to consider collecting associated organisms. For example, it should be standard procedure to collect root and/or stem nodules from leguminous species (Date 1995). Sampling or photographing the organisms associated with species may be helpful to provide information on pests, diseases or pollinators

2.3.1.2 Documentation

Unfortunately early and very important germplasm collections in genebanks were undertaken before the "computer age". At that time passport data was generally limited to a few notes in a field note book. The most important passport data is exact location, that today should if possible be recorded with a Global Positioning System (GPS), as well as the "unique" number – passport number that identifies the collected germplasm. *The only unique identifier of a collected population is the passport number* and it is essential that this remains associated with the sample at all times. Much germplasm is currently duplicated in the global genebank system; however, because germplasm is exchanged using genebank number that is not associated with original passport number the extent of duplication is unknown.

There are many examples of plant collecting data sheets in the literature (e.g. Zippel et al. 2010) a comprehensive review of data to record in the field is provided by Moss and Guarino (1995). The data to collect in the field as follows:

(a) A preliminary description of the locality this can be elaborated later after consulting maps;
(b) GPS location;
(c) Habitat (site) data including landform, slope, dominant plant species, degree of disturbance etc.
(d) Information on species/populations collected such as color, height, relative abundance, diseases observed etc.;
(e) Ethnobotanical information if travelling with knowledgeable residents of the area.

Collectors can combine information gathered by visual observation onto data sheets with GPS readings and (digital) photographs of locality, site, plants and plant parts. Additionally simple sketch maps of a collecting site and locality that includes, if possible, a permanent marker such as road mile marker. These simple sketch maps are particularly helpful in the situation where recollecting/resurveying is conducted, for example, to monitor genetic erosion. An example of a collecting site map redrawn from a sketch maps is shown (Fig. 2.2). An example of the type of passport data that should be collected when sampling wild species is provided (Appendix 2).

Combining passport data with other available databases such as climatic data and soil type data can lead to learning much about germplasm. When samples have been characterized and evaluated, characterization and evaluation databases can also be used to analyze collections. A simple example for Japan is shown that illustrates combining location of collecting site with seed colour (Fig. 2.3) (Takeya et al. 2009). Such information can be displayed on Google Earth® (Takeya et al. 2009). DIVA-GIS, a freely available software, is available that allows analysis of genebank and herbarium databases to elucidate genetic, ecological and geographic patterns in the distribution of crops and wild species (Hijmans et al. 2001). Analysis of passport data can provide insights into which collected samples should be targeted for specific evaluation. While cold tolerance would not be expected to be associated with germplasm growing close to the equator that is where the cold tolerant rice variety called Silewah from Indonesia was found. The clue to why this rice variety was cold tolerant was the altitude it was growing, 1,300 m.

Of particular help in ensuring that collected germplasm is used after collecting is to obtain ethnobotanical information from local informants. In rural areas people tend to have knowledge about a wide variety of both cultivated and wild plant species. This information can provide valuable insights into the potential wider use of the germplasm. Information related to gathering ethnobotanical data can be found in Cotton (1996) and Martin (2004).

2.3.2 Post Field Collection

On returning to base there are a series of activities that need to be undertaken by the collector(s). These include securing the collection, writing a report on the collecting mission and ensuring all data that was collected is entered into the appropriate database.

How the collection is secured will depend on the material collected. Quarantine clearances maybe required for material that crosses borders. The collection should be duplicated for safety at separate locations. The collection should also be put into appropriate storage conditions or planted to ensure the collection maintains its genetic integrity. Herbarium materials, DNA samples, soil and/or rhizobia samples should be deposited appropriately.

A trip report is needed to summarize all aspects of the trip and should include:

- The objectives of the mission;
- The environment of the target region;

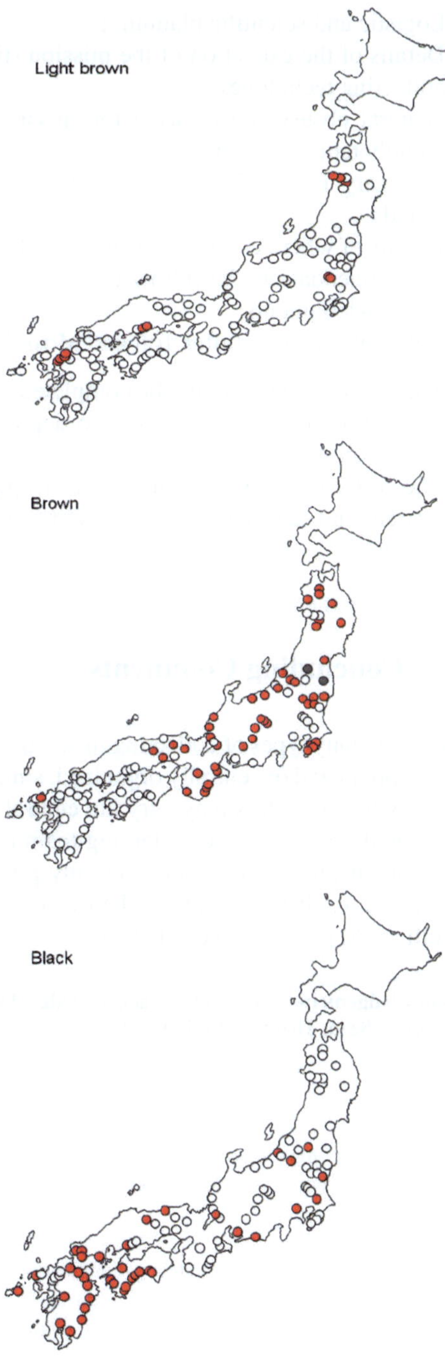

Fig. 2.3 Variation in seed color of *Glycine soja* across Japan shows clinal variation for *black* and *brown* seed color but not for light brown seed color (Modified from Takeya et al. 2009)

Appendix 2

Appendix 2.

Names and Addresses of collectors in Local language

Names and Addresses of collectors in English

Basic data

Scientific name		**Date**	dd	mm	yy
Local name		Collecting no.			
Plant no. codes		Site no.			

Location

Latitude				Longitude	
Geocode method	GPS	DGPS	Estimate	Map	Google Earth
Land holder	(In local language)				(In English)
Village	(In local language)				(In English)
Nearest town	(In local language)				(In English)
State / Country	(In local language)				(In English)

Site

Topography	Hill	Plain	Mountains		
Altitude (m)					
Altitude method	Altimeter	DEM	GPS	Estimate	Map
Land use					
Soil characteristics*				Geology*	
Climate*				Aspect / Slope	

Habitat

Associated vegetation type		Forest	Bushes	Cultivated	Grassland	Other specify ( )
Associated plants specific		Dominant sp.				
		Other spp.				
Shading (%)	Heaby	Medium	Light	Open (none)		
Degree of disturbance	High	Medium	Low	None		
State	Vegetative	Flowering	Mature	Past maturity		
Status	Wild	Weedy	Cultivated	Mixed		
Introgression	Yes (extent)			No		
Disease assessment	Leaf				Pod / seed	
Pest assessment	Leaf				Pod / seed	

* Information may be obtained from maps

Names and Addresses of collectors in Local language

Names and Addresses of collectors in English

Plant characteristics

Leaf pub.	High	Medium	Low	None	
Viable seeds/pod (10)					
Ovules/pod(10)					
Flower color					
Phenology status	More flowers than fruits Only fruits			More fruits than flowers Fruits already dispersed	
Frequency	Rare	Few	Frequent	Very frequent	Highly frequent
Species characters					

Population characteristics

Population variation			
Population size (m^2)		Sampling area (m^2)	
Number of plants sampled		Seed collected from	Plant Ground Both

Check list

Photo numbers	Site	Habitat	Plants
Herbarium voucher	Yes / No, Number:		
Other samples taken (e.g. rhizobium / soil)	Yes / No, What was sample?		
Sampling methods	Random Core of population	Regular Edge of population	Transect (linear) Other
Plant no. codes		Collecting no.	
Ethnobotanical information			

Sketch Map

References

Brown ADH, Marshall DR (1995) A basic sampling strategy: theory and practice. In: Reid R, Guarino L, Ramanatha Rao V (eds) Collecting plant genetic diversity. CAB International, Wallingford, pp 75–91

Cheyne P (2003) Access and benefit-sharing agreements: bridging the gap between scientific partnerships and the convention of biological diversity. In: Smith RD, Dickie JB, Linington SH, Pritchard HW, Probert RJ (eds) Seed conservation: turning science into practice. Royal Botanic Gardens, Kew, pp 5–26

Cotton CM (1996) Ethnobotany: principles and applications. Wiley, New York

Date RA (1995) Collecting Rhizobium, Frankia and mycorrhizal fungi. In: Guarino L, Ramanatha Rao V, Reid R (eds) Collecting plant genetic diversity. CAB International, Wallingford, pp 551–560

Day Rubenstein K, Heisey P, Shoemaker R, Sullivan J, Frisvold G (2005) Crop genetic resources: an economic appraisal. Economic Research Service/USDA Economic information bulletin no. (EIB2), 47 pp

Dransfield J (1986) A guide to collecting palms. Ann Missouri Bot Gard 73:166–176

Engels JMM, Arora RK, Guarino L (1995) An introduction to plant germplasm exploration and collecting: planning, methods and procedures, follow-up. In: Guarino L, Ramanatha Rao V, Reid R (eds) Collecting plant genetic diversity. CAB International, Wallingford, pp 31–63

ENSCONET (2009) ENSCONET seed collecting manual for wild species, 1st edn. Available online as pdf file at http://www.google.com/search?hl=en&source=hp&q=ensconet+seed+collecting+manual&aq=2v&aqi=g-v3g-bs1&aql=&oq=ENSCONET+

Evenson RE, Gollin D (1997) Genetic resources, international organizations and improvement in rice varieties. Econ Dev Cult Change 45(3):471–500

FAO (1993) Code for collecting PGRFA is found in the report of the conference of FAO – twenty-seventh session 1993 as XIV. Appendix E – international code of conduct for plant germplasm collecting and transfer

Fay MF (2003) Using genetic data to help guide decisions about sampling. In: Smith RD, Dickie JB, Linington SH, Pritchard HW, Probert RJ (eds) Seed conservation: turning science into practice. Royal Botanic Gardens, Kew, pp 91–96

Guarino L, Ramanatha Rao V, Reid R (1995) Collecting plant genetic diversity. CAB International, Wallingford

Hammer K, Teklu Y (2008) Plant genetic resources: selected issues from genetic erosion to genetic engineering. J Agric Rural Dev Trop Subtrop 109(1):15–50

Hay FR, Smith RD (2003) Seed maturity: when to collect seeds from wild plants. In: Smith RD, Dickie JB, Linington SH, Pritchard HW, Probert RJ (eds) Seed conservation: turning science into practice. Royal Botanic Gardens, Kew, pp 99–133

Hijmans RJ, Guarino L, Cruz M, Rojas E (2001) Computer tools for spatial analysis of plant genetic resources data: 1. DIVA-GIS. Plant Genet Res Newsl 127:15–19

ITPGRFA (2009) International treaty of plant genetic resources for food and agriculture. FAO, Rome. Available online as a pdf file from site http://www.planttreaty.org/texts_en.htm

Martin GJ (2004) Ethnobotany: a methods manual. People and plants international conservation. Earthscan, Oxford

Maxted N, Guarino L (2003) Planning plant genetic conservation. In: Smith RD, Dickie JB, Linington SH, Pritchard HW, Probert RJ (eds) Seed conservation: turning science into practice. Royal Botanic Gardens, Kew, pp 39–78

Morishima H, Shimamoto Y, Sato YI, Chitrakon S, Sano Y, Barbier P, Sato T, Yamagishi H (1996) Monitoring wild rice populations in permanent study sites in Thailand. In: Rice genetics III. IRRI, Los Banos, pp 377–380 (available online)

Moss H, Guarino L (1995) Gathering and recording data in the field. In: Guarino L, Ramanatha Rao V, Reid R (eds) Collecting plant genetic diversity. CAB International, Wallingford, pp 367–417

Smith RD, Dickie JB, Linington SH, Pritchard HW, Probert RJ (2003) Seed conservation: turning science into practice. Royal Botanic Gardens, Kew. Available online at: http://www.kew.org/science-research-data/kew-in-depth/msbp/publications-data-resources/technical-resources/seed-conservation-science-practice/

Takeya M, Yamasaki F, Tomooka N (2009) A web-based search and map display system for the integration of collection sites of plant genetic resources with geographic, climatic and plant characteristic data. Agric Info Res 18:82–90 (in Japanese with an English abstract)

Toll JA, Moss H (1995) Reporting on germplasm collecting missions. In: Guarino L, Ramanatha Rao V, Reid R (eds) Collecting plant genetic diversity. CAB International, Wallingford, pp 597–613

Tomooka N, Kaga A, Isemura T, Vaughan D (2011) *Vigna,* chapter 15. In: Kole C (ed) Wild relatives of crops – genomic and breeding resources. Springer, Heidelberg

Vaughan DA, Sitch LA (1991) Gene flow from the jungle to farmers: wild-rice genetic resources and their uses. Bioscience 41:22–28

Vitha S, Yoder DW (2005) High throughput processing of DNA samples on FTA filter paper for PCR analysis. Microscopy and Imaging Center, Texas A and M University, College Station. http://www.tamu.edu/mic

Von Bothmer R, Seberg O (1995) Strategies for collecting of wild species. In: Guarino L, Ramanatha Rao V, Reid R (eds) Collecting plant genetic diversity. CAB International, Wallingford, pp 93–111

Way MJ (2003) Collecting seed from non-domesticated plants for long-term conservation. In: Smith RD, Dickie JB, Linington SH, Pritchard HW, Probert RJ (eds) Seed conservation: turning science into practice. Royal Botanic Gardens, Kew, pp 165–201, Pages 39–78

Young JA, Young CG (1986) Collecting, processing, and germinating seeds of wildland plants. Timber Press, Portland

Zippel E, Wilhalm T, Thiel-Egenter C (2010) Manual on vascular plant recording techniques in the field and protocols for ATBI+M sites – inventory and sampling of specimens, chapter 14. In: Eymann J, Degreef J, Häuser Ch, Monje JC, Samyn Y, VandenSpiegel D (eds) Manual on field recording techniques and protocols for all taxa biodiversity inventories and monitoring, vol 8, part 2. Available in pdf format at http://www.abctaxa.be/volumes/volume-8-manual-atbi/

Chapter 3
Seed Banks for Future Generation

H.F. Chin, P. Quek, and U.R. Sinniah

3.1 Introduction

The world leaders of the United Nations convened a conference in Rio de Janeiro for the Earth Summit in 1992. The important outcome of this conference was a comprehensive Blueprint – Agenda 21 which highlighted biodiversity conservation. Biodiversity became a "buzz word" used by policy makers, politicians, scientist and the public. This was followed by the Global Strategy for plant conservation which was adopted by the Parties to the Convention on Biological Diversity (CBD) in 2002, thus contributing to the biodiversity target agreed to at the World Summit in Johannesburg 2002. Plants are among our most valuable resources for our survival and well being. When disaster strikes, causing paralysis to agriculture, seed banks serve as "insurance policy" and come to the rescue. In situ conservation is the most appropriate method to maintain species of wild animals and plants in their natural habitats. However, it is impractical to conserve the large number of plant species and their wild relatives in their natural habitats. Therefore ex situ method of conservation using seed banks is a popular, efficient and economical method. The important role played by seed banks in the conservation of plant genetic resources is now globally recognized.

H.F. Chin (✉)
Regional Office for Asia, The Pacific and Oceania, Bioversity International,
UPM BOX 236, Serdang, Selangor, Malaysia

Department of Crop Science, Universiti Putra Malaysia, Serdang, Selangor, Malaysia
e-mail: h.chin@cgiar.org

P. Quek
Regional Office for Asia, The Pacific and Oceania, Bioversity International,
UPM BOX 236, Serdang, Selangor, Malaysia

U.R. Sinniah
Department of Crop Science, Universiti Putra Malaysia, Serdang, Selangor, Malaysia
e-mail: umarani@putra.upm.edu.my

M.N. Normah et al. (eds.), *Conservation of Tropical Plant Species*,
DOI 10.1007/978-1-4614-3776-5_3, © Springer Science+Business Media New York 2013

The idea of seed bank is credited to Nicolai Vavilov, a Russian botanist who collected 200,000 cultivated plants from around the world (Cohen 1991). Today, the oldest recognized seed bank is located at the Vavilov Institute in Russia and the newest is Svalbard Global Seed Vault (SGSV) located on the Norwegian Island of Spitsbergen in Arctic region, established in 2008. This will be one of the largest seed banks in the world with a capacity of over four million accessions. To date the International Rice Research Institute (IRRI) has deposited the largest number of accessions, amounting to 112,807 for a single crop and its wild relatives (Barona 2011). Presently there are 1,750 genebanks worldwide, storing 7.4 million accessions (Dulloo et al. 2010). This number is the seed banks officially recognized in institutions, research organizations and government departments and does not include private collections from farmers, hobbyists and societies. Thus the real number of accessions will surely exceed the 7.4 million recorded.

A seed bank deals with things which are alive; hence we must handle them with great care. Seeds are collected, processed and only those with good quality are stored and catalogued. Seeds deposited in gene banks do not bring immediate benefits or pay dividends. They are only utilized in times of need or when plant breeders use them to select traits from existing collections and use them to breed a new variety for the farmers (Chin 1994). Viability of seeds stored in seed banks is of utmost importance. To maintain the viability of seeds, they must be of the orthodox type whereby based on the understanding of the physiology of seeds, ideal conditions of low moisture of seeds, relative humidity and low temperature, can be provided to extend the storage life of such seeds, thus reducing cost involved in regeneration of the collection.

This chapter will discuss the physiology of seed in relation to storage; leading to the way a typical seed bank functions, with adequate facilities for data processing, seed testing and regeneration. These functions are needed to ensure routine management of seed banks and to facilitate distribution of seed samples. Awareness of the existence of seed banks locally and worldwide must be created and an easily accessible global system of information must be made available to the plant breeders and others who make use of them.

3.2 Seed Biology

Kesseler and Stuppy (2006), describe seed as the time capsule of life. Seeds are living things and have to be handled with great care to ensure their survival. It is important that we know and understand their biology, diversity and longevity. In the past, poets described seed as the awesome vessel of power and wherein all the mystery is unfolded. Guest (1933) in a poem professed that seed is "the miracle of life and is a dime's worth of power that no man can create and a dime worth of mystery, destiny and fate". The biologists regard the seed of any plant as one of the most fascinating and puzzling mysteries of biology.

The seed, an embryonic plant, is well protected by the testa or seed coat. Its metabolic activities are at an extremely low rate while dormant, yet after a long period it can be re-awakened by the stimuli of a favourable environment. Nature's own seed banks are found buried in soil and lake beds. Millions of seed are found buried in the fields or farms. On tillage and excavation seeds are found or seedlings found sprouting. From this, man learned to store dry seeds in seed banks, with a suitable environment of low humidity and temperature. The orthodox way for seeds to be stored is for them to be dried to a very low moisture content around 5% of the dry weight of the seed; at times even as low as 1% and kept in sealed containers at −18°C, which allows storage for a few decades (IBPGR 1976). This is the unique ability bestowed upon seeds. A very important and basic question which always escapes our notice and curiosity is what mechanism allows seeds to withstand desiccation, avoid death and survive the stress that kills the great majority of organism?

3.3 Seed Longevity and Deterioration

The longevity of seeds is beyond that ever imagined; in some rare cases wheat seeds have been found to be viable despite being buried in the Egyptian tombs for centuries (Barton and Polunin 2005). Lotus seeds buried in the Manchurian lakes for hundreds of years have been known to sprout (Shen-Miller 2002). Porsild et al. (1967) reported that seeds of arctic Cupine, at least 10,000 years old, found in lemming burrows deeply buried in permanently frozen silt of the Pleistocene age, germinated and grew into normal healthy plants. Experimental data of Beal reported by Darlington (1951) showed after 70 years of storage, seeds of *Oenothera biennis* and *Rumex crispus* were capable of germination. Under natural conditions weed seeds buried in soil under pastures for 32 years were known to germinate and grow into seedlings (Brenchley 1918). A seed does not die suddenly, it undergoes a process of seed deterioration which is not visible and what really causes a seed to deteriorate are not fully understood (Narayana Murthy et al. 2003). Recent reviews indicate free radical-mediated lipid peroxidation, enzyme inactivation, protein degradation, membrane disruption, and DNA and RNA damage as major causes of seed ageing (Priestley 1986; Smith and Berjak 1995; Walters 1998; McDonald 1999). Seed deterioration leads to the death of seeds, but at the stage of deterioration, seeds are still germinable but they affect the yield of a crop in two ways: there will be fewer plants per unit area also poorer performance by the surviving plants. Therefore, the method of handling and storage of seeds will influence the rate of deterioration thus limiting or promoting longevity.

3.4 Seed Storage Principles and Practices

An old famous quotation "Care with the seeds joy with the harvest" is true till today. Seeds are living materials and have to be handled with care. There is a great diversity in seed types and the needs and conditions of storage will differ depending on period of storage required. Often seeds are stored for food as grains or planting materials for the next season which is relatively easy to achieve while the concern here is long term storage as breeding materials or genetic resource materials in which maintenance of viability is essential, especially for very long periods in seed banks.

Seeds have been classified into (1) orthodox and (2) recalcitrant (Roberts 1973), according to tolerance to drying and chilling. Hanson et al. (1984) called them non-desiccation sensitive and the desiccation sensitive which is more descriptive and later Berjak et al. (1990) referred recalcitrant seeds as homoiohydrous group which are mainly tropical species with many important plantation crops such as rubber, cocoa and forestry species (Chin 1978); Chin et al. 1981, 1984). A third class of seed storage behavior termed intermediate has been recognized, whereby seeds are more tolerant of desiccation than the recalcitrant type but the tolerance is much limited as compared to an orthodox species, however, they are said to lose viability more rapidly at low temperature (Ellis et al. 1990, 1991). According to Berjak and Pammenter (2004), a distinct difference between these categories is not present, rather an extended continuum of seed storage behavior from the most desiccation tolerant of orthodox species to the intermediate seed and finally of the minimally recalcitrant to those extremely desiccation sensitive types are present. To date, both recalcitrant and intermediate seed types cannot be stored in conventional seed banks. In vitro technology is utilized to conserve such materials using either slow growth storage or using cryopreservation i.e. storage at ultra-low temperature of liquid nitrogen (−196°C) (Benson 1999).

Orthodox seeds comply with the general rule-of-thumb for seed storage, which state that for each 1% reduction in seed moisture, the life of the seed is doubled and for each 5°C reduction in seed temperature the life of the seed is doubled. These rules- of-thumb are not as accurate as those of Roberts and Abdalla (1968) formulae, but they do stress the importance of low seed moisture and low seed temperature in preserving high germination. In the case of orthodox seeds, the best method of storage is first to dry the seed so as to obtain as low a moisture level as possible (in the region of 5–6%). The moisture must be kept at this level through-out the storage period or else the benefit of drying will be lost. The maintenance of seeds in a dry condition can be achieved in three ways. First, seeds should be stored in the moisture-proof room or stores which have been dehumidified to maintain the desired low relative humidity. Secondly, they must be packed in moisture-proof containers i.e. tins or bags, and thirdly, seeds should be placed in air-tight containers with desiccant. The store should be air-conditioned and in case of long term storage the temperature should be as low as −20°C. Under these conditions rice seeds of 4–6% moisture in air tight tins at −20°C can have a storage life of about 75 years. In general, to preserve viability of orthodox seeds, seed moisture and seed temperature should be reduced as soon as possible to a desirable level, and maintained at this level. In addition, cryogenic storage as a method of long term storage was found to be successful and recommended as a method of storage for orthodox seeds. Freezing seeds at −196°C was conducted about 100 years ago by Brown and Escombe (1897–98). They found that seeds of 12 species with low moisture content survived without damage after 110 days storage. This study was later followed by other researchers. Lipman (1936) exposed seeds of barley, corn, sweet clover and vetch to −273.1°C, with no impairment of viability. The superiority of subfreezing temperatures over higher temperatures for seed storage has well been established for many kinds of seeds (Ellis et al. 1996; Pukacka and Ratajczak 2005; Concellon et al. 2007).

3.5 Future Trends in Seed Banking

3.5.1 Ultra Low Moisture

Refrigeration technology has advanced tremendously, however, availability and reliability of electricity in developing countries is still prohibitive, so new storage technology has to be developed. One, which has already been initiated, is the study on the possibility of using ultra low moistures, i.e. drying seeds to very low moisture content of 1–2% (Pérez-García et al. 2007). At such low moisture content it may not be necessary to have refrigeration at all; studies have shown at low moisture content orthodox seeds can remain viable even at ambient temperature for years, so we look forward to this new method. If further studies show no detrimental effects, i.e. genetic changes or loss of viability over long periods, then developing countries will opt for this method.

3.5.2 Cryogenic Storage

In the last decade studies have shown that hundreds of crop seeds can be stored at ultra low temperatures of −196°C. At such low temperatures metabolic activities are minimal and deterioration is practically non-existent. This method is proven to be successful and cost effective, therefore it will be an ideal method for long-term seed conservation. Stanwood and Bass (1981) estimated that the cost of storing onion seeds over 100 years is reduced to a quarter over the conventional method. Cryogenic storage is in practice at Fort Collins Colorado where the National Center for Genetic Resources Preservation of the USA is located. Further tests on a wider range of crops and their cost have to be re-evaluated. In future we expect to see more seed banks using cryogenic storage.

Cryopreservation has been hailed as the only viable option for storage of recalcitrant species (using excised embryos) and those that are propagated vegetatively. Bioversity (formerly known as IPGRI) in collaboration with other institutes is actively involved with cryopreservation work of cassava, coconut and banana. In the meantime, it is important to carry out studies on the effect of storage under cryopreserved conditions might have on the stability of the genome. Many new inputs on optimization of cryoprotocols and improvement after cryogenic storage is being established (Hajari et al. 2011; Berjak et al. 2011). With all the recent developments and advances in biotechnology, many of the problems will be solved and we can expect to see more germplasm either in seeds or in other forms, such as tissues, meristems, embryos, and artificial seeds cryopreserved for long-term conservation in gene banks.

3.5.3 *Permafrost*

In nature there are many suitable cold places for the storage of germplasm without involving high cost in refrigeration. Some of these places are located in old mines, caves and tunnels, the extremely cold Arctic and Antarctic circles. A very good example is the Nordic Gene bank located in Svalbard in the Island of Spitzbergen. A group of Scandinavian countries agreed to store their germplasm in this common gene bank. The samples are stored in glass ampoules and moisture content of seeds is around 2–4%. The temperature in the mine is constant from −2°C to 3°C. An international permafrost seed bank named Svalbard Global Seed Vault is now established.

3.6 Development and Types of Seed Banks

There are many types of seed banks, which can be classified into short, mid-or long-term according to their storage period. There are also base collections which do not distribute seed, and active collections that make the seed available to breeders or farmers. They are also classified as single or multi species such as rice or mixture of beans or legumes in the big family and its wild relatives or collection of a mixture of vegetables.

Nature has conserved plant species through various mechanisms inbuilt in seeds through millions of years of evolution. Man has learnt the necessity to conserve as far back as 700 B.C. in parts of China and India (Plucknett et al. 1987). Modern genetic conservation was pioneered by Vavilov (1940, 1957). Vavilov in 1920 started storing seed samples in metal cases at ambient temperature. This is the humble beginning of a seed bank. In 1992 it was estimated there were 993 gene banks in 121 countries holding about four million accessions. Today the CGIAR centres house the world's largest ex situ collection of some 600,000 accessions of major food and forage crops (CGIAR 1996). This is followed in the twenty-first century by the Millennium Seed Bank in Royal Botanic Gardens in Kew, United Kingdom. The Millennium Seed Bank Project (MSBP) comprises over 100 organisations in more than 50 countries. The main aim is to collect and safely store 24,200 threatened species by 2010. Now over 18,000 species have been secured all over the world for such species. A new programme known as MSBP-2 is established to address the oncoming challenges of the next decade. At the beginning of the twenty-first century concurrently with the Millennium Seed Bank Project in Kew the Svalbard Global Seed Vault was established. The Svalbard Global Seed Vault was opened officially in 2008. It is designed to store duplicate of seeds from seed collections of banks all around the world including many from developing countries. The aim of this bank is to ensure that they can replace them for countries in case they lost them due to natural disasters, war or in events they cannot afford to maintain them. These countries may later make use of their duplicates stored in Svalbard. This particular seed bank in Norway is very unique, making use of green technology depending on permafrost to keep the seeds at very low temperature, it is one of the world's largest seed bank with a capacity of four million accessions, presently only holding 30,000 new seed samples from India, Africa and Korea. Recently 112,807 accessions of rice seeds from IRRI have been deposited into the Svalbard seed bank.

Seed banks in the past century have grown and advanced tremendously with over 1,400 genebanks. The trends have been traced from the primitive to the ultra modern ones operated by robots and using green technology. Examples of genebanks include the Asian Vegetable Research Development Centre (AVRDC), houses 57,000 accessions of 408 different vegetable species and varieties collected from 154 countries. Seed banks are also grouped into national, regional and international banks. Even in some countries, a national bank is further divided into state banks, for instance in USA, the National Center for Genetic Resource Preservation (NCGRP) is the base collection, but several states have seed banks for active distribution of seed. From the national seed banks of individual countries now they joined hands to become regional and international seed banks. By the twenty-first century a number of international centres have joined hands and linked together as a network. One outstanding example is that in 2006, eleven International Agricultural Research Centres (IARCs) of the CGIAR holding ex situ germplasm collection, signed agreements with the Governing Body of the International Treaty on Plant Genetic Resources for Food and Agriculture (ITPGRFA) placing the collections they hold under the Treaty (Article 15). These agreements placed the ex situ collections of PGRFA held by those Centres (some 650,000 accessions of the world's most important crops) within the purview of the Treaty. Under these agreements, the Centres recognize the authority of the Governing Body of the ex situ collections. CGIAR collections held in-trust for the world community based on agreements with FAO (2002).

The CGIAR seed collections in the Centres are unique resources to be made available to all researchers. These seeds are invaluable and have been proven to help countries in need for agricultural development and growth in countries emerging from wars and those that are recovering from natural disasters. In the past decade over a million seed samples have been distributed mainly to universities and national agricultural research systems where scientists need to develop new crop varieties with higher yield, more nutrition and resistant to pests and diseases. Recently breeding for climate change such as drought and flood are top in priority. The seed banks of the (CGIAR) centres in Table 3.1 are the main sources of seeds for breeders to work on in case of emergencies.

Table 3.1 CGIAR seed collections held in the centers

Center	Crops	Number of accessions
International Center for Tropical Agriculture (CIAT), Cali, Columbia	Forages	18,138
	Bean	31,718
International Maize and Wheat Improvement Center (CIMMYT) Mexico	Maize	20,411
	Wheat	95,113
International Center for Agriculture in the Dry Areas (ICARDA), Allepo, Syria	Barley	24,218
	Chickpea	9,116
	Faba bean	9,074
	Wheat	30,270
	Forages	24,581
	Lentil	7,827
International Crops Research Institute for the Semi-Arid Tropics (ICRISAT), Patancheru, India	Chickpea	16,961
	Groundnut	14,357
	Pearl millet	21,250
	Pigeonpea	12,698
	Sorghum	35,780
International Institute for Tropical Agriculture (IITA), Ibadan, Nigeria	Bambara groundnut	2,029
	Cowpea	15,001
	Soybean	1,909
	Wild Vigna	1,634
International Livestock Research Institute (ILRI). Nairobi, Kenya	Forages	11,537
International Rice Research Institute (IRRI). Los Banos, Philippines	Rice	80,617

3.7 Establishing a Seed Bank

Seed bank can range from a collection of seed vessels housed in simple structures where communities store their seeds till the next season to complex structures with cold room and laboratory facilities. A better understanding of a seed bank function is to look at examples of processes and inputs that goes into a seed bank managed by National Systems and by International Centres. Many International Centres have seed banks and an example is the International Rice Research Institute (IRRI) rice genebank (IRRI 2000). The aims of a seed bank should be well spelled out and it has to be crop specific. For example the role of a rice genebank is to assist in the ex situ conservation of the cultivated rice varieties and the wild species of rice. Justification for it includes: (1) Traditional varieties tend to be lost through genetic erosion in places where farmers adopt new varieties through replacement of traditional varieties grown over many generations, (2) Wild varieties threats can come from habitat changes due to human disturbances, development pressure and climate change and (3) Plants with biotic and abiotic stress traits that is important for breeding of resistance and adaptation to climate changes.

Facilities in a seed genebank include facilities for seed processing, regeneration and characterization. Most information provided are based on the IRRI manual (IRRI 2000).

3.7.1 Drying Rooms

An important facility for seed processing has to ensure that seeds can be dried properly. Seeds collected or harvested from the field are first cleaned of diseased grains, pest, debris and empty grains beside seed contaminants. These are then dried in a drying room where the temperature and relative humidity are kept low. For rice seed a drying temperature of 15°C and 15% relative humidity are used by IRRI. This is within the range recommended by the FAO/IPGRI Genebank Standards (1994) of 10–15% RH and a temperature of 10–25°C for drying seeds. The amount of time seeds are kept in the drying room are estimated from moisture isotherms graph to determine the seed equilibrium moisture content. This graph is prepared for different species by measuring the moisture content over time within a specific drying condition. Other drying methods include using salt solutions, oven drying and using silica gel. The drying process is to achieve the critical moisture content before storage and differs in different species, from about 6% for pea and mung bean (protein rich) to 4.5–5% for rice, wheat and barley (starch rich), and 3.3% for soybean (oil rich) (Rao et al. 2006). The seeds can then be checked for damage and removed if necessary before packing into containers or bags for storage. Drying rooms should be well built and managed to avoid pest attacks during drying.

3.7.2 Storage Rooms

As seeds are sealed in containers or bags the humidity is not subject to change and the temperature used for storage will determine the longevity of the seeds. Storage periods are generally divided into short term, medium term and long term. Long term storage is for base collection maintaining longevity of greater than 50 years and up to 100 years when kept at −20°C. Medium term storage is for active collection and is maintained for 20–40 years in storage. These are seeds that genebank and breeders would be expected to use in the near future. They are kept at +2°C which requires less energy to maintain than −20°C. Working collection are usually maintained in short term storage maintaining at ambient or slightly below in air-conditioned rooms. These standards on storage determine the goal for genebank curators to achieve. In reality, where resources are limited, some genebanks may only be able to opt for medium term storage and use deep freezers instead. Deep freezers are feasible only for seed bank with a small base collection and assuming stable electrical supply. For long term storage, seed banks can also opt for "black box" storage hosted by genebanks with excess long-term and medium-term storage facilities. The Svalbard Global Seed Vault located on the Norwegian island of Spitsbergen in the remote Arctic Svalbard archipelago, about 1,300 km from the North Pole is an example of long-term "black box" storage at the global level (http://en.wikipedia.org/wiki/Svalbard_Global_Seed_Vault). Another option is the use of long-term ultra dry storage where seeds moisture content are maintained at 3% or less and maintained at –5°C to –10°C in storage. Brassica seeds have been shown to germinate after 40 years (Pérez-García et al. 2009) in this condition. The impact of the method for genebank is the lower energy cost in maintaining long-term storage, improving energy efficiency.

3.7.3 *Screenhouses*

Screen houses facilitate control pollination and pests. This is an important facility for regeneration of seed accessions that are not obligate in-breeders.

3.7.4 *Fields*

Outside the screenhouses, the fields are the main areas for regeneration, multiplication and characterization of germplasm. Characterization involves the traits and character of the plant that is not affected by the environment. Characters that are influenced by the environment like yield and resistant to pest and diseases are part of the evaluation work.

3.7.5 *Laboratories*

It is an integral component of seed bank for seed testing, germplasm characterization and data management. Tasks performed in laboratories include viability tests (seed germination and dormancy breaking), seed moisture determination, determining the presence of transgenes (to prevent unintentional geneflow from genetically modified organisms), cytogenetical study and tissue culture.

3.8 Seed Bank Functions

3.8.1 *Maintaining Collection*

Seed genebanks carry out the basic function of acquiring new germplasm accessions through collecting missions and exchange of germplasm. Example in the IRRI rice genebank, their collection consists of cultivated materials, mostly landrace or breeding materials of *O. sativa, O. glaberrima*, wild species, and representative species from eight genera in the tribe Oryzeae found in Asia, Africa, Australia, and the Americas (IRRI 2000). The amount of rice seeds recommended for keeping in IRRI genebank is 120 g or more for rice seeds kept in active and base collections involving *O. sativa* and *O. glaberrima.* This amount includes samples for viability testing, seed health evaluation, sample for base collection (60 g) and sample for duplicate storage.

To reduce duplication further and to make the collection more useful, a core collection could be developed. Core collection is a subset of an entire collection that represents with a minimum of repetition, the genetic diversity of a crop species and its wild relatives (Johnson and Hodgkin 1999). Putting available diversity into a smaller subset of samples help in the utilization of accessions conserved in genebanks. The Rural Development Administration (RDA) genebank has developed the Powercore software to help genebanks to identify accessions for core collections (Kim et al. 2007). The software works on identifying all useful alleles or characteristics in accessions so that these can be retained in genebank. It helps in reducing the redundancy of useful alleles and thus enhancing the richness of the core collection. The tool is available from the RDA genebank and runs on MS Windows Platform.

3.8.2 Germplasm Acquisition and Conservation

The role of seed genebanks is to ensure the conservation and continued availability of genetic resources for crop improvement, to restore valuable germplasm which has been lost in the country, and as an important conduit for germplasm exchange. The acquisition of accessions into the genebank has to consider the regeneration capacity of the seed genebank. This is one consideration that is frequently overlooked as regeneration is not due until 10 or more years down the road. Let's assume a seed genebank with short term storage of 10 years and has regeneration capacity for 1,000 accessions per year (Table 3.2)

Table 3.2 shows us the impact of acquiring accessions without consideration of the regeneration capacity of the genebank. If the regeneration needs from year 11 to year 20 is taken into consideration the following will be the impact.

Year 12 – there is lack of regeneration capacity of 500 which can be accommodated in year 11 if this lack is anticipated. **Year 14** – This huge increase for regeneration cannot be accommodated in the year before and after. This situation can result in loss of accessions if genebank does not have alternative arrangement for regeneration. **Year 18** – This will be another problem year but with experience from year 14 genebank may be able to manage.

3.8.3 Viability Testing

Seeds held in genebank are subjected to changes in the environment which affects the seed viability. Genebank carries out viability testing at regular intervals to check the seed viability. In Fig. 3.1 the initial viability test result provides an indication of

Table 3.2 Accessions acquisition example

Year	Accessions acquired	Regeneration due	Regeneration capacity	Excess/lack of capacity
1	500			
2	1,500			
3	1,000			
4	4,000			
5	200			
6	500			
7	100			
8	2,000			
9	0			
10	200			
11		500	1,000	500
12		1,500	1,000	–500
13		1,000	1,000	0
14		4,000	1,000	–3,000
15		200	1,000	800
16		500	1,000	500
17		100	1,000	900
18		2,000	1,000	–1,000
19		0	1,000	1,000
20		200	1,000	800

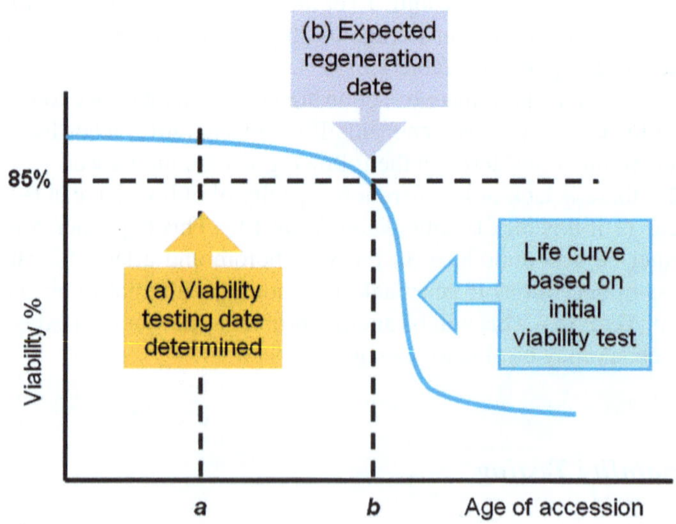

Fig. 3.1 Life curve

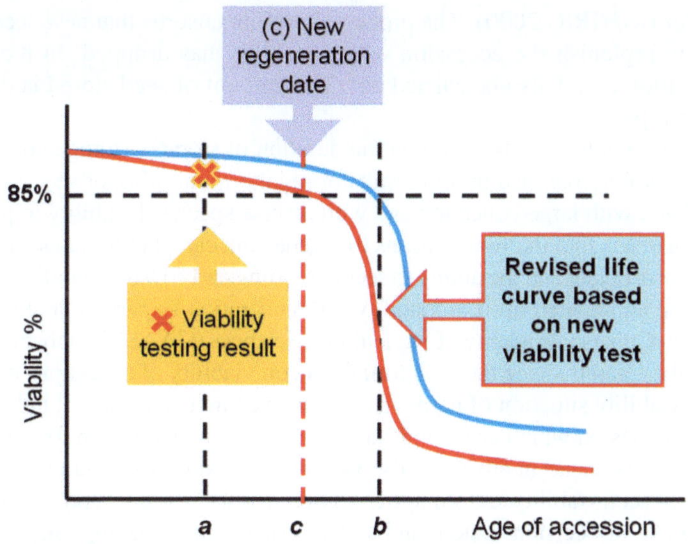

Fig. 3.2 New life curve

the expected regeneration date (point ***b***). A monitoring date can then be set (point ***a***). Based on new viability test at point ***a*** (Fig. 3.2), a new life curve can be drawn and the new regeneration date determined (point ***c***). The actual viability also determines the schedule and frequency of future monitoring where needed. The importance of the viability test in determining the regeneration date of an accession is necessary so that, accessions that loose viability faster than anticipated can be identified and regenerated accordingly to prevent loss of the accession. Adding this to the ability to predict the future regeneration timing will help genebank to plan in case of regeneration capacity over run.

3.8.4 Regeneration

Regeneration takes place for the purpose of; (a) increasing seed numbers, (b) replenishing stocks in active and base collections. Regeneration is a procedure that is avoided as much as possible as it increases the risk of change in the genetic make-up of the accession due to selection pressure, mechanical mixtures and others. This risk of loss of genetic integrity is high for genetically heterogeneous germplasm accessions. Regeneration is also an expensive procedure to undertake and the ability to predict accessions due for regeneration becomes very important so that facilities and funds can be sourced when needed. In a seed bank regeneration is carried out when the accession seed viability drops below the agreed standard percentage e.g. 85% for rice (IRRI 2000). The process therefore ensures that new seeds can be harvested to replenish the accession where viability has dropped. In the process, characterization could also be carried out. The amount of seed stored is dependent on the crop type.

Regeneration is not solely based on the duration of storage, storage condition and seed initial viability but also on the results of regular monitoring of seed viability. In large genebank with large collection and with diverse species the ability to predict the regeneration needs into the future means that genebank can plan for alternative sites if capacity is insufficient and the ability to secure funding early. Using viability test results to predict the anticipated drop in viability a Regeneration Expert System (RES) was developed by Chinese Academy of Agricultural Sciences (CAAS) Genebank, as a platform to analyse and manage the monitoring data on viability of accessions and predict the overall viability situation of germplasm conserved in a genebank. It helps curators to make decisions on which accessions need regeneration and to plan for regeneration work many years ahead in advance. The system is based on formula presenting relations between seed viability and storage conditions for specific crop and species (Zhang 2002). The use of decision system and expert systems in managing large genebanks will be the norm in future as information technology is enhanced.

3.9 Intellectual Property, Germplasm Collection and Exchange

Development in the global policy instruments such as the Convention of Biodiversity and the ITPGRFA has changed the process of germplasm acquisition and exchange. Example, IRRI and the other 10 CGIAR centres have placed their genebank ex situ collections under the ITPGRFA. On the IRRI website (http://irri.org/our-science/genetic-diversity/sharing-rice) they provide small quantities of rice free of charge on demand to any individual or organization anywhere in the world for the purposes of research, breeding, or training for food and agriculture in accordance to the ITPGRFA. They would also accept contributions of seeds to add to their collection. Their Information products are released under suitable Open Content License such as Google Books.

3.9.1 Distribution, Exchange and Uses

The distribution and exchange of seeds from genebanks are subjected to issues of Intellectual Property Rights (IPR) and international treaties. For the IRRI genebank and those of international centres, all international germplasm exchanges, with the exception of restoring germplasm to the country of origin, must now be accompanied by a Material Transfer Agreement (MTA) appropriate to the germplasm. The standard MTA is used for the germplasm governed by the Multilateral System (MLS) of the ITPGRFA. For germplasm governed by the Convention on Biological Diversity, exchange is based on a non-standard MTA that must be negotiated each time between the governments of the donor and the recipient. However, breeder or farmer has authority to determine the conditions of exchange without involvement of governmental officials for germplasm where they hold a form of Intellectual Property Rights. With all those changes breeders now have to be exceptionally careful to ensure they have the right to use germplasm in their breeding programmes, and to ensure that they use the material legally. In all these exchanges the plant quarantine rules and regulations of importing and exporting countries should be adhered to.

3.9.2 Procedures for Germplasm Exchange with IRRI

Here is an example on the seed exchange procedure used in IRRI. There are two major sources of rice germplasm those from the Genetic Resources Centre (GRC) which are basically genebank materials and those from the Plant Breeding, Genetics, and Biotechnology (PBGB) division which are improved germplasm with desired traits such as greater yields developed by IRRI.

In PBGB, the International Network for Genetic Evaluation of Rice (INGER) handles most seed requests. INGER distributes materials developed by IRRI, WARDA, CIAT, IITA and NARES. The exchange process requires the seed importer to accept the terms and conditions of the MTA associated with the seeds. The seed importer has to provide documents related to the plant quarantine rules of the importing country. This will include an import permit/licence (if required by importing country), requirements for phyto-sanitary certificate (seed treatment needed), shipping instructions and custom declarations requirements (if needed). Importer should acknowledge the receipt of seeds based on shipping information provided by the exporter.

3.10 Genebank Management and Information System

Most large genebanks have their own genebank management system software (GMS) developed based on various database platforms. The GMS is improved as IT evolved over time. In the 1980s GMS were database system that stored data in text. Today's data includes multimedia besides the traditional text data. As Geographical Information System (GIS) evolved and with faster computer processor speed and larger storage, satellite images and visualization of accession data on maps became common (Fig. 3.3). The internet now provides the means to update data on line and for all users to query the GMS directly on line. The changes in types of data, information and media that can be stored essentially have not changed the way these GMS are structured. They are still based on the categories that are in the descriptor lists. Besides, the GMS developed by genebanks, there are GMS applications that were developed for smaller genebanks and network of genebanks to adopt. A better understanding of the various aspects of genebank documentation can be found in the guidebook for genetic resources documentation (Painting et al. 1995).

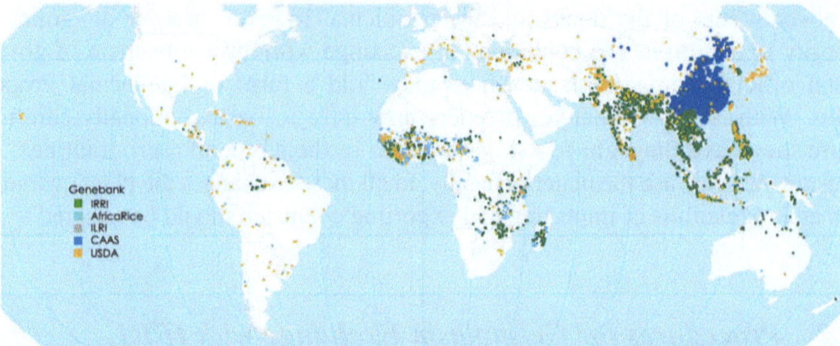

Fig. 3.3 Distribution of rice accessions from five genebanks. *IRRI* International Rice Research Institute, *AfricaRice* Africa Rice Center (formerly known as WARDA), *CAAS* Chinese Academy of Agricultural Sciences, *USDA* United States Department of Agriculture, *ILRI* International Livestock Research Institute (Source: Paule et al. 2010)

Examples of documentation system and genebank related applications that had been promoted to genebanks are;

1. **pcGRIN and GRIN-Global** (Agriculture Research Services, USDA)
2. **BG-Base,** a Collection Management Software
3. **DIVA-GIS** (Geographical Information System)

3.11 Descriptor List

Each seed bank needs a documentation system to ensure that information regarding the accession in storage are maintained and can be accessible to the users. The information stored can be found in the crop descriptor lists. A descriptor list for a crop offers the possibility to facilitate the international exchange and use of plant resources, uniformity in data description, collection and record-keeping. The descriptor list is a set of documentation standards that assist countries to improve their capacity to store, manage and share information about their conserved biodiversity.

Descriptors are the basis of major information platforms such as GENESYS EURISCO, and the FAO World Information and Early Warning System. Descriptors is also an important tool to enable information sharing for crops covered under ANNEX 1 of the International Treaty of Plant Genetic Resources for Food and Agriculture.

Bioversity in collaboration with international partners have developed many crop descriptors and uses the following definitions in genetic resources documentation. These are categories to which descriptors are assigned (Bioversity International and WARDA 2007).

(a) **Passport descriptors:** These provide the basic information used for the general management of the accession (including registration at the genebank and other identification information) and describe parameters that should be observed when the accession is originally collected.
(b) **Management descriptors:** These provide the basis for the management of accessions in the genebank and assist with their multiplication and regeneration.
(c) **Environment and site descriptors:** These describe the environmental and site-specific parameters that are important when characterization and evaluation trials are held. They can be important for the interpretation of the results of those trials. Site descriptors for germplasm collecting are also included here.
(d) **Characterization descriptors:** These enable an easy and quick discrimination between phenotypes. They are generally highly heritable, can be easily seen by the eye and are equally expressed in all environments. In addition, these may include a limited number of additional traits thought desirable by a consensus of users of the particular crop.
(e) **Evaluation descriptors:** The expression of many of the descriptors in this category will depend on the environment; consequently, special experimental designs and techniques are needed to assess them. Their assessment may also require complex biochemical or molecular characterization methods. These types of descriptors include characters such as yield, agronomic performance, stress susceptibilities and biochemical and cytological traits. They are generally the most interesting traits in crop improvement. Characterization will normally be the responsibility of genebank curators, while evaluation will typically be carried out elsewhere (possibly by a multidisciplinary team of scientists). The evaluation data should be fed back to the genebank, which will maintain a data file based on five categories of descriptors.

3.12 Information Sharing Networks

The exchange of seeds is now a complicated process and the ITPGRFA is a process to simplify the exchange of crops listed in its ANNEX 1. Information from genebanks are now more available resulting in development of information sharing networks. Information networks among genebanks can be grouped into international centers, regional and national information networks. Examples of such information networks are listed below.

3.12.1 International Centres Genebank Network

The System-wide Information Network for Genetic Resources (SINGER) is the germplasm information exchange network of the Consultative Group on International Agricultural Research (CGIAR) and its partners (http://singer.cgiar.org/). SINGER members collectively hold more than half a million accessions of crop, forage and tree diversity in their germplasm collections. SINGER provides easy access to information about this diversity from 11 CGIAR centres. The interactive website allows for searching and ordering samples from the various genebanks.

3.12.2 Regional Genebank Network

A European Genebank Integrated System, or AEGIS for short, aims to establish a European Collection, which would be a virtual European Genebank, to be maintained in accordance with agreed quality standards, and to be freely available in accordance with the terms and conditions set out in the International Treaty on Plant Genetic Resources for Food and Agriculture (http://aegis.cgiar.org/). The AEGIS initiative was developed by the network, the European Cooperative Programme for Plant Genetic Resources (ECPG) to bring together information of crops conserved in 650 institutions scattered over about 43 European countries.

3.12.3 National Genebank Networks

The NIAS (National Institute of Agrobiological Sciences) Genebank is an example of a central coordinating institute in Japan for conservation of plants, microorganisms, animals and DNA related to agriculture. The NIAS Genebank coordinates this activity in collaboration with a network of institutes throughout Japan (http://www.gene.affrc.go.jp/about_en.php). An on line search is available for germplasm related to agriculture within the NIAS Genebank system. The databases of the NIAS Genebank include information on passport data, evaluation as well as more general information on genetic resources.

3.13 Benefits and Uses for the Future

Seed stored in gene banks can be regarded as an agriculture insurance policy and are the foundation of humans' food. These are species that have useful traits such as being more drought or disease resistant as well as varieties undergoing genetic erosion. They also provide genetic resources for countries to recover after natural or man-made catastrophes. For example, when the American corn leaf blight visited United States in 1970, the farmers, food manufacturers, and consumers found out just how important genetic diversity is. A hybrid corn developed through artificial breeding with desirable characteristics was used widely by the agricultural community. The shift to an extensive monoculture system resulted in the destruction of the crop due to attack by a fungus called *Bipolaris maydis*. The losses to corn leaf blight were about 710 million bushels and the epidemic illustrates the vulnerability of food crops to pests especially those in monoculture (Tatum 1971). Another example of the use of genebanks can be seen after the tsunami in 2004, farmers in Malaysia and Sri Lanka whose rice fields were badly affected by the salt left behind were provided with salt-tolerant varieties not normally grown in their areas (Shetty 2005). This illustrates the usefulness of seed banks in conserving crop diversity so that in times of need seeds with resistance and partial resistance could be made available to save the food supply. A recent example on the importance of germplasm conservation is "scuba rice" (Barclay 2009). Scuba rice is an IRRI bred variety with origins from an Indian low yielding rice variety called FR13A. It attracted attention due to a flood tolerant trait. The gene responsible for this trait was found and was named gene "SUB1". Insertion of this gene into high-yielding varieties has allowed many rice farmers to cope with frequently flooded rice fields.

The awareness of the importance of genetic diversity and conservation needs to be enhanced. To date, wars and need for housing land or development can make seed banks susceptible to destruction. For example, seed banks in Afghanistan, Egypt and Iraq were ransacked and in 2010, the Pavlovsk seed bank in Russia faced threats from housing development (Vidal 2010). Today the world is confronted by increased population growth leading to food scarcity. In addition we are faced with climate change and other natural disasters. It cannot be disputed from the above mentioned examples that back up and protection of the world's diverse agricultural heritage is of utmost importance not only for this generation but for generations to come as it provides options to deviate from roadblocks placed by nature.

References

Barclay A (2009) Scuba rice: Stemming the tide in flood-prone South Asia. Rice Today 8(2): 26–31

Barona MJ (2011) Mankind takes a giant leap – again. Rice Today 10(3):30–31, International Rice Research Institute

Barton LV, Polunin N (2005) SEEDS: their preservation and longevity. Asiatic Publishing House, Delhi, 216 pp

Benson EE (1999) Cryopreservation. In: Benson EE (ed) Plant conservation biotechnology. Taylor & Francis, London, pp 83–95

Berjak P, Pammenter NW (2004) Recalcitrant seeds. In: Benech-Arnold RL, Sanchez RA (eds) Handbook of seed physiology: applications to agriculture. Haworth Press, New York, pp 305–345

Berjak P, Farant JM, Mycock DJ, Pammenter NW (1990) Recalcitrant (homoiohydrous) seeds: the enigma of their desiccation sensitivity. Seed Sci Technol 18:297–310

Berjak P, Boby Varghese S, Pammenter NW (2011) Cathodic amelioration of the adverse effects of oxidative stress accompanying procedures necessary for cryopreservation of embryonic axes of recalcitrant-seeded species. Seed Sci Res 21:187–203

Bioversity International, IRRI, WARDA (2007) Descriptors for wild and cultivated rice (*Oryza* spp.). Bioversity International/International Rice Research Institute/WARDA, Africa Rice Center, Rome/Los Banos/Cotonou

Brenchley WE (1918) Buried weed seeds. J Agric Sci 9:1–31

Brown HT, Escombe F (1897–1898) Note on the influence of very low temperatures on the germinative power of seeds. Proc R Soc Lond B 62:160–165

Chin HF (1978) Production and storage of recalcitrant seeds in the tropics. Acta Hortic 83:17–21

Chin HF (1994) Seedbanks: conserving the past for the future. Seed Sci Technol 22:385–400

Chin HF, Aziz M, Ang BB, Hamzah S (1981) The effect of moisture and temperature on the ultrastructure and viability of seeds of *Hevea brasiliensis*. Seed Sci Technol 9:411–422

Chin HF, Hor YL, Mohd Lassim MB (1984) Identification of recalcitrant seeds. Seed Sci Technol 12:429–436

Cohen BM (1991) Nikolai Ivanovich Vavilov: the explorer and plant collector. Econ Bot 45(1): 38–46

Concellon A, Anon MC, Chaves AR (2007) Effect of low temperature storage on physical and physiological characteristics of eggplant fruit (*Solanum melongena* L.). LWT- Food Sci Technol 40:389

Consultative Group on International Agricultural Research (1996). Report of the CGIAR Genetic Resources Policy Committee on facing the poverty challenge

Darlington HT (1951) The seventy year period for Dr. Beal's seed viability experiment. Am J Bot 38:379–381

Dulloo ME, Hunter D, Borelli T (2010) Ex situ and in situ conservation of agricultural biodiversity: major advances and research needs. Not Bot Horti Agrobot Cluj 38(2) special issue:123–135

Ellis RH, Hong TD, Roberts EH (1990) An intermediate category of seed storage behaviour? I. Coffee. J Exp Bot 41:1167–1174

Ellis RH, Hong TD, Roberts EH (1991) Effect of storage temperature and moisture on the germination of papaya seeds. Seed Sci Res 1:69–72

Ellis RH, Hong TD, Astley D, Pinnegar AE, Kraak HL (1996) Survival of dry and ultra-dry seeds of carrot, groundnut, lettuce, oilseed rape, and onion during five years; hermetic storage at two temperatures. Seed Sci Technol 24:347–358

FAO/IPGRI (1994) Genebank standards. FAO and IPGRI, Rome

Guest EA (1933) A package of seeds. http://www.goantiques.com/detail,package-seeds-edgar, 765874.html. Accessed on 15 November 2011

Hajari E, Berjak P, Pammenter NW, Paula Watt M (2011) A novel means for cryopreservation of germplasm of the recalcitrant-seeded species, *Ekebergia capensis*. CryoLetters 32(4):308–316

Hanson J, Williams JT, Freund R (1984) Institutes conserving crop germplasm: the IBPGR global network of genebanks. IBPGR, Rome, 25 pp

IBPGR (1976) Report of IBPGR working group on engineering, design and cost aspects of long-term seed storage facilities. International Board for Plant Genetic Resources, Rome

IRRI (2000) Rice genebank operation manual http://www.knowledgebank.irri.org/extension/index.php/grcmanual

Johnson RC, Hodgkin T (1999) Core collections for today and tomorrow. IPGRI, Rome, 81 pp

Kesseler RL, Stuppy W (2006) Seeds time capsule of life. Papadaley, London, 264 pp

Kyu-Won K, Chung H-K, Cho G-T, Ma K-H, Chandrabalan D, Gwag J-G, Kim T-S, Cho E-G, Park Y-J (2007) PowerCore. Bioinformatics, vol 23, issue 16. Oxford University Press, Oxford, pp 2155–2162

Lipman CB (1936) Normal viability of seeds and bacterial spores after exposure to temperatures near absolute zero. Plant Physiol 11:201–205

McDonald MB (1999) Seed deterioration: physiology, repair and assessment. Seed Sci Technol 27:177–237

Narayana Murthy UM, Kumar PP, Sun WQ (2003) Mechanisms of seed ageing under different storage conditions for *Vigna radiata* (L.) Wilczek: lipid peroxidation, sugar hydrolysis, Maillard reactions and their relationship to glass state transition. J Exp Bot 54(384):1057–1067

Painting KA, Perry MC, Denning RA, Ayad WG (1995) Guidebook for genetic resources documentation. IPGRI, Rome

Paule Ma C, Cuerdo JG, Reyes MA, Rala A, Van Etten J, Nelson A, Hijmans RJ (2010) Mapping genebank collections. Rice Today 9(2):30–31

Pérez-García F, González-Benito ME, Gómez-Campo C (2007) High viability recorded in ultra-dry seeds of 37 species of Brassicaceae after almost 40 years of storage. Seed Sci Technol 35:143–153
Pérez-García F, Gómez-Campo C, Ellis RH (2009) Successful long-term ultra dry storage of seed of 15 species of Brassicaceae in a genebank: variation in ability to germinate over 40 years and dormancy. Seed Sci Technol 37(3, 10):640–649
Plucknett DL, Smith NJH, Williams JT, Anishetty NM (1987) Genebanks and the World's food. Princeton University Press, Princeton
Porsild AE, Pharington CR, Mulligan GA (1967) *Lupinus arcticus'* Wats. grown from seeds of pleistocene age. Science 158:113–114
Priestley DA (1986) Seed aging: implications of seed storage and persistence in the soil. Cornell University Press, Ithaca, 304 pp
Pukacka S, Ratajczak E (2005) Production and scavenging of reactive oxygen species in *Fagus sylvatica* seeds during storage at varied temperature and humidity. J Plant Physiol 162: 873–885
Rao NK, Hanson J, Dulloo ME, Ghosh K, Nowell A, Larinde M (2006) Manual of seed handling in genebanks. Handbooks for genebanks no. 8. Bioversity International, Rome
Roberts EH (1973) Predicting the storage life of seeds. Seed Sci Technol 1:499–514
Roberts EH, Abdalla FH (1968) The influence of temperature, moisture, and oxygen on the period of seed viability in barley, broad beans, and peas. Ann Bot 32:97–117
Shen-Miller J (2002) Sacred lotus the long-living fruits of China Antique. Seed Sci Res 12:131–143
Shetty P (2005) Tsunami-hit farmers to grow salt-tolerant rice. http://www.scidev.net/en/news/tsunamihit-farmers-to-grow-salttolerant-rice.html. Accessed 2 Feb 2005
Smith MT, Berjak P (1995) Deteriorative changes associated with the loss of viability of stored desiccation-tolerant and desiccation-sensitive seeds. In: Kigel J, Galili G (eds) Seed development and germination. Marcel Dekker, New York, pp 701–746
Stanwood PC, Bass CN (1981) Seed germplasm preservation using liquid nitrogen. Seed Sci Technol 9:423–427
Tatum LA (1971) The southern corn leaf blight epidemic. Science 171(3976):1113–1116
Vavilov NI (1940) The new systematic of cultivated plants. In: Huxley J (ed) The new systematics. Clarendon Press, Oxford, pp 549–566
Vavilov NI (1957) World resources of cereals, leguminous seed crops and flax, and their utilization in plant breeding. Academy of Sciences of the USSR, Moscow
Vidal J (2010) Pavlovsk seed bank faces destruction. guardian.co.uk, Sunday 8 Aug 2010. Article history: http://www.guardian.co.uk/environment/2010/aug/08/pavlovsk-seed-bank-russia
Walters C (1998) Understanding the mechanisms and kinetics of seed aging. Seed Sci Res 8:223–244
Zhang Z (2002) Plant genetic resources data and information management and public awareness. In: Zhang Z, Zhou M, Ramanatha Rao V (eds) Plant genetic resources network in East Asia. Proceedings of the meeting for the regional network for conservation and use of plant genetic resources in East Asia, 13–16 Aug 2001, Ulaanbaatar, Mongolia. IPGRI Office for East Asia, Beijing

Chapter 4
Pollen Cryobanking for Tropical Plant Species

P.E. Rajasekharan, B.S. Ravish, T. Vasantha Kumar, and S. Ganeshan

4.1 Introduction

Tropical plant species hybridize to create diversity, which is often represented as a hierarchy of discrete units such as species, ecosystems, and landscapes. Pollen has a vital role to play in hybridization, which results in seed production. Conserving pollen ex situ for future use is an important process in tropical species that are wild relatives of crops, especially for conservation of tropical plant genetic resources. The mature male gametophyte of most plant species is desiccation tolerant which render them ideally suitable for storage, especially under cryogenic conditions. A major impact of pollen cryopreservation will be to maintain genetically diverse stocks of pollen collected from wild for future conservation and use, a strategy which can also accomplish conservation of Nuclear Genetic Diversity (NGD) to conserve genetic variation expressed through pollen. This is the only variability that could be tapped under natural situations, especially when information on their breeding behavior is less known. Long term pollen cryopreservation has been attempted successfully in many species for conservation of NGD. Conservation status, methods to conserve NGD has been adequately reviewed, there are some problems related to monocots and recalcitrant species which require more attention.

The life span of pollen at ambient temperature varies with species, ranging from a few hours to several months. Application of cryogenic techniques for conservation of NGD of tropical plant species sourced from wild habitats would enable extended

P.E. Rajasekharan • B.S. Ravish • S. Ganeshan (✉)
Division of Plant Genetic Resources, Indian Institute of Horticultural Research,
Hessaraghatta Lake P.O, Bangalore 560 089, India
e-mail: ganeshans77@vsnl.com

T.V. Kumar
Section of Medicinal crops, Indian Institute of Horticultural Research,
Hessaraghatta Lake P.O, Bangalore 560 089, India

M.N. Normah et al. (eds.), *Conservation of Tropical Plant Species*,
DOI 10.1007/978-1-4614-3776-5_4, © Springer Science+Business Media New York 2013

use of the male gametophyte for providing access to conserved nuclear genetic variability, biotechnology research, besides genetic enhancement of derived crops, which require introgression of pollen transmitted genes from wild sources. Long term ex situ conservation through cryopreservation eliminates the need for frequent sampling of pollen from the wild for assisted pollination needs, leading to sustainable use of tropical plant resources available in situ, and allowing normal evolutionary processes to operate. A ‘Pollen Cryobank’ can maintain genetically diverse stocks of pollen collected from such plants and provide the required male parent in a viable and fertile form for primary and supplementary pollination needs, to improve seed set for species amplification, thereby increasing the chance of species recovery for eco- rehabilitation/eco-restoration programs. Cryopreservation is one of the methods used to conserve plant gene pool components in the form of seeds, pollen and other tissue materials. At this temperature of liquid nitrogen (−196°C), all the metabolic activities is kept under suspended animation resulting in long term conservation (Engelmann 2004).

Pollen is small in size and their tolerance to desiccation makes them suitable for storage. Pollen can be used for international exchange of germplasm, in a Plant Genetic Resources (PGR) management program. Cryopreserved pollen is required for studies in both applied and fundamental aspects of pollen biology (Rajasekharan and Ganeshan 2003). Viability of pollen at ambient temperature varies with time in different species. Hence, viability of pollen needs to be assessed before cryopreservation and at regular intervals, especially after long term cryopreservation. Moisture content plays a major role during cryopreservation, excess moisture in pollen forms ice crystals damaging pollen membranes irreversibly affecting viability (Towill 1985). Alexander and Ganeshan (1993) reviewed the work on pollen storage in fruit crops. Hoekstra (1995) assessed the merits and demerits of pollen as a genetic resource. Ganeshan and Rajasekharan (1995) reviewed work on pollen storage in ornamental crops. Grout (1995) detailed the methodology for pollen cryopreservation. Barnabas and Kovcas (1997) and Berthoud (1997) stressed the importance and need for pollen conservation. Hanna and Towill (1995) reviewed the importance of pollen as a genetic resource for breeding. The techniques for pollen collection, viability assessment (pre- and post –storage), processing for cryopreservation, retrieval and fertility assessment followed for different species has been recently reviewed and presented in detail (Ganeshan et al. 2008). The response to cryopreservation obtained with pollen of some tropical species is presented in this chapter as case studies. Breeders routinely store pollen in liquid nitrogen in the framework of their improvement programs (Towill and Walters 2000).

Pollen, which is an interesting material for genetic resource conservation of various species, is stored by several institutes. In India, the NBPGR conserves cryopreserved pollen of 65 accessions belonging to different species (Mandal 2000) and the Indian Institute for Horticultural Research (IIHR, Bangalore) conserves pollen of 650 accessions belonging to 40 species from 15 different families, some of which have been stored for over 15 year (Ganeshan and Rajashekaran 2000). In the USA, the NCGRP conserves pollen of 13 pear cultivars and 24 Pyrus species (Reed et al. 2000).

4.2 Need and Importance of Pollen Cryopreservation-Breeding Perspective

There are several areas cryopreserved pollen can be of use in crop breeding; these include:

4.2.1 Crossing Involving Wild Relatives

In most instances, the wild relatives of crops or species in general do not adopt well in different climatic conditions due to biotic and abiotic stress under ex situ situations resulting in poor pollen production. The species in question is grown as male parent under ideal environmental conditions and pollen is collected at the time of flowering and stored in liquid nitrogen. Pollen cryopreservation may thus help to overcome this environmental specificity. Asynchronous flowering can also be overcome by cryo-storing pollen of the early or late flowering male parent in liquid nitrogen.

4.2.2 Shuttle Breeding

In order to speed up variety/hybrid development, depending on seasonal advantages, the breeding generations are forwarded to locations across the world and the common conserved/cryopreserved male is used in crosses to develop advanced pre-bred lines/varieties/hybrids in the seed industry. Cryo-biological systems with required quantity of the pollen for crosses could be located in production areas and the pollen used for hybrid seed production.

4.2.3 Maintenance of B Line in Male Sterile Based Seed Production

Often, the maintainer line is required to be grown to restore the fertility of male sterile populations developed by breeders. Pollen cryopreservation of the maintainer lines can be accomplished by growing them under ideal environmental conditions and pollen is collected at the time of flowering and stored in liquid nitrogen for use in different seed producing locations. This would enable increasing the seed production by planting more female parents. When MS line is used, parental increase 'A' line can be accomplished by storing the pollen of 'B' line. Under normal circumstances, appropriate male to female ratio is required to be maintained which could be labor intensive. An increase in seed parent 'A' line can avoid growing of 'B' line which could result in reduction of 15% time required for pollen collection and crossing.

4.2.4 General Crosses

Crosses with cryopreserved male parent can be accomplished repeatedly over time and space in the form of cumulative pollinations to obtain enhanced seed set especially in multi-carpellary fruiting species. Also, to overcome unfavorable weather conditions as in the rainy season, pollen from the common or frequently used parent can be collected, cryostored and used in crosses. Sometimes in fruit tree species, It is easier to import pollen from other established orchards/locations or exotic lines which has a distinct advantage over seed as introductions, reducing the breeding time.

4.2.5 Pollen Storage/Cryopreservation in Cross-Pollinated Species

Pollen storage helps reduction in genetic drift and splitting of adaptive gene complexes, since a temporal genetic snapshot of the genotype x environment variability can be captured and conserved. In most cases the security of male parent is not compromised, ensuring a steady supply of pollen at appropriate times for crossing.

4.3 Extent of Value Addition Contributed by Cryopreserved Pollen in Genetic Enhancement

The availability of pollen with good quality in a pollen Cryobank will provide a constant supply for extended durations. Pollen in such a state can be termed “value added” by virtue of its potentially extended life, for having been able to kept viable and fertile for extended.

At the Pollen Cryobank of IIHR, (Ganeshan and Rajasekharan 2000) information pertaining to 650 pollen samples of more than 45 species belonging to 15 families is stored in a database. The software support provided for this database was FoxPro version –2. For easy management of data one main database and three sub-databases were created. Genus, species and cultivar information is available in a separate database. Information pertaining to longevity, viability and fertility of pollen after cryopreservation, media used for pollen germination, year wise listing of collected samples, year-wise list of samples cryopreserved in different years are available within the database, which is menu driven and user friendly (Ganeshan et al. 2008).

Although genetic conservation through pollen storage does not accomplish whole genome conservation, a breeder involved in genetic enhancement can have access to a Pollen Cryobank facility, for sourcing nuclear genetic variability in his genetic amelioration program. Besides the already existing role of pollen cryobanks in breeding, there are many promising applications, which have come to focus with the recent advances in allied bio-scientific areas. The response to cryopreservation obtained with pollen of some tropical species is presented in Table 4.1.

Pollen of tropical Rare, Endangered &Threatened (RET) plants *Celastrus paniculatus, Oroxylum indicum, Cayratia pedata, Decalepis hamiltonii* and *Holostemma adakodien* (Rajasekharan and Ganeshan 2002), are selected as case studies for investigating their cryogenic behavior. Information on pollen cryopreservation for the above mentioned RET plant species are not available as published literature and no attempts have been made to store pollen even at low temperatures; no studies have been carried out with regard to pollen longevity under ambient conditions. Hence preliminary investigations were attempted and results presented.

4.4 Description of RET Species

4.4.1 *Celastrus paniculatus*

Is a woody climber distributed in Western Ghats. Twigs are brown with peal lenticels. Leaves are alternate, variable in shape; elliptic, oval, broadly obovate or suborbicular. Flowers are white or yellow to green arranged in cyme. Flowers are functionally unisexual and pentamerous; with five stamens; petals are inserted on the margin of cup shaped disk. Fruits are globose orange to red aril covering seeds. Seeds are used in the treatment of paralysis, abdominal disorders, leprosy, skin diseases, fever, beriberi, sores. Seed oil is used in stimulating the intellect. Root powder is used in cancerous tumor treatment (Anonymous 1986).

4.4.2 *Oroxylum indicum*

Is a small to medium sized deciduous tree found in Western Ghats with bipinnate or tripinnate, ovate elliptic leaves, purple fleshy campanulate flowers and long flat sword shaped fruits (capsule) with seeds round, compressed, surrounded by transparent butterfly shaped wing. Bark of the tree is used in treatment of diarrhea, dysentery and rheumatism (Ravikumar and Ved 2000).

4.4.3 *Cayratia pedata*

Is a climber and found distributed in the Western Ghats. Leaves are three or five foliate. Flowers are arranged in Corymbose. Flowers are tetramerous four petals, four stamen, superior ovary. Fruit is a globose berry with 1–4 seeds. Seeds are semiglobose. Leaves are used as astringent and refrigerant (Ravikumar and Ved 2000).

Table 4.1 Pollen cryopreservation in plant species

Botanical name/common name	Duration	Percentage of germination: Before cryostorage	After cryostorage	Reference
Citrus limon	3.5 years	51	52	Ganeshan and Sulladmath (1983)
Juglans regia	1 year	91	65	Alexander and Ganeshan (1993)
Olea europea	Brief	54	51	Alexander and Ganeshan (1993)
Phoenix dactylifera	Brief	56	53	Alexander and Ganeshan (1993)
Prunus amygdalus	Brief	71	72	Alexander and Ganeshan (1993)
Prunus armeniaca	Brief	35	36	Alexander and Ganeshan (1993)
Prunus persica	Brief	55	58	Alexander and Ganeshan (1993)
Annona cherimola	3 months	57.1	13.6	Lora et al. (2006)
Asclepiadaceae	Brief	[a]	[a]	Shashikumar (2006)
Vanilla	Brief	[a]	[a]	Anonymous (2006)
Pecteilis	Brief	[a]	[a]	Anonymous (2006)
Habenaria	Brief	[a]	[a]	Anonymous (2006)
Satyrium	Brief	[a]	[a]	Anonymous (2006)
Solanaceae	Various periods	45.96	20.70	Rajasekharan et al. (1998); Rajasekharan and Ganeshan (2003)
S.melongena 2 y (capsicum, tomato, *Solanum* 10 species)				
Allium sepa	360 days	45.13	46.90	Ganeshan (1986a)
Catharanthus roseus	Brief	[a]	[a]	Shashikumar (2006)
Punica granatum	Brief	[a]	[a]	Ganeshan and Rajasekharan (2000)
Citrus aurantifolia	1 year	37.5	27.4	Rajasekharan et al. (1995)
C. limon				
Poncirus trifoliata		40.1	34.9	
Gladiolus	10 years	56.37	54.07	Rajasekharan et al. (1994)
Rosa sp.	1 year	62.5	63	Rajasekharan and Ganeshan (1994)
Carica papaya	5 years	65	52	Ganeshan (1985a, 1986)
Vitis vinifera L.	5 years	73	76	Ganeshan (1985)
Mangifera indica	5.5 years	[a]	[a]	Rajasekharan and Ganeshan (2002a)
Oil palm	8 years	[a]	[a]	Tandon et al. (2007)

[a]Tested through field pollinations

4.4.4 *Decalepis hamiltonii*

Is a climbing shrub found in open rocky slopes and rocky cervices of dry to moist forest region of Western Ghats. The plant exudes latex when cut, which is sticky and milky; leaves are distinctly greenish pink, opposite orbicular or elliptic, obovate. Flowers are arranged in cymes. Stigma is flat pentangular with five depressions one at each angle in which pollenia is located. Fruit follicles are cylindrically oblong and woody when dried. Seeds are egg shaped with long silky hair. Roots are highly aromatic. It is used as appetizer, blood purifier and preservative besides serving as a source of bioinsecticide for stored food (Ravikumar and Ved 2000).

4.4.5 *Holostemma adakodien*

A climber occurring in Western Ghats. Leaves are ovate, coriaceous, and finely pubescent. Flowers are arranged in cyme, pinkish purple fleshy actinomorphic, hermaphrodite polygamous, pentamerous and epigynous. Calyx has five inconspicuous teeth or a narrow circular ridge at the top of the ovary five petals and five sepals Stamen arises from epigynous disc. Filaments are inflexed fruits are thick boat shaped. Seeds are many, oval, winged along margin with silky white hair at apex, used in the treatment of ophthalmopathy, orichitis, cough, fever, tuberculosis and also as expectorant, tonic stimulant (Ravikumar and Ved. 2000).

4.5 Protocols for RET Species

Pollen collection is the critical step wherein care should be taken to collect it as soon as anthesis takes place, before pollinators are attracted and it is free from contaminants. On the day of collection, the flowers were harvested and brought to the laboratory. Petals were carefully separated and pollen grains were extracted by scraping the mature anthers which were about to dehisce.

4.5.1 *Flowering Time*

In *O. indicum*, the flower blooms at night and fall by morning indicative of anthesis, these flowers were used for pollen extraction. Extracted pollen was dried to attain optimum moisture content. In *C. paniculatus* flowers bloom in the morning and in *C. pedata* by afternoon. Since the pollen size is small in these species,

Table 4.2 Flowering period, blooming, moisture content and germination times for RET plant species

Species	[a]Flowering period	Blooming time	Pollen/ pollenia	Moisture content (%)	Hours for optimum germination	
					Fresh (h)	Cryostored
C. paniculatus	April–May	Morning	Pollen	72.5	1.5	1.5
O. indicum	May–August	Night	Pollen	65.5	2	2
C. pedata	June–July	Afternoon	Pollen	75	1	1
D. hamiltonii	April–May	Morning	Pollenia	72	2	3
H. adakodien	July–August	Morning	Pollenia	72	2	4

[a]In hours

freshly dehisced anthers were collected. In *D. hamiltonii* and *H. adakodien* pollen aggregates in the form of pollenia, were collected for the present study. Pollen/pollenia samples were obtained from plants collected from wild and domesticated at Indian Institute of Horticultural Research field gene bank (located at an elevation of 890 M from mean sea level at latitude of 13°N and longitude of 77°37E. The average rainfall is about 1,400 mm per annum). The details of flowering period, blooming time and form of pollen in the RET plant species, along with initial germination time needed in presented in Table 4.2.

4.5.2 Pollen Samples

Viability was assessed by initial germination using Brewbaker and Kwack medium, consisting of 15% sucrose, supplemented with 100 μg/ml Ca $(NO_3)_4$ by the hanging drop method. Pollen dispersed in hanging drop position was incubated in a moist chamber at $25 \pm 2°$ C (Ganeshan et al. 2008). After germination pollen was stained with Alexander stain (Alexander 1980) and viewed under microscope. Germinated pollen was counted and percentage germination computed. Five replicates were assessed with ten microscopic fields taken for count.

4.5.3 Cryopreservation

Fresh pollen was enclosed in gelatin capsule which were sealed in aluminum pouches and stacked in canisters. These canisters were completely immersed in liquid nitrogen, and frozen at −196°C in a Mach SM-43, MVE USA make cryobiological system. Care was taken to minimize the moisture content in pollen before cyropreserving, since presence of moisture in pollen may form ice crystals and cause irreversible damage to membranes, (Ganeshan et al. 2008) affecting pollen viability. Pollen was cryopreserved for 1 week before viability was assessed, samples were taken out of the cryobiological system and were thawed to ambient temperature and germination was carried out similar to fresh pollen.

Table 4.3 Mean, standard deviation and calculated T – values for RET medicinal plants

Species	Fresh pollen germination (mean)	Cryopreserved pollen germination (mean)	T value
C. paniculatus	75.91 (60.58)	52.15 (46.21)	18.942[a]
O. indicum	82.55 (65.22)	70.62(57.12)	15.190[a]
C. pedata	91.58 (73.19)	93.20 (75.24)	1.485 NS
D. hamiltonii	90.18 (71.72)	89.00 (70.61)	2.751 NS
H. adakodien	89.41 (71.02)	87.00 (68.97)	1.043 NS

Figures in *parenthesis* indicate corresponding transformed values
[a]Highly significant at 1%; table T value 5% = 2.776: 1%4.604

4.5.4 Statistical Analysis

The data was subjected to't-test', after converting the percentage germination values to angular transformed values (Arcsine conversions) (Panse and Sukhatme 1989). The details of flowering period, blooming time and form of pollen in the five different RET plant species along with initial germination time needed in presented in Table 4.2.

4.6 Response to Cryopreservation

Fresh pollenia of *D. hamiltonii* and *H. adakodien* showed germination within 2 h of incubation in Brewbaker and Kwack medium where as cryopreserved pollenia of *D. hamiltonii* germinated after 3 h and *H. adakodien* after 4 h of incubation (Table 4.2). In the other three species, there was no change in time taken for germination.

Pollen grains from the five RET plant species can be recovered viable after cryogenic storage in various proportions. There was a variable response to germination of cryopreserved pollen among RET species (Table 4.3). There was a significant (P = 0.05) decline in germination profiles, in *O. indicum* and *C. paniculatus* after 1 week of cryopreservation (Vikas et al. 2009); pollen of *C. pedata* showed a slight increase in germination, which however, was not statistically significant. Similarly, pollenia of *H. adkodien* and *D. hamiltonii* retained their germination profiles after 1 week of cryostorage, although the initiation of the same was considerably delayed by 1–2 h.

The differential responses to cryostorage among pollen of the five tropical RET species, could be attributed to their inherent genetic and physiological status and the level of maturity among the samples collected for the study. Such responses are commonly encountered with species of tropical origin (Towill 1985).

4.7 Conclusions

Pollen of tropical RET plant species are amenable to cryopreservation; impact of such a storage over long term duration needs to be investigated. Such studies are useful for establishing pollen cryobanks for RET species, for assisted pollinations and conserving nuclear genetic variability, maintaining genetically diverse stocks of pollen collected from such species for undertaking region specific ex situ conservation programs. This technique conserves variability in the male gametophyte perhaps is the only available genetic variation expressed through pollen of these tropical RET species, especially when their breeding systems are less understood. Further, it is essential to ascertain the fertilizing ability of cryopreserved pollen in these species, which could facilitate primary and supplementary pollination needs for improving seed set under species amplification projects, attempting to enhance the probability of species recovery for their eco-rehabilitation/eco-restoration in different tropical habitats. A pollen Cryobank can provide a constant supply of viable and fertile pollen to allow supplementary pollinations for improving seed set (Ganeshan and Rajasekharan 2000)

References

Alexander MP (1980) A versatile stain for pollen, fungi, yeast and bacteria. Stain Technol 53:13–18

Alexander MP, Ganeshan S (1993) Pollen storage. In: Chadha KL, Pareek OP (eds) Fruit crops Part I, vol I, Advances in horticulture. Malhotra Publishing House, New Delhi, pp 481–486

Anonymous (1986) The wealth of India: a dictionary of Indian raw materials and industrial products. Publication and information Directorate/CSIR, New Delhi

Anonymous (2006) Establishment of seed and pollen Cryobank for ex situ conservation and sustainable utilization of orchids of Western Ghats – (2005). Project report finding submitted to the Department of Biotechnology, Government of India by William Decruze Nair GM and Ganeshan S

Barnabas B, Kovcas G (1997) Storage of pollen. In: Shivanna KR, Sawney VK (eds) Pollen biotechnology for crop production and improvement. Cambridge University Press, Cambridge, UK, pp 293–314

Berthoud J (1997) Strategies for conservation of genetic resources in relation with their utilization. Euphytica 96:1–12

Engelmann F (2004) Plant cryopreservation: progress and prospects. In Vitro Cell Dev Biol Plant 40:427–433

Ganeshan S (1985a) Cryogenic preservation of Grape (*Vitis vinifera* L.) Pollen. Vitis 24:169–173

Ganeshan S (1985b) Storage and longevity of papaya (*Carica papaya* L. 'Washington') Pollen effect of low temperature and humidity. Gartenbauwissenchaf 50:227–230

Ganeshan S (1986) Viability and fertilizing capacity of onion pollen (Allium cepa L.) stored in liquid nitrogen. Trop Agric (Trinidad) 63:46–48

Ganeshan S, Rajasekharan PE (1995) Genetic conservation through pollen storage in ornamental crops Chadha KL, Bhattacharjee SK (Eds.) Malhotra Publishing House New Delhi. Advances in Horticulture Part-1 Ornamental Crops, pp 87–108

Ganeshan S, Rajasekharan PE (2000) Current Status of pollen cryopreservation research: relevance to tropical horticulture. Cryopreservation of tropical plant germplasm programs and application JIRCAS/IPGRI publication (Eds: Engelmann and H.Takagi), pp 360–365

Ganeshan S, Rajasekharan PE, Shashikumar S, Decruze W (2008) Cryopreservation of pollen. In: Reed B (ed) Plant cryopreservation: a practical guide. Springer, Berlin, pp 443–464

Grout BWW (Ed.) (1995) Genetic Preservation of Plant Cells In Vitro Springer Verlag, Berlin

Hanna WW, Towill LE (1995) Long term pollen storage. In: Janick J (ed) Plant breeding reviews. Wiley, New York, pp 179–207

Hoekstra FA (1995) Collecting pollen for genetic resources conservation. In: Rao VR, Reid R, Guarino L (eds) Collecting plant genetic diversity. Technical guidelines. CAB International, Wallingford, pp 527–550

Lora J, Perez de Oteyza MA, Fuentetaja P, Hormaza JI (2006) Low temperature storage and in vitro germination of cherimoya (*Annona cherimola* Mill.) pollen. Scientia Horticulturae 108: 91–94

Mandal BB (2000) Cryopreservation research in India: current status and future perspectives. In: Engelmann F, Takagi H (eds) Cryopreservation of tropical plant germplasm – current research progress and applications. IPGRI/JIRCAS, Rome/Tsukuba, pp 282–286

Panse VG, Sukhatme PV (1989) Statistical methods for agricultural workers. ICAR, New Delhi, 359

Rajasekharan PE, Ganeshan S (1994) Freeze preservation of rose pollen in liquid nitrogen: Feasibility viability and fertility status after long term storage. J Hot Sci 69:565–569

Rajasekharan PE, Rao TM, Janakiram T, Ganeshan S (1994) Freeze preservation of Gladiolus pollen in liquid nitrogen: Feasibility, viability and fertility status after long-term storage. Euphytica 80:115–119

Rajasekharan PE, Alexander MP, Ganeshan S (1998) Long term pollen preservation of wild tomato, brinjal species and cultivars in liquid nitrogen. Indian J Pl Gen Resour 11:117–120

Rajasekharan PE, Ganeshan S (2002) Conservation of medicinal plant biodiversity – an Indian perspective. J Med Arom Plant Sci 24:132–147

Rajasekharan PE, Ganeshan S (2003) Feasibility of pollen cryopreservation in Capsicum sp. Capsicum eggplant News Letter 22:87–90

Ravikumar K, Ved DK (2000) 100 red-listed medicinal plants of conservation concern in Southern India, 1st edn. Foundation for Revitalization of Local Health Traditions (FRLHT), Bangalore, December

Reed BM, DeNoma J, Chang Y (2000) Application of cryopreservation protocols at a clonal genebank. In: Engelmann F, Takagi H (eds) Cryopreservation of tropical plant germplasm – current research progress and applications. IPGRI/JIRCAS, Rome/Tsukuba, pp 246–249

Shashikumar S (2006) Pollen biology of few horticulturally important plants. PhD Thesis. Bangalore University Bangalore

Tandon R, Chaudhury RK, Shivanna R (2007) Cryopreservation of oil palm pollen. Current Science L 92(2):181–183

Towill LE (1985) Low temperature and freeze/vacuum-drying preservation of pollen. In: Kartha KK (ed) Cryopreservation of plant cells and organs. CRC Press, Boca Raton, pp 171–198

Towill LE, Walters C (2000) Cryopreservation of pollen. In: Engelmann F, Takagi H (eds) Cryopreservation of tropical plant germplasm – current research progress and applications. IPGRI/JIRCAS, Rome/Tsukuba, pp 115–129

Vikas M, Gautam R, Tandon HY, Mohan Ram (2009) Pollination ecology and breeding system of *Oroxylum indicum*(Bignoniaceae) in the foothills of the Western Himalaya. J Trop Ecol 25:93–96

Chapter 5
In Vitro Genebanks for Preserving Tropical Biodiversity

Barbara M. Reed, Sandhya Gupta, and Esther E. Uchendu

5.1 Introduction

Much of the world's germplasm of vegetatively propagated crops is maintained as clonal collections in field genebanks, plantations or orchards, such as seedlings of rubber, coconut and oil palm or species of tropical fruit such as citrus, cacao and banana, or root and tuber crops. Clonal collections also exist for important staple food crops such as cassava, sweet potato and yams, and aroids such as *Colocasia* and *Xanthosoma* or spices, plantation crops or medicinal and aromatic plants. These field genebanks often do not represent the entire range of genetic variability within the respective crop genepool and most of them represent only a fraction of the variability which should be conserved (Reed et al. 2011; Withers and Williams 1986). The need for increased diversity in genebanks was highlighted by inclusion of a statement on the importance of ex situ conservation in the ninth article of the Convention on Biological Diversity (Pennisi 2010). In vitro methods are well documented as being beneficial for secondary storage and conservation planning (Benson 1999; Engelmann 1997; Sarasan 2011; Sarasan et al. 2006). In vitro propagation and conservation also contribute to the maintenance of natural populations through

B.M.Reed (✉)
United States Department of Agriculture-Agricultural Research Service,
National Clonal Germplasm Repository, 33447 Peoria Rd, Corvallis,
OR 97333-2521, USA
e-mail: Barbara.Reed@ars.usda.gov

S. Gupta
Tissue Culture and Cryopreservation Unit, National Bureau of Plant
Genetic Resources, Pusa Campus, New Delhi 110 012, India
e-mail: sandhya@nbpgr.ernet.in

E.E. Uchendu
Formerly Department of Plant Agriculture,
University of Guelph, Guelph ON N1G 2W1, Canada

M.N. Normah et al. (eds.), *Conservation of Tropical Plant Species*,
DOI 10.1007/978-1-4614-3776-5_5, © Springer Science+Business Media New York 2013

the reintroduction of preserved material to the original habitat (Pence 2011). A wide range of in vitro conservation tools are available for conserving plant biodiversity (Noor et al. 2011; Reed et al. 2011).

5.2 In Vitro Culture for Conservation

In vitro culture is an effective method for ex situ conservation of plant genetic diversity, allowing multiplication from small amounts of plant material, with low impacts on wild populations (Fay 1994). Methods used for in vitro germplasm storage must meet two basic criteria; first, minimal maintenance for an extended period of time and second, stability of the genetic integrity of the stored material. Low maintenance in vitro germplasm collections provide an alternative to growing plants as field collections, in nurseries or in greenhouses (Ashmore 1997; Reed 1999; Taylor 2000). The advantages of using in vitro methods include (a) conservation of: vegetatively propagated plants, plants with recalcitrant seeds, as well as endangered species, and (b) aid in the establishment of germplasm banks, storage equipments and cryopreserved collections and also (c) exchange of germplasm at the international level. Conservation of plant genetic resources using these in vitro procedures is in progress in many countries (Table 5.1; Sect. 5.5 this chapter). In vitro storage provides a platform for the collection, conservation, documentation, distribution, exchange of germplasm information and technology across the world. This results in capacity building in the area of tropical crop management and utilization, and increases the preservation of valuable genes which are needed for plant improvement.

5.2.1 Medium-Term In Vitro Storage

Short- or medium-term storage encompasses slow-growth strategies ranging from temperature reduction and other environmental manipulation to chemical additions in the growth medium. Slow-growth strategies are used in a number of institutions throughout the world for the preservation and distribution of clonally propagated plant germplasm (Ashmore 1997). Large collections of tropical crops held in vitro include germplasm at some international centers: cassava at Centro International de Agricultura Tropical (CIAT) in Columbia, banana and plantain at Bioversity International's Transit Center (ITC) in Belgium, potato at the International Potato Center (CIP) in Peru, taro and yam at the Centre for Pacific Crops and Trees (CePaCT), the Secretariat of the Pacific Community in Fiji and yam at the International Institute of Tropical Agriculture (IITA) in Nigeria . Viable tissue culture techniques are the main requirement for this form of storage. Techniques for some crops require additional research, but methods for others are well defined. Slow growth conditions provide a secondary storage method for clonal field collections, a storage mode for

Table 5.1 Some of the in vitro tropical plant collections conserved in genebanks

Genebank/location	Plant taxa	Common name	Storage form	Reference/websites
Argentina	*Solanum tuberosum*	Potato	In vitro cryopreservation	http://inta.gob.ar/proyectos/planes-tecnologicos
Instituto Nacional de Tecnologia Agropecuaria (INTA) Balcarce and Castelar	*Ipomoea batatas*	Sweet potato		
	Manihot esculenta	Cassava		
Australia	*Musa*	Banana	In vitro	https://sites.google.com/a/cgxchange.org/musanet/musa-collections/collections-in-asia-and-pacific/australia---qdpi
Queensland Horticulture Institute, Maroochy Research Station, Nambour	*Carica* spp., *Interspecific hybrids*	Papaya		
Australia	*Conostylis dielsia*	Flora of Western Australia	In vitro	http://www.bgpa.wa.gov.au/
The Plant Science Laboratory at Kings Park and Botanic Garden, Perth	*C. micrantha*	Endangered species		
	C. wonganensis, *Anigozanthos viridis*			
	A. humilis			
	A. kalbar-riensis			
Australia	*P. saxicola*	Orchid	seed	http://www.rbgsyd.nsw.gov.au/
The Royal Botanic Gardens and Domain Trust, Sydney	*D. arenaria*			
	D. flavescens			
	D. bracteata			
Barbados	*Saccharum officinarum*	Sugar cane	In vitro	http://wicscbs.org/
West Indies Central Sugar Cane Breeding Station				

(continued)

Table 5.1 (continued)

Genebank/location	Plant taxa	Common name	Storage form	Reference/websites
Caribbean Agricultural Research and Development Institute (CARDI)	*Ipomoea batatas*	Sweet potato	Field and some in vitro	www.cardi.org
Ministry of Agriculture and Rural Development (MARD)		Cassava Yam Pineapple Onion Pigeon-pea Maize Hot peppers Fruit crops Cut flowers	In vitro and field	www.fao.org/fileadmin/templates/agphome/…/**BARBADOS**.PDF
Belgium	*Musa* spp.	Banana plantain	In vitro	http://bananas.bioversityinternational.org/es/partnerships-mainmenu-34/58-partnerships/98-inibap-.html
Katholieke Universiteit Leuven; Bioversity International			Apical meristem	
Bolivia	*Ananas comosus, 6, Ipomoea batatas, 35*	Fruit and vegetable crops	In vitro	http://www.fao.org/docrep/013/i1500e/Bolivia.pdf
Banco de Germoplasma, Santa Cruz de la Sierra	*Manihot esculenta 49* *Musa* spp. *19 Solanum tuberosum* *Stevia rebaudiana*			
Bolivia	*Canna edulis 2*	Root and tuber crops	In vitro	Biodiversity in trust: conservation and use of plant genetic resources in CGIAR Centres. By Dominic Fuccillo, Linda Sears, Paul Stapleton

Sistema Nacional de Conservación y Desarrollo de Recursos Genéticos de Bolivia, Cochabamba	*Oxalis tuberosa*, 463 *Pachyrhizus ahipa 2* *Polymnia sonchifolia* 6 *Solanum* spp.1394 *Tropaeolum tuberosum*, 64 *Ullucus tuberosus 114*			
Brazil	*Bromeliad*	Bromeliads	In vitro backup for field collections	http://www.cenargen.embrapa.br/
EMBRAPA Mandioca e Fruticultura Tropical, Cruz das Almas	*Ananas* (220)	Pineapple		
Brazil	*Ananas121*	Fruit and vegetable crops	In vitro	
EMBRAPA Recursos Genéticos e Biotecnologia, Brasília	*Rubus5* *Asparagus11* *Arachis10* *Vanilla 15* *Solanum 339* *Ipomoea batatas20* *Bromelia 39* *Dioscorea 13* *Stevia rebaudiana12* *Pfaffia glomerata 329* *Lippia 40* *Manihot 549* *Mentha 67* *Fragaria 1* *Vitis 28* *Syngonanthus 1* *Schumbergera 1* *Pereskia1*			

(continued)

Table 5.1 (continued)

Genebank/location	Plant taxa	Common name	Storage form	Reference/websites
	Lavandula1			
	Costus1			
	Sinningia1			
	Musa 8			
	Phaseolus15			
	Uncaria1			
	Piper2			
	Psychotria 9			
	Solanum 162			
	Pilocarpus 30			
China Institute of Crop Germplasm Resources, CAAS and Keshan National Gene Bank for Potato	*Solanum* (600 accessions)	Potato	In vitro	http://icgr.caas.net.cn/cgris_english.html
China Institute of Crop Germplasm Resources CAAS and Xuzhou National Gene Bank for Sweet Potato	*Ipomoea batatas* (1,000 accessions)	Sweet potato	In vitro	http://icgr.caas.net.cn/cgris_english.html
Guangdong Academy of Agricultural Sciences (GAAS)/Research Institution of Fruit Tree (FTRI)	*Musa* (210 accessions)	Banana	In the field or in vitro	
Colombia Centro International de Agricultura Tropical (CIAT), Cali	*Manihot* spp. (>6,500 accessions)	Cassava	In vitro	http://www.ciat.cgiar.org/Paginas/index.aspx

Costa Rica	*Dioscorea alata*	Yam	In vitro	www.catie.ac.cr/
	Dioscorea esculent	Cassava		
	Dioscorea dumetotum	Fruit		
	Dioscorea cayenensis	Coffee		
	Dioscorea pentaphylla	Cocoa		
	Dioscorea trífida			
Centro Agronómico Tropical de Investigación y Enseñanza (CATIE)	*Dioscorea bulbifera* *Manihot esculenta*			
Ethiopia	*Brachiaria*	Grasses	In vitro	http://www.fao.org/wairdocs/ilri/x5491e/x5491e0b.htm
Forage Agronomy Group (FLAG) at International Livestock Center for Africa (ILCA), Addis Ababa				
Fiji	*Colocasia esculenta*	Sweet potato, taro, banana, cassava, yam, vanilla	In vitro	www.apaari.org/wp-content/plugins/download…/download.php?id…
PRAP (Pacific Regional Agricultural Programme), SPC, Suva	*Ipomoea batatas*			
	Dioscorea esculenta			
	Manihot esculenta			
	Musa			
France	*Dioscorea* spp.	Yam	In vitro	http://en.ird.fr/
GeneTrop, GAP unit, IRD, Montpellier			Meristem	
Germany	*Allium* spp.	Garlic	In vitro	http://www.ipk-gatersleben.de/Internet/Forschung/Genbank
IPK Leibniz Institute of Plant Genetics and Crop Plant Research, Gatersleben	*Capsicum* spp.	Pepper		

(continued)

Table 5.1 (continued)

Genebank/location	Plant taxa	Common name	Storage form	Reference/websites
Germany	*Allium*	Garlic	In vitro	http://www.ipk-gatersleben.de/Internet/Forschung/Genbank/InvitroErhaltung
Leibniz Institute of Plant Genetics and Crop Plant Research (IPK), Gatersleben	*Solanum tuberosum*	Potato		
Ghana	*Dioscorea* spp.	Yam	In vitro	http://www.cropsresearch.org/
CSIR-Crops Research Institute, Kumasi				
Ghana	*Dioscorea* spp.	Yam	In vitro	http://www.fao.org/docrep/013/i1500e/i1500e01.pdf
Tissue Culture Laboratory, University of Ghana	*Manihot* spp.	Banana Plantain Cassava, Cocoyam		
	Musa spp.			
	Xanthosoma spp.			
India	*Dioscorea*, *Colocasia*, *Alocasia*, *Xanthosoma*, *Musa*, *Morus*, *Aegle marmelos*, *Zingiber*, *Curcuma*, *Piper*, *Vanilla*, *Simmondsia*	Sweet potato, yam, taro, banana, mulberry, *bael*, piper, ginger, turmeric, elephant foot yam, jojoba, vanilla	In vitro	http://www.nbpgr.ernet.in/
National Bureau of Plant Genetic Resources, Delhi				
India	*Bambusa*	Trees	In vitro	http://www.tbgri.in/
Tropical Botanic Gardens and Research Institute in Trivandrum, Kerala		Bamboo	Meristem	
		Orchid		
India, National Research Centre on Banana (NRCB), Trichy, Tamil Nadu	*Musa*, 145 acc	Banana	In vitro	http://www.nrcb.res.in/index.html
Japan	*Morus*	Mulberry	Cryopreservation (dormant buds)	http://www.gene.affrc.go.jp/plant/SEARCH/outline.html
NIAR (National Institute of Agrobiological Resources)		Taro	Meristems	
		Yam		

Kenya	*Cenchrus*	Grass	In vitro	http://www.pgrfa.org/gpa/ken/nsc.html
Kenya Agricultural Research Institute's National Genebank (KARI-NGBK)	*Pyrethrum*	Medicinal plants, trees		
	Citrus	Citrus		
Malaysia	*Shorea leprosula*	Tropical forest trees	Embryo rescue	http://www.frim.gov.my/
Forest Research Institute				
Malaysia, Kuala Lumpur				
Mexico	*Opuntia* spp.	Cactus pear	In vitro backup for some	http://cactus-congress.manesk-ovtravel.com/files/Posters/A%20Global%20Perspective%20Prelim_Enza.pdf?PHPSESSID=b5dc3614d59f507d66b7b1be083e316c
National Center for Genetic Resources CENARGEN				
Tepatitlan, Jal				
Nigeria	*Ananas comosus*	Pineapple	In vitro	Personal communication
National Center for Genetic Resources and Biotechnology (NACGRAB)	*Phoenix dactilifera*	Date palm		Deputy Director, Dr. Aladele http://www.nacgrab.gov.ng/biotechnology
	Saccharum spp.	Sugar cane		
Nigeria	*Dioscorea* spp.	Yam	In vitro	http://www.iita.org/genetic-resources-center
International Institute of Tropical Agriculture (IITA) Ibadan	*Manihot esculenta*	Cassava	Meristem	
Pakistan	*Solanum tuberosum*	Potato	Buds	http://www.parc.gov.pk/pgri.html
Plant Genetics Resources Program, National Agricultural Research Center, Islamabad	*Ipomoea batatas*	Sugarcane	In vitro	
	Musa	Banana		
	Piper betle	Betel leaf		
	Saccharum spp.	Sweet potato		

(continued)

Table 5.1(continued)

Genebank/location	Plant taxa	Common name	Storage form	Reference/websites
New Guinea, Papua	*Ipomoea batatas*	Sweet potato	Backup in vitro	www.pgrfa.**org**/gpa/png/docs/PNG_report_GPA_monitoring.pdf
National Agriculture				
Research Institute				
Coffee Research Institute	*Coco nucifera*	Coconut	From embryos	
Kainantu; Cocoa and Coconut Research Station, Madang	*Coffea*	Coffee	In vitro	
Peru	*Solanum tuberosum*	Potato	In vitro	http://www.cipotato.org/genebank/
International Potato Research Institute, Lima	*Ipomoea batatas*	Sweet potato		
Philippines	*Musa* spp.	Banana	In vitro	https://sites.google.com/a/cgxchange.org/musanet/musa-collections/collections-in-asia-and-pacific/philippines---npgrl-uplb
National Plant Genetic Resources Laboratory IPB/UPLB				
Romania	*Solanum tuberosum*	Potato	In vitro	http://www.svgenebank.ro/activi-ties_invitro.asp
SVGB, Biodiversity for Food Security, Suceava			Node cuttings	
Russian Federation Timiryazev Institute of Plant Physiology, Russian Academy of Sciences, Moscow	*Solanum tberosum*	Potato	In vitro	http://www.ippras.ru/
	Rosa spp.	Rose	Shoot tips	http://translate.google.com/translate?sl=ru&tl=en&js=n&prev=_t&hl=en&ie=UTF-8&layout=2&eotf=1&u=http%3A%2F%2Fwww.ippras.ru%2F&act=url
	Rubus spp.	Raspberry		
South Africa	*Solanum tuberosum*	Potato	In vitro	http://www.arc.agric.za/
Agricultural Research Council, Pretoria				
Spain	*Solanum tuberosum*	Potato	In vitro	http://www.neiker.net/neiker/germoplasma/ingles/ingles.html
NEIKER-Instituto Vasco de Investigacion y Desarrollo Agrario, Vitoria				

Tanzania	*Manihot* spp.	Cassava	In vitro	http://www.spgrc.org.zm/index.php?option=com_content&view=article&id=21&Itemid=34
National Plant Genetic Resources Center (NPGRC), Arusha				
United Kingdom	*Angraecum*	Cactus	In vitro	http://www.kew.org/science/micro_cons_prog.html
	Paphiopedilum	Palm		
	Phragmipedium	Orchid		
	Cypripedium	Tree species		
USA, Georgia	*Ipomoea batatas*	Sweet potato	Field and in vitro	http://www.ars.usda.gov/main/site_main.htm?modecode=66-07-00-00
		Sorghum		
Plant Genetic Resources Conservation Unit (PGRCU), Griffin		Peanut Subtropical and Tropical legumes Warm-season grass		
USA, Hawaii	*Ananas comosus*	Pineapple	*Ananas* in vitro	http://www.ars.usda.gov/main/site_main.htm?modecode=53-20-03-65
	Artocarpus altilis spp.	Breadfruit	Field collections of other tropical fruit	
	Averrhoa carambola	Starfruit		
	Bactris gasipaes spp.	Peach palm		
	Canarium ovatum spp.	Pili nut		
	Carica papaya spp.	Papaya		
	Dimocarpus longan	Longan		
	Litchi chinensis	Lychee		
	Macadamia integrifolia	Macadamia		
	Malpighia glabra	Acerola		
	Nephelium lappaceum	Rambutan		
	Nephelium rambutan-ak	Pulasan		

(continued)

Table 5.1(continued)

Genebank/location	Plant taxa	Common name	Storage form	Reference/websites
	Passiflora edulis f. flavicarpa and spp.	Passion fruit		
	Psidium guajava	Guava		
	Camellia sinensis	Tea		
USA, Hawaii	*Artocarpus altilis*	Breadfruit	In vitro backup	www.ntbg.org/breadfruit/resources/
National Tropical Botanic				
Garden (NTBG), Maui				
USA, Ohio	*Saintpaulia*	African violet	In vitro	http://cincinnatizoo.org/conservation/crew/crew-plant-research/
Cincinnati Zoo and Botanic Garden	*Asimina tetramera*	Pawpaw		
	Hedeoma todsenii	Native plants		
USA, Puerto Rico	*Musa, Bambusa*	Banana	Field and in vitro	http://www.reeis.usda.gov/web/crisprojectpages/403046.html
USDA, Agricultural Research Service, Tropical Agriculture Research Station Mayaguez	*Cacao, Atemoya, Sugar Apple, Custard Apple, Sapodilla, Mamey Sapote, and Garcinia* spp.	Bamboo		
		Tropical fruit crops		
USA, Wisconsin	*Solanum* spp. (~6,000 accessions)	Potato	In vitro	http://www.ars-grin.gov/nr6/
US Potato Genebank USDA, Agricultural Research Service, Sturgeon Bay NRSP-6			Seed	
			Field	

experimental material, or a reserve of germplasm for plant distribution. In vitro storage is not often the only form of germplasm held in genebanks, but when it is, it is a versatile tool that can provide readily available, disease indexed plants for distribution to scientists or farmers. Medium-term conservation is obtained under slow-growth conditions by considering: (1) physiological stage of the explant, (2) osmotic agents and growth inhibitors, (3) reduced temperature, (4) medium alterations such as reduced mineral or sucrose concentrations, (5) reduced oxygen or (6) alginate encapsulation (Dulloo et al. 1998; Engelmann 1997, 1999; Harding et al. 1997; Rai et al. 2008; Reed et al. 2005; Sarasan 2011; Zee and Munekata 1992). Storage techniques vary with the crop, the facility and its mandate.

5.2.2 *Advantages of In Vitro Storage*

5.2.2.1 Germplasm Collection, Exchange and Distribution

In vitro collections play an increasingly important role in storing and distributing germplasm throughout the world (Engelmann 2011; Sarasan 2011). In vitro plantlets are often a preferred form of plant distribution, as they are more likely to be free of insect-pests and many of the disease problems of field plants. In addition the in vitro plantlets are available at any time of year for distribution. Collections of yam exist in over 40 countries, and in vitro collections are being used for exchange of germplasm and as secured backup collections in many countries (FAO 2010). It is ideal to initiate the collections from virus-tested, disease-free plants for this purpose. As a minimum screen, the cultures should be indexed for bacterial and fungal contaminants at the initiation phase and again during culture to ensure the most phytosanitary propagule possible (Leifert et al. 1994; Reed and Tanprasert 1995; Viss et al. 1991). Virus indexing and cleanup provide additional value to the germplasm and allow for easier international exchange (Malaurie et al. 1998). In some cases germplasm can be collected from the wild as in vitro plants. This is often useful when the plants have a long transit time or only very small populations are available for collection (Pence 2005).

5.2.2.2 Plant Quarantine

Many plant certification programs incorporate in vitro meristem culture as a standard technique for producing virus-negative and phytoplasma-negative plants from stock collections (Malaurie et al. 1998; Reed et al. 2005). Tropical crops have specific diseases that are common in the field but often can be eliminated in vitro. Thermotherapy and meristem-tip culture is used for producing cassava collections free of viruses and bacterial diseases (IPGRI/CIAT 1994). Similar procedures are available for *Musa* species and cultivars (Van de Houwe 1999; Van den Houwe and Panis 2000).

5.2.2.3 Protection from Disease and Environmental Threats

Field collections are at risk from infection by various pathogenic organisms or attack by insect pests. In addition, floods, droughts, fire and weather conditions may destroy whole collections in the field. In vitro collections can be duplicated at remote sites so they are not affected by local disasters.

5.2.2.4 Secondary Metabolite Production

In vitro collections of tissues and organs of medicinal plants are used for production of phytochemicals (Heine-Dobbernack et al. 2008; Nigro et al. 2004).

5.3 Managing an In Vitro Collection

5.3.1 Replicates

Adequate replicate cultures are needed to avoid loss of accessions. This may require backup collections at another site, or holding more individual plantlets at one site. Replicates range from 5 to 20 in various collections (Malaurie et al. 1998; Reed et al. 2005; Van de Houwe 1999). Exchange of materials between laboratories in a "black box" arrangement could be used to provide a secure backup for materials. These materials are held at another lab without any manipulation and could be returned to the owner if needed.

5.3.2 Evaluation and Subculture

A number of protocols are used for evaluation and timing of subculture. In some cases all of the collection is repropagated once per year; in other cases, this is done only as often as needed. Cultures in the *Musa* collection at ITC are repropagated yearly and half of each accession is subcultured in a separate operation to avoid transfer errors. Other laboratories evaluate the cultures at set intervals and repropagate those that are declining in storage (Aynalem et al. 2006; Reed et al. 2005).

5.3.3 Genetic Erosion

As with any collection, there is concern about genetic erosion, but the occurrence with an in vitro collection is less than that of a field collection. The loss of some accessions during the storage period could result in loss of those accessions in the entire field collection, resulting in reduced diversity for the collection (Dussert et al. 1997). In the case of in vitro collections, if they are a secondary backup collection, genetic erosion would be reduced because a dead or diseased field plant could be replaced with one from the in vitro collection. This is especially pertinent if the most at risk accessions are backed up in vitro. Since in vitro cultures are clonally propagated, it is important to handle them in a manner that will reduce the possibility of somaclonal variation. Culture protocols that reduce or eliminate callus production and workers who carefully evaluate plants during culture to avoid mutations are key points for maintaining the genetic integrity of the accessions.

5.3.4 Culture Indexing

An important but often ignored step during preparation for in vitro germplasm storage is indexing the cultures for latent bacterial and fungal infections (Kane 1995, 2000; Reed and Tanprasert 1995). Contaminants often become evident after a period of storage and this latent infection may contribute to the death of stored cultures (Wanas et al. 1986). Indexing procedures using bacteriological media to detect contaminants are more effective than visual examination of cultures (Kane 2000; Viss et al. 1991). Improved detection of bacterial contaminants will provide healthier cultures, longer storage times, and safer materials for distribution. Proper laboratory protocols are important to maintaining the sterility of the cultures. Every step of the culture process should be evaluated and designed to reduce the chance of carrying or introducing contaminants to the cultures (Cassells 1991; Gunson and Spencer-Phillips 1994; Leifert and Woodward 1997; Reed and Tanprasert 1995).

5.3.5 Database Management

Management of data associated with the in vitro collection should be maintained in a readily available form. Identification numbers, date of initiation into culture, storage date, growth and storage media, subculture interval and growth characteristics should be maintained for each accession. A secure backup of the database information is also important. As more data are amassed about plants held in national and international genebanks, it is increasingly difficult to manage the data and make it widely accessible. Smaller genebanks may have a more difficult time and may lack the resources to develop their systems. Because of the importance of information sharing, the Agricultural Research Service (ARS) of the United States Department of Agriculture, the Bioversity International and the Global Crop Diversity Trust are developing an easy-to-use, Internet-based information management system (GRIN-Global) that will be available to all of the world's plant genebanks. The system will be based on the ARS Germplasm Resources Information Network (GRIN), a database that holds information on over 480,000 accessions (distinct varieties of plants) in the ARS National Plant Germplasm System (NPGS). GRIN has also been adopted by the National Genebank System of Canada for germplasm information management.

Bioversity International, on behalf of the CGIAR System-wide Genetic Resources Program, the Global Crop Diversity Trust and the Secretariat of the International Treaty on Plant Genetic Resources for Food and Agriculture has developed GENESYS (www.genesys-pgr.org), that provides access to data associated with millions of accessions of food crops, wild crop relatives and underutilized species worldwide. This is a step toward establishing a global information system of plant genetic resources. There are currently 24 million records accessed from databases in the CGIAR system-wide information network for genetic resources of crop, forage and tree germplasm (SINGER), the European Cooperative Program for Plant Genetic Resources (ECPGR) and the US Germplasm Resources Information Network (GRIN) computer systems. These records include passport, phenotypic and environmental data, and may include information such as optimal growing climates, nutritional content and resistance to pests and diseases.

5.3.6 Emergency Priority

If accessions are to be saved during any emergency situations, they should be categorized on the basis of available information on their characterization and evaluation. Priority should be given to those accessions that: (1) have rare genes or gene combinations, (2) are difficult to maintain in the field, (3) are not readily available at the active genebank sites, and/or (4) are exotic and have been introduced with great efforts. This priority should be set when the accession is accepted into the genebank and should be used as priority for in vitro backup collections or cryopreservation (Mandal 2003).

5.4 Factors Involved in Slow Growth

5.4.1 Temperature

A number of factors are involved in the successful in vitro storage of germplasm collections; however temperature is the most studied parameter (Engelmann 2011; Sarkar et al. 2001; Van den Houwe and Panis 2000). Lowering the temperature of growth by 10–15°C is effective for many tropical plants. Species such as *Musa* and *Coffea* do not survive below 15–20°C, but at temperatures slightly higher, the growth is reduced and culture time is increased (Bessembinder et al. 1993; Dulloo et al. 1998; Dussert et al. 1997; Van den Houwe and Panis 2000; Zandvoort et al. 1994). Storage under growth room conditions can be used to extend the subculture interval of some species to a year or more (Gupta and Mandal 2003). Medium-term conservation at growth room temperatures is used with several crop species. Shoot cultures of banana and mulberry are maintained for 12 months at the normal culture-room temperature (25±2°C), while sweet potato has a shelf life of about 12–16 months with the inclusion of mannitol in culture medium at moderate levels (1–1.5%) under same culture conditions (Mandal 1997, 2003). Slow growth at room temperature is economical for laboratories where it is difficult to maintain cold rooms and is useful for plants that cannot tolerate cold temperatures. Several tropical crops such as sweet potato, taro, yams, ginger, turmeric, pepper, banana, mulberry, and some medicinal and aromatic plants are being conserved under normal growing conditions at 25°C in the in vitro repository of the National Bureau of Plant Genetic Resources (NBPGR), New Delhi (Gupta and Mandal 2003). Conservation at growth room or slightly lower temperatures (15–24°C) can be effective, especially where the summer temperatures reach 45–50°C. Six potato genotypes had moderate to good survival (55.5–77.8%) when stored at 24°C for 12 months with 20 g/L sucrose and 40 g/L sorbitol. *Vanilla planifolia* shoot cultures were held for more than 1 year at 22±2°C without subculture, on medium with sucrose and mannitol and were maintained with yearly subculture for more than 7 years (Divakarana et al. 2006). Chilling sensitivity greatly varies between species, however, banana, ginger and sweet potato cultures could not survive at temperatures below 15°C (Mandal 1997). For *Coffea*, cultures were extended through changes in storage temperature, sugar, mineral salts, growth regulators, growth retardants or oxygen tension (Dulloo et al. 1998; Dussert et al. 1997; Engelmann 1991). Reduced sucrose was not effective for *C. arabica* microcuttings, but reduced temperature could be used successfully. *C. canephora* microcuttings did not survive low temperatures (Bertrand-Desbrunais et al. 1992). An in vitro collection of yam germplasm is maintained at Institut de Recherche pour le Developpement, Montpellier, France (IRD), with 6 test tubes of each accession at growth room temperature. The minimal growth conditions allow most of the accessions to remain for 2 years, but technical constraints could reduce subcultures to a low of 6–8 months (Malaurie et al. 1998).

5.4.2 Growth Medium

Changes in growth regulators (Dussert et al. 1997; Van den Houwe and Panis 2000), increases in osmotic pressure (Dekkers et al. 1991; Zandvoort et al. 1994), mineral nutrient changes (Reed 1993; Zee and Munekata 1992) and chemical growth inhibitors are also used to reduce plant growth and increase storage time (Bertrand-Desbrunais et al. 1992; Dekkers et al. 1991; Gupta and Mandal 2003; Jarret 1997). Sugarcane grown with 1% or 3% mannitol survived for 105 days without transfer while cultures on 2% mannitol lasted 165 days. Sugarcane germplasm was stored as somatic embryos for 8 months at 18°C on half strength Murashige and Skoog medium (MS) (1962) with increased sucrose or sorbitol (Watt et al. 2009). *Ensete ventricosum* clones collected from southwestern Ethiopia stored well for 6 months at slightly lower than normal temperature (15°C or 18°C) with mannitol in the medium (Negash et al. 2001). *Curcuma* shoots of 8 wild species could be stored for 10 months to 3 years at room temperature when cytokinins were optimized (Tyagi et al. 2004). At IRD, the in vitro yam collection contains over 21 accessions and is maintained for 6 months to 2 years in medium with low mineral nutrient concentrations and a low sucrose concentration (Malaurie et al. 1998). Shoot cultures of tropical *Rubus* germplasm could be stored for 6–9 months on medium with 25% MS medium nitrogen (Reed 1993).

5.4.3 Plant Materials

The size and type of propagule stored is very important. Shoot cultures are most commonly used, and the size, shape and condition of the starting material is very important to successful storage. The most commonly stored propagules are shoot cultures or buds and seed; however embryos, cotyledons, rhizomes, tubers and corms could also be stored. The size of shoots for storage varies greatly with the plant type and should be evaluated with the germplasm in question. Normally a shoot of 2–4 cm is preferred and often clumps of shoots provide better survival in storage than single shoots.

In vitro rhizome formation in ginger allowed cost-efficient conservation as the rhizome cultures could be stored for up to 20 months at 25 ± 2°C (Tyagi et al. 2006). Rhizomes can withstand power failures for short duration and also can be transplanted to the soil directly (without hardening) in case of emergency. In vitro culture also allows for exchange of germplasm and production of disease-free rhizomes in ginger (Tyagi et al. 2006). Corm forming cultures of taro could be conserved at 25 ± 2°C for up to 15 months, but shoot-forming cultures could last for only 6 months (Hussain and Tyagi 2006).

Synthetic seeds, propagules gelled in a sodium alginate matrix, can be stored for several months. Shoot buds and protocorms of *Vanilla* in alginate could be stored for 10 months at 22 ± 2°C but lost viability in 1 month at 5°C and 15°C (Divakarana et al. 2006). Sucrose-dehydrated synseeds of ginger stored better than air-dehydrated or fresh synseeds (Sundararaj et al. 2010). Guava shoot tips encapsulated in alginate could be stored for several months at low temperatures or with reduced sucrose (Rai et al. 2008). Encapsulated *Rauvolfia serpentine* shoot tips survived for 14 weeks at 4°C and could be grown into plants (Ray and Bhattacharya 2008).

The Svalbard seed vault in Norway opened in 2008 is one of the centers that guarantee future availability and diversity of various seed producing plant accessions. Exceptions to seed storage occurs in situations where plants do not produce viable seeds, seeds are sterile, short lived, does not tolerate water loss or when the clones are heterozygous and do not produce true-to-type seeds. Seed storage may also not be an option in some cases where the selected clonal material are more productive than seed-derived lines (Towill 1988). Many tropical seeds are used as explant materials for in vitro germplasm preservation because they have short-lived seeds.

Orthodox seeds can withstand dehydration to less than 5% moisture content (of seed dry weight) during storage and thus are more predisposed to safe storage at low temperature and moisture conditions compared to recalcitrant seeds such as fruit trees and palms which are high in moisture content and sensitive to low moisture and temperature (Engelmann 2004; Normah and Vengadasalam 1992). Seeds of high moisture content often face a serious challenge during storage. Seeds of many tropical and sub-tropical plant species are recalcitrant in nature. Examples include rubber, coconut, lemon, mango, citrus, cocoa and coffee. They are mostly stored under slightly wet conditions so as to avoid drying injury but under this condition their viability may be reduced. Maintenance would be only for a short duration. However, as a substitute for seed storage, embryonic axes may be excised and stored for some recalcitrant seeds or vegetatively propagated species (Normah and Makeen 2008; Walters et al. 2002). Embryonic axes are suitable for maintenance of genetic diversity of a given population especially in cases where seed storage is not an option (Makeen et al. 2005; Normah and Makeen 2008).

5.4.4 Culture Conditions

Culture conditions before storage and duration of growth after subculture have important effects on length of storage (Reed and Aynalem 2005; Zee and Munekata 1992). Light quality and intensity, both before and during storage, need further investigation (Reed 2002). Ginger stored under 16- and 24-h light survived up to 14 months compared to 20% cultures held in the dark (Tyagi et al. 2006). Some limitations on research are due to the need for varied growth room conditions and the cost related to this requirement. The storage container used also affects the duration and security of storage. Standard culture tubes with polypropylene caps are widely used for conservation of *Morus* spp. up to 12–15 months without subculturing (Gupta et al. 2002). The use of polypropylene caps had added advantage over the cotton plugs enclosure as it reduces the evaporation of water from the media therefore reducing the drying of media in the culture tubes. This helps to increase shelf life of cultures. Glass test tubes with screw caps enclosures are used to reduce fungal contamination in cultures of *Morus* to about 90%. Medium browning resulting from exudation of phytochemicals from plant tissues is a major problem affecting the shelf life of in vitro cultures in storage. This could be controlled by use of medium additives including activated charcoal, antioxidants and phenolic reducing compounds. Low temperature condition if not carefully managed can expose stored tissues to hyperhydricity (Uchendu et al. 2011).

5.4.5 Contamination Control

Contamination by either endogenous or exogenous microorganisms can affect the establishment of in vitro cultures (Bausher and Niedz 1998; Langens-Gerrits et al. 1997;

Skirvin et al. 1993, 1999). Fungal, bacterial and insect contamination in storage can be minimized by adopting appropriate sterile techniques at initiation and during transfer of axenic cultures and also by using heat-sealed tissue culture bags for germplasm storage (Bausher and Niedz 1998; Reed 1992; Reed et al. 2005). Watt et al. (2003) sprayed long leafless stems of *Eucalyptus* sp. with 70% (v/v) ethanol and also added 1 g calcium hypochlorite solution into the first culture medium to control growth of bacteria with high success. Many new decontamination agents are available for surface sterilization, such as 8-hydroxyl-quinolinol-sulfate (Cassells 1997; Laimer Da Camara Machado et al. 1991; Leifert and Woodward 1997).

Aegle marmelos and mulberry cultures with fungal contamination could be sterilized with 0.1% sodium hypochlorite or 0.1% mercuric chloride with 80–100% recovery depending on the genus. Sodium dichloroisocyanurate was reported as an effective sterilizing agent (Niedz and Bausher 2002; Parkinson et al. 1996). Fumigation of culture room, inoculation room and media preparation room or the whole laboratory becomes necessary in the tropical countries after rainy season to kill spore built up due to high humidity.

Mites are very harmful for any tissue culture facility (Blake 1988; Cassells 1997). Mites should be avoided by treating new explants carefully. Oil or miticide treatment of parent materials may be needed to avoid mite outbreaks. Explants from materials with likely mite infestations should be isolated from other cultures and sealed so any hatching mites cannot move into the room. Mites can contaminate cultures as they move rapidly from one container to another leaving a trail of fungal contamination. Mite contamination in the cultures can be controlled by adding few drops of filter sterilized insecticides in the test tubes or culture containers. Any infected cultures should be autoclaved immediately when noticed, if left unautoclaved, mites can move to healthy cultures (Leifert and Waites 1994).

5.5 International In Vitro Storage Facilities

In vitro genebanks are a valuable adjunct to a field genebank. They are also valuable for cryopreservation research and to provide initial explants or stock cultures for a cryo genebank. Significant progress has been made in this area and the impact can be accessed by the increasing number of accessions in the genebanks. In the following section, global and national in vitro genebanks are briefly discussed (Table 5.1). A report on the state of the world's genetic resources was produced by the Food and Agriculture Organization of the United Nations (FAO 2010) and includes some references to in vitro storage. Numerous problems, including reliability of electricity supplies, pests and disease related problems as well as lack of staff, equipment, or funds limit the amount of in vitro storage that is possible in tropical countries (FAO 2010). The listings are as comprehensive as we were able to determine through a wide range of sources.

5.5.1 Africa

Republic of Benin, Cameroon, Republic of Congo, Ghana, Kenya, Mali, Nigeria and Uganda mentioned having in vitro storage facilities at their websites (FAO 2010).

Egypt: In 2004, the National Genebank of Egypt became operational with a storage capacity for 200,000 accessions (15% of capacity was being used by the end of 2006) as well as facilities for in vitro conservation and cryopreservation; however the laboratories were looted during political unrest in 2011 and only the seed bank remains.

Ethiopia: The Forage Agronomy Group (FLAG) at International Livestock Center for Africa (ILCA), Addis Ababa holds in vitro collections of clonally propagated grasses.

Ghana: CSIR-Crops Research Institute, Kumasi and Tissue Culture Laboratory, University of Ghana maintains some vegetable and fruit crops in vitro.

Kenya: The Kenya Agricultural Research Institute's National Genebank (KARI-NGBK) holds in vitro accessions of some grasses, medicinal plants and trees.

Nigeria: The International Institute of Tropical Agriculture (IITA) is using cassava cryopreservation as means of longer-term preservation of cassava in vitro collections (Dumet et al. 2011). The National Center for Genetic Resources and Biotechnology (NACGRAB) holds in vitro collections of pineapple, date palm and sugar cane.

South Africa: A potato collection is held at the South Africa Agricultural Research Council, Pretoria.

Tanzania: a cassava collection is held at the Tanzania National Plant Genetic Resources Center (NPGRC), Arusha.

5.5.2 Australia, Asia and the Pacific

Australia: Queensland Horticulture Institute, Maroochy Research Station, Nambour, holds banana and papaya. The Plant Science Laboratory at Kings Park & Botanic Garden, Perth maintains endangered native flora.

China: Institute of Crop Germplasm Resources, CAAS The National Sweet Potato in vitro Storage Unit is in Xuzhou, Jiangsu Province and the National Potato in vitro Storage Unit in Keshan, Heilongjiang Province (Ren 1994). In addition there are in vitro collections at many universities (Gu 1998). Guangdong Academy of Agricultural Sciences (GAAS)/Research Institution of Fruit Tree (FTRI) has about 210 accessions of banana in the field or in vitro as a national collection including 120 cultivated species.

Fiji: The Centre for Pacific Crops and Trees (CePaCT), the Secretariat of the Pacific Community holds collections of about 1,800 accessions from several crops including banana, cassava, potato, sweet potato, taro and yam. CePaCT maintains safety duplicates of the national vegetatively propagated crop collections from the Pacific islands. These are held on MS medium at 20°C. These include: *Alocasia* (11), banana (158), breadfruit (13), cassava (24), coconut (1), *Cyrtosperma* (29), potato (11), kava (23), sweet potato (225), vanilla (7), *Xanthosoma* (9), yam (247), taro (1053).

India: At the Tropical Botanical Garden and Research Institute (TBGRI), 38 accessions of 26 medicinal plant species are maintained as shoot cultures in the in vitro bank. These are: *Acorus calamus*, *Adhatoda beddomei*, *Alpinia calcarata*, *Baliospermum solanifolium*, *Celastrus paniculatus*, *Geophila reniformis*, *Holostemma annulare*, *Plumbago rosea*, *Rauvolfia micrantha*, *R. serpentina*, *Rubia cordifolia* L., *Utleria salicifolia* (TBGRI 2010). The National Bureau of Plant Genetic Resources, New Delhi, India holds about 40,000 in vitro cultures of more than 2,000 accessions belonging to six crop groups (temperate fruit, tropical fruit, tuber, bulbous, medicinal/aromatic and spices/industrial crops) including 48 genera and about 133 species. Main tropical species include sweet potato (255), yams (140), taro (191), banana (411), mulberry (61), ginger (181), turmeric (162), pepper (8), cardamom (5), vanilla (4), jojoba (12), *Xanthosoma* (9) and *Aegle marmelos* (2). These accessions (8–12 tubes per accession) are maintained at 25°C under normal growth conditions with the subculture period ranging from 8 to 22 months. Tropical medicinal plants maintained as shoot cultures include: *Acorus calamus*, *Centella asiatica*, *Holostemma ada-kodien*, *Coleus* spp., *Costus speciosus*, *Curculigo orchioides*, *Plumbago* spp., *Rauvolfia* spp., and *Tylophora indica* (Sharma and Pandey, Chap. 12 this volume).

Indonesia: In Indonesia some endangered medicinal plants like *Alyxia reinwardtii*, *Rauvolfia serpentine*, *Ruta graveolens* and *Pimpinella purpruatian* have been conserved using in vitro culture technique and cryopreservation (Anonymous 2010).

Japan: The National Institute of Agrobiological Sciences (NIAS) genebank of Japan holds tropical accessions in three genera: mulberry (70 accessions), taro (25 accessions) and yam (34 accessions).

New Guinea, Papua: Has several research sites holding sweet potato, coffee and coconut.

Pakistan: Plant Genetic Resources Institute of the National Agricultural Research Center Pakistan has an in vitro laboratory for conservation of the germplasm of vegetatively propagated species including tropical species, sweet potato, banana and sugarcane.

Taiwan: The World Vegetable Center-AVRDC maintains crops that are vegetatively propagated such as shallot and garlic as living plants in field genebanks, and there is a tissue culture laboratory for embryo rescue activities but no in vitro collection.

5.5.3 *Central America and the Caribbean*

There are many tissue culture facilities in the Central America—Caribbean area. *Caribbean*: The Caribbean Agricultural Research and Development Institute (CARDI), formerly had a micropropagation program that distributed virus-free in vitro propagated germplasm but that program was discontinued and was taken over by individual countries.

Costa Rica: The Center for Tropical Agricultural Research and Education (CATIE) in Costa Rica maintains in vitro holdings of yams (58) in seven species and cassava (*Manihot esculenta*) (40). Thermotherapy is used to eliminate viruses. Cultures are held at 27°C in 20 cm tubes with 10 tubes per accession. Growth medium includes standard growth regulators. Fruit crops and coffee are also held in vitro.

Cuba: In vitro plants of tropical roots and tubers, plantains and bananas are maintained at Instituto Nacional de Investigaciones de Viandas Tropicales (INIVT) Cuba (Roca and Tay 2007) and also a sugarcane collection is also held in vitro. *The West Indies*: The West Indies Central Sugarcane Breeding Station (WICSBS) in Barbados conserves about 3,500 accessions of sugarcane (FAO 2010). The International Cocoa Genebank (ICGT) at the University of the West Indies conserves about 2,300 accessions of cocoa germplasm.

5.5.4 *Europe*

There are over 600 genebanks and plant collections in the European Union that are now being coordinated into 'A European Genebank Integrated System' (AEGIS). England, Belgium, Germany, Poland, Romania and the Russian Federation maintain cryopreservation facilities and conserve some germplasm in vitro, mainly potato. Germany has more than 3,200 in vitro accessions including potato. Belgium holds in vitro collections of banana and plantain germplasm at the International Network for Improvement of Banana and Plantain (INIBAP) center in Leuven with more than 1,500 in vitro accessions. Large potato collections are held at the Institut National de la Recherche Agronomique (INRA)-Rennes (France) and at the Russian Federation (VIR) as well as in Germany (Leibniz Institute of Plant Genetics and Crop Plant Research [IPK]) and in Spain at the Basque Institute of Agricultural Research and Development [BIARD]-NEIKER (Instituto Vasco de Investigacion y Desarrollo Agrario-NEIKER) (Barandalla et al. 2003).

5.5.5 *North America*

In North America, Canada, Mexico and the United States operate long- and medium-term genebanks.

Canada: The Canadian genebank holds in vitro collections of 180 potato genotypes at the Potato Genebank, Fredericton, New Brunswick. Also, the Plant Cell Technology Laboratory, University of Guelph holds in vitro collections of breadfruit and tobacco. All other in vitro collections in Canada are of temperate fruit crops.

Mexico: The National Center for Genetic Resources (CENARGEN) Tepatitlan, holds a collection of *Opuntia* cactus in vitro. Most other collections are held in the field.

United States: The National Plant Germplasm System of the US Department of Agriculture operates a system of germplasm conservation with 31 genebanks within the country and conserves 7% of the germplasm holdings, including more than 50% of the genera conserved worldwide in genebanks. Tropical and subtropical germplasm is held in vitro at genebanks in Hawaii [*Ananas* (217) (Zee and Munekata 1992), *Artocarpus* (16), Carica (1), Camellia (5), *Passiflora* (7), *Vaccinium* (15) (Zee et al. 2008), *Vasconcellea* (2), *Zingiber* (9)]. These shoot cultures are maintained on Woody Plant Medium (WPM) (Lloyd and McCown 1980), MS or ½ MS medium without plant growth regulators (PGR) in tubes at 20°C for 3–18 months. In Puerto Rico there are mostly *Musa* (150 accessions) stored in tubes or tissue culture bags at 28°C (Irish et al. 2009), but also some *Bambusa*, *Guadua*, and *Theobroma*. About 700 accessions of sweet potato are in vitro at USDA-ARS Plant Genetic Resources Conservation unit (PGRCU) Griffin, Georgia (Jarret 1997; Jarret and Florkowski 1990; Jarret and Gawel 1991). The USDA potato (*Solanum*) collection is held at Sturgeon Bay, Wisconsin, with in vitro collections of 400 varieties of potato from around the world, 250 mapping populations, and 250 wild species/genetic stocks. The potato collections (five tubes per accession) are held at 10°C in 20 × 150 mm test tubes on MS medium with 10–30 g D-sorbitol. Shoots are repropagated at 20°C as needed.

5.5.6 South America

There are many germplasm collections throughout South America, but not many include in vitro cultures.

Argentina: The National Institute of Agricultural Technology (INTA) implemented a Plant Genetic Resources Network (PGRN) with nine Active Banks and 11 collections distributed in diverse ecological areas and a Base Bank that maintains a duplicate of the active collections. The PGRN is charged with the ex situ conservation of crop species and their wild relatives; including morphological, genetic, agronomic, biochemical and molecular characterization, evaluation and documentation. The goal is to make the genetic diversity available for research and plant breeding programs. In 1990 the Institute of Biological Resources developed an in vitro conservation bank, with collections of vegetatively propagated sweet potato and cassava. The sweet potato collection includes 379 introduced cultivars, landraces and commercial materials derived from meristems. This collection is conserved under slow growth conditions (17–19°C, light intensity of 3,000 lux and 16-h photoperiod). Three replicates of each accession are maintained for 3 month intervals in tubes (18 × 150 mm) on MS medium with 20 g/L sucrose, 20 g/L mannitol and 8 g/L agar. The cassava collection includes 72 landraces conserved under the same slow growth conditions as sweet potato. For each accession three to six replicates are maintained for 3 month intervals in tubes (24 × 150 mm) on MS medium with 20 g/L sucrose, 0.01 mg/L NAA, 0.02 mg/L BAP and 8 g/L agar. INTA at Balcarce has an extensive in vitro collection of potato cultivars.

Bolivia: Banco de Germoplasma, Santa Cruz de la Sierra and the Sistema Nacional de Conservación y Desarrollo de Recursos Genéticos de Bolivia, Cochabamba hold in vitro collection of fruit and tuber crops.

Brazil: Brazilian Agricultural Research Corporation (EMBRAPA) Recursos Genéticos e Biotecnologia, Brazilia holds a wide range of in vitro collections. In vitro collections are in most cases subsets of the field collections. The 1,636 accessions are held in three to five tubes per accession at 20°C on mostly MS or ½ MS medium. At Mandioca e Fruticultura Tropical, Cruz das Almas, there are collections of pineapple held on ½ x MS medium at 21°C for 230–770 days.

Columbia: The genus *Manihot* is in collection at the in vitro conservation laboratory of the Genetic Resources Program of the International Center for Tropical Agriculture (CIAT). It has the greatest genetic diversity of any collection and includes a wide range of geographic areas. More than 6,500 accessions from 28 countries are represented for *M. esculenta* and nearly 1,000 wild species. Materials are available from Colombia, Brazil and other South American countries as well as from Central America and the Caribbean, Asia and several other areas. This was one of the first in vitro genebanks established (IPGRI/CIAT 1994). The collection is held under slow growth conditions at 23–24°C, with 18.5 μmol m^{-2} s^{-1} light and 12 h photoperiod. Five tubes of each genotype are held and subcultured once each year. Materials are initiated after virus elimination procedures and are free of the Cassava Common Mosaic Virus (CsCMV), Cassava virus X (CsXV) and frog skin disease (FSD). Procedures are available at www.ciat.cgiar.org/urg.

Ecuador: The Departamento Nacional de Recursos Fitogenéticos y Biotecnología Instituto Nacional Autonomo de Investigaciones Agropecuarias (DENAREF-INIAP), is the most important institution for the conservation of Equador's plant genetic resources. The gene bank holds 20,000 accessions of Andean tubers, quinoa, *Amaranthus*, cucurbits, tree tomato and peppers. There are also collections held at INIA- Maracay, the Bolivarian Republic of Venezuela, INIA Carillanca (Chile), INIAP, Centro de Bioplantas (Cuba). This collection consists mainly sugar cane, but no specific information is available.

Peru: International Potato Research Institute (CIP), Lima, Peru holds the world's largest sweet potato (*Ipomea batatas*) collection with more than 6,400 accessions. It also has the third largest potato collection (about 8% of total world holdings) in the field genebank. CIP maintains 2,800 in vitro accessions in its sweet potato germplasm collection. Potato and sweet potato plantlets are held in test tubes for about 2 years at 6–8°C under low light and on an osmotic regulator to slow the growth.

5.6 Conclusions

In vitro techniques are complementary to other strategies of plant genetic resources conservation. These techniques are used in a wide range of germplasm laboratories from single genus collections to very diverse gene banks. In vitro conservation is not intended to replace conventional approaches for in situ and ex situ conservation, instead it provides security and flexibility for a collection. In vitro conservation offers researchers and genebank curators a set of additional tools to allow them to improve the conservation of germplasm collections placed under their responsibility. Although there remain species that cannot be cultured, great strides are being made to improve tissue culture methods/techniques for the culture of both crop and endangered plants.

Dramatic progress has been made during the last 30 years in the development of in vitro culture techniques for the conservation of plant germplasm. Slow-growth techniques are immediately applicable in many cases and techniques are under research for additional species. In vitro conservation techniques offer advantages of rapid multiplication, ready availability and secure backup capabilities for germplasm laboratories. Basic laboratory infrastructure already exists in many genebanks and can be utilized to provide backup to field collections. It is also important to develop in vitro protocols that are adaptable to low technology laboratories.

The main challenge to progress for in vitro collections is the introduction of contaminant-free plants into culture and finding the appropriate growth medium for suitable micropropagation. This is a challenging project due to the wide diversity of plants involved in genebanks and the complexity of disease mechanisms, but advances are being made in both field and in vitro collection process to facilitate healthy plant production. The use of microbiological media to screen explants for bacteria and fungi early in the culture process will result in healthier collections and can be easily implemented. Developing more customized growth media for unique plant groups should also result in improved culture response and better storage in vitro.

Acknowledgements The following are gratefully acknowledged for helping us in retrieval of in vitro genebank information for the chapter. **Claire Arakawa**, USDA-ARS NCGR-Hilo Hawaii; **Ariana Digilio**, INTA Castelar, Argentina; **Stefano Diulgheroff**, AGPM, FAO; **Natalie Feltman,** Deputy Director: Plant Genetic Resources, Department of Agriculture, Fisheries and Forestry, Republic of South Africa; **Luigi Guarino**, Crop Diversity Trust; **Brian M. Irish**, USDA-ARS, Tropical Agriculture Research Station, Mayaguez, PR.; **Max Martin**, USDA-ARS, US Potato Genebank, Sturgeon Bay, WI.; **Luis Mroginski**, Instituto de Botánica del Nordeste, Universidad Nacional del Nordeste, Corrientes, Argentina; **Prem Narain Mathur,** South Asia Coordinator and Senior Scientist, Diversity Assessment and Use, Bioversity International, Office for South Asia, New Delhi, India; **Juliano Gomes Pádua**, EMBRAPA Recursos Genéticos e Biotecnologia,

Brasília, Brasil; **Victoria Rivero**, Instituto de Recursos Biológicos, INTA Balcarce, Argentina; **Ericson Aranzales Rondon,** International Center for Tropical Agriculture (CIAT), Cali, Colombia; **William Solano**, investigador en Recursos Fitogenéticos y Biotecnología del CATIE en Costa Rica; **Mary Taylor,** Genetic Resources Coordinator/Centre of Pacific Crops and Trees, Suva, Fiji; **Fernanda Vidigal**, Embrapa Mandioca e Fruticultura Tropical, Cruz das Almas, Brazil; our appologies to any other contributors that we may have overlooked.

References

Anonymous (2010) Country report on the state of plant genetic resources for food and agriculture: Indonesia. FAO, Rome

Ashmore SE (1997) Status report on the development and application of in vitro techniques for the conservation and use of plant genetic resources. International Plant Genetic Resources Institute, Rome

Aynalem HA, Righetti TL, Reed BM (2006) Nondestructive evaluation of in vitro-stored plants: a comparison of visual and image analysis. In Vitro Cell Dev Biol Plant 42(6):562–567

Barandalla L, Sanchez I, Ritter E, Ruiz de Galerreta JI (2003) Conservation of potato (*Solanum tuberosum* L.) cultivars by cryopreservation. Spanish J Agri Res 1:9–13

Bausher MG, Niedz RP (1998) A discussion of in vitro contamination control of explants from greenhouse and field grown trees. Proc Fla State Hort Soc 111:260–263

Benson EE (1999) Plant conservation biotechnology. Taylor and Francis, London

Bertrand-Desbrunais A, Noirot M, Charrier A (1992) Slow growth in vitro conservation of coffee (*Coffea* spp.) 2: influence of reduced concentrations of sucrose and low temperature. Plant Cell Tiss Organ Cult 31:105–110

Bessembinder JJE, Staritsky G, Zandvoort EA (1993) Long-term in vitro storage of *Colocasia esculenta* under minimal growth conditions. Plant Cell Tiss Organ Cult 33(2):121–127

Blake J (1988) Mites and thrips as bacterial and fungal vectors between plant tissue cultures. Acta Hortic 225:163–166

Cassells AC (1991) Problems in tissue culture: culture contamination. In: Debergh PC, Zimmerman RH (eds) Micropropagation technology and application. Kluwer, Dordrecht, pp 31–44

Cassells AC (1997) Pathogen and microbial contamination management in micropropagation, vol 12. Kluwer, Dordrecht

Dekkers AJ, Rao AN, Goh CJ (1991) In vitro storage of multiple shoot cultures of gingers at ambient temperatures of 24–29°C. Sci Hortic 47:157–167

Divakarana M, Nirmal K, Babua K, Peterb V (2006) Conservation of *Vanilla* species, in vitro. Sci Hortic 110:175–180

Dulloo ME, Guarino L, Engelmann F, Maxted N, Newbury JH, Atterc F, Ford-Lloyd BV (1998) Complementary conservation strategies for the genus *Coffea*: a case study of Mascarene *Coffea* species. Genet Resour Crop Evol 45:565–579

Dumet D, Korie S, Adeyemi A (2011) Cryobanking cassava germplasm at IITA. Acta Hortic 908, ISHS 2011 439–446

Dussert S, Chabrillange N, Anthony F, Engelmann F, Recalt C, Hamon S (1997) Variability in storage response within a coffee (*Coffea* spp.) core collection under slow growth conditions. Plant Cell Rep 16(5):344–348

Engelmann F (1991) In vitro conservation of horticultural species. Acta Hortic 298:327–332

Engelmann F (1997) In vitro conservation methods. In: Callow JA, Ford-Lloyd BV, Newbury HJ (eds) Biotechnology and plant genetic resources. Conservation and use. CAB International, Rome, pp 119–160

Engelmann F (1999) Management of field and in vitro germplasm collections. In: Proceedings of a consultation meeting, CIAT/International Plant Genetic Resources Institute, Cali/Rome, 15–20 Jan 1996

Engelmann F (2004) Plant cryopreservation: progress and prospects. In Vitro Cell Dev Biol Plant 40:427–433

Engelmann F (2011) Use of biotechnologies for the conservation of plant biodiversity. In Vitro Cell Dev Biol Plant 47(1):5–16
FAO (2010) The second report on the state of the world's plant genetic resources for food and agriculture. FAO, Rome. ISBN 978-92-5-106534-1
Fay MF (1994) In what situations is in vitro culture appropriate to plant conservation? Biodivers Conserv 3:176–183
Gu J (1998) Conservation of plant diversity in China: achievements, prospects and concerns. Biol Conserv 85:321–327
Gunson HE, Spencer-Phillips PTN (1994) Latent bacterial infections: epiphytes and endophytes as contaminants of micropropagated plants. In: Nicholas JR (ed) Physiology growth and development of plants in culture. Kluwer, Dordrecht, pp 379–396
Gupta S, Mandal BB (2003) In vitro methods for PGR conservation: principles and prospects. In: Chaudhury R, Pandey R, Malik SK, Mal B (eds) In vitro conservation and cryopreservation of tropical fruit species. IPGRI, Rome, pp 71–80
Gupta S, Mandal B, Gautam P (2002) In vitro and cryorepository of NBPGR. In: Kumar N, Negi P, Singh N (eds) Plant biotechnology for sustainable hill agriculture. Defence Agricultural Research Laboratory, Pithoragarh, pp 20–25
Harding K, Benson EE, Clacher K (1997) Plant conservation biotechnology: an overview. Agro Food Ind Hi Tech 8:24–29
Heine-Dobbernack E, Kiesecker H, Schumacher HM (2008) Cryopreservation of dedifferentiated cell cultures. In: Reed BM (ed) Plant cryopreservation: a practical guide. Springer Science and Business Media LLC, New York, pp 141–176
Hussain Z, Tyagi RK (2006) In vitro corm induction and genetic stability of regenerated plants in taro [*Colocasia esculenta* (L.) Schott]. Indian J Biotechnol 5:535–542
IPGRI/CIAT (1994) Establishment and operation of a pilot in vitro active genebank. Report of a CIAT-IBPGR collaborative project using cassava (*Manihot esculenta* Crantz) as a model. International Plant Genetic Resources Institute and International Center for Tropical Agriculture, Rome
Irish BM, Goenaga RJ, Reed BM (2009) Amending storage vessel and media improves subculture interval of *Musa* sp. tissue culture plantlets. HortScience 44(4):1103
Jarret RL (1997) Effects of chemical growth retardants on growth and development of sweet potato (*Ipomoea batatas* (L.) Lam.) in vitro. Plant Growth Reg 16:227–231
Jarret RL, Florkowski WJ (1990) In vitro active vs. field genebank maintenance of sweet potato germplasm: major costs and considerations. HortScience 25(2):141–146
Jarret RL, Gawel N (1991) Chemical and environmental growth regulation of sweet potato (*Ipomoea batatas* (L.) Lam.) in vitro. Plant Cell Tiss Organ Cult 25:153–159
Kane ME (1995) Bacterial and fungal indexing of tissue cultures. In Vitro Cell Dev Biol Plant 31:25A
Kane ME (2000) Culture indexing for bacterial and fungal contaminants. In: Gray DJ, Trigiano RN (eds) Plant tissue culture concepts and laboratory exercises. CRC Press, Boca Raton, pp 427–431
Laimer Da Camara Machado M, Da Camara MA, Hanzer V, Kalthoff B, Weib H, Mattanovich D (1991) A new, efficient method using 8-hydroxyl-quinolinol-sulfate for the initiation and establishment of tissue cultures of apple from adult material. Plant Cell Tiss Organ Cult 27:155–160
Langens-Gerrits M, Albers M, De Klerk GJ (1997) Hot-water treatment before tissue culture reduces initial contamination in *Lilium* and *Acer*. In: Cassells AC (ed) Pathogen and microbial contamination management in micropropagation, vol 12. Kluwer, Dordrecht, p 219
Leifert C, Waites WM (1994) Dealing with microbial contaminants in plant tissue and cell culture: hazard analysis and critical control points. In: Lumsden PJ, Nicholar JR, Davies WJ (eds) Physiology growth and development of plants in culture. Kluwer, Dordrecht, pp 363–378
Leifert C, Woodward S (1997) Laboratory contamination management; the requirement for microbiological quality assurance. In: Cassells AC (ed) Pathogen and microbial contamination management in micropropagation, vol 12. Kluwer, Dordrecht, pp 237–244
Leifert C, Waites B, Keetley JW, Wright SM, Nicholas JR, Waites WM (1994) Effect of medium acidification on filamentous fungi, yeasts and bacterial contaminants in *Delphinium* tissue cultures. Plant Cell Tiss Organ Cult 36:149–155
Lloyd G, McCown B (1980) Commercially feasible micropropagation of mountain laurel, *Kalmia latifolia*, by use of shoot-tip culture. Comb Proceed Int Plant Prop Soc 30:421–427

Makeen AM, Normah MN, Dussert S, Clyde MM (2005) Cryopreservation of whole seeds and excised embryonic axes of *Citrus suhuiensis* cv. limau lang-kat in accordance to their desiccation sensitivity. Cryo Lett 26:259–268

Malaurie B, Trouslot M-F, Berthaud J, Bousalem M, Pinel A, Dubern J (1998) In vitro storage and safe international exchange of yam (*Dioscorea* spp.) germplasm. EJB Electron J Biotechnol 1(3):103–117

Mandal BB (1997) Application of in vitro/ cryopreservation techniques in conservation of horticultural crop germplasm. Acta Hortic 447: 483–489

Mandal BB (2003) Management of in vitro garmplasm collections: Practical approaches. In: Mandal BB, Chaudhury R, Engelmann F, Bhag Mal, Tao KL, Dhillon BS (eds) Conservation biotechnology of plant germplasm. NBPGR, New Delhi, India/ IPGRI, Rome, Italy/ FAO, Rome, Italy, pp 131–140.

Murashige T, Skoog F (1962) A revised medium for rapid growth and bio assays with tobacco tissue cultures. Physiol Plant 15:473–497

Negash A, Krens F, Schaart J, Visser B (2001) In vitro conservation of enset under slow-growth conditions. Plant Cell Tiss Organ Cult 66:107–111

Niedz RP, Bausher MG (2002) Control of in vitro contamination of explants from greenhouse- and field-grown trees. In Vitro Cell Dev Biol Plant 38:468–471

Nigro SA, Makunga NP, Grace OM (2004) Medicinal plants at the ethnobotany–biotechnology interface. S African J Bot 70:89–96

Noor N, Kean C, Vun Y, Mohamed-Hussein Z (2011) In vitro conservation of Malaysian biodiversity—achievements, challenges and future directions. In Vitro Cell Dev Biol Plant 47(1):26–36

Normah MN, Makeen AM (2008) Cryopreservation of excised embryos and embryonic axes. In: Reed BM (ed) Plant cryopreservation: a practical guide. Springer Science and Business Media LLC, New York, pp 211–240

Normah MN, Vengadasalam M (1992) Effects of moisture content on cryopreservation of *Coffea* and *Vigna* seeds and embryos. Cryo Lett 13:199–208

Parkinson M, Prendergast M, Sayegh AJ (1996) Sterilisation of explants and cultures with sodium dichloroisocyanurate. Plant Growth Reg 20:61–66

Pence VC (2005) In vitro collecting (IVC). I. The effect of collecting method and antimicrobial agents on contamination in temperate and tropical collections. In Vitro Cell Dev Biol Plant 41(3):324–332

Pence VC (2011) Evaluating costs for the in vitro propagation and preservation of endangered plants. In Vitro Cell Dev Biol Plant 47(1):176–187

Pennisi E (2010) Tending the global garden. Science 329:1274–1277

Rai MK, Jaiswal VS, Jaiswal U (2008) Encapsulation of shoot tips of guava (*Psidium guajava* L.) for short-term storage and germplasm exchange. Sci Hortic 118(1):33–38

Ray A, Bhattacharya S (2008) Storage and plant regeneration from encapsulated shoot tips of Rauvolfia serpentina—an effective way of conservation and mass propagation. S African J Bot 74(4):776–779

Reed BM (1992) Cold storage of strawberries in vitro: a comparison of three storage systems. Fruit Var J 46(2):98–102

Reed BM (1993) Improved survival of in vitro-stored *Rubus* germplasm. J Am Soc Hortic Sci 118:890–895

Reed BM (1999) The in vitro genebank of temperate fruit and nut crops at the National Clonal Germplasm Repository-Corvallis. In: Engelmann F (ed) Management of field and in vitro germplasm collections. International Plant Genetic Resources Institute, Rome, pp 132–135

Reed B (2002) Photoperiod improves long-term survival of in vitro-stored strawberry plantlets. HortScience 37(5):811–814

Reed BM, Aynalem HA (2005) Iron formulation affects in vitro cold storage of hops. Acta Hortic 668:257–262

Reed BM, Tanprasert P (1995) Detection and control of bacterial contaminants of plant tissue cultures. A review of recent literature. Plant Tiss Cult Biotechnol 1:137–142

Reed BM, Engelmann F, Dulloo E, Engels J (eds) (2005) Technical guidelines for the management of field and in vitro germplasm collections. IPGRI/FAO/SGRP, Rome

Reed B, Sarasan V, Kane M, Bunn E, Pence V (2011) Biodiversity conservation and conservation biotechnology tools. In Vitro Cell Dev Biol Plant 47(1):1–4

Ren QM (1994) Advances in the research of fruit germplasm resources in China. In: Rao VR (ed) Proceedings of East Asia coordinators' meeting on plant genetic resources, Beijing. IPGRI, Singapore. Regional Office for Asia, the Pacific and Oceania, pp 99–102

Roca WM, Tay D (2007) Global strategy for ex-situ conservation of sweet potato genetic resources. CIP, Peru

Sarasan V (2011) Importance of in vitro technology to future conservation programmes worldwide. Kew Bull 65(4):549–554

Sarasan V, Cripps R, Ramsay MM, Atherton C, McMichen M, Prendergast G, Rowntree JK (2006) Conservation in vitro of threatened plants—progress in the past decade. In Vitro Cell Dev Biol Plant 42(3):206–214

Sarkar D, Chakrabarti SK, Naik PS (2001) Slow-growth conservation of potato microplants: efficacy of ancymidol for long-term storage in vitro. Euphytica 117:131–142

Skirvin RM, McMeans O, Wang WL (1993) Storage water is a source of latent bacterial contamination in vitro. Plant Pathol 1:63–65

Skirvin RM, Motoike S, Norton MA, Ozgur M, Al-Juboory K, McMeans O (1999) Workshop of micropropagation establishment of contaminant-free perennial plants in vitro. In Vitro Cell Dev Biol Plant 35:278–280

Sundararaj SG, Agrawal A, Tyagi RK (2010) Encapsulation for in vitro short-term storage and exchange of ginger (*Zingiber officinale* Rosc.) germplasm. Sci Hortic 125:761–766

Taylor M (2000) New regional genebank in Fiji was made-to-order for Pacific island nations. Diversity 16:19–21

TBGRI (2010) Annual report of the Tropical Botanical Garden and Research Institute. Tropical Botanical Garden and Research Institute, Thiruvananthapuram

Towill LE (1988) Genetic considerations for germplasm preservation of clonal materials. HortScience 23:91–95

Tyagi RK, Yusuf A, Dua P, Agrawal A (2004) In vitro plant regeneration and genotype conservation of eight wild species of *Curcuma*. Biol Plant 48(1):129–132

Tyagi RK, Agarwal A, Yusuf A (2006) Conservation of Zingiber germplasm through in vitro rhizome formation. Sci Hortic 108:210–219

Uchendu EE, Paliyath G, Brown DCW, Saxena PK (2011) In vitro propagation of the North American ginseng (*Panax quinquefolius* L.). In Vitro Cell Dev Biol Plant 47:710–718

Van de Houwe I (1999) INIBAP germplasm Transit Centre: managing the in vitro medium term genebank for *Musa* spp. In: Engelmann F (ed) Management of field and in vitro germplasm collections. International Plant Genetic Resources Institute, Rome, pp 127–131

Van den Houwe I, Panis B (2000) In vitro conservation of banana: medium-term storage and prospects for cryopreservation. In: Razadan MK, Cocking EC (eds) Conservation of plant genetic resources in vitro, vol 2. Science, Enfield, p 17

Viss PR, Brooks EM, Driver JA (1991) A simplified method for the control of bacterial contamination in woody plant tissue culture. In Vitro Cell Dev Biol 27P:42

Walters C, Touchell DH, Power P, Wesley-Smith J, Antolin MF (2002) A cryopreservation protocol for embryos of the endangered species *Zizania texana*. Cryo Lett 23:291–298

Wanas WH, Callow JA, Withers LA (1986) Growth limitations for the conservation of pear genotypes. In: Withers LA, Alderson PG (eds) Plant tissue culture and its agricultural applications. Butterworths, London, pp 285–290

Watt MP, Berjak P, Makhathini A, Blakeway F (2003) In vitro field collection techniques for *Eucalyptus* micropropagation. Plant Cell Tiss Organ Cult 75:233–240

Watt MP, Banasiak M, Reddy D, Albertse EH, Synyman SJ (2009) In vitro minimal growth storage of *Saccharum* spp. hybrid (genotype 88 H0019) at two stages of direct somatic embryogenic regeneration. Plant Cell Tiss Organ Cult 96:263–271

Withers LA, Williams JT (1986) In vitro conservation. IBPGR, Rome

Zandvoort EA, Hulshof MJH, Staritsky G (1994) In vitro storage of *Xanthosoma* spp. under minimal growth conditions. Plant Cell Tiss Organ Cult 36:309–316

Zee FT, Munekata M (1992) In vitro storage of pineapple (*Ananas* spp.) germplasm. HortScience 27(1):57–58

Zee F, Strauss A, Arakawa C (2008) Propagation and cultivation of 'Ohelo'. Coop Ext Serv Bull Fruits Nuts 13:1–6

Chapter 6
Cryopreservation

Florent Engelmann and Stéphane Dussert

6.1 Introduction

Seeds of many important food plants undergo a dehydration period at the end of their maturation process, and are tolerant to extensive desiccation. They can be stored at low moisture content and low temperature for long periods of time. Such seeds, which are mainly, but not exclusively, found in temperate regions of the globe, are termed orthodox (Roberts 1973). Over 90% of the 7.3 million accessions currently stored ex situ in genebanks worldwide are composed of orthodox seeds (FAO 2009).

By contrast, in subtropical and tropical regions, numerous species such as cacao, coconut, coffee, citrus, rubber, oil palm and many forest and fruit tree species, produce seeds, which do not undergo maturation drying and are shed at relatively high moisture content (Chin 1988). Such seeds are unable to withstand desiccation and are often sensitive to chilling. Therefore, they cannot be maintained under the conventional seed storage conditions described above, i.e. low moisture content and low temperature. Seeds of this type are called recalcitrant (Roberts 1973; Chin and Roberts 1980) or intermediate (Ellis et al. 1990, 1991), depending on the extent of their desiccation sensitivity and have to be kept in moist, relatively warm conditions to maintain viability. Even when such seeds are stored in an optimal manner, their lifespan is limited to days, occasionally months.

F. Engelmann (✉)
IRD, UMR DIADE, BP 64501, 911 avenue Agropolis,
Montpellier cedex 5 34394, France

Bioversity International, Via dei Tre Denari 472/a, Maccarese (Fiumicino),
Rome 00057, Italy
e-mail: florent.engelmann@ird.fr

S. Dussert
IRD, UMR DIADE, BP 64501, 911 avenue Agropolis,
Montpellier cedex 5 34394, France

M.N. Normah et al. (eds.), *Conservation of Tropical Plant Species*,
DOI 10.1007/978-1-4614-3776-5_6, © Springer Science+Business Media New York 2013

There are other species whose conservation in the form of seeds is impossible. Some species, such as banana and plantain (*Musa* spp.) do not produce seeds. Some crops such as potato (*Solanum tuberosum*), other root and tuber crops such as yams (*Dioscorea* spp.), cassava (*Manihot esculenta),* sweet potato (*Ipomoea batatas)* and sugarcane (*Saccharum* spp.) have either sterile genotypes and/or some, which produce orthodox seeds. However, because these species are allogamous, their seeds cannot be employed for the conservation of particular genotypes. These crops are thus propagated vegetatively to maintain genotypes as clones.

The field genebank is the traditional ex situ storage method for the problem materials mentioned above (Engelmann and Engels 2002). The First Report on the State of the World's Plant Genetic Resources for Food and Agriculture (FAO 1996) indicated that 527,000 accessions were maintained in field genebanks at that time. No updated data were provided in the second report (FAO 2009) but this number can only have increased during this period. One advantage of this storage method is that genetic resources under conservation can be readily accessed and observed, thus permitting detailed characterisation and evaluation. However, some drawbacks limit its efficiency and threaten its security (Withers and Engels 1990; Engelmann 1997a). The plants remain exposed to pests, diseases and other natural hazards such as drought, weather damage, human error and vandalism. In addition, they are not in a condition allowing germplasm exchange because of the risks of disease transfer through the exchange of vegetative material. Field genebanks are costly to maintain and require considerable inputs in the form of land, labour, management and materials, and, in addition, their capacity to ensure the maintenance of much diversity is limited.

Other categories of plant materials are in urgent need of improved storage technologies. These include biotechnology products such as clones obtained from elite genotypes, cell lines with special attributes and genetically transformed material (Engelmann 1991). This new germplasm is often of high added value and very difficult to produce. The last category of species urgently requiring conservation actions comprises rare and endangered plant species, the number of which is increasing rapidly.

In the light of the difficulties presented by the categories of problem materials outlined above, efforts have been made to improve the quality and security of conservation offered by field genebanks, and to understand seed recalcitrance to make seed storage more widely available. It has also been recognized that alternative approaches to genetic conservation were needed for these problem materials and, since the early 1970s attention has turned to the possibilities offered by biotechnology, specifically in vitro or tissue cultures (Engelmann 2012). Tissue culture techniques are of great interest for storage of plant germplasm (Engelmann 1991; Bunn et al. 2007) as they allow conserving virus-free, miniaturized explants in an aseptic environment, which can be propagated with high multiplication rates for rapid distribution. In vitro propagation protocols have been established for several thousands plant species (George 1996).

In vitro techniques are efficiently employed for medium term storage of numerous plant species (Engelmann 2012, Reed et al. Chap. 5 this volume). However, for long-term storage, cryopreservation, i.e. storage at ultra-low temperature, usually that of liquid nitrogen (−196°C), is the only current method. At this temperature, all cellular divisions and metabolic processes are stopped. The plant material can thus be stored without alteration or modification for extended periods of time. Moreover, cultures are stored in a small volume, protected from contamination, requiring very limited maintenance. It is important to realize that cryopreservation is the only technique currently available to ensure the safe and cost-efficient long-term conservation of the germplasm of problem species.

For cryopreservation, a range of in vivo and in vitro explants can be employed including seeds and dormant buds for in vivo materials and cell suspensions, calluses, shoot tips, somatic embryos, zygotic embryos and embryonic axes for in vitro materials. In this chapter, we describe the various cryopreservation techniques developed for the different materials available and present the current and future utilizations of cryopreservation for conservation of tropical plant germplasm.

6.2 Cryopreservation Protocols

6.2.1 Principle of Cryopreservation

There are materials, such as orthodox seeds or pollen, which display natural dehydration processes and can be cryopreserved without any pretreatment. However, most biological materials employed in cryopreservation, such as cell suspensions, calluses, shoot tips, embryos contain high amounts of cellular water and have thus to be dehydrated artificially to protect them from the damages caused by the crystallization of intracellular water into ice (Mazur 1984). The techniques employed and the physical mechanisms upon which they are based vary between classical and new cryopreservation techniques (Withers and Engelmann 1998). In classical techniques, dehydration of samples takes place both before and during cryopreservation (freeze-induced dehydration), whereas in new techniques, dehydration takes place only before cryopreservation. In optimal conditions, all freezable water is removed from the cells during dehydration and the highly concentrated internal aqueous compartment vitrifies. Vitrification can be defined as the transition of water directly from the liquid phase into an amorphous phase or glass, whilst avoiding the formation of crystalline ice (Fahy et al. 1984).

6.2.2 Controlled Rate Cooling Techniques

Controlled cooling cryopreservation techniques involve slow cooling down to a defined prefreezing temperature, followed by rapid immersion in liquid nitrogen. With temperature reduction during slow cooling, the cells and the external medium initially supercool, followed by ice formation in the medium (Mazur 1984). The cell membrane acts as a physical barrier and prevents the ice from seeding the cell interior and the cells remain unfrozen but supercooled. As the temperature is further decreased, an increasing amount of the extracellular solution is converted into ice, thus resulting in the concentration of intracellular solutes. Since cells remain supercooled and their aqueous vapour pressure exceeds that of the frozen external compartment, cells equilibrate by loss of water to external ice. Depending upon the rate of cooling and the prefreezing temperature, different amounts of water will leave the cell before the intracellular contents solidify. In optimal conditions, most or all intracellular freezable water is removed, thus reducing or avoiding detrimental intracellular ice formation upon subsequent immersion of the specimen in liquid nitrogen, during which vitrification of internal solutes occurs. However, too intense freeze-induced dehydration can incur different damaging events due to concentration of intracellular salts and damages to the cell membrane (Meryman et al. 1977). Rewarming should be as rapid as possible to avoid the phenomenon of recrystallization in which ice melts and reforms at a thermodynamically favourable, larger and more damaging crystal size (Mazur 1984).

Controlled cooling procedures include the following successive steps: pregrowth of samples, cryoprotection, slow cooling (0.1–2.0°C/min) to a determined prefreezing temperature (usually around –40°C), rapid immersion of samples in liquid nitrogen, storage, rapid rewarming, and recovery. Controlled cooling techniques are generally operationally complex since they may require the use of sophisticated programmable freezers.

Controlled cooling cryopreservation techniques are generally employed for dormant buds (Sakai and Nishiyama 1978; Towill and Ellis 2008), apices of cold-tolerant species (Reed and Uchendu 2008) and undifferentiated culture systems such as cell suspensions and calluses (Withers and Engelmann 1998; Kartha and Engelmann 1994). They are thus of limited interest for cryopreservation of tropical plant germplasm.

6.2.3 New Cryopreservation Techniques

In new techniques, cell dehydration is performed prior to cryopreservation by exposure of samples to highly concentrated cryoprotectant solutions and/or air desiccation, until most or all freezable water has been extracted from the cells, which results in vitrification of the aqueous compartment. Intracellular ice formation is thus avoided. This dehydration step is generally followed by direct immersion of samples in liquid nitrogen. These new procedures offer practical advantages compared to controlled cooling freezing techniques. They are more appropriate for complex organs such as shoot-tips, embryos or embryonic axes. By precluding ice formation in the system, these procedures are operationally less complex than controlled cooling ones (e.g. they do not require the use of controlled freezers). This is particularly important for cryopreservation of tropical plant germplasm, which can thus be implemented in a larger number of tropical countries, provided that basic tissue culture facilities are present and that a reliable source of liquid nitrogen is available.

In all these new protocols, the critical step is dehydration, and not cooling, as in controlled cooling protocols (Engelmann 2000). Therefore, if samples to be cryopreserved can be desiccated to sufficiently low water contents with no or little decrease in survival in comparison to non-dehydrated controls, no or limited further drop in survival is generally observed after cryopreservation (Engelmann 2009). Seven different cryopreservation procedures can be identified: (1) encapsulation-dehydration; (2) vitrification; (3) encapsulation-vitrification; (4) dehydration; (5) pregrowth; (6) pregrowth-dehydration; and (7) droplet-vitrification.

6.2.3.1 Encapsulation-Dehydration

Encapsulation-dehydration is based on the technology developed for the production of artificial seeds. Explants are encapsulated in calcium alginate beads, pregrown in liquid medium enriched with sucrose for 1–7 days, partially desiccated with silica gel to a water content around 20% (fresh weight basis), then cooled rapidly. Survival is high and growth recovery of cryopreserved samples is generally rapid and direct, without callus formation. This technique has been applied to apices of numerous species from temperate and of tropical origin as well as to cell suspensions and somatic embryos of several species (Gonzalez-Arnao and Engelmann 2006; Engelmann et al. 2008). A simplification of this technique has been proposed recently by Bonnart and Volk (2010), which involves encapsulation of samples in alginate medium containing 2 M glycerol + 0.5 M sucrose, immediately followed by air-dehydration.

6.2.3.2 Vitrification

Vitrification involves treatment of samples with progressively more concentrated cryoprotectant solutions. Samples are firstly exposed to a loading solution with intermediate concentration, generally 2 M glycerol + 0.4 M sucrose (Matsumoto et al. 1994). Kim et al. (2009a) have devised a series of alternative loading solutions, which proved highly efficient with sensitive species such as chrysanthemum. They are then dehydrated with highly concentrated vitrification solutions. The most commonly employed vitrification solutions are the Plant Vitrification Solutions PVS2 (Sakai et al. 1990) and PVS3 (Nishizawa et al. 1993) developed by the group of Prof. Sakai in Japan, which contain (w/v) 15% ethylene glycol + 30% glycerol + 15% DMSO + 13.7% sucrose and 50% glycerol + 50% sucrose, respectively. Kim et al. (2009b) have developed a series of alternative vitrification solutions, derived from the original PVS2 and PVS3, some of which produced higher survival and recovery compared to the original vitrification solutions, for cryopreservation of sensitive species such as chrysanthemum. Explants are placed in cryotubes, in small quantities of vitrification solution (0.5–1.0 ml), which are cooled rapidly by direct immersion in liquid nitrogen. After rapid warming in a water-bath at 37–40°C, the highly toxic vitrification solutions are removed and replaced by an unloading solution containing 0.8–1.2 M sucrose, thereby allowing progressive rehydration of the samples. Explants are then transferred on culture medium in the dark for 1 week, then transferred to standard culture conditions for recovery. Vitrification protocols have been developed for numerous plant species, both from temperate and tropical origin (Sakai and Engelmann 2007; Sakai et al. 2008).

6.2.3.3 Encapsulation-Vitrification

Encapsulation-vitrification is a combination of encapsulation-dehydration and vitrification. Samples are encapsulated in alginate beads, then treated following a standard vitrification protocol, as described previously. The major interest of this technique is that encapsulated explants are not in direct contact with the highly concentrated vitrification solutions, thereby decreasing their toxicity. This technique has been applied to apices of a number of temperate and tropical plant species (Sakai and Engelmann 2007; Sakai et al. 2008).

6.2.3.4 Dehydration

Dehydration is the simplest cryopreservation procedure. It consists in dehydrating explants, then cooling them rapidly by direct immersion in liquid nitrogen, except for oily seeds, which may require to be slowly precooled prior to cryoexposure (e.g. *Coffea arabica* seeds). This technique is mainly used with seeds, zygotic embryos or embryonic axes extracted from seeds. With seeds, desiccation is usually performed by equilibration drying in controlled relative humidity using saturated salt solutions. In contrast, with zygotic embryos or embryonic axes, the air current of a laminar airflow cabinet is the most common procedure, but more precise and reproducible dehydration conditions can be achieved by using a flow of dry, sterile compressed air, silica gel or saturated salt solutions. These last three dehydration procedures are of particular relevance to tropical countries where the very high air humidity does not allow dehydration of samples down to moisture contents compatible with survival after cryopreservation. Ultra-rapid drying in a stream of compressed dry air (a process called flash drying developed by Berjak's group in South Africa) allows cryopreserving samples with relatively high water content, thus reducing desiccation injury (Berjak et al. 1989). Dehydration has been applied to seeds, embryos and embryonic axes of a large number of recalcitrant and intermediate tropical species (Engelmann 1997b). Optimal survival is generally obtained when samples are cryopreserved with a water content comprised between 10% and 20% (fresh weight basis). The use of Differential Scanning Calorimetry in a wide range of non-orthodox seed species allowed to better understand the basis of the optimal hydration status for seed cryopreservation (Dussert et al. 2001; Hor et al. 2005). In particular, it was shown that it can be predicted as a function of the seed lipid content and that the seed unfrozen water content, which is the optimal hydration status for cryopreservation in coffee seeds (Dussert et al. 2001), is always achieved using a desiccation RH of ca. 80%.

6.2.3.5 Pregrowth

The pregrowth technique consists of cultivating samples on medium with cryoprotectants (generally sucrose), then cooling them rapidly by direct immersion in liquid nitrogen. The pregrowth technique has been applied for cryopreservation of *Musa* meristematic cultures (Panis et al. 2002).

6.2.3.6 Pregrowth-Dehydration

In a pregrowth-dehydration protocol, explants are pregrown in the presence of cryoprotectants, dehydrated under the laminar airflow cabinet or with silica gel, and then cryopreserved rapidly. This method has been applied to various explants of different species, including asparagus stem segments (Uragami et al. 1990), oil palm somatic embryos (Dumet et al. 1993), coconut zygotic embryos (Assy-Bah and Engelmann 1992) and, more recently, to coriander somatic embryos (Popova et al. 2010).

6.2.3.7 Droplet-Vitrification

Droplet-vitrification is the latest technique developed (Panis et al. 2005). Samples are treated following a standard vitrification protocol, except for the cooling step, for which explants are placed on an aluminum foil in minute droplets of vitrification solution, which are immersed rapidly in liquid nitrogen. The main advantage of this technique lies with the fact that explants are in direct contact with liquid nitrogen during cooling and with the unloading solution during rewarming, thereby ensuring very high cooling and rewarming rates. The number of species to which it has been successfully applied is increasing rapidly (Sakai and Engelmann 2007).

6.3 Current Application of Cryopreservation to Tropical Plant Species

6.3.1 Cryopreservation of Vegetatively Propagated and Recalcitrant Seed Species

In the case of vegetatively propagated tropical species, cryopreservation protocols have been established for root and tubers, fruit and forest trees, ornamentals and plantation crops, and often applied to large numbers of accessions within species

(Engelmann 2009). Vitrification-based protocols have been employed in most cases, often producing high survival and recovery percentages. Different reasons can explain these good results. The meristematic zone of apices, from which organized growth originates, is composed of a relatively homogenous population of small, actively dividing cells, with little vacuoles and a high nucleo-cytoplasmic ratio. These characteristics make them more susceptible to withstand desiccation than highly vacuolated and differentiated cells. The whole meristem is generally preserved when using vitrification-based techniques, thus allowing direct, organized regrowth after cryopreservation. These good results are also linked with the availability of tissue culture protocols for vegetatively propagated plants. Indeed, many vegetatively propagated species for which cryopreservation protocols have been developed are cultivated crops. Cultural practices, including in vitro micropropagation, are well established for such species. In addition, in vitro material is "synchronized" by the tissue culture and pregrowth procedures. Relatively homogenous samples in terms of size, cellular composition, physiological state and growth response are employed for cryopreservation, thus increasing the chances of positive and uniform response to treatments. Finally, vitrification-based procedures allow using samples of relatively large size (shoot tips of 0.5 to 2–3 mm), which can resume growth directly without any difficulty.

By contrast, only limited success has been achieved until now with cryopreservation of non-orthodox species (Engelmann 2009). The desiccation is generally employed for freezing embryos or embryonic axes. Survival is variable, but generally low. Various reasons can explain this situation. Many recalcitrant and intermediate species are wild species, and consequently, there is no or little information on their biology and seed storage behaviour. Moreover, only very few teams worldwide are working on conservation and cryopreservation of recalcitrant and intermediate seeds. In vitro culture protocols are often non-existent or not fully operational. Seeds and embryos of such species are often very large, have a complex tissue composition and display very large variations in moisture content between and within seed lots, making the successful application of a single protocol difficult.

Various options can be explored to improve cryopreservation of non-orthodox seed species. In the case of intermediate seeds, very precisely controlled desiccation may allow cryopreserving whole seeds as demonstrated recently with various coffee species (Dussert et al. 1997; Dussert and Engelmann 2006). Additional cryopreservation techniques should also be tested with embryos and embryonic axes. In the case of highly recalcitrant seed species, for which it will be impossible to cryopreserve whole seeds, embryos or embryonic axes due to their high desiccation sensitivity, alternative explants, such as shoot apices sampled on embryos, adventitious buds or somatic embryos induced from the embryonic tissues represent the only alternative option. A good illustration of such a situation is provided by the work performed at the International Intermediate Cocoa Quarantine Centre at Reading, UK, where floral-derived somatic embryos are cryopreserved to back-up the 600 accessions maintained as whole plants in the greenhouse (Wetten et al. 2011).

Finally, analytical techniques, which allow describing and understanding the biological and physical processes, which take place during cryopreservation can be highly instrumental for establishing more efficient cryopreservation protocols, as demonstrated in several cases (Engelmann 2012).

6.3.2 Large Scale Utilization of Cryopreservation

Even though its routine use is still limited, there is a growing number of genebanks and botanic gardens where cryopreservation is employed on a large scale for different types of materials, which are, or not, tolerant to dehydration (Engelmann 2012). However, only very few cryopreserved collections of tropical plant species currently exist, including coffee, *Musa* and cassava. In the case of coffee, using a protocol including controlled dehydration and cooling (Dussert and Engelmann 2006), cryopreserved collections of seeds are being established in CATIE (Tropical Agricultural Research and Higher Education Center, Costa Rica), and in IRD Montpellier (France), where it already includes over 200 accessions. Cryopreservation is also systematically employed for storing all the new embryogenic cell lines of coffee and cacao produced by the Biotechnology Laboratory of the Nestlé Company based in Tours, France (Florin et al. 1999). Finally, cryopreserved collections are under development for long-term storage of tropical plants: 630 banana accessions have been cryopreserved at the INIBAP International Transit Center (Panis et al. 2007) and 540 cassava accessions at the International Center for Tropical Agriculture (CIAT, Cali, Colombia) (Gonzalez-Arnao et al. 2008).

Cryopreservation often involves tissue culture, which is susceptible of inducing modifications in cryopreserved cultures and regenerated plants. It is thus necessary to verify that the genetic stability of the cryopreserved material is not altered before routinely using this technique for the long-term conservation of plant genetic resources. There is no report of modifications at the phenotypical, biochemical, chromosomal or molecular level that could be attributed to cryopreservation (Engelmann 1997a, 2009).

The few studies performed on the cost of cryopreservation confirm the interest of this technique from a financial perspective, in view of its low utilization cost (Engelmann 2012). Recently, a detailed study compared the costs of establishing and maintaining a coffee collection in the field and under cryopreservation, using the CATIE collection in Costa Rica as example (Dulloo et al. 2009). The results indicate that cryopreservation costs less (in perpetuity per accession) than conservation in field genebanks. A comparative analysis of the costs of both methods showed that the more accessions there are in cryopreservation storage, the lower the per accession cost. In addition to cost, the study examined the advantages of cryopreservation over field collection and showed that for non-orthodox seed species that can only be conserved as live plants, seed cryopreservation may be the method of choice for long-term conservation of genetic diversity.

6.3.3 Additional Use of Cryopreservation

A new utilization of cryopreservation has been proposed recently, viz. the elimination of viruses from infected plants, a process called cryotherapy (Brison et al. 1997; Helliot et al. 2002; Wang et al. 2008). In cryotherapy, plant pathogens such as viruses, phytoplasmas and bacteria are eradicated from shoot tips by exposing them briefly to liquid nitrogen. The cryo-treatment destroys the more differentiated cells in which viruses are located, whereas the pathogen-free meristematic cells withstand cryopreservation and can regenerate healthy plants. Thermotherapy followed by cryotherapy of shoot tips can be used to enhance virus eradication. To date, severe pathogens in several tropical plants including banana (*Musa* spp.), *Citrus* spp., sweet potato (*Ipomoea batatas*) have been eradicated using cryotherapy. These pathogens include viruses (banana streak virus, sweet potato feathery mottle virus and sweet potato chlorotic stunt virus), sweet potato little leaf phytoplasma and Huanglongbing bacterium causing 'citrus greening'.

6.4 Cryopreservation: Progress and Prospects

Cryopreservation is still routinely employed in a limited number of cases only. However, the development of the new vitrification-based techniques has made its application to a broader range of species possible. A very important advantage of these new techniques is their operational simplicity, since they will be mainly applied in tropical countries where the largest part of genetic resources of problem species is located, and where, in many instances, only basic infrastructures are available. Another one is their broad applicability, which is of particular relevance to the conservation of wild species, for which large amounts of genetic diversity need to be conserved. In the case of vegetatively propagated species, cryopreservation techniques are well advanced and they are already applied on a large scale for several tropical plants such as *Musa*, sweet potato and cassava. Research is much less advanced for recalcitrant seed species. This is due to the large number of such, mainly wild, species, with very different characteristics, which fall within this category and to the comparatively limited level of research aiming at improving the conservation of these species. However, various technical approaches, which need to be explored, have been identified to improve the efficiency and increase the applicability of cryopreservation techniques to recalcitrant species. In addition, research is actively performed by various groups in universities, research institutes, botanic gardens and genebanks worldwide to improve knowledge of biological mechanisms underlying seed recalcitrance. It is hoped that new findings on critical issues such as understanding and control of desiccation sensitivity will contribute significantly to the development of improved cryopreservation techniques for recalcitrant seed species. It can thus be realistically expected that in the coming years, our understanding of the biological mechanisms involved in desiccation tolerance as well as in cryopreservation will increase and that cryopreservation will become more frequently employed for long-term conservation of biodiversity of tropical plant species.

References

Assy-Bah B, Engelmann F (1992) Cryopreservation of mature embryos of coconut (*Cocos nucifera* L.) and subsequent regeneration of plantlets. Cryo Lett 13:117–126

Berjak P, Farrant JM, Mycock DJ, Pammenter NW (1989) Homoiohydrous (recalcitrant) seeds: the enigma of their desiccation sensitivity and the state of water in axes of *Landolphia kirkii* Dyer. Planta 186:249–261

Bonnart R, Volk GM (2010) Increased efficiency using the encapsulation-dehydration cryopreservation technique for *Arabidopsis thaliana*. Cryo Lett 31:95–100

Brison M, de Boucaud MT, Pierronet A, Dosba F (1997) Effect of cryopreservation on the sanitary state of a cv Prunus rootstock experimentally contaminated with Plum Pox Potyvirus. Plant Sci 123:189–196

Bunn E, Turner SR, Panaia M, Dixon KW (2007) The contribution of in vitro technology and cryogenic storage to conservation of indigenous plants. Aust J Bot 55:345–355

Chin HF (1988) Recalcitrant seeds: a status report. International Plant Genetic Resources Institute, Rome

Chin HF, Roberts EH (1980) Recalcitrant crop seeds. Tropical Press Sdn. Bhd, Kuala Lumpur

Dulloo ME, Ebert AW, Dussert S, Gotor E, Astorga C, Vasquez N, Rakotomalala JJ, Rabemiafara A, Eira M, Bellachew B, Omondi C, Engelmann F, Anthony F, Watts J, Qamar Z, Snook L (2009) Cost efficiency of cryopreservation as a long term conservation method for coffee genetic resources. Crop Sci 49:2123–2138

Dumet D, Engelmann F, Chabrillange N, Duval Y (1993) Cryopreservation of oil palm (*Elaeis guineensis* Jacq.) somatic embryos involving a desiccation step. Plant Cell Rep 12:352–355

Dussert S, Engelmann F (2006) New determinants of coffee (*Coffea arabica* L.) seed tolerance to liquid nitrogen exposure. Cryo Lett 27:169–178

Dussert S, Chabrillange N, Anthony F, Engelmann F, Recalt C, Hamon S (1997) Variability in storage response within a coffee (*Coffea* spp.) core collection under slow growth conditions. Plant Cell Rep 16:344–348

Dussert S, Chabrillange N, Roquelin G, Engelmann F, Lopez M, Hamon S (2001) Tolerance of coffee (*Coffea* spp.) seeds to ultra-low temperature exposure in relation to calorimetric properties of tissue water, lipid composition and cooling procedure. Physiol Plant 112:495–505

Ellis RH, Hong T, Roberts EH (1990) An intermediate category of seed storage behaviour? I. Coffee. J Exp Bot 41:1167–1174

Ellis RH, Hong T, Roberts EH, Soetisna U (1991) Seed storage behaviour in *Elaeis guineensis*. Seed Sci Res 1:99–104

Engelmann F (1991) In vitro conservation of tropical plant germplasm—a review. Euphytica 57:227–243

Engelmann F (1997a) In vitro conservation methods. In: Ford-Lloyd BV, Newburry JH, Callow JA (eds) Biotechnology and plant genetic resources: conservation and use. CABI, Wellingford, pp 119–162

Engelmann F (1997b) Importance of desiccation for the cryopreservation of recalcitrant seed and vegetatively propagated species. Plant Genet Res Newsl 112:9–18

Engelmann F (2000) Importance of cryopreservation for the conservation of plant genetic resources. In: Engelmann F, Takagi H (eds) Cryopreservation of tropical plant germplasm—current research progress and applications. JIRCAS/IPGRI, Tsukuba/Rome, pp 8–20

Engelmann F (2009) Use of biotechnologies for conserving plant biodiversity. Acta Hort 812:63–82
Engelmann F (2012) Germplasm collection, storage and preservation. In: Altman A, Hazegawa PM (eds) Plant biotechnology 2010: basic aspects and agricultural implications, Elsevier, pp 255–268
Engelmann F, Engels JMM (2002) Technologies and strategies for ex situ conservation. In: Engels JMM, Rao VR, Brown ADH, Jackson MT (eds) Managing plant genetic diversity. CABI/IPGRI, Wallingford/Rome, pp 89–104
Engelmann F, Gonzalez-Arnao MT, Wu WJ, Escobar RE (2008) Development of encapsulation-dehydration. In: Reed BM (ed) Plant cryopreservation: a practical guide. Springer, Berlin, pp 59–76
Fahy GM, MacFarlane DR, Angell CA, Meryman HT (1984) Vitrification as an approach to cryopreservation. Cryobiology 21:407–426
FAO (1996) Report on the state of the world's plant genetic resources for food and agriculture. Food and Agriculture Organization of the United Nations, Rome
FAO (2009) Draft second report on the world's plant genetic resources for food and agriculture. Food and Agriculture Organization of the United Nations, Rome
Florin B, Brulard E, Lepage B (1999) Establishment of a cryopreserved coffee germplasm bank. In: Abstracts of the CRYO'09 annual meeting of the society for cryobiology, Tsukuba, 21–26 Jul 2009, p 67
George EF (1996) Plant propagation by tissue culture. Part 2—in practice, 2nd edn. Exegetics, Edington
Gonzalez-Arnao MT, Engelmann F (2006) Cryopreservation of plant germplasm using the encapsulation-dehydration technique: review and case study on sugarcane. Cryo Lett 27:155–168
Gonzalez-Arnao MT, Panta A, Roca WM, Escobar RH, Engelmann F (2008) Development and large scale application of cryopreservation techniques for shoot and somatic embryo cultures of tropical crops. Plant Cell Tiss Org Cult 92:1–13
Helliot B, Panis B, Poumay Y, Swennen R, Lepoivre P, Frison E (2002) Cryopreservation for the elimination of cucumber mosaic and banana streak viruses from banana (*Musa* spp.). Plant Cell Rep 12:1117–1122
Hor YL, Kim YJ, Ugap A, Chabrillange N, Sinniah UR, Engelmann F, Dussert S (2005) Optimal hydration status for cryopreservation of intermediate oily seeds: *Citrus* as a case study. Ann Bot 95:1153–1161
Kartha KK, Engelmann F (1994) Cryopreservation and germplasm storage. In: Vasil IK, Thorpe TA (eds) Plant cell and tissue culture. Kluwer, Dordrecht, pp 195–230
Kim HH, Lee YG, Ko HC, Park SU, Gwag JG, Cho EG, Engelmann F (2009a) Development of alternative loading solutions in droplet-vitrification procedures. Cryo Lett 30:291–299
Kim HH, Lee YG, Shin DJ, Kim T, Cho EG, Engelmann F (2009b) Development of alternative plant vitrification solutions in droplet-vitrification procedures. Cryo Lett 30:320–334
Matsumoto T, Sakai A, Yamada K (1994) Cryopreservation of in vitro grown apical meristems of wasabi (*Wasabia japonica*) by vitrification and subsequent high plant regeneration. Plant Cell Rep 13:442–446
Mazur P (1984) Freezing of living cells: mechanisms and applications. Amer J Physiol 247:C125–C142, Cell Physiol 16
Meryman HT, Williams RJ, Douglas MSJ (1977) Freezing injury from solution effects and its prevention by natural or artificial cryoprotection. Cryobiology 14:287–302
Nishizawa S, Sakai A, Amano AY, Matsuzawa T (1993) Cryopreservation of asparagus (*Asparagus officinalis* L.) embryogenic suspension cells and subsequent plant regeneration by vitrification. Plant Sci 91:67–73
Panis B, Strosse H, Van den Henda S, Swennen R (2002) Sucrose preculture to simplify cryopreservation of banana meristem cultures. Cryo Lett 23:375–384
Panis B, Piette B, Swennen R (2005) Droplet vitrification of apical meristems: a cryopreservation protocol applicable to all *Musaceae*. Plant Sci 168:45–55
Panis B, Van den houwe I, Piette B, Swennen R (2007) Cryopreservation of the banana germplasm collection at the ITC (INIBAP Transit centre). In: Proceedings of the 1st meeting of COST 871

working group 2: technology, application and validation of plant cryopreservation, Florence, 10–13 May 2007, pp 34–35

Popova E, Kim HH, Paek KY (2010) Cryopreservation of coriander (*Coriandrum sativum* L.) somatic embryos using sucrose preculture and air desiccation. Sci Hortic 124:522–528

Reed BM, Uchendu E (2008) Controlled rate cooling. In: Reed BM (ed) Plant cryopreservation: a practical guide. Springer, Berlin, pp 77–92

Roberts HF (1973) Predicting the viability of seeds. Seed Sci Technol 1:499–514

Sakai A, Engelmann F (2007) Vitrification, encapsulation-vitrification and droplet-vitrification: a review. Cryo Lett 28:151–172

Sakai A, Nishiyama Y (1978) Cryopreservation of winter vegetative buds of hardy fruit trees in liquid nitrogen. Hortscience 13:225–227

Sakai A, Kobayashi S, Oiyama IE (1990) Cryopreservation of nucellar cells of navel orange (*Citrus sinensis* Osb. *var. brasiliensis* Tanaka) by vitrification. Plant Cell Rep 9:30–33

Sakai A, Hirai D, Niino T (2008) Development of PVS-based vitrification and encapsulation-vitrification protocols. In: Reed BM (ed) Plant cryopreservation: a practical guide. Springer, Berlin, pp 33–58

Towill LE, Ellis DD (2008) Cryopreservation of dormant buds. In: Reed BM (ed) Plant cryopreservation: a practical guide. Springer, Berlin, pp 421–442

Uragami A, Sakai A, Magai M (1990) Cryopreservation of dried axillary buds from plantlets of *Asparagus officinalis* L. grown in vitro. Plant Cell Rep 9:328–331

Wang Q, Panis B, Engelmann F, Lambardi M, Valkonen JPT (2008) Elimination of plant pathogens by cryotherapy of shoot tips: a review. Ann Appl Biol 154. http://www3.interscience.wiley.com/journal/119879031/issue. Accessed on 15 Nov 2011

Wetten A, Adu-Gyamfi R, Rodriguez-Lopez C (2011) Apples and cocoa: distinct challenges in cryopreservation of two germplasm collections. In: COST action 871 cryopreservation of crop species in Europe final meeting, Agrocampus Ouest INPH, Angers, 8–11 Feb 2011, pp 10–11

Withers LA, Engelmann F (1998) In vitro conservation of plant genetic resources. In: Altman A (ed) Biotechnology in agriculture. Marcel Dekker, New York, pp 57–88

Withers LA, Engels JMM (1990) The test tube genebank—a safe alternative to field conservation. IBPGR Newsl Asia Pac 3:1–2

Chapter 7
Biomarkers from Molecules to Ecosystems and Biobanks to Genebanks

Keith Harding and Erica E. Benson

7.1 Introduction

Major advances have been made regarding the in vitro conservation of economically important tropical plants (Benson et al. 1996; Benson 1999, 2004; Coates and Dixon 2007; Garming et al. 2010; Lakshmanan et al. 2011; Noor et al. 2011) culminating in the establishment of In Vitro Genebanks (IVGBs) as reviewed by Benson (2008) and other contributors to this book. However, there is a significant part of tropical plant diversity for which in vitro preservation is an aspiration rather than an actuality. Tropical species present exacting challenges for their conservation ranging from understanding their complex life cycles to ameliorating the effects of stress during storage recalcitrance (Benson 2008; Berjak et al. 2011) and optimising cryostorage regimes (Nadarajan et al. 2008; Nashatul et al. 2007). The motivation for this chapter is provided by the vulnerability of tropical species to extinction proneness in their natural environment (Ghazoul and Sheil 2010) and the need to step up the pace of developing integrated conservation measures (Coates and Dixon 2007). The revised 2011–2020 targets of the Convention on Biological Diversity's (CBD) Global Strategy for Plant Conservation are timely actions in response to the rapid rate of species decline. This review concerns Target 8 (Jackson and Kennedy 2009; Sharrock et al. 2010): "Objective 2: Plant diversity is urgently and effectively conserved; Target 8: At least 75% of threatened plant species in ex situ collections, preferably in the country of origin and at least 20% available for recovery and restoration programmes". The practical realisation of in vitro plant conservation has been mainly achieved on a 'case-by-case' basis using empirical approaches underpinned by fundamental research (Benson 2008, 2004; Benson and Harding 2012; Litz et al. 2004; Lakshmanan et al. 2011; Nashatul et al. 2007; Reed 2008). For tropical species, it is crucial to generate technically-relevant knowledge concerning

K. Harding (✉) • E.E. Benson
Damar, Drum Road, Cuparmuir, Cupar, Fife KY15 5RJ, Scotland, UK
e-mail: k.harding-damar@tiscali.co.uk; e.benson-damar@tiscali.co.uk

M.N. Normah et al. (eds.), *Conservation of Tropical Plant Species*,
DOI 10.1007/978-1-4614-3776-5_7, © Springer Science+Business Media New York 2013

the stresses incurred by different in vitro conservation and seed storage manipulations (Benson 2008; Berjak et al. 2011; Noor et al. 2011). Thus, research will need to take into account the practical and logistical constraints of conserving diverse and problematic genetic resources in seed banks and in vitro genebanks (Higa et al. 2011; Nadarajan et al. 2006, 2007). Furthermore, a range of storage options may need to be tested for the preservation of different types of germplasm from the same species (Costa Nunes et al. 2003).

Biomarkers are used in both fundamental and applied plant conservation research to aid the development of preservation protocols and improve storage outcomes (Benson 2008; Harding et al. 2009; Laamanen et al. 2008). Their wider applications range from investigating the molecular basis of adaptive stress physiologies to assessing genetic diversity in genebank accessions and their exploitation as ecological indicators in environmental impact monitoring (Fang et al. 2008; Kranner et al. 2006; Tausz et al. 2004; Karp et al. 1997; Kosakivska 2008). Knowledge of species variation and their genetic structure is essential for making informed decisions regarding effective conservation measures. Consequently, molecular technologies are increasingly applied to evaluate the extent of biodiversity and genetic diversity within a species (Ayad et al. 1997; de Vicente 2004; de Vicente and Andersson 2006; Spooner et al. 2005) and assess their potential to adapt and survive the challenges of human impacts and climate change (Ahuja et al. 2010; Kramer and Havens 2009; Nicotra et al. 2010). This chapter considers how the Barcode of Life initiative and biomarkers can be used to assist the conservation of tropical plants and their genetic resources.

7.2 The Barcode of Life: BOLI, iBOL, BOLD, CBOL

A new approach is being explored to assist evaluations of biodiversity. DNA barcoding is a taxonomic tool that uses a genetic marker in a plant genome to identify that it belongs to a specific species. It tests the hypothesis that all biological species may be identified using a short DNA sequence of a gene from a standardised position in the genome to produce a DNA barcode characteristic of a species. The concept is analogous to the universal product code of optical, machine-readable black bar stripes used to trace retail consumer products. DNA barcoding of life on Earth is a global scientific initiative that utilises genetic markers as taxonomic tools to identify species from DNA derived from a specimen (Ratnasingham and Hebert 2007). Since 2003, researchers around the world have joined together to form the Barcode of Life Initiative (BOLI). The International Barcode of Life (iBOL) project is a Canadian-led research alliance involving some 26 countries. It enables scientists to collect specimens, obtain their DNA barcode records and build an informatics platform to store and share the information for the identification of species. The progressive input of barcode data is creating a global resource for the identification of species moving towards a barcode reference library for all life. The barcoding technique uses minute amounts of tissue derived from a specimen. The iBOL project aims to collect specimens from a broad range of taxa to produce DNA barcode records for all species. DNA barcoding utilises the natural DNA sequence diversity in a standardised gene region to identify species with the potential to discover unknown species. Barcode sequences are uploaded on an online informatics platform, the Barcode of Life Data Systems (BOLD). This information becomes part of a growing reference library of validated DNA barcodes for all life on Earth. To unify these activities the Consortium for the Barcode of Life (CBOL) has been established as an international collaboration devoted to developing DNA barcoding as a research tool in taxonomy and a global standard for species identification (CBOL 2009).

Information emerging from BOLI will be used to inform both in situ and ex situ conservation strategies. As a growing biodiversity database, this huge resource provides information that is likely to facilitate tropical plant conservation, research and activities related to species and biomarker discovery. Sass et al. (2007) have tested the potential of using barcoding markers for species identification of cycads which may have future applications in their conservation (Donaldson 2004). DNA barcodes have been used for the identification of Amazonian trees (Gonzalez et al. 2009) for which the process can assist the creation of inventories to inform conservation strategies where the workload for classic taxonomic surveys is high as is the case for tropical forests. DNA barcode markers can also help identify plants, detect taxonomic errors and enable the identification of plants in their juvenile stages (Gonzalez et al. 2009). All these applications will have relevance for ex situ conservation, particularly as genotype differences and the adaptive behaviours of tropical plants can greatly impact their responses to preservation and tissue culture treatments.

7.3 Biomarker Definitions

There are numerous situations where biomarkers have a specific application and utility within a given context. A generic definition of a biological marker or 'biomarker' is a substance or descriptor that is used to indicate the 'state' of a given biological system. Therefore, biomarkers may be a characteristic, attribute, feature or quality that can be quantitatively determined as an indicator of any biological process. They may be indicators of:

1. Normal biological systems that reflect complex interactions as detected by various molecular biological 'omics' platforms;
2. Abnormal biological processes reflecting disease states in pathology, hypersensitive reactions and age-related degeneration and
3. Biological response mechanisms to environmental triggers, chemical agents or treatments.

As indicators of a given state within a biological system, a wide range of generic biomarkers exist that fall within this general description. Some of the more common include: DNA sequence fragments, RNA molecules, proteins and metabolites reflecting respectively the following omics platforms: genomics, epigenomics, transcriptomics, proteomics and metabolomics (Ahuja et al. 2010; Carpentier et al. 2007, 2008; Miguel and Marum 2011; Shulaev and Oliver 2006; Volk 2010; Zhu et al. 2006).

7.4 Biomarkers for Tropical Plant In Vitro Conservation

Biomarkers are an integral part of Biological Resource Centres (BRCs), biobank, biorepository and genebank processes for which they are used diversely in genetic and biospecimen-related research as diagnostic tools and performance indicators for the purposes of quality and risk management. Examples that may be applied to tropical plant in vitro conservation can be placed into two categories, the first concerns those that are used for conservation research and the second for conservation operations with overlap in both contexts. Biomarkers may be used in research to explore the basis of different storage responses in order to improve conservation outcomes. They may be used in the operational framework of a 'working' genebank, as quality performance indicators or as diagnostic tools in accordance with the BRCs guiding principles: purity, authenticity and stability (Benson 2008; ISBER 2008; OECD 2007). The following sections illustrate the main types of biomarkers used in biobanking and genebanking contexts while exploring their potential applications for tropical plant conservation and related environmental studies.

7.5 On/Off Response Biomarkers

In the clinical context, the concept of biomarkers describes the molecular events characteristic for various stages between exposure to an adverse agent and the manifestation of a disease (Holland et al. 2003). Biomarkers have proven to be valuable tools for molecular epidemiology; where studies require the careful handling and storage of precious biological samples (Holland et al. 2005). There are many factors which affect the quality of the samples and the stability of biomarkers that are influenced by the initial collection stage (Balasubramanian et al. 2010). These are termed preanalytical variables and they include: the tissue type, time of collection, containers used, preservatives and other additives, type of transport and length of transit time (Betsou et al. 2010). The Standard PREanalytical Code (SPREC) was thus developed by the medical/clinical biobanking sector motivated by the need to harmonize biospecimen traceability in preanalytical processes and enable interconnectivity and interoperability between different biobanks, research consortia and infrastructures (Betsou et al. 2010). The concept has obvious utility in biorepositories and culture collections that service environmental and biodiversity communities. The clinical SPREC has been recalibrated as a putative code that may be adopted for biobanks holding different types of biodiversity. This 'tool' presents a new paradigm for environmental biobanking sectors and has been recently explored in algal culture collections (Benson et al. 2011). Equally so, the adoption of SPREC would be a timely advancement to assist progress in the conservation of tropical plant germplasm, particularly where it is to be used for molecular, genetic and systematic studies. It is prudent to evaluate the potential impact that preanalytical factors have on the down-stream treatment and use of genetic resources, especially where the original samples require further processing, for example cryopreservation of isolated cells/tissue types, DNA/RNA and preparation of specimens for cytogenetic, immunological, biochemical and omics analyses (Palmirotta et al. 2011; Kugler et al. 2011).

Relevant to in vitro conservation, Riegman et al. (2008) suggest that an ideal biomarker has a ubiquitous expression and shows 100% loss, or vice versa, gain of activity upon inadequate processing, storing conditions and temperature variations. Thus, consistent with the generic definition, a biomarker with an on/off response could serve as quality indicator if it shows a sample has the required quality for a given assay or diagnostic technique (Bank and Schmehl 1989; Verleysen et al. 2004; Yin et al. 2011). Biomarkers applied in this situation form part of an evidence-based test that is used to determine the quality and usability of a specimen especially during its storage life time (Balasubramanian et al. 2010; Kugler et al. 2011). For instance, they may also serve as indicators of variation in storage conditions revealing departures from: (1) biobanking Best Practices (BPs) which are protocols, methods or processes that are the most effective in realising a particular outcome compared with any other approach; they are checked and tested for compliance with required outcome(s) and should be achieved without complications, changes or unforeseen problems and (2) Standard Operation Procedures (SOPs) which comprise a portfolio of technically detailed, documented standard procedures that are consistently and commonly used by the personnel in a biobank or biorepository.

Biomarkers can be applied to many different types of biospecimens as specific or generic indicators of connecting biological processes. As preanalytical procedures and low temperature storage are vital components of biobanking operations; Riegman et al. (2008) advocate that biomarkers should be found for all types of samples and their derivative samples and products held by biobanks. Examples of on/off biomarker responses in plant genebanks include viability stains (dead/alive) and indicators of morphogenetic competency (totipotent/non-totipotent) and in certain circumstances the presence or absence of specific secondary metabolites (e.g. medicinal compounds).

7.6 Transitional and Temporal Biomarkers

Patterns of gene expression and steady state levels of gene products for metabolic pathways and cellular 'house keeping' activities conceivably provide a basis for the discovery of potentially useful biomolecules as biomarkers in clinical biobanking. However, other types of biomarker may be useful as applied for the conservation of plants in vitro. The notion of on/off switching mechanisms that regulate patterns of gene expression may be less obvious in an organism's adaptive responses to environmental stimuli, chemical triggers and changing habitats. The stochastic nature of gene expression is all too evident as demonstrated by the out-of-steady-state variability of the cyanobacterial circadian clock (Chabot et al. 2007). Simply having the entire genome switched into full expression bioenergetically limits an organism's ability to regulate and conserve its resources. The complex patterns of differential gene expression are vital sequences in developmental pathways and growth of tropical plant life cycles. While some genes may be switched on, the dynamics of their expression may be considered more as a phasing process, where the levels of various gene products independently oscillate between the lows and highs of their potential limits allowing the plant to adapt coping strategies for environmental challenges. Understanding these mechanisms is relevant to the development of in vitro conservation protocols for tropical plants, especially as their complicated life cycles and adaptive strategies impinge upon their responses to storage treatments. The complexity of the dynamic regulatory processes is indicated by the intricate patterns of expression for an array of genes as revealed by the application of functional genomics and proteomics in plant cryobiology (Volk 2010). As in vitro plant conservation, cryostorage and related subjects increasingly enter into the genomics arena, it is likely to lead to the further identification and quantification of gene candidates as potential biomarkers in the temporal dynamics of tropical plant cryopreservation (Noor et al. 2011). Biomarker candidates are beginning to emerge from studies of the interactions between mRNA-microRNAs (Rodriguez-Enriquez et al. 2011) and protein-metabolite molecules (Ahuja et al. 2010; Shulaev and Oliver 2006) and other biomolecular changes involving chemical modifications to DNA, RNA and the modulation of chromatin structure over time (Harding et al. 2000; Johnston et al. 2005, 2007, 2009, 2010; Miguel and Marum 2011). The implications of other transitional and temporal biomarkers is further shown by investigations into the role(s) of small RNA molecules which are beginning to reveal the complexity of microRNA networks and developmental plasticity in plants (Rubio-Somoza and Weigel 2011) and short interfering (si)RNAs in the regulation of gene expression from plant to progeny (Mosher and Melnyk 2010).

7.7 Ecosystems, Ecological Indicators and Stress Biomarkers

As the generic definition prescribes, biomarkers are not necessarily restricted to molecular interactions at the cellular level they can also be indicators of the dynamics in a given situation and as such are equally applicable to the macro-level changes that occur in ecosystems (Kosakivska 2008). In the context of biological resources, conservation and plant diversity studies, the systems analysis of the biome has considerable importance in developing the discipline of biomics (Cramer 2002). In the broader sense and for the purposes of conservation, biomarkers can also incorporate ecological indicators, for example, the use of indicator species for biodiversity assessments and environmental change (Lindenmayer et al. 2000). One of the more compelling applications of biomarkers concerns the study and mitigation of the impacts of climate change (Ahuja et al. 2010) in the context of both ecological (in situ) and ex situ (in vitro) conservation research (Berjak et al. 2011).

The protection of species *and* their habitats is key to in situ and ex situ conservation and critical for the maintenance of biodiversity. Therefore, the wider application of biomarkers in tropical ecosystem conservation i.e. the ecological context is supported by the concept and practice of bionomics which deals with plant habitats and life cycle strategies with respect to the changes that occur in the natural environment and how plants respond to them (Kosakivska 2008). The importance of understanding these interactions is well recognised and forms a vital component of the Global Strategy for Plant Conservation; Target 8 which aims to conserve 75% of the threatened plant species in ex situ collections where 20% are available for recovery restoration programmes. Coates and Dixon (2007) highlight the importance of the linkage between ex situ conservation and reintroduction programmes regarding effective off-site (ex situ) protection of germplasm. This link involves the concepts and practical principles of ex situ conservation technologies that relate to the preservation of germplasm using traditional seed banking methods and the more challenging biotechnological techniques of in vitro culture and cryopreservation. These approaches support conservation science by providing an alternative means to protect tropical plant germplasm that cannot be effectively stored as seed (Berjak et al. 2011). In this scenario, biomarkers are indicators of sensitivity and tolerance to stressful storage treatments (chilling, freezing, desiccation) and can be used to help develop and optimise cryostorage protocols. For example, storage-related stress biomarkers have been applied to study tropical plant species that produce recalcitrant seeds for which the cryopreservation of excised zygotic embryos/embryonic axes or vegetative propagules is a practical, but technically limiting option (Berjak et al. 2011; Naidoo et al. 2011; Pilatti et al. 2011).

Biomarkers may also assist the development of in vitro and cryostorage methods for tropical species represented by a critically low number of rare plants and for which the sacrifice of precious germplasm for storage protocol development experiments is restrictive (Nadarajan et al. 2006, 2007, 2008; Staines et al. 1999).

7.8 Biomarkers, Quality and Risk Management

The guiding principles of BRCs, biobanks, biorepositories and genebanks are authenticity, purity and stability (OECD 2007). Molecular genetic and other biochemical biomarkers are used for the authentication of tropical plant germplasm in in vitro genebanks. The detection of microbial contamination of germplasm in in vitro collections and phytosanitary regulation using diagnostics in disease indexing may also be considered as a category of biomarker (Diekmann and Putter 1996;

Leifert and Cassells 2001; Lievens et al. 2005; Rao 2004; Sugii 2011; Van den Houwe and Swennen 2000). Biomarkers also have an application in the monitoring of stability for the purposes of quality assurance and quality control in genebanks. One of the principal concerns regarding the protection of plant germplasm by in vitro technologies is the question of genetic stability and maintenance of trueness-to-type (Harding 1996, 1999, 2004; Harding et al. 2000, 2009). These issues arise from the phenomenon of somaclonal variation (Rodriguez-Enriquez et al. 2011) and the production of undesirable 'off types' which are counter to the objectives of plant conservation (Scowcroft 1984; Msogoya et al. 2011). Regarding biospecimen science, a process or technique that may produce genetically dissimilar material is an issue of quality; the preservation and provision of material by a repository therefore requires stringent evaluation of the risks that could potentially compromise stability. Biomarkers have therefore a key role in risk assessment, quality control and management, especially in the identification and evaluation of quality performance indicators.

7.9 Biomarkers and Cryobionomics

In the context of in vitro conservation, the use of cryopreservation for the long-term storage of tropical plant germplasm is ultimately the process that enables the restoration and sustainable utilisation of a species. To assist this process, the concept of cryobionomics has been developed and explored as the biological science dealing with a cryopreserved organisms' behaviour and its performance in a habitat following its reintroduction back into the natural environment (Harding 2004, 2010; Harding et al. 2005, 2009). Cryobionomics provides a framework for hypothesis driven research to critically examine the linkage between the induction of cryoinjury and germplasm stability and it can be used to assess the risks of cryopreservation and its associated in vitro manipulations and help to mitigate against them. This interdisciplinary subject, as defined by Harding (2004) requires phenotypic, cytological, biochemical and molecular biological knowledge of the organism to assess possible cellular/biochemical damage, impairment of metabolism and loss of reproductive functions. It examines the potential temporal shifts in gene expression causing disruption of normal regulatory mechanisms, growth and development. Analogous to the requirements of clinical biospecimen science, biomarkers have an application in the predetermination and evaluation of events that give rise to cryoinjury along with those epigenetic and genetic factors that alter stability thus affecting the quality of preserved germplasm (Côte et al. 2000; Johnston et al. 2005, 2010; Kaity et al. 2008, 2009; Martínez-Montero et al. 2002; Okere and Adegeye 2011; Sahijram et al. 2003; Scocchi et al. 2004; Scowcroft 1984). This is relevant to tropical species following their reintroduction where evidence of the impact of ex vitro physiological performance and/or stress tolerance affecting the functionality of recalcitrant plant species recovered from cryopreserved samples is beginning to emerge (Sershen et al. 2010).

7.10 Conclusions

The realisation of in vitro conservation has been mainly achieved on a case-by-case basis using empirical approaches for protocol development informed by fundamental research largely aimed at tropical crops. Omics research is providing valuable insights into how some tropical plants respond to different preservation treatments. However, empirical methods of testing and fundamental research are time and resource intensive and may not be feasible for the routine conservation of a wider range of tropical plant diversity for which a basic knowledge of in vitro responses is sparse. A more cohesive approach in the use of biomarkers for genebank operations and biopreservation and biospecimen science research may help to resolve some of the critical issues affecting tropical plant conservation.

Acknowledgements The authors gratefully acknowledge Dr Fay Betsou and colleagues in the ISBER Biospecimen Science Working Group.

References

Ahuja I, de Vos RCH, Bones AM, Hall RD (2010) Plant molecular stress responses face climate change. Trends Plant Sci 15:664–674

Ayad WG, Hodgkin T, Jaradat A, Rao VR (1997) Molecular genetic techniques for plant genetic resources. Report of an IPGRI workshop, IPGRI, Rome, 9–11 Oct 1995

Balasubramanian R, Müller L, Kugler K, Hackl W, Pleyer L, Dehmer M, Graber A (2010) The impact of storage effects in biobanks on biomarker discovery in systems biology studies. Biomarkers 15:677–683

Bank HL, Schmehl MK (1989) Parameters for evaluation of viability assays: accuracy, precision, specificity, sensitivity, and standardization. Cryobiology 26:203–211

Benson EE (1999) Plant conservation biotechnology. Taylor and Francis, London

Benson EE (2004) Cryo-conserving algal and plant diversity: historical perspectives and future challenges. In: Fuller B, Lane N, Benson EE (eds) Life in the frozen state. CRC Press, Florida

Benson EE (2008) Cryopreservation of phytodiversity: a critical appraisal of theory & practice. Crit Rev Plant Sci 27:141–219

Benson EE, Harding K (2012) Cryopreservation of shoots and meristems: an overview of contemporary methodologies. In: Loyola-Vargas VM, Ocho-Alejo N (eds) Plant cell culture protocols, 3rd edn. Humana Press (in press)

Benson EE, Krishnapillay B, Mansor M (1996) The potential of biotechnology in the in vitro conservation of Malaysian forest germplasm: an integrated approach. In: Hussein N, Bacon PS, Choon KK (eds) ODA proceedings of the 3rd conference forestry forest products research, vol I. FRIM Publication, Kuala Lumpur

Benson EE, Betsou F, Amaral R, Santos LMA, Harding K (2011) Standard preanalytical codes: a new paradigm for environmental biobanking sectors explored in algal culture collections. Biopres Biobank 9:1–12

Berjak P, Bartels P, Benson EE, Harding K, Mycock D, Pammenter N, Sershen, Wesley-Smith J (2011) Cryo-conservation of South African plant genetic diversity. In Vitro Cell Dev Biol Plant 47:65–81

Betsou F, Lehmann S, Ashton G, Barnes M, Benson EE, Coppola D, DeSouza Y, Eliason J, Glazer B, Guadagni F, Harding K, Horsfall DJ, Kleeberger C, Nanni U, Prasad A, Shea K, Skubitz A,

Somiari S, Gunter E (2010) Standard preanalytical coding for biospecimens: defining the sample PREanalytical code. Cancer Epidemiol Biomarkers Prev 19:1004–1011

Carpentier SC, Witters E, Laukens K, Van Onckelen H, Swennen R, Panis B (2007) Banana *Musa* spp. as a model to study the meristem proteome: acclimation to osmotic stress. Proteomics 7:92–105

Carpentier SC, Coemans B, Podevin N, Laukens K, Witters E, Matsumura H, Terauchi R, Swennen R, Panis B (2008) Functional genomics in a non-model crop: transcriptomics or proteomics? Physiol Plant 133:117–130

CBOL Plant Working Group (2009) A DNA barcode for land plants. Proc Natl Acad Sci USA 106:12794–12797

Chabot JR, Pedraza JM, Luitel P, Oudenaarden AV (2007) Stochastic gene expression out-of-steady-state in the cyanobacterial circadian clock. Nature 450:1249–1252

Coates DJ, Dixon KW (2007) Current perspectives of plant conservation biology. Aust J Bot 55:187–193

Costa Nunes DA, Benson EE, Oltramari AC, Araujo PS, Moser JR, Viana AM (2003) In vitro conservation of *Cedrela fissilis* (Meliaceae) a native tree of the Brazilian Atlantic forest. Biodivers Conserv 12:837–848

Côte FX, Goue O, Domergue R, Panis B, Jenny C (2000) In-field behaviour of banana plants (*Musa* AA sp.) obtained after regeneration of cryopreserved embryonic cell suspensions. Cryo Lett 21:19–24

Cramer W (2002) Biome models. In: Mooney HA, Canadell JG (eds) The earth system: biological and ecological dimensions of global environmental change, vol 2, Encyclopedia of global environmental change. Wiley, Chichester

de Vicente MC (2004) The evolving role of genebanks in the fast-developing field of molecular genetics. Issues in genetic resources. No. 11, IPGRI, Rome

de Vicente MC, Andersson MS (2006) DNA banks-providing novel options for genebanks? Topical reviews in agricultural biodiversity. IPGRI, Rome

Diekmann M, Putter CAJ (1996) FAO/IPGRI technical guidelines for the safe movement of germplasm. *Musa*, vol 15, 2nd edn. IPGRI, Rome

Donaldson JS, IUCN/SSC Cycad Specialist Group (2004) Cycads. Status survey and conservation action plan. IUCN, Gland

Fang JY, Wetten A, Johnston J (2008) Headspace volatile markers for sensitivity of cocoa (*Theobroma cacao* L.) somatic embryos to cryopreservation. Plant Cell Rep 27:453–461

Garming H, Roux N, Van den Houwe I (2010) The impact of the *Musa* international transit centre. Review of its services and cost-effectiveness and recommendations for rationalization of its operations. Bioversity International, Montpellier

Ghazoul J, Sheil D (2010) Tropical rain forest ecology, diversity, and conservation. Oxford University Press, New York

Gonzalez MA, Baraloto C, Engel J, Mori SA, Petronelli P, Riera B, Roger A, Thebaud C, Chave J (2009) Identification of Amazonian trees with DNA barcodes. PLoS One 4:e7483. doi:10.1371/journal.pone.0007483

Harding K (1996) Approaches to assess the genetic stability of plants recovered from in vitro culture. In: Normah MN, Narimah MK, Clyde MM (eds) Proceedings of the international workshop: in vitro conservation of plant genetic resources. Plant Biotechnology Laboratory, University Kebangsaan Malaysia, Kuala Lumpur

Harding K (1999) Stability assessments of conserved plant germplasm. In: Benson EE (ed) Plant conservation biotechnology. Taylor and Francis, London

Harding K (2004) Genetic integrity of cryopreserved plant cells: a review. Cryo Lett 25:3–22

Harding K (2010) Plant and algal cryopreservation: issues in genetic integrity, concepts in 'cryobionomics' and current applications in cryobiology. In: Proceedings of the Asia Pacific conference on plant tissues cultures and agrobiotechnology (APaCPa) 2007, Kuala Lumpur. Asia Pac J Mol Biol Biotech 18:151–154

Harding K, Marzalina M, Krishnapillay B, Nashatul ZNA, Normah MN, Benson EE (2000) Molecular stability assessments of trees regenerated from cryopreserved Mahogany seed germplasm using non-radioactive techniques to examine chromatin structure and DNA methylation status of the ribosomal genes. J Trop Forest Sci 12:149–163

Harding K, Johnston J, Benson EE (2005) Plant and algal cell cryopreservation: issues in genetic integrity, concepts in 'Cryobionomics' and current European applications. In: Benett IJ, Bunn E, Clarke H, McComb JA (eds) Contributing to a sustainable future. Proceedings of the Australian branch of the IAPTC & B, Perth, Western Australia, pp 112–119

Harding K, Johnston JW, Benson EE (2009) Exploring the physiological basis of cryopreservation success and failure in clonally propagated in vitro crop plant germplasm. Agr Food Sci 18:3–16

Higa TC, Paulilo MTS, Benson EE, Pedrotti E, Viana AM (2011) Developing seed cryobanking strategies for *Tabebuia heptaphylla* (Bignoniaceae) a hardwood tree of the Brazilian South Atlantic forest. Cryo Lett 32:329–338

Holland NT, Smith MT, Eskenazi B, Bastaki M (2003) Biological sample collection and processing for molecular epidemiological studies. Mutat Res 543:217–234

Holland NT, Pfleger L, Berger E, Ho A, Bastaki M (2005) Molecular epidemiology biomarkers-sample collection and processing considerations. Toxicol Appl Pharmacol 206:261–268

ISBER (2008) Best practices for repositories: collection, storage, retrieval and distribution of biological materials for research. Cell Preserv Technol 6:3–58

Jackson WP, Kennedy K (2009) The global strategy for plant conservation: a challenge and opportunity for the international community. Trends Plant Sci 14:578–580

Johnston JW, Harding K, Bremner DH, Souch G, Green J, Lynch PT, Grout B, Benson EE (2005) HPLC analysis of plant DNA methylation: a study of critical methodological factors. Plant Physiol Bioch 43:844–853

Johnston JW, Harding K, Benson EE (2007) Antioxidant status and genotypic tolerance of *Ribes* in vitro cultures to cryopreservation. Plant Sci 172:524–534

Johnston J, Benson EE, Harding K (2009) Cryopreservation of in vitro *Ribes* shoots induces temporal changes in DNA methylation. Plant Physiol Bioch 47:123–131

Johnston J, Pimbley I, Harding K, Benson EE (2010) Detection of 8-hydroxy-2′-deoxyguanosine a marker of DNA damage in germplasm and DNA exposed to cryogenic treatments. Cryo Lett 31:1–13

Kaity A, Ashmore SE, Drew RA, Dulloo ME (2008) Assessment of genetic and epigenetic changes following cryopreservation in papaya. Plant Cell Rep 27:1529–1539

Kaity A, Ashmore SE, Drew RA (2009) Field performance evaluation and genetic integrity assessment of cryopreserved papaya clones. Plant Cell Rep 28:1421–1430

Karp A, Kresovich S, Bhat KV, Ayad WG, Hodgkin T (1997) Molecular tools in plant genetic resources conservation: a guide to the technologies. IPGRI Technical Bulletin, no. 2, IPGRI, Rome

Kosakivska IV (2008) Biomarkers of plants with different types of ecological strategies. Gen Appl Plant Physiol 34:113–126

Kramer AT, Havens K (2009) Plant conservation genetics in a changing world. Trends Plant Sci 14:599–607

Kranner I, Birtić S, Anderson KM, Pritchard HW (2006) Glutathione half-cell reduction potential: a universal stress marker and modulator of programmed cell death? Free Radical Biol Med 40:2155–2165

Kugler KG, Hackl WO, Mueller LAJ, Fiegl H, Graber A, Pfeiffer RM (2011) The impact of sample storage time on estimates of association in biomarker discovery studies. J Clin Bioinform 1:1–9

Laamanen J, Uosukainen M, Häggman H, Nukari A, Rantala S (2008) Cryopreservation of crop species in Europe. In: Proceedings of the CRYOPLANET COST action 871, Agrifood research working papers 153, MTT Agrifood Research, Oulu

Lakshmanan P, Reed BM, Sarasan V (2011) Special issue on biodiversity. In Vitro Cell Dev Biol Plant 47:1–200

Leifert C, Cassells AC (2001) Microbial hazards in plant tissue and cell cultures. In Vitro Cell Dev Biol Plant 37:133–138

Lievens B, Grauwet TJMA, Cammue BPA, Thomma BPHJ (2005) Recent developments in diagnostics of plant pathogens: a review. Recent Res Dev Micro 9:1–23
Lindenmayer DB, Margules CR, Botkin DB (2000) Indicators of biodiversity for ecologically sustainable forest management. Conserv Biol 14:941–950
Litz R, Moon P, Benson EE, Stewart J, Chavez VM (2004) A biotechnology strategy for the medium and long-term, conservation of cycads. (New York Botanical Garden). Bot Rev 70:47–53
Martínez-Montero ME, Ojeda E, Espinosa A (2002) Field performance of sugarcane (*Saccharum* sp.) plants derived from cryopreserved calluses. Cryo Lett 23:21–26
Miguel C, Marum L (2011) An epigenetic view of plant cells cultured in vitro: somaclonal variation and beyond. J Exp Bot 62:3713–3725. doi:10.1093/jxb/err155
Mosher RA, Melnyk CW (2010) siRNAs and DNA methylation: seedy epigenetics. Trends Plant Sci 15:204–210
Msogoya TJ, Grout BW, Maerere AP (2011) Performance of micropropagation-induced off-type of East African highland banana (*Musa* AAA—East Africa). J Animal Plant Sci 10:1334–1338
Nadarajan J, Staines HJ, Benson EE, Marzalina M, Krishnapillay B, Harding K (2006) Optimization of cryopreservation protocol for *Sterculia cordata* zygotic embryos using Taguchi experiments. J Trop Forest Sci 18:222–230
Nadarajan J, Staines HJ, Benson EE, Marzalina M, Krishnapillay B, Harding K (2007) Optimization of cryopreservation for *Sterculia cordata* zygotic embryos using vitrification techniques. J Trop Forest Sci 19:79–85
Nadarajan J, Mansor M, Krishnapillay B, Staines HJ, Benson EE, Harding K (2008) Applications of differential scanning calorimetry developing cryopreservation strategies for *Parkia speciosa* a tropical tree producing recalcitrant seeds. Cryo Lett 29:95–110
Naidoo C, Benson EE, Berjak P, Goveia M, Pammenter NW (2011) Exploring the use of DMSO and ascorbic acid to promote shoot development by excised embryonic axes of recalcitrant seeds. Cryo Lett 32:166–174
Nashatul Zaimah NA, Benson EE, Marzalina M (2007) Viability of *Elateriospermum Tapos* (Perah) embryos after storage. J Trop Forest Sci 19:1–5
Nicotra AB, Atkin OK, Bonser SP, Davidson AM, Finnegan EJ, Mathesius U, Poot P, Purugganan MD, Richards CL, Valladares F, van Kleunen M (2010) Plant phenotypic plasticity in a changing climate. Trends Plant Sci 15:684–692
Noor NM, Kean CW, Vun YL, Zeti AMH (2011) In vitro conservation of Malaysian biodiversity—achievements, challenges and future directions. In Vitro Cell Dev Biol Plant 47:26–36
OECD (2007) OECD best practice guidelines for biological resource centres. OECD, Paris
Okere AU, Adegeye A (2011) In vitro propagation of an endangered medicinal timber species Khaya grandifoliola C. Dc. Afr J Biotechnol 10:3335–3339
Palmirotta R, Ludovici G, De Marchis ML, Savonarola A, Leone B, Spila A, De Angelis F, Morte DD, Ferroni P, Guadagni F (2011) Preanalytical procedures for DNA studies: the experience of the interinstitutional multidisciplinary biobank (BioBIM). Biopres Biobank 9:35–45
Pilatti FK, Aguiar T, Simões T, Benson EE, Viana AM (2011) In vitro and cryogenic preservation of plant biodiversity in Brazil. In Vitro Cell Dev Biol Plant 47:82–98
Rao NK (2004) Plant genetic resources: advancing conservation and use through biotechnology. Afr J Biotechnol 3:136–145
Ratnasingham S, Hebert PDN (2007) BOLD: The barcode of life data system. Mol Ecol Notes 7:355–364 (www.barcodinglife.org)
Reed BM (2008) Plant cryopreservation: a practical guide. Springer, New York
Riegman PHJ, Morente MM, Betsou F, de Blasio P, Geary P (2008) Biobanking for better healthcare. Mol Oncol 2:213–222
Rodriguez-Enriquez J, Dickinson HG, Grant-Downton RT (2011) MicroRNA misregulation: an overlooked factor generating somaclonal variation? Trends Plant Sci 16:242–248
Rubio-Somoza I, Weigel D (2011) MicroRNA networks and developmental plasticity in plants. Trends Plant Sci 16:258–264

Sahijram L, Soneji JR, Bollamma KT (2003) Invited review: analyzing somaclonal variation in micropropagated bananas (*Musa* spp.). In Vitro Cell Dev Biol Plant 39:551–556

Sass C, Little DP, Stevenson DW, Specht CD (2007) DNA barcoding in the cycadales: testing the potential of proposed barcoding markers for species identification of cycads. PLoS One 2(11):e1154. doi:10.1371/journal.pone.0001154

Scocchi A, Faloci M, Medina R, Olmos S, Mroginski L (2004) Plant recovery of cryopreserved apical meristem-tips of *Melia azedarach* L. using encapsulation/dehydration and assessment of their genetic stability. Euphytica 135:29–38

Scowcroft WR (1984) Genetic variability in tissue culture: impact on germplasm conservation and utilisation, Report (AGPG: IBPGR/84/152), IBPGR, Rome

Sershen, Berjak P, Pammenter NW (2010) Effects of cryopreservation of recalcitrant *Amaryllis belladonna* zygotic embryos on vigor of recovered seedlings: a case of stress 'hangover'? Physiol Plant 139:205–219

Sharrock S, Hird A, Kramer A, Oldfield S (2010) Saving plants, saving the planet: botanic gardens and the implementation of GSPC target 8. Botanic gardens conservation international, Richmond

Shulaev V, Oliver DJ (2006) Metabolic and proteomic markers for oxidative stress. New tools for reactive oxygen species research. Plant Physiol 141:367–372

Spooner D, van Treuren R, de Vicente MC (2005) Molecular markers for genebank management. IPGRI technical bulletin, no. 10, IPGRI, Rome

Staines HH, Marzalina M, Krishnapillay B et al (1999) Using Taguchi experimental design for developing cryopreservation strategies for recalcitrant seeds. In: Marzalina M, Khoo KC, Jayanthi N et al (eds) Proceedings of the IUFRO seed symposium 1998—recalcitrant seeds. FRIM Publication, Kuala Lumpur

Sugii NC (2011) The establishment of axenic seed and embryo cultures of endangered Hawaiian plant species: special review of disinfestation protocols. In Vitro Cell Dev Biol Plant 47: 157–169

Tausz M, Sircelj H, Grill D (2004) The glutathione system as a stress marker in plant ecophysiology: is a stress-response concept valid? J Exper Bot 55:1955–1962

Van den Houwe I, Swennen R (2000) Characterization and control of bacterial contaminants in in vitro cultivars of banana (*Musa* spp.). Acta Hortic 530:69–79

Verleysen H, Samyn G, Van Bockstaele E, Debergh P (2004) Evaluation of analytical techniques to predict viability after cryopreservation. Plant Cell Tiss Org Cult 77:11–21

Volk GM (2010) Application of functional genomics and proteomics to plant cryopreservation. Curr Genomics 11:24–29

Yin LL, Poobathy R, James J, Julkifle AL, Subramaniam S (2011) Preliminary investigation of cryopreservation by encapsulation-dehydration technique on *Brassidium* Shooting Star orchid hybrid. Afr J Biotech 10:4665–4672

Zhu G-Y, Geuns JMC, Dussert S, Swennen R, Panis B (2006) Change in sugar, sterol and fatty acid composition in banana meristems caused by sucrose-induced acclimation and its effects on cryopreservation. Physiol Plant 128:80–94

Section II
Current Status

Chapter 8
Conservation of Tropical Fruit Genetic Resources

M.N. Normah, S.K. Malik, R. Chaudhury, I. Salma, and M.A. Makeen

8.1 Introduction

There is a vast diversity of tropical fruits in Asia, America and Africa. Tropical fruit tree genetic resources in the Asian region alone include more than 500 species of edible tropical fruits (Ramanatha Rao and Bhag Mal 2002). Fruits under cultivation include banana, citrus, mango, pineapple, papaya, durian, rambutan, jackfruit, litchi, longan, tamarind, chempedak, carambola, langsat, guava, sour sop, custard apple and salak (Verheij and Coronel 1991; Singh 1993; Arora and Ramanatha Rao 1995), the predominant fruits being banana, pineapple, citrus, mango and papaya. Other than these cultivated species many are considered rare or underutilized. These rare fruits are generally neglected because they have not been exploited commercially and there is a lack of improved varieties. As far as their potential is concerned, many of these are important for the well being of the people in the different regions as

M.N. Normah (✉)
Institute of Systems Biology, Universiti Kebangsaan Malaysia,
Bangi, UKM, Selangor 43600, Malaysia
e-mail: normah@ukm.my

S.K. Malik • R. Chaudhury
Tissue Culture and Cryopreservation Unit, National Bureau of Plant
Genetic Resources, Pusa Campus, New Delhi 110 012, India
e-mail: skm@nbpgr.ernet.in; rekha@nbpgr.ernet.in

I. Salma
Malaysian Agriculture Research Institute (MARDI),
P.O. Box 12301, General Post Office, 50774 Kuala Lumpur, Malaysia
e-mail: salma@mardi.gov.my

M.A. Makeen
Department of Crop Sciences, University of Kordofan,
P.O. Box 160, Elobeid 51111, Sudan
e-mail: makeen_makeen2004@yahoo.com; makeenabdalla@kordofan.edu.sd

M.N. Normah et al. (eds.), *Conservation of Tropical Plant Species*,
DOI 10.1007/978-1-4614-3776-5_8,

supplemental food, nutritionally balanced diets and enhancing household incomes. In addition, fruit species, both cultivated and wild, contribute substantially towards sustainability of the ecosystems (Ramanatha Rao and Bhag Mal 2002). They are also a source of useful genes for related crop species. Some of the genera that include these rare species are *Garcinia*, *Lansium*, *Baccaurea*, *Artocarpus* and *Nephelium*.

The genetic diversity of tropical fruit trees is increasingly threatened; in the case of cultivated species by specialization of production systems in a few varieties and by land use changes, and in the case of wild relatives due to habitat loss. The recognition of the value of the genetic diversity of tropical fruits trees and the threats that it faces has led to conservation efforts by countries within tropical regions (Bhag Mal et al. 2010). Conservation however is difficult since tropical fruit trees generally possess recalcitrant seeds that cannot be stored in conventional seed genebanks. Conservation efforts therefore have focused on in situ including on-farm and ex situ strategies involving the maintenance of germplasm in field genebanks, as well as conserving species in botanical gardens, recreational parks and agrotourism parks. In vitro conservation is another alternative method of conservation for many of these fruit species (Drew et al. 2007). Two reviews in particular, Ramanatha Rao and Bhag Mal (2002) and Bhag Mal et al. (2010) have covered many conservation efforts especially in the Asian region. In this chapter, conservation efforts will be covered for the whole tropical region in general and emphasis will be given to three main areas; Malaysia, South Asia (India), and tropical Africa, occasionally touching tropical America, depending on the availability of information.

8.2 In Situ Conservation

In situ and on-farm conservation and management strategies provide an important complement to ex situ approaches because in these approaches the diversity is maintained in farms, orchards and home gardens by farmers, who derive benefits from their cultivation in the case of domesticated species, and from their harvest in natural forests in the case of their wild relatives. Under the Asian Development Bank (ADB) funded project on tropical fruits, Bioversity International (formally known as the International Plant Genetic Resources Institute, IPGRI) promoted the concept of on-farm conservation of fruit genetic resources (Nares et al. 2001). The results of the ADB project helped in planning and developing larger actions under United Nations Environment Programme (UNEP)/Global Environment Facility (GEF) project "Conservation and Sustainable Use of Cultivated and Wild Tropical Fruit Diversity: Promoting Sustainable Livelihood, Food Security and Ecosystem Services" started in January 2009. Bhag Mal et al. (2010) described in detail the efforts done in the Asia, the Pacific and Oceana. The project focuses on four tropical fruit tree species, citrus (*Citrus* spp.), mango (*Mangifera indica*), mangosteen (*Garcinia mangostana*), rambutan (*Nephelium lappaceum*) and their wild relatives (some of which are edible). These are commercially important species in the region with high diversity levels, both at intraspecific and interspecific levels. This GEF funded project is being implemented by UNEP and executed by Bioversity International in India, Indonesia, Malaysia and Thailand. These four countries were selected, as they are located in the centres of diversity of these species (Bhag Mal et al. 2010) (Ramanatha Rao and Sthapit, Chap. 1 this volume).

8.3 Ex Situ Conservation

Ex situ approach comprises storage outside the natural habitat/environment i.e. storing the seed in cold storage in a seed bank, maintaining plants in field genebanks or botanical gardens and in vitro storage of plantlets, cells, tissues or pollen (under slow growth conditions or cryopreservation). Conservation of seeds is an efficient and reproducible technique in which orthodox seeds are dried to appropriate moisture content and stored at a low temperature and regenerated when needed. However, for many tropical fruit species with recalcitrant seeds they must be conserved as plants/trees in field genebanks, which also provide easy and ready access to conserved material for research and use. Nevertheless, field genebanks tend to be expensive and have problems of losses of material from diseases, natural disasters and require continued funding for management. For a number of plant species the alternative methods (in vitro and cryopreservation) are not yet fully developed to play a major role in conservation strategy (Ramanatha Rao and Sthapit, Chap. 1 this volume)

8.3.1 Botanical Gardens

Botanical gardens play an important role in plant ex situ conservation, taxonomic research, horticultural and economic botany and public education and nature appreciation. Large botanical gardens such as the Royal Botanic Gardens, Kew, commonly have seed banks and herbaria. Traditionally, the activities of botanical gardens were horticultural and taxonomic research but the urgent need for biodiversity conservation, public education and technical outreach have become an increasingly important focus (Chen et al. 2009). Irawati (Chap. 9 this volume) describes the role of botanical gardens in detail for the conservation of orchids. Two examples of a conservation effort for tropical fruits in botanical gardens/agrotourism parks are Mekarsari Fruit Garden Indonesia (MFGI), Bogor, West Java, Indonesia and the Fairchild Tropical Botanic Garden (FTBG), Florida, USA.

Conservation of tropical fruit germplasm at MFGI is described in detail by Tirtawinata (2002). The number of plants collected ex situ at the garden includes 43 families, 202 species, and 672 varieties, which comprise a total of approximately 37,000 plants (Tirtawinata 2002). Research activities include breeding and selection programs. The garden has attracted many local and oversea visitors and the facilities, activities and collections support agrotourism of the country.

Fairchild's Center for Tropical Plant Conservation is dedicated to conserving tropical plants, driven by the imperative to avoid the extinction of species and their habitats. These activities are measured by the delivery of quantifiable conservation benefits to Fairchild's priority geographic investment regions (South Florida, Caribbean, oceanic islands, tropical Africa, and Madagascar) and plant groups (palms, cycads, tropical fruit and tropical trees). Main activities include field exploration of important plant areas, conservation assessments, species recovery and direct support to in situ conservation. The Fairchild Farm serves as the permanent home for the living genetic collections of the Tropical Fruit Program of the Fairchild's Tropical Botanical Garden (FTBG). The living genetic collections are the cornerstone of the program and serve a role for the conservation of clones of key tropical fruit species, investigation into applied horticulture, and for distribution of plants to the local, national and international community. The primary collections are of avocado, mango, jackfruit, mamey sapote, sapodilla, canistel, abrico (*Mammea americana*), caimito, spanish lime (*Melicoccus bijugatus*) and tamarind. (http:www.fairchildgarden.org/livingcollections/tropicalfruitprogram).

8.3.2 Field Collection/Genebanks

According to Bhag Mal et al. (2010), 52 field genebanks were identified/established for a range of crops, viz., mango (21), citrus (13), rambutan (8), jackfruit (4), litchi (4) and mangosteen (2) under the national programs in collaborating countries under the ADB and UNEP/GEF projects in Asia, the Pacific and Oceana regions. These field genebanks were enriched with more accessions collected during the ADB project, thus enhancing the accessible genetic diversity. This included mango (734 accessions), citrus (511), rambutan (195), jackfruit (126), litchi (60) and mangosteen (59). Out of a total of 2,184 accessions collected, 1,685 accessions of six target crops were added to field genebanks in several countries (Bhag Mal et al. 2010). Ex situ conservation of tropical fruit tree genetic resources is an expensive and labour intensive programme requiring considerable land resources. Hence, any effort in developing and managing a fruit tree genebank should be based on sound strategy. The guidelines for field genebank management developed by Bioversity International (Saad and Ramanatha Rao 2001) and guidelines for field and in vitro genebanks (Reed et al. 2005) may be used to promote improved establishment and management of field genebanks. Currently there are efforts being made by Bioversity International and the FAO to develop new standards for establishing and managing field genebanks as well as in vitro genebanks.

As for tropical America, Brazil represents a model for the conservation of plant genetic resources. Ferreira and Pinto (2010) describe these efforts in detail. Brazil has a structure for the conservation of plant genetic resources comprising the National Center for Genetic Resources and Biotechnology Research – CENARGEN, located in Brasilia, Federal District, as well as a network of active genebanks distributed all over the country in research units, universities and state research institutions. CENARGEN is one of the 39 research units of the Brazilian Agriculture Research Corporation (EMBRAPA), which coordinates the National Network of

Table 8.1 Genetic resources of tropical and subtropical fruit crops in Brazil

Fruit crops	Species	Number of collections	Number of accessions
Avocado	*Persea americana*	6	216
Pineapple	*Ananas comosus* and other species	3	823
Acerola	*Malpighia glabra*	9	309
Anonacea	*Annona* spp. and *Rollinia*	9	223
Banana	*Musa* spp.	6	554
Cashew	*Anacardium occidentale* and other species	4	505
Citrus	*Citrus* spp. and other genus	4	3,023
Guava	*Psidium guajava* and other species	6	428
Papaya	*Carica papaya* and other species	5	382
Mango	*Mangifera indica* and other species	7	458
Passion fruit	*Passiflora edulis* and other species	8	305
Cupuassu	*Theobroma grandiflorum*	4	508
Assai	*Euterpe oleraceae*	1	130
Camu-camu	*Myrciaria dubia*	2	140
Guarana	*Paullinia cupana*	1	270
Other fruit crops	Several (±200 species)	17	2,300
Total	±300 species	92	10,574

Source: Ferreira and Pinto (2010)

Active Genebanks (RENARGEN), of the Embrapa Research System (SEG). The program has approximately 160 Active Germplasm Banks (BAGs), and 24 out of them are for tropical and subtropical fruit tree species. There are about 300 species and 10,574 accessions under conservation, including duplications (Table 8.1).

All the documentation of the material under conservation is regularly updated via a computerized system as part of the new Brazilian Genetic Resources Information System (SIBRARGEN) (Ferreira and Pinto 2010).

The current collection at the National Clonal Germplasm Repository for Tropical Fruit and Nut Crops U.S. Department of Agriculture's Agricultural Research Service (USDA, ARS 2011a) in Florida is approximately 863 (statistics generated 4 Sept 2011), consisting of 25 genera, 72 taxa of 65 species. The genera include *Ananas, Artocarpus, Averrhoa, Dimorcarpus, Durio, Carica, Nephelium and Psidium.* Crops are maintained as living plants in 33 acres of field planting and selected germplasm are grown in tissue culture and glasshouses (USDA, ARS 2011a). ARS facilities in Hawaii hold *Ananas* (217), *Artocarpus* (16), *Carica* (1), *Passiflora* (7); in Puerto Rico, mostly *Musa* (150 accessions). The Germplasm Resources Information Network (GRIN) web server provides germplasm information about the collections.

8.3.3 In Vitro Genebanks and Cryopreservation

Various efforts for development of protocols for example through IPGRI (Ramanatha Rao and Bhag Mal 2002) involved Citrus Research Institute of China and National Bureau of Plant Genetic Resources (NBPGR), India. The work on cryopreservation of citrus was undertaken in collaboration with Universiti Kebangsaan Malaysia (UKM) and Universiti Putra Malaysia (UPM) in Malaysia with support from the Rural Development Administration (RDA), Republic of Korea. These studies revealed that seed survival decreased below 16% seed moisture content (MC) in *Citrus aurantifolia* (Cho et al. 2002) and that seeds cryopreserved without testa survived better than those with testa at about 7% seed MC. Protocol for cryopreservation of zygotic embryonic axes of *C. madurensis* was developed (Cho et al. 2001). Subsequently, to promote the work on in vitro conservation and cryopreservation, a 3-year project on 'Development of Advanced Technologies for Germplasm Conservation of Tropical Fruit Species' funded by the Australian Centre for International Agricultural Research (ACIAR) was implemented in Australia, Malaysia, the Philippines, Thailand and Vietnam (Drew et al. 2007). The ACIAR funded project was very successful in developing and standardizing the techniques for in vitro conservation and cryopreservation of several fruit species which greatly helped in developing suitable strategies/plans for conservation of tropical fruit genetic resources. These projects/programs were able to put fruit crops on a higher development agenda of the national governments and assisted the national agricultural research systems in tropical fruit tree germplasm in collection, characterization and evaluation, conservation, human resource development and in strengthening collaboration among the partners. Nevertheless, research is still needed to refine and standardize the existing cryopreservation protocols so that these can be effectively used for conservation of tropical fruit tree genetic resources.

Elsewhere, tropical fruit germplasm is held at genebanks in Brazil, *Ananas* (121 accessions) and Belgium holds the international collection of banana and plantain germplasm (International Network for the Improvement of Banana and Plantain, INIBAP) with more than 1,500 in vitro accessions (Reed et al., this volume). Wild *Citrus* species of India namely *C. indica* and *C. macroptera* (Malik and Chaudhury 2006) and Australia namely *C. australasica, C. inodora* and *C. garrawayi* (Hamilton et al. 2009) have been successfully cryopreserved using seeds, embryos and embryonic axes. In China, cryopreservation of seeds and embryonic axes of *C. grandis* was undertaken at different moisture contents (Wen et al. 2010) and successful cryopreservation of commercial *Citrus* cultivars using vitrification technique was reported by Wang and Deng (2004).

8.4 Genetic Diversity for Utilization

Much of tropical fruit tree germplasm collecting work focused on collecting of elite material, as is common for horticultural species, especially tree species. Hence, characterization of tropical fruit genetic resources usually lagged behind relative to annual crop species. In Asia, this problem was remedied somewhat under the UNEP/GEF project 'Conservation and Sustainable Use of Cultivated and Wild Tropical Fruit Diversity: Promoting Sustainable Livelihood, Food Security and Ecosystem Services'. The number of accessions characterized in the 10 countries is 3,359 with

190 elite lines identified (Bhag Mal et al. 2010). Conservation and use of tropical fruit species diversity in the Philippines is jointly funded by IPGRI and Department of Agriculture – Bureau of Agricultural Research (DA-BAR), the project systematically collects, documents new and existing collections, provides improved guidelines for their management, and characterizes and evaluates the germplasm so that useful traits as well as valuable accessions are identified. The crops covered by the project include jackfruit and other *Artocarpus* species, pili (*Canarium ovatum*), mangosteen and other *Garcinia* species, and durian (*Durio zibethinus*). Collecting and exploratory expeditions resulted in the collection of 33 new durian accessions; likewise, 40 tropical fruit species were found during a qualitative survey, 11 fruit species were evaluated using the descriptors for tropical fruits, and 20 pili accessions in Albay were characterized morphologically, using biochemical and molecular markers (Ilao 2007). Molecular characterization is now being carried out to understand the extent and distribution of genetic diversity in tropical fruit species. Molecular markers such as simple sequence repeat (SSR), amplified fragment length polymorphic (AFLP), and random amplification of polymorphic DNA (RAPD) are being developed.

Morphological and molecular characterization and evaluation of the most important tropical and subtropical fruit germplasm are being conducted in the active genebanks, and also at CENARGEN (Ferreira and Pinto 2010). Exploitation of SSR, sequence related amplified polymorphism (SRAP) and cleaved amplified polymorphic sequences – single nucleotide polymorphism (CAPS-SNP) markers for genetic diversity of *Citrus* germplasm collection was reported by Amar et al. (2011) to provide basis for future efficient use of these molecular markers in the genetic analysis of *Citrus*. Genetic diversity within a jackfruit (*Artocarpus heterophyllus* Lam.) germplasm collection in China was investigated using AFLP markers (Li et al. 2010) and assessment of genetic diversity among mango (*Mangifera indica* L.) genotypes was studied using RAPD markers (Rajwana et al. 2008). Others include Barkley et al. (2006), assessing genetic diversity and population structure in a *Citrus* germplasm collection utilizing simple sequence repeat markers (SSRs) and citrus microsatellite (Barkley et al. 2009) and a molecular phylogeny of the orange subfamily (Rutaceae: Aurantioideae) using nine cpDNA sequence (Bayer et al. 2009). Molecular markers have also been developed for genetic diversity analysis and for the production of molecular genetics linkage maps at USDA-ARS. Families of avocado, mango, and jackfruit have been produced that should allow the mapping of Quantitative Trait Loci (QTL) involved with disease resistance, fruit quality, and yield. A Candidate Gene Approach (CGA) is also being used to find genes involved with disease resistance and for control of flowering (USDA, ARS 2011b).

8.5 Fruit Species Conservation in Malaysia

Many tropical fruits originated in Malaysian forests, which 'constitute one of the richest floras in the world' (Knight 1983). As some of these indigenous species are vulnerable to genetic erosion, conservation of genetic resources of these species is

Table 8.2 Status of fruit genebanks in Malaysia

Institutions	No. of species	No. of accessions
MARDI	164	3,500
DOA, Peninsular Malaysia	109	900
DOA, Sabah	204	997
DOA, Sarawak	75	7,118

a priority. With the recalcitrant nature of the seeds of many tropical fruit species at present, establishment of field genebanks or field collections is the most suitable method for conserving these genetic resources.

8.5.1 Field Genebank

In Malaysia, the ex situ collections of fruit species are kept by various government agencies such as Malaysian Agricultural Research Institute (MARDI), Department of Agriculture (DOA), Peninsular Malaysia, Sabah and Sarawak, Forest Research Institute of Malaysia (FRIM) (Table 8.2), Universiti Putra Malaysia (UPM) and Universiti Kebangsaan Malaysia (UKM).

The collection of major fruit tree species started in the early 1900s by DOA Peninsular Malaysia with the aim of selecting accessions with high fruit quality and yield to be recommended for planting, and tied to the crop improvement program. During the period of 1993–1995 there were about 456 clones from 27 fruit species registered (Wan Darman 2008). These fruit species are maintained at various DOA stations in Peninsular Malaysia (Table 8.3). Collection of the rare fruit species started recently and about 84 species are maintained in the field genebank at DOA, Ulu Paka, Terengganu.

MARDI started to collect and conserve major fruit species in 1972. A systematic collection of landraces of *Durio zibethinus* (durian), *Nephelium lappaceum* (rambutan), *Mangifera indica* (mango), *Artocarpus heterophllus* (jackfruit), *Artocarpus integer* (cempedak), *Garcinia mangostana* (mangosteen) and *Lansium domesticum* (langsat) was carried out in collaboration with International Board of Plant Genetic Resources (IBPGR) in 1992 and maintained at MARDI Kemaman, Terengganu. These fruit species were originally established as working collections with the main objective of characterizing the germplasm for further use in breeding. Currently, comprehensive fruit germplasm collections comprised of major fruits such as durian, mango, rambutan, cempedak, nangka, mangosteen, langsat, bananas, citrus and germplasm for the exotic fruits such as carambola, papaya, pineapple and ciku (zapota) (*Manilkara zapota*) as well as underutilized fruits are maintained in the field genebanks situated at various MARDI Stations throughout the country. The collection of wild relatives of fruit species started in 1981 and since then expeditions to collect fruit tree species were continuously carried out with the Forest Departments. Accessions of the wild fruit species are maintained in field genebanks situated at various MARDI stations such as Serdang, Kemaman, Jerangau, Bukit Tangga, and Jelebu. There are about 2,000 accessions representing 150 species of the rare and wild fruit species established

Table 8.3 Status of some fruit species in field genebanks at DOA, Peninsular Malaysia

No.	Location	Fruit species	No. of species/ clones/accessions
1.	DOA, Gajah Mati, Kedah	Mango	32 clones
2.	DOA, Lekir, Perak	Guava	9 clones
		Pomelo	13 clones
3.	DOA, Serdang, Selangor	Rambutan	11 clones
		Durian	52 clones
4.	DOA, Sedayan, Negeri Sembilan	Jackfruit	6 clones
5.	DOA, Pulau Gadong, Melaka	Sapodilla	5 clones
6.	DOA, Ayer Hitam, Johor	Star fruit	18 clones
7.	DOA, Kg. Awah, Pahang	Lemon	3 clones
		Tangerine	3 clones
8.	DOA, Bukit Goh Pahang	Mangosteen	3 accessions
9.	DOA, Hulu Paka Terengganu	Durian	62 clones
			67 accessions
		Rambutan	32 clones
		Cempedak	8 clones
		Nangka	2 clones
			3 accessions
		Langsat	1 accession
		Dokong	5 accessions
		Duku Langsat	10 clones
		Mangosteen	1 accession
		Sapodilla	3 accessions
		Water rose apple	3 clones
		Rare fruits	84 species
10.	DOA, Batang Merbau Kelantan	Duku	5 accessions
		Dokong	3 accessions

Source: Wan Darman (2008)

in MARDI field genebanks (Table 8.4) (Salma et al. 2005) and in few other genebanks (Fig. 8.1). Moderately comprehensive collections were recently established for *Mangifera odorata* (kuini), *Mangifera foetida* (bacang), *Mangifera caesia* (binjai), *Garcinia prainiana* (cerapu), *Baccaurea motleyana* (rambai), *Garcinia atroviridis* (asam gelugor), *Artocarpus odorotissimus* (terap) and *Nephelium ramboutan-ake* (pulasan) throughout the country to conserve their intraspecific variation.

The Fruit Germplasm Collection at UKM is managed by the Faculty of Science and Technology. The collection was started in 1992 to facilitate research on indigenous rare fruit species. The collection covers an area of about 2 ha within the Greenhouse and Experimental Plot Complex. There are also a small number of pre-existing trees at Taman Pantun, a garden on the campus developed from an orchard established before the University was established. In addition, some species are also maintained around the campus for landscape purposes; at present there are 36 species representing 22 genera from 14 families collected from Peninsular Malaysia. The main genera are *Garcinia*, *Citrus*, *Baccaurea, Lansium, Artocarpus*, *Nephelium*, *Annona* and *Durio* (Normah et al. 2002). The fruit germplasm collection is now

Table 8.4 Status of some fruit species in MARDI field genebanks

No.	Location	Fruit species	No. of species/ clones/accessions
1.	MARDI Bt Tangga	Durian	20 acc
	MARDI Kemaman		695 acc
	Jerangau		26 acc and 10 spp.
	Kuala Kangsar		56 acc
2.	MARDI Bukit Tangga	Pulasan	15 acc
	MARDI Bukit Redan		41 acc
	MARDI Serdang		35 acc
3.	MARDI Bukit Tangga	Rambutan	19 acc
	MARDI Bukit Redan		41 acc
	MARDI Kemaman		235 acc
4.	MARDI Kemaman	Cempedak	110 acc
	MARDI Kluang		137 acc
5.	MARDI Kluang	Belimbing	116 acc
6.	MARDI Serdang	Underutilized fruits	1,200 acc
	MARDI Jelebu		200 acc

Source: Salma et al. (2005)

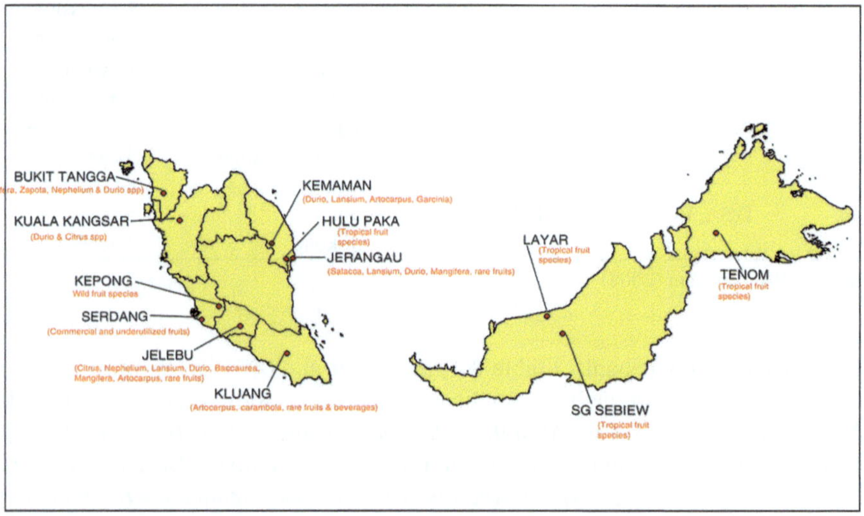

Fig. 8.1 Field genebanks in Malaysia (Source: Salma et al. 2010; Wong et al. 2008; Wan Darman 2008; Chai et al. 2008)

used as a source of materials for research, especially for seed studies, micropropagation and cryopreservation.

In Sabah the collection of fruit tree species are focused for the improvement program and presently the DOA Sabah has at least 200 species of fruit trees in the genebank namely at the Agriculture Research Station, Ulu Dusun, Sandakan (UDARS) and ARS, Tenom (Table 8.5) (Wong et al. 2008). While in Sarawak there are 7,118 accessions representing 75 fruit species are conserved in the field

Table 8.5 Status of some fruit species in field genebanks at DOA Sabah

No.	Location	Fruit species	No. of species/ clones/accessions
1.	DOA Sabah	*Mangifera*	19 spp., 15 var
2.		*Durio*	9 spp., 23 clones
3.		*Lansium*	4 clones
4.		*Artocarpus*	11 spp., 5 var
5.		*Citrus*	7 spp., 4 var
		Nephelium	6 spp., 14 var
6.		*Persea*	70 acc, 4 var
7.		Underutilized fruits	~150 spp.

Source: Wong et al. (2008)

genebanks at Bintulu Agricultural Park, Betong Layar Agriculture Station and at the Agriculture Research Centre Sarawak (Chai et al. 2008).

Many major fruit species are vegetatively propagated and maintained in the genebank. However, the underutilized fruit species are planted from seeds, except for kuini and pulasan where they are vegetatively propagated. Each clone or accession is represented by three to six plants. The trees are planted at a distance of 20×20 m and a hectare of land can accommodate about 48 trees. For some of the fruit species, they are duplicated at a different site.

8.5.2 Characterization, Evaluation and Utilization

The main objective of genetic resource conservation is for utilization in crop improvement. As such characterization and evaluation of all the accessions are the main activities of the germplasm collection. Most of the accessions in the field genebanks are characterized and evaluated using morpho-agronomic characters. Characterization and evaluation provide diagnostic information for identification of accession as well as information on desirable traits for crop improvement. Evaluation of the accessions is normally carried out by multidisciplinary team consisting of breeders, entomologists, pathologists, agronomists and others. Mohd Shukor et al. (2007) reported that in most of the germplasm collections in the country, the accessions or species are partially (75–90%) characterized or evaluated.

The nutritional composition of the fruits was analyzed for the proximates and antioxidants (Mohd Shukri et al. 2009). In addition, another important activity of the genetic resource conservation is to determine the genetic diversity present in the genepool and the genebank. A study on DNA polymorphism in 211 accessions of *Nephelium* showed that the degree of genetic variation for six primers varied from 1.24 to 4.02 with the mean value for Shannon Diversity Index (no diversity is 0 and more diversity is shown with higher numbers) of 2.25 (Chew et al. 2002). Similarly, Shannon Diversity Index of 85 *Lansium domesticum* accessions was 2.32 (Song et al. 2000). Salma (2005) who studied the genetic variation in 176 *Durio* accessions using molecular markers showed that durian genotype exhibited a high level of genetic variation with the mean degree of polymorphism of 90.38%. Whereas recently Elcy et al. (2012) reported that four groups were identified from *Citrus reticulata* accessions in Malaysia with a mean genetic distance of 0.170. Low genetic variability within species was probably due to vegetative propagation. These studies highlight the need for genetic diversity for local germplasm collection. Only a small portion of the total collections in the institutions was evaluated for biotic and abiotic stress reactions. Wild *Durio* species were evaluated for resistance to stem canker and found that accessions from *D. lowianus, D. kutejensis* and *D. graveolens* were susceptible to canker (Nik Masdek et al. 2008).

Fruit breeding at MARDI started with the selection and cloning by vegetative propagation of individual trees with emphasis on high fruit quality. Selection from farmers' fields also produced elite accession e.g. Masmerah pineapple, Subang 6 papaya, R3 rambutan and D 24 durian. Systematic conventional breeding was adopted for improvement of fruit varieties. Varieties such as MDUR 78, MDUR 79 and MDUR 88 were produced from crossings of D24 and D7 (Salma et al. 2010). The crossing of local accession Subang 6 with the exotic accession Hawaii Sunrise Solo produced Eksotika papaya, and an improved Eksotika II was released from an F1 hybridization (Chan 2002). The various hybridization programmes successfully yielded new hybrids of pineapple (Hybrid 1, Josapine), starfruit, rambutan and watermelon. Josapine was the first successfully commercialized pineapple hybrid . Mutation breeding as a tool was adopted for improvement of banana and mangosteen. United Plantations Berhad (a private company), in collaboration with University of Malaya, developed somaclonal variants of banana such as Intan (Berangan) and Mutiara (Rastali) that are now commercially cultivated. Genetic engineering permits the insertion of specific genes into the genomes of target host and a project on genetic engineering of papaya for resistance to papaya ringspot virus disease is underway at MARDI.

8.5.3 Documentation

In line with the commitment made under the Convention of Biodiversity (CBD), International treaty of Plant Genetic Resource (ITGRFA) and National Policy on Biological Diversity (NPBD), MARDI has developed its Agricultural biodiversity information system or AgrobIS (www.mardi.my) to meet the need for easy access to information on biological resources currently held at various genebanks, microbial resource collection centre and arthropod repository facility in MARDI. This is to encouraging the exchange of information on biological diversity at local and international levels to enhance its sustainable utilization.

8.5.4 Constraints

Conservation of fruit trees in the field genebank requires relatively large plots of land, and growing trees are relatively expensive to maintain. In addition, field genebanks are exposed to natural calamities or adverse environmental conditions such as droughts, floods, pests and diseases. In some cases, field genebanks may be competing with land development for the same piece of land. As such, field genebank may need to be relocated and replanted to give way to other priority activity. Other methods of conservation such as in vitro and cryopreservation methods can be alternatives to extensive duplication in field genebanks.

8.5.5 *In Vitro Conservation and Cryopreservation*

Normah et al. (2011) reviewed the work done in Malaysia on development of cryopreservation protocols for tropical fruit species. Existing cryopreservation procedures developed for temperate or non-woody tropical species are not readily adaptable to woody tropical species with recalcitrant seeds. Besides a few species of *Citrus*, attempts to develop protocols for successful cryopreservation of woody tropical fruit species such the *Garcinia* species and *Lansium domesticum* using vitrification and encapsulation–dehydration techniques resulted in failure due to the highly sensitive characteristics of the embryos/embryonic axes and shoot tips to dehydration, or to various steps of the vitrification technique and LN exposure. Recently a first success was noted with *Nephelium ramboutan-ake* shoot tips (Chua and Normah 2011). The addition of Vitamin C (ascorbic acid) as an antioxidant to the improved preculture and PVS2 treatments resulted in the first case of survival of shoot tips after LN exposure. Survival after exposure to liquid nitrogen is an important finding, suggesting that cryopreservation is possible for this tropical recalcitrant tree species with the right modification and improvement of the technique.

8.6 Fruit Species Conservation in South Asia (India)

India is endowed with a rich genetic diversity of fruits. Tropical fruits constitute a major portion of the spectrum of fruit diversity available in India. Important tropical fruits mango, banana, citrus, papaya and guava are grown in 4.29 m ha out of the total area of 6.32 m ha of fruit crops and account for 81% of the total annual fruit production (Anonymous 2010a). These and several other tropical fruit species not only meet the needs of local and export markets for fresh fruit, but also contribute substantially to the fruit processing industry. Mango, citrus, banana and guava, due to the presence of vast diversity and favorable flavor, were supported and improved by local fruit growers and horticulturists for wide adoption. The less important, so-called underutilized fruits, remain confined mainly to natural wild, semi-wild and semi-domesticated conditions, albeit with large ever increasing variability. Besides their importance as potential horticulture species, these plants are incidentally store houses of genes for adaptation to hot and hardy climates, salt tolerance, diseases tolerance and several essential nutritional values. However, ensuring continued growth in fruit production and to have vast diversity of fruits would require continuous selection and development of desired genotypes adapted to diverse agro-climatic conditions of the country. In this regard the role of genetic resources as the base material for crop improvement is invaluable and hence, their collection, evaluation, documentation, conservation and utilisation are of prime importance. Systematic exploration and collection of germplasm in major fruit crops from indigenous and exotic sources resulted in widening the genetic base and selection of suitable genotypes. Efforts for conservation of existing diversity are being made using various in situ and ex situ methods. Selection of suitable conservation strategies depends upon reproductive and breeding mechanisms and the physiology of suitable plant propagules. These factors determine the sample size of the propagules to be stored, and the appropriate conservation technologies to be applied. Accordingly, several conservation strategies are suggested and utilized by conservation biologists for achieving successful conservation of targeted species.

Conservation of horticulture genetic resources (HGR) and specifically the under-utilized fruit species and wild species of major fruits which are still grown as natural wild and in semi-wild conditions would require adoption of complementary conservation strategies where, suitable in situ and ex situ conservation methods are to be employed to achieve the successful conservation goals. Within fruit crops, specific conservation strategy is to be developed and adopted based on extent of genetic diversity available, mechanism of propagation, reproductive biology and present biological status of the species. In situ conservation involves promoting growth of plant species in their natural habitats where evolutionary processes continue to operate, making it a dynamic system. A majority of the lesser-known tropical fruit species in India grow in diverse climatic and edaphic conditions and are adapted to arid and semi-arid climates. The Indian National PGR System advocates for the complementary conservation strategy using more than one method to achieve comprehensive conservation of species. The Ministry of Environment and Forest is mandated with in situ conservation while ex situ conservation activities are looked after by NBPGR. Horticulture genetic resources of fruits comprise candidates for conservation based on their biological status and propagation method (Fig. 8.2).

8.6.1 Mango

Mango (*Mangifera indica* L.) is the most important fruit crop of India. It occupies the largest area (2.3 m ha) among fruit crops and production-wise, mango ranks second (15.0 m) with nearly 21% contribution to the total fruit production of the country (Indian Horticulture Data Base 2010). The nearly 1,000 varieties of mango cultivated in India provide an unusual diversity of flavours and tastes; managing them is a challenge for farmers and PGR managers.

The origin of mango is suggested to be Eastern India to Myanmar and possibly further in the Malay region (Popenoe 1920; Vavilov 1951). *M. indica* originated in the Assam-Myanmar (Assam-Burma) region Mukherjee (1951). To date, available taxonomic and molecular evidence suggests that mango probably evolved within a

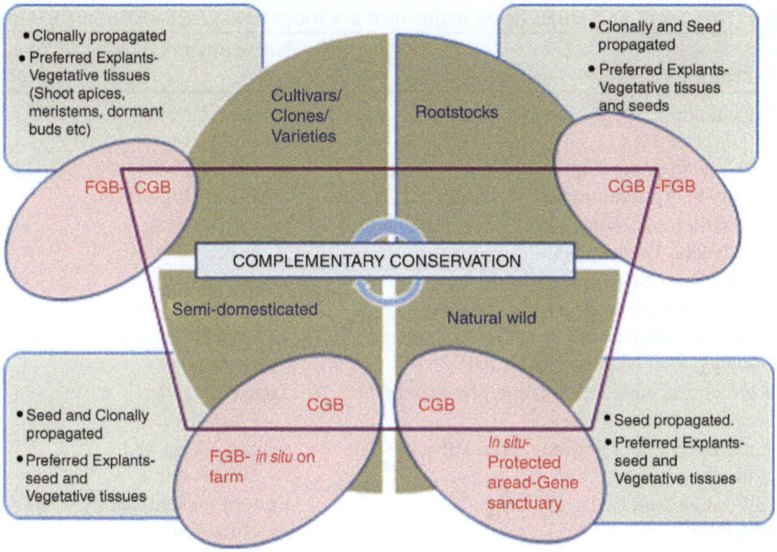

Fig. 8.2 Complementary conservation strategies for conservation of fruit genetic resources. *FGB* Field Gene Bank, *CGB* Cryo Gene Bank

large area comprising north-western Myanmar, Bangladesh and north-eastern India (Mukherjee 1997).

The Indian peninsula is the most important centre of diversity for *M. indica* having a large number of cultivated varieties as well as wild types. Truly wild mango trees are found in north-eastern India, sub-Himalayan tracts, deep gorges of the Bahraich and Gonda hills in Uttar Pradesh and in the outer hills in Kumaon and Garhwal in Uttaranchal (Brandis 1874; Kanjilal et al. 1982). Semi-wild trees are present in the forests of almost all parts of the Indian subcontinent. Other *Mangifera* species, namely *M. andamanica* King, *M. khasiana* Pierre, *M. sylvatica* Roxb. and *M. camptosperma* Pierre are also reported from India (Mukherjee 1985; Yadav and Rajan 1993). *M. sylvatica* occurs up to 1,300 masl in the eastern Himalayas in Sikkim, Darjeeling district in west Bengal, Khasi Hills in Meghalaya, Upper Assam and Surma valley (Assam) and in the Andaman Islands (Mukherjee 1949; Agharkar and Roy 1951). *M. khasiana* is an endemic but insufficiently known species from Khasi hills and is distinguished from *M. sylvatica* in having smaller and narrower leaves, smaller flowers, inflorescence with fasciculate divergent branching, petals ovate lanceolate and disc glandular or with distinct lobes (Mukherjee 1985). *M. andamanica* is widespread in tropical, wet evergreen forests of Andaman Islands.

In India there are nearly 1,000 varieties of mango grown, however, only 20 varieties are commercially grown due to their specific ecogeographic requirements for optimum performance (www.mangifera.org). Mango being a cross-pollinated tree is highly heterozygous and naturally established seedlings are the most important source of its diversity in India. Almost all the commercial cultivars of mango have arisen as a result of selection from these seedlings

Table 8.6 Number of accessions being maintained at various field genebanks of mango in India

Name of centre	Name of agroclimatic zone	No. of accessions
Bihar Agricultural College, Sabour, Bihar	Middle Gangetic Plains Region	157
Rajasthan College of Agriculture, Maharana Pratap University of Agriculture and Technology, Udaipur	Middle Gangetic Plains Region	16
Bidhan Chandra Krishi Viswa Vidyalaya, Mohanpur, District-Nadia, West Bengal	Lower Gangetic Plains Region	94
Central Institute for Subtropical Horticulture, Lucknow, Uttar Pradesh	Upper Gangetic Plains Region	735
Govind Bullabh Pant University of Agriculture and Technology, Pant Nagar, Uttaranchal	Western Himalayan Region	202
Indian Agricultural Research Institute, New Delhi	Trans-Gangetic Plains Region	48
Agricultural Experiment Station, Paria, Valsad, Gujarat	Gujarat Plains and Hills Region	177
Regional Fruit Research Station, Vengurla, Maharashtra	Eastern Plateau and Hills Region	379
Fruit Research Station, Sangareddy, Andhra Pradesh	Southern Plateau and Hills Region	510
Horticulture College and Research Institute, Periyakulam, Tamil Nadu	Southern Plateau and Hills Region	75
Indian Institute of Horticultural Research, Bangalore, Karnataka	Southern Plateau and Hills Region	269

Source: www.mangifera.org

8.6.1.1 Conservation

Mangifera species occur mainly as complex biotic communities in tropical humid forests, subtropical rain forests and dry forests/woodlands of Indo-Malayan biogeographic realm (Mukherjee 1985). For in situ conservation, the hilly regions of East Orissa and forests bordering Myanmar in Manipur valley are suggested as the appropriate locations (Rajan et al. 1999; Yadav 1999).

Systematic attempts of ex situ conservation of mango genetic resources were made in the past few decades and more than 1,200 accessions were collected and maintained in various field genebanks (Chadha 1976, 1996; Yadav and Rajan 1993). However, much of the diversity in the field genebanks is limited to only cultivated types and does not contain variability in the wild types. Major germplasm as active and base collections of *M. indica* are listed in Table 8.6. Besides these centres, germplasm is also being maintained at the Agricultural Research Institute, Pusa, Bihar; Horticulture and Agroforestry Research Programme, ICAR Research Complex for Eastern Region, Ranchi, Jharkhand; Fruit Research Station, Krishanagar, West Bengal; Horticultural Experiment and Training Centre, Basti, Uttar Pradesh; Horticulture Experiment and Training Centre, Saharanpur, Uttar Pradesh; Fruit Research Station, Aurangabad, Maharashtra; Gujarat Agricultural University, Shahi Bagh, Ahmedabad, Gujarat; Maharashtra Krishi Vishav Vidhalaya

KV, Parbani, Maharashtra; MPKVV, Rahuri, Maharashtra; Fruit Research Station, Kodur, Andhra Pradesh; and University of Agricultural Sciences, Dharwad, Karnataka. One of the largest field genebanks of mango was established at the Central Institute for Subtropical Horticulture, Lucknow, which has more than 735 accessions from many parts of the country as well as from exotic sources (Anonymous 2010b).

In several clonally propagated crops, in vitro method is a successful complementary strategy of germplasm conservation. But in mango no viable regeneration protocol has been developed so far. Pollen storage of important mango varieties following dehydration and freezing was demonstrated by Iyer and Subramanian (1989). This method has utility in gene pool conservation and in hybridisation programmes. Pollen of 249 accessions of mango belonging to several important indigenous cultivars available in the field genebanks at Central Institute of Sub-tropical Horticulture, Lucknow and other institutes were successfully cryopreserved in the National Cryogenebank. Pollen viability was tested after 4 years of cryostorage using in vitro germination, the flurochromatic reaction (FCR) method and by fruit set following field pollination (Chaudhury et al. 2010). Wu et al. (2007) reported successful cryopreservation of embryogenic cultures induced from nucelli and cotyledon cuts and also the direct somatic embryogenesis in mango (*Mangifera indica* L. Var Zihua). In vitro conservation and cryopreservation of mango germplasm still requires much effort due to the highly recalcitrant nature of explants for in vitro establishment, desiccation and freezing.

8.6.2 Banana

Banana (*Musa* spp.), including plantain, is the third most important fruit crop of India in terms of the area under cultivation (0.77 m ha). However, with respect to production, it tops the list with an annual production of 26.47 m (Anonymous 2010a). The term banana is given to varieties that are used for fruit and eaten raw whereas plantain is used for those that are edible only when cooked. Besides its use as fruit and vegetable, banana has several other uses because of which it is often termed as 'Kalpatharu', plant with all imaginable uses. The fruit is rich in phosphorus, calcium and potassium and does not contain any fat. The stem is taken as a medicine for eliminating stones in kidney and bladder. Further, banana plant has an important role in many social and religious ceremonies (Amalraj et al. 1993). Reference to plantain in ancient Indian literature like Valmiki's Ramayana (2029 BC), Kautilya's Arthasastra (250–300 BC) and the Tamil Classic Silappadikaram (500–600 AD) suggests very early domestication of the plant in India (Krishnamurthi and Seshadri 1958).

The wild bananas are distributed in South-east Asia and the Pacific (Simmonds 1962). The cultivated bananas originated from *M. acuminata* Colla (2n = 22, *AA* genome) and *M. balbisiana* Colla (2n = 22, *BB* genome). The wild types are seed bearing with poor edibility; *Musa acuminata* was the first species to develop edible traits. It is probable that human selection favoured parthenocarpy, structural heterozygosity and seed sterility, leading to the production of edible seedless fruits (Simmonds 1994). Most cultivated bananas are triploid comprising three major groups, *AAA*, *AAB* and *ABB*. Diploids *AA*, *AB* and tetraploids *AAAB*, *AABB*, *ABBB* are less abundant.

India possesses rich diversity in wild *Musa* spp. in the north-eastern region while variability of cultivars, especially belonging to the genomic groups *AA*, *AB*, *AAB* and *ABB* exists in the southern region. Maximum genetic variability of *M. acuminata* and *M. balbisiana* occur in north-east India. *Musa flaviflora* Simmonds is localised to Manipur and Meghalaya. Several wild species are recorded from southern and western regions of India. These include *M. acuminata* subspecies *burmannica*, *M. balbisiana*, *M. nagensium* Prain, *M. sikkimensis* Kurz, *M. cheemanii* Simmonds, *M. ornata*, *M. laterita*, *M. velutina* Wendl. Ex Drude and *M. sanguinea*. Banana cultivation is predominantly practiced in Tamil Nadu, Karnataka, Kerala, Andhra Pradesh, Maharashtra, Bihar, West Bengal and north-eastern states. The variability in cultivars decreases as one proceeds from extreme south. It is also interesting to note that all the Indian *AA* and *AB* cultivars occur only in the extreme south. The most common commercial varieties grown in different states are; Poovan (*AAB*) in Bihar; Pachabale (*AA*) in Karnataka; Nendran (*AAB*) in Kerala; Kapuravalli (*ABB*), Monthan (*AAB*), Morris (*AAA*), Mysore Poovan (*AAB*), Nendran (*AAB*), Pachanadan (*AAB*), Rasthali (*AAB*) and Robusta (*AAA*) in Tamil Nadu; and Dwarf Cavendish (*AAA*), Poovan (*AAB*), and Rasthali/ Morthaman (*AAB*) in West Bengal.

8.6.2.1 Conservation

The largest field collection of banana, comprising 685 accessions, is maintained at National Research Centre for Banana, Trichy, Tamil Nadu. Other centres where significant collections are available includes Indian Institute of Horticultural Research, Bangalore, Karnataka (241), Tamil Nadu Agricultural University, Coimbatore, Tamil Nadu (125), Banana Research Station, Kannara (190), Central Horticultural Experiment Station, Hajipur, Bihar (115) and Gujarat Agricultural University Research Station, Kovuur, Gujarat (51).

In vitro conservation of banana was initiated at Tissue Culture and Cryopreservation Unit of NBPGR in 1986. Presently, 415 accessions belonging to 13 *Musa* species are maintained in the in vitro repository. These collections include banana and plantain varieties, wild species and synthetic hybrids. The average storage period varies from 6 to 12 months at 25°C and 6/16 months, depending on the genotypes (Anonymous 2010c). Included in these collections are 168 category 1 (virus tested) accessions received from the International Network for the Improvement of Banana and Plantain (INIBAP), Belgium. Germplasm received so far includes accessions resistant to some common insect-pests and diseases of banana. Cryopreservation of banana germplasm has also been initiated and presently 45 accessions are being maintained as cryocollections at NBPGR.

8.6.3 *Citrus*

Citrus is the second most important fruit crop of India with an estimated production of 9.63 m from an area of 0.98 m ha (Anonymous 2010a). Mandrin (*Citrus reticulata* Blanco), sweet orange (*C. sinensis* Osbeck), acid lime (*C. aurantifolia* Swing) and lime (*C. limon* Burn. f.) are the major cultivated species of the country. Other species that are cultivated to a lesser extent include seedless lime (*C. latifolia* Tan.), pummelo (*C. grandis* Osbeck), grapefruit (*C. paradisi* Macf.) and belladikithuli (*C. maderaspatana* Tan.). Citrus fruits have numerous uses though primarily they are eaten fresh or prepared into juice concentrate. Pulp and seed are used for cattle feed and molasses as also for flavouring and for the production of pharmaceuticals, soaps and perfumes. Fermented orange juice produces vinegar and alcohol. Some species are known to have properties of curing fever and colic (Solley 1997). Quite a few species are grown for ornamental fruits and flowers.

South-east Asia, Australia, the intervening islands between Asia and Australia and central Africa are recognised as important centres of origin of *Citrus* and related genera (Swingle and Reece 1967). *Citrus reticulata, C. sinensis* and *C. aurantifolia*, the commercially important species grown in India, originated in India and China (McPhee 1967). In India, north-eastern Himalayan region and foothills of the central and western Himalayan tracts are rich sources of *Citrus* diversity. As many as 17 *Citrus* species, their 52 cultivars and 7 probable natural hybrids are reported to have originated in the north-eastern region (Bhattacharya and Dutta 1956). A number of commercial varieties were developed from native germplasm through selection and crop improvement.

8.6.3.1 Conservation

Presently most of the germplasm of *Citrus* species and allied genera in India is maintained and conserved in field genebanks. The largest collection of 614 accessions of *Citrus* and related genus *Poncirus* and *Severinia* including 62 exotic and 552 indigenous collections are maintained at National Research Centre, Nagpur (Anonymous 2010b). The indigenous collections include 26 Rangpur Lime, 24 Rough Lemon, 12 Cleopatra Mandarin, 14 Trifoliate Orange, 16 Trifoliate Orange hybrids and 18 other rootstocks. Forty scion cultivars were introduced from exotic sources. Besides these, germplasm of rare, wild and endangered species namely *C. indica*, *C. macroptera*, *C. latipes*, *C. assamensis*, *C. ichangensis*, *C. megaloxycarpa*, *C. rugulosa* and some probable natural hybrids collected from north-eastern India is maintained in the field genebanks. Other centres where *Citrus* germplasm is maintained include, Regional Fruit Research Station, Punjab Agricultural University, Abohar, Punjab; Central Horticultural Experiment Station, Chethalli, Karnataka; Indian Institute of Horticultural Research, Bangalore, Karnataka; Horticultural Experiment Station, Bhatinda, Punjab; Indian Agriculture Research Institute, New Delhi; Mahatama Phule Krishi Vidyapeeth, Rahuri, Maharashtra; Citrus Improvement

Project, Tirupati, Andhra Pradesh; Citrus Experiment Station, Katol, Maharashtra; Horticultural Experiment Station, Periyakulam, Tamil Nadu; Citrus Research Station, Petlur, Venkatagiri, Andhra Pradesh and Citrus Experiment Station, Tinsukia, Assam (Karihaloo et al. 2005; Singh et al. 2010).

Long-term field maintenance of citrus is difficult since the plants are highly prone to bacterial and viral diseases. In India in vitro conservation and cryopreservation of Citrus germplasm was taken up at NBPGR, New Delhi and at limited scale at NBRI, Lucknow. In vitro long-term preservation was attempted in *C. aurantifolia* by root apices (Bhat et al. 1992) and in *C. grandis* using shoot apices (Chaturvedi 2002). In the case of *C. aurantifolia,* root cultures retained their ability to regenerate shoot buds even after 3 years of storage and regenerated plants with normal diploid chromosomes which were established successfully in soil (Bhat et al. 1992). However, the low frequency of plant regeneration limits the use of root cultures for germplasm conservation until the frequency of shoot regeneration can be improved (Bhat et al. 1992). Similarly, shoot cultures of *C. grandis* were successfully preserved by in vitro method for more than 32 years (Chaturvedi 2002).

The presence of a high degree of polyembryony in many *Citrus* species provides the opportunity to conserve the original genotype as seed despite high levels of heterozygosity. However, seeds of many *Citrus* species display recalcitrant or intermediate storage behaviour and, therefore, cannot be stored using conventional –20°C storage methods. Cryopreservation of embryos and embryonic axes provides the most successful and reliable method for long term conservation of citrus germplasm; high recovery rate of genetically stable plantlets with normal growth has prompted the initiation of efforts for establishment of base collection of several indigenous and exotic citrus species in National Cryogenebank at NBPGR. To date, successful cryopreservation was achieved in *Citrus* by using a wide range of explants namely embryogenic callus, cell suspensions, nucellar callus, somatic embryos, shoot apices, zygotic embryos and embryonic axes in several *Citrus* species and cultivars (Malik et al. 2010). Approximately 25 *Citrus* species and allied genera were tested for long, medium and short-term conservation in various laboratories around the world (Malik et al. 2010). At NBPGR, seed storage behaviour of different *Citrus* species was studied and seeds and embryonic axes of more than 650 accessions comprising 20 species are conserved in the cryogenebank (Table 8.7). Pollen of 64 accessions belonging to *C. limon, C. aurantifolia* and *P. trifoliata* was cryopreserved at Indian Institute of Horticultural Research, Bangalore (Ganeshan and Rajasekharan 2000).

8.6.4 Jackfruit

Jackfruit (*Artocarpus heterophyllus* Lamk.) is a tropical tree indigenous to South and South-east Asia. Wild forms of jackfruit are commonly found in India, Thailand, the Philippines, Myanmar, Indonesia, Malaysia and to a considerable extent in Brazil. The prominent and economically important species of genus *Artocarpus*

Table 8.7 Seed storage behaviour and successful cryopreservation of *Citrus* species undertaken at Cryogenbank at NBPGR

Species	Seed storage behaviour	Explant stored	Method of cryostorage	Recovery (%)	Total accessions
C. reticulata	Intermediate	Seed	SDFF	93.3	53
		EA	DFF, ED, VT	75, 80, 70	
C. sinensis	Intermediate	Seed	SDFF	70	75
		EA	DFF, ED, VT	90, 80, 90	
C. aurantifolia	Intermediate	Seed	SDFF	55	51
		EA	DFF, VT	57, 67	
C. limon	Intermediate	Seed	SDFF	88	55
C. medica	Intermediate	Seed	SDFF	80	32
		EA	DFF, VT	85, 77	
C. grandis	Intermediate	Seed	SDFF	85	60
		EA	DFF, ED, VT	95, 100, 90	
C. paradisi	Intermediate	Seed	SDFF	88	18
		EA	DFF, ED, VT	75, 100	
C. jambhiri	Recalcitrant	Seed	SDFF	68.7	62
		EA	DFF, ED, VT	96, 100, 90	
C. karna	Recalcitrant	Seed	SDFF	60.5	19
		EA	DFF, ED, VT	73, 70, 70	
C. latipes	Intermediate	Seed	SDFF	60	2
		EA	DFF, ED, VT	64, 45, 77	
C. macroptera	Intermediate	Seed	SDFF	70	7
		EA	DFF, ED, VT	87, 62, 92	
C. indica	Intermediate	Seed	SDFF	86	27
		EA	DFF, ED, VT	90, 87, 90	
C. aurantium	Intermediate	Seed	SDFF	90	10
C. limetta	Intermediate	Seed	SDFF	80	6
C. limettioides	Intermediate	Seed	SDFF	80	5
C. limonia	Intermediate	Seed	SDFF	85.6	26
C. pseudolimon	Intermediate	Seed	SDFF	74	10
C. ambelycarpa	Intermediate	Seed	SDFF	90	1
C. maderaspantna	Intermediate	Seed	SDFF	50	1
C. madurensis	Intermediate	Seed	SDFF	52	4
C. myritifoila	Intermediate	Seed	SDFF	60	2
C. pectinifera	Intermediate	Seed	SDFF	56.6	5
C. reshni	Intermediate	Seed	SDFF	87.5	10
C. regulosa	Intermediate	Seed	SDFF	71	2
C. samperflorens	Intermediate	Seed	SDFF	70	2
C. taiwanica	Intermediate	Seed	SDFF	90	3
C. tangerina	Intermediate	Seed	SDFF	90	3
Poncirus trifoliata	Intermediate	Seed	SDFF	88.3	5
		EA	DFF	70	

EA embryonic axes, *SDFF* silica drying followed by fast freezing, *DFF* desiccation in sterile conditions followed by fast freezing, *ED* encapsulation-dehydration, *VT* vitrification include *A. heterophyllus* and *A. altilis* (Parkinsons) Forsberg (breadfruit). Jackfruit is a tetraploid with 2n=56.

A. heterophyllus is reported to have originated in Western Ghats of India and later was distributed to South-east Asia and other tropics; it has a high degree of variability in South-east Asia. Being protandrous and mainly cross-pollinated, the plants show great variability in many characters including texture, odour and flavour of the pulp.

8.6.4.1 Conservation

The 62 germplasm accessions of jackfruit are maintained in the field genebanks at NBPGR Regional Station, Thrissur, Kerala, 44 accessions at Horticulture and Agroforestry Research Programme, ICAR Research Complex for Eastern Region, Ranchi, Jharkhand; other institutes in India where germplasm is maintained are Indian Institute of Horticulture Research, Bangalore and Horticulture Experiment and Training Center, Saharanpur, Uttar Pradesh.

Jackfruit seeds lose viability within 15–30 days of harvest and below the moisture content of 26%, thus the species is regarded as truly recalcitrant (Chandel et al. 1995; Chaudhury and Malik 2004). Attempts are underway in different laboratories to conserve the embryonic axes through cryopreservation where limited success (30% survival) was reported after desiccation to 13–14% moisture content (Chaudhury and Malik 2004) and 50% survival using vitrification (Thammasiri 1999; Krishnapillay 2000) Experiments to recover viable embryonic axes desiccated to below 0.42 g water g^{-1} dry weight using air desiccation were unsuccessful (Dumet and Berjak 1995). However, on pretreatment of axes with 0.75 M sucrose, viable axes with lower moisture content of 0.21 g water g^{-1} dry weight could be obtained. These desiccated axes, however, failed to survive liquid nitrogen exposure.

8.6.5 Underutilized Fruits

Tropical underutilized fruits are a vast genetic wealth in the Indian sub-continent, however, desired attention for their genetic resource management is lacking due to their comparatively low commercial importance and limited research on genetic improvement of cultivars. In India several fruit species have local importance and are considered as important underutilized fruits: *Aegle marmelos* (Bael), *Buchanania lanzan* (Chironji), *Capparis decidua* (Ker), *Carissa* species (Karonda), *Cordia* species (Lasora), *Diospyros melanoxylon* (Tendu), *Emblica officinalis* (Aonla), *Feronia limonia* (Wood apple), *Ficus palmata* (Wild Fig), *Garcinia* species (Kokam, Malabar Tamarind and Mysore Gamboge), *Grewia* species (Phalsa), *Madhuca indica* (Mahua), *Manilkara hexandra* (Khirni), *Phoenix sylvestris* (Wild date), *Pithecellobium dulce* (Manila Tamarind), *Salvadora* species (Pilu and Meswak),

Syzygium cumini (Jamun), *Tamarindus indica* (Tamarind) and *Ziziphus* species (Ber). Most of these fruits originated in India and are still found wild or semi wild in limited areas with local socio-economic importance.

8.6.5.1 In Situ Conservation

In situ conservation is important for underutilized fruit species still occurring as natural wild or in the semi-domesticated conditions using following two approaches:

1. In natural habitats like protected areas and national reserves: Fruit species specific area based on presence of natural diversity are to be identified for species found as only natural wild namely *Buchanania lanzan, Capparis decidua, Diospyros melanoxylon, Manilkara hexandra, Salvadora oleodis, S. persica, Tamarindus indica, Pithecellobium dulce*. For species where both natural wild and cultivated genotypes are available wild populations are to be protected immediately. Such species are *Aegle marmelos, Emblica officinalis, Grewia* species, *Carissa* species, *Cordia* species, *Madhuca* species and *Ziziphus* species. Fruit species and possible protected areas for in situ conservation are to be identified based on socio-economic importance, population size, available diversity and biodiversity conservation policy of respective state government.
2. On-farm conservation for local natural selections/cultivars/farmer's varieties. In some underutilized fruits, local selections or farmers varieties have been developed or identified since time immemorial. These local selections are grown as isolated plants or in small numbers in the homestead gardens, farmers fields, backyards or in the village community lands. Such selections need urgent attention for further characterization, evaluation and on-farm conservation. Underutilized fruits where such selections are indentified and available are *Syzygium cumini, Cordia myxa, Tamarindus indica, Aegle marmelos, Embelica officinalis* and *Ziziphus* species.

8.6.5.2 Ex Situ Conservation

Field Genebanks

Field genebank conservation of underutilized fruits has been recently undertaken especially under the AICRP (All India Coordinated Research Project) on Arid Fruits at various ICAR (Indian Council of Agricultural Research) institutes namely Central Institute of Arid Horticulture, Bikaner; Central Institute of Sub-tropical Horticulture, Lucknow; Central Arid Zone Research Institute, Jodhpur; NBPGR Regional Station, Jodhpur; Indian Institute of Horticulture Research, Bangalore and at State Agricultural Universities namely Chaudhary Charan Singh Agricultural University (CCSHAU), Hisar and Regional Station, Bawal, Haryana; Mahatma Phule Krishi Vidyapeeth (MPKV), Rahuri, Maharashtra; Dantiwada Agricultural University

(DAU), Sardarkrushinagar; Sri Karan Narendra (SKN) College of Agriculture, Rajasthan Agriculture University (RAU), Jobner, Rajasthan; Maharana Partap University of Agriculture and Technology (MPUAT), Udaipur, Rajasthan; Narendra Deva University of Agriculture and Technology (NDUAT), Faizabad, Uttar Pradesh; Rajasthan Agricultural University (RAU), Bikaner, Rajasthan; and other state horticulture stations at Tamil Nadu and Andhra Pradesh.

Ex situ conservation of underutilized fruits is important to safeguard the genetic wealth and to use germplasm for genetic improvement to develop desirable cultivars or varieties. Field genebanks have an important place in conservation and maintenance of such species. Nevertheless, it must be noted that important agronomic practices and reproductive biology for successful management of such underutilized fruit species are not understood, and some increased attention needs to be paid on these issues so that these potentially useful species could be conserved for the benefit of future generations (Malik et al. 2011b).

Genebank and Cryogenebank Conservation

Seed storage behaviour, germination characteristics and cryopreservation of several tropical underutilized have been studied and cryobanking of germplasm has been undertaken in India (Malik et al. 2011a). Conservation of germplasm in the form of seeds for underutilized fruit species that are predominantly cross pollinated, only ensures the genepool conservation of these species due to the heterozygous nature of seeds. As most of these species are found natural wild or semi-wild and propagated through seeds in nature, conservation of available genetic variability essentially required for the selection of desired genotypes needs to be protected safely and timely. In most of these fruit species farmers or local people are propagating progenies of these fruits using seeds as no commercial cultivars are available and even if a few have been identified, planting material is not available. Once the promising genotypes or cultivars are identified in these species, conservation of their vegetative tissues to achieve true-to-type conservation can be attempted using in vitro methods. It is to emphasize here that conservation of vegetative tissues in these tropical woody species would be an enormous task as most of the species are known for their recalcitrance as far as in vitro establishment is concerned and equally difficult task would be to successfully cryopreserve the vegetative tissues excised from in vitro cultures. It is, therefore, recommended to conserve the available genetic diversity of such economically important species in the best possible way to fulfill the objective of safe guarding these species from genetic erosion. For genetic improvement and genotype conservation, collected and characterized elite genotypes are presently being conserved in the field genebanks at various horticultural organizations. It is, therefore, emphasized that a complementary conservation strategy (Ramanatha Rao 1998) involving the use of more than one relevant approach would be the best option for achieving safe conservation of these underutilized fruit species facing severe threat of extinction. Table 8.8 shows seed storage behaviour and accessions of underutilized fruits cryopreserved.

Table 8.8 Seed longevity and seed storage behaviour of tropical underutilized fruit species

Species	Shelf life (longevity at ambient temperature 25–34°C)	Seed storage behaviour	Accessions cryostored	Accessions cryostored in the form of seed/embryo/embryonic axes
Aegle marmelos	36 months	I	80	Seed and embryo
Buchanania lanzan	11 months	I	127	Seed
Capparis decidua	14 months	I[a]	88	Seed
Carissa carandas	5–6 months	I	9	Seed
Cordia myxa	3–4 months	I	24	Seed
Diospyros melanoxylon	15 months	I	16	Seed
Emblica officinalis	19 months	O	31	Seed
Garcinia indica	30–45 days	R	0	Nil
Grewia subinaequalis	6–7 months	I	13	Seed
Madhuca indica	7 days	R	2	Embryonic axis
Manilkara hexandra	4 months	I	46	Seed and embryonic axis
Pithecellobium dulce	24 months	O	14	Seed
Salvadora oleoides	14–21 days	R	23	Seed and embryonic axis
Syzygium cumini	30 days	R	0	Nil
Tamarindus indica	20 months	O	10	Seed
Ziziphus nummularia	22 months	O[b]	15	Seed

O orthodox, *I* Intermediate, *R* recalcitrant

[a]Also reported by ICRAF Agroforestry database

[b]Also reported by Hong et al. (1998), Malik et al (2011b)

8.7 Conservation of Tropical Fruit Trees in Africa

Africa is richly endowed with various plants genetic resources in terms of cultivated field crops, forest trees and horticultural crops, namely fruits, medicinal and ornamental plants. Additionally, many plant species, crop relatives and land races occur in the wild.

Agro-biodiversity in Africa is literally immense owing to the variety of agro-ecological zones across the continent. Most African countries have strategies and plans to assess diversity of plant genetic resources for food and agriculture. However, one can state that no tangible actions were taken to support and encourage in situ and ex situ conservation methods, especially for the local fruit trees. Assessment of this diversity in terms of characterization, evaluation and setting up procedures to monitor and measure the level of diversity per se and to examine status of genetic erosion remains meager. Owing to lack of efforts to conserve, it could be assumed that the genetic resources of most species, including local fruit species, are undergoing deleterious genetic erosion. Control of genetic erosion in Africa seems difficult due to lack of policies, and a low level of public cognizance on conservation of their genetic resources, in addition to a lack of technical and financial assistance. Fortunately, rural people in Africa very much value their indigenous fruit trees and other genetic resources; they preferentially conserve them in their home yards although these genotypes are in too small numbers to give viable populations.

The wide range of indigenous fruit trees in many areas of Africa can enable farmers to meet their varied household needs for food, nutrition, medicines, etc. These species are often part of the traditional diet and culture and the subject of a body of indigenous knowledge regarding their management and use. The role of indigenous fruit trees in poverty alleviation was well recognized (Schreckenberg et al. 2006). Africa's fruits have not been brought up to their potential in terms of quality, production, and availability. Geographically speaking, few have moved beyond Africa's shores; horticulturally speaking, most remain poorly known (Lost crops of Africa, Vol. III-Fruits 2008).

Exotic fruit trees (e.g. mangoes, avocado, Citrus, bananas) have received considerable research work for their breeding, cultural practices and conservation efforts. Substantial number of tropical rare and underutilized fruit trees in Africa remains understudied in the sense of being neglected in exploration, collection, evaluation, particularly their nutritive value, documentation and conservation. In this context, the medicinal uses of bush mango or African mango *Irvingia gabonesis* are well recognized as a result of extensive research revealing its usage in fighting obesity, lowering blood cholesterol, controlling appetite and reducing blood hypertension and sugar in addition its uses as food fruits (Oguntola 2011). Most recently (Raebild et al. 2011), domestication of important indigenous fruit trees species has started in dry West Africa using *Adansonia digitata* (baobab tree), *Parkia biglobosa*, *Tamarindus indica*, *Vitellaria paradoxa* (shea nut tree) and *Ziziphus mauritiana*. Knowledge of genetic parameters such as fruit traits is almost absent, and characterization of these genotypes is still ongoing. Conservation of these species is poor due to the few planted trees at sporadic trial sites and those grown in their natural habitat where they are subject to biotic and a biotic hazards. It is worth noting that these species and many other local fruit species are shared by most African countries. Therefore, the research work of Raebild and co-workers (2011) embarked primarily on germplasm collection and conservation based on provenance trials to cater for the species distribution area in various African countries.

In situ and ex situ conservation methods applied for some field crops and few fruit species were firstly documented by the report of the first State of the World's Plant Genetic Resources for Food and Agriculture (SoW) that was presented to the Fourth International Technical Conference on Plant Genetic Resources held in Leipzig, Germany, in 1996. The first SoW report highlighted the poor documentation available on much of the world's ex situ plant genetic resources. This problem continues to be a substantial obstacle to the increased use of plant genetic resources for food and agriculture in crop improvement and research. Frequent problems exist in standardization and accessibility, even for basic passport information. The report revealed that major African countries have already established natural reserves, botanical gardens, arboreta usually maintained by universities and other tertiary education institutes whose mandates tend to neglect indigenous crops; and individual researchers tend to embark on their own research programs. No actions were reported to encourage and support field maintenance of traditional fruit varieties and other crops as well, owing to lack of operational and capital funds. National genebanks in many African countries were also established with the same fate of inadequacy in maintenance, seed processing, viability, replications, testing prior storage and other important characteristics.

However, the second SoW report indicated growing interest in neglected and under-utilized crops in recognition of their potential to produce high-value niche products and as novel crops for the new environment conditions that are expected to result from climate change (FAO 2009). According to country reports, data on storage facilities in Africa are less complete than for other continents. Most countries reported having seed and field genebanks, but only Benin, Cameroon, Congo Brazzaville, Ghana, Kenya, Mali, Nigeria and Uganda reported having in vitro storage facilities (Reed et al., this volume). No country specified having an ability to conserve germplasm cryogenically. Seed genebanks are generally much more important and widespread in the continent than field genebanks. There were many problems posed in use of storage facilities, including reliability of electricity supplies, pests and disease problems, and lack of staff, equipment, or funds. Guinea reported the loss of its entire ex situ collection as a result of a failure in the electricity supply. Regular viability testing was carried out in Madagascar, Nigeria, Uganda and Zambia, but generally not elsewhere. The systematic regeneration of stored material appears sporadic. Funding, staffing and facilities were frequently reported to be inadequate to allow the necessary germplasm regeneration to be undertaken.

Most African nations reported having characterization and evaluation data on their collections, but with some exceptions (e.g. Ethiopia, Kenya and Mali), it was generally incomplete and not standardized. Togo indicated that its documentation was in a rudimentary state and several other countries reported serious weaknesses.

Kenya reported its intention to develop national documentation systems. While three countries reported that they still maintain some records on paper and eight use spreadsheets, at least eight others have dedicated electronic systems and Kenya, Ghana and Togo reported using generic databases to manage information on ex situ collections. Nevertheless, the research on tropical fruit trees in Africa recently started to pick up momentum in terms of novel efforts to undertake successful cryopreservation of zygotic embryos and embryonic axes of recalcitrant and intermediate seeded local fruit trees as a means for ex situ conservation of long term germplasm storage in the face of inevitable climate change hazards that might lead to their being endangered. For instance, in South Africa, embryonic axes of the local palm *Phoenix reclinata*, which is heavily utilized as food by local community, were successfully cryopreserved (Ngobese et al. 2010). Most recently, zygotic embryos and embryonic axes of recalcitrant seeded fruit species *Ekebergia capensis*, *Strychnos gerrardii* and *Boophane disticha* were cryopreserved with successful regeneration of adventitious buds and shoots and the plantlets were successfully acclimated (Hajari et al. 2011; Berjak et al. 2011), Such research work will unequivocally pave the way for reliable long term conservation of the tropical fruit trees in Africa rather than in situ conservation where these species are subject to biotic and a biotic deleterious hazards.

Yet, in order to set up sound genetic resource conservation in Africa, liaison and coordination amongst countries should be effected and national and regional conservation strategies and working plans should be scheduled.

8.8 Challenges and Future Directions

To establish and maintain viable ex situ populations for even a small fraction of the overall plant diversity would demand enormous resources, in terms of land area, funds and human expertise (Oldfield 2009). Many tropical countries do not maintain a botanical garden or field genebanks and the financial support, both public and private, is often limited. This also applies for funding in research; fruit crops are usually given low priority in granting research funds, thus limiting the amount of research that can be carried out. For conservation to be truly effective, an integrated approach is required, in research, maintaining and supplementing conservation efforts. The challenges are particularly great in tropical regions that have the greatest species diversity, a high proportion of trees with recalcitrant seeds and relatively few field genebanks, botanic gardens and arboreta. Another need or a challenge for near future is rationalization of collections and gap filling. Despite recent efforts, many tropical fruit genetic resources remain incomplete in terms of genetic diversity as most of the collection in the past was focused on elite genotypes, those that the horticulturalists could put in immediate efforts for selection and commercialization. This needs to be corrected through enriching extensive collections with genetically diverse accessions. Additionally, the use of available geographic information and mapping, the distribution and gaps in collections may be identified and targeted collecting may be carried out.

The challenge in obtaining cryopreservation protocols for these species involves not only the recalcitrant nature of the materials but also in the availability of materials. Tropical recalcitrant species have additional phenological considerations such as seasonality or irregular and unpredictable seed production; some species flower once in 7 years and even the clonal varieties do not produce fruits and seeds at regular yearly intervals. Hence, there is scarcity and shortage of fruit for experimental purpose (Normah et al. 2011). Shoot tips from in vitro plants are usually used for cryopreservation to overcome this problem. Thus in vitro shoot tips are produced from induction of multiple or adventitious shoots from seeds/embryos or from other parts of a seedling.

Understanding the phenomenon of seed recalcitrance and consequently developing sound conservation practices for these species is of major scientific and practical importance. The integrative approach of systems biology using detailed analyses of data from genomics and related areas of transcriptomics, proteomics, and metabolomics may shed light on the interacting networks of genes and proteins downstream from the response signals, leading to understanding seed or explant recalcitrance (Normah et al. 2011). Along with this, in order to formulate and develop appropriate and more effective conservation strategies as well as ways to efficiently utilize the available genetic diversity of these important resources, information on reproductive biology, improved morphological data along with information on the genetic diversity in tropical fruit species is necessary. Partnerships need to be strengthened between all parts of the world and, increasingly, researchers need to work together with policy makers and local communities to develop integrated conservation solutions for tropical fruit species.

8.9 Conclusions

Conservation efforts for tropical fruit species are being undertaken by many countries within the region with the support of various organizations such as Bioversity International and ACIAR. This is especially so in South East Asia and South Asia (India). Unlike Malaysia, India has a nationally managed PGR system and hence is less decentralized. Both of these approaches have their merits and demerits. Brazil as a representative of tropical America has done commendable efforts in conservation of these species. USDA, ARS, (USA) is also carrying out similar efforts. Tropical Africa however has many challenges and greater efforts and commitment by the countries involved are needed.

Due to the seed recalcitrance nature of many tropical fruit species, conservation efforts are mainly carried out ex situ, in field genebanks and botanical gardens. Genetic diversity using a range of genetic markers is being investigated at various regions to assist in breeding and improvement of the crops. Documentation in the form of passport, characterization and evaluation data are kept in various information systems such the AgroBIS, MARDI, Malaysia, the new Brazilian Genetic Resources Information System (SIBRARGEN) and the USDA, ARS Germplasm Resources Information Network (GRIN).

Field genebanks have disadvantages: require a lot of space, and is relatively expensive to maintain; are exposed to natural calamities or adverse environmental condition such as drought, flood, pest and diseases. In vitro and cryopreservation are alternative techniques for conservation of tropical fruit species with recalcitrant seeds. Nevertheless, in vitro genebanks established so far are mainly for *Musa* and *Ananas* species while cryopreservation methods are still being developed and improved for many of the species. However, for *Citrus* species known to have a continuous range of seed characteristics, cryopreservation for several of these species is being implemented in India, China and Australia. Underutilized fruit germplasm are also cryoconserved. Conservation of tropical fruit species faces many challenges ranging from the nature of the species, funding and national policies. Efforts and support need to be given to assist research for improvement of protocol development with further understanding of recalcitrant nature of the plant. It is necessary to increase awareness of the importance of conservation of these species for the policy makers and funding bodies in order to have successful programs for the conservation of these precious tropical fruit species.

References

Agharkar SP, Roy N (1951) On the origin and distribution of cultivated mangoes. Indian J Genet 11:48

Agroforestry Database (World Agroforestry Centre) www.worldagroforestrycentre.org/sites/treeDBS/AFT

Amalraj VA, Velayudan KC, Agrawal RC, Rana RS (1993) Banana genetic resources. Scientific monograph no. 3, National Bureau of Plant Genetic Resources, New Delhi

Amar MH, Biswas MK, Zhang Z, Guo W-W (2011) Exploitation of SSR, SRAP and CAPS-SNP markers for genetic diversity of citrus germplasm collection. Sci Hortic 128:220–227

Anonymous (2010a) Indian horticulture database. National Horticulture Board, Ministry of Agriculture, Gurgaon

Anonymous (2010b) Annual progress report of National Bureau of Plant Genetic Resources-2010–11. NBPGR, Pusa campus, New Delhi

Anonymous (2010c) Annual progress report of National Bureau of Plant Genetic Resources-2010–11. NBPGR, Pusa campus, New Delhi

Arora RK, Ramanatha Rao V (1995) Proceedings of expert consultation on tropical fruit species of Asia. Malaysian Agricultural Research and Development Institute, Serdang, Kuala Lumpur, 17–19 May 1994. IPGRI Office for South Asia, New Delhi, 116 p

Barkley NA, Roose ML, Krueger RR, Federici CT (2006) Assessing genetic diversity and population structure in a citrus germplasm collection utilizing simple sequence repeat markers (SSRs). Theor Appl Genet 112:1519–1531

Barkley NA, Krueger RR, Federici CT, Roose ML (2009) What phylogeny and gene genealogy analyses reveal about homoplasy in citrus microsatellite alleles. Plant Syst Evol 282:71–86

Bayer RJ, Mabberley DJ, Morton C, Miller CH, Sharma IK, Pfiel BE, Rich S, Hitchock R, Sykes S (2009) A molecular phylogeny of the orange subfamily (Rutaceae: Aurantioideae) using nine cpDNA sequence. Am J Bot 96:668–685

Berjak P, Varghese SB, Pammenter NW (2011) Cathodic amelioration of the adverse effects of oxidative stress accompanying procedures necessary for cryopreservation of embryonic axes of recalcitrant-seeded species. Seed Sci Res 21:187–203

Bhag Mal, Ramanatha Rao V, Arora RK, Percy E, Sajise SBR (2010) Conservation and sustainable use of tropical fruit species diversity: bioversity's efforts in Asia, the Pacific and Oceania. Indian J Plant Genet Resour 24(1):1–22

Bhat SR, Chitralekha P, Chandel KPS (1992) Regeneration of plants from long-term root culture of lime, *C. aurantifolia* (Christm.) Swing. Plant Cell Tiss Organ Cult 29:19–25

Bhattacharya SC, Dutta S (1956) Classifications of *citrus* fruits of Assam. ICAR Sci Monogr 20:1–110

Brandis DD (1874) Forest flora of north west and central India. Allen, London

Chadha KL (1976) Germplasm of mango, grape and guava in India. Tec. Doc. No.13. All India co-ordinated fruit improvement project, Lucknow, 89

Chadha KL (1996) Status report on genetic resources of mango in India. (Synthesised and edited by Arora RK, Ramanatha Rao V, Rao AN). IPGRI-APO, Singapore

Chai CC, Teo GK, Lau CY, Pawozen AMA (2008) Conservation and utilisation of indigenous vegetables in Sarawak. In: Mohd Shukor N, Salma I, Mohd Said S (eds) Agrobiodiversity in Malaysia. MARDI, Kuala Lumpur, pp 47–55

Chan YK (2002) Fruit breeding at MARDI: a retrospect over three decades. In: Drew R (ed) Proceedings of the international symposium on tropical and subtropical fruits. Acta Hortic 575:279–286

Chandel KPS, Chaudhury R, Radhamani J, Malik SK (1995) Desiccation and freezing sensitivity in recalcitrant seeds of tea, cocoa and jackfruit. Ann Bot 76:443–450

Chaturvedi HC (2002) Conservation of phytodiversity through in vitro morphogenesis. In: Nandi SK, Palni LMS, Kumar A (eds) Role of plant tissue culture in biodiversity conservation and economic development. Gyanodaya Prakashan, Nainital, India, pp 503–511

Chaudhury R, Malik SK (2004) Desiccation and freezing sensitivity during seed development in jackfruit. Seed Sci Technol 32:785–795

Chaudhury R, Malik SK, Rajan S (2010) An improved pollen collection and cryopreservation method for highly recalcitrant tropical fruit species of mango (*Mangifera indica* L.) and Litchi (*Litchi chinensis* Sonn.). Cryo Lett 31:268–278

Chen J, Cannon CH, Hu H (2009) Tropical botanical gardens: at the in situ ecosystem management frontier. Trends Plant Sci 14(11):584–589

Chew PC, Clyde MM, Normah MN, Salma I (2002) DNA polymorphism in accessions of *Nephelium lappaceum* L. In: Engels JMM, Ramanatha Rao V, Brown AHD, Jackson MT (eds) Managing plant genetic diversity. CABI, New York, pp 57–60

Cho EG, Hor YL, Kim HH, Ramanatha Rao V, Engelmann F (2001) Cryopreservation of *Citrus madurensis* zygotic embryonic axes by vitrification: importance of pregrowth and preculture conditions. Cryo Lett 22:391–396

Cho EG, Normah MN, Kim HH, Ramanatha Rao V, Engelmann F (2002) Cryopreservation of *Citrus aurantifolia* seeds and embryonic axes using a desiccation protocol. Cryo Lett 23:309–316

Chua SP, Normah MN (2011) Effects of preculture, PVS2 and vitamin C on the survival of recalcitrant *Nephelium ramboutan-ake* shoot tips after cryopreservation by vitrification. Cryo Lett 32(6):506–515

Drew R, Sarah A, Somsri S, Normah MN, Hoa TT, Damasco O, Ramanatha Rao V (2007) Advanced technologies for germplasm conservation of tropical fruit species. Acta Hortic 760:91–97

Dumet D, Berjak P (1995) Desiccation tolerance and cryopreservation of embryonic axes of recalcitrant species. In: Ellis RH, Black M, Murdoch AJ, Hong TD (eds) Basic and applied aspects of seed biology. Proceedings of the 5th international workshop on seeds, Reading

Elcy GPS, Mahani MC, Park Y-J, Normah MN (2012) Simple sequence repeats (SSR) profiling of cultivated Limau Madu (*Citrus reticulata* Blanco) in Malaysia. Fruits 67:67–76

FAO (2009) The second report on the state of the world's plant genetic resources for food and agriculture. Commission on Genetic Resources for Food and Agriculture, Rome

Ferreira FR, Pinto ACQ (2010) Germplasm conservation and use of genebanks for research purposes of tropical and subtropical fruits in Brazil. Acta Hortic 864:21–27

Ganeshan S, Rajasekharan PE (2000) Conservation of nuclear genetic diversity (NGD) in *Citrus* genepool: implications in citrus breeding and genetic conservation. In: Singh S, Ghosh SP (eds) Hi tech citrus management, Proceedings of international symposium on citriculture. ISC/ICAR/NRC for Citrus, Nagpur, 23–27 Nov 1999, pp 85–90

Hajari E, Berjak P, Pammenter NW, Watt MP (2011) A novel means for cryopreservation of germplasm of the recalcitrant-seeded species, *Ekebergia capensis*. Cryo Lett 32:308–316

Hamilton KN, Ashmore SE, Pritchard HW (2009) Thermal analysis and cryopreservation of seeds of Australian wild *Citrus* species (Rutaceae): *Citrus australasica, C. inodora and C. garrawayi*. Cryo Lett 30:268–279

Hong TD, Linington SH, Ellis RH (1998) Compendium of information on seed storage behaviour, Volumes I and II. Royal Botanic Gardens, Kew, UK

Ilao SSL (2007) Conservation and utilization of plant genetic resources in the Philippines: status and directions. International training-workshop. The conservation and utilization of tropical/subtropical plant genetic resources, pp 91–116

Iyer CPA, Subramanian TR (1989) Genetic resources activities concerning tropical fruit plants. In: Paroda RS, Arora RK, Chandel KPS (eds) Plant genetic resources: Indian perspective. NBPGR, New Delhi, pp 310–319

Kanjilal UN, Kanjilal PC, Das A (1982) Mangifera indica L. Flora of Assam (Reprint). Vol 1, A Von Book Company, Delhi, India, pp 335–336

Karihaloo JL, Malik SK, Rajan S, Pathak RK, Gangopadhyay KK (2005) Tropical fruits. In: Dhillon BS, Tyagi RK, Saxena S, Randhawa GJ (eds) Plant genetic resources: horticultural crops. Narosa Publishing House, New Delhi, pp 121–145

Knight RJ Jr (1983) Tropical fruits of Asia with potential for expanded world production. Proc Trop Region Am Soc Hortic Sci 27(A):41–93

Krishnamurthi S, Seshadri VS (1958) Origin and evolution of cultivated banana. Indian J Hortic 15:135–145

Krishnapillay B (2000) Towards the use of cryopreservation as a technique for conservation of tropical recalcitrant seeded species. In: Razdan MK, Cocking EC (eds) Conservation of plant genetic resources in vitro, vol 2, Applications and limitations. Science//IBH, USA/Oxford/New Delhi, pp 139–166

Li Y, Mao Q, Feng F, Ye C (2010) Genetic diversity within a jackfruit (*Artocarpus heterophyllus* Lam.) germplasm collection in China using AFLP markers. Agri Sci China 9(9):1263–1270

Lost Crops of Africa, Vol. III-Fruits (2008) National academy of sciences, Washington, USA. http://www.nap.edu/catalog/11879.html

Malik SK, Chaudhury R (2006) The cryopreservation of embryonic axes of two wild and endangered *Citrus* species. Plant Genet Resour 4:1–7

Malik SK, Chaudhury R, Dhariwal OP, Bhandari DC (2010) Genetic resources of tropical underutilized fruits in India. NBPGR, New Delhi

Malik SK, Kumar S, Pal D, Choudhary R, Uchoi A, Rohini MR, Chaudhury R (2011a) Cryopreservation and in vitro conservation of *Citrus* genetic resources. In: Citrus biodiversity: proceedings of national seminar on "citrus biodiversity for livelihood and nutritional security". National Research Centre on Citrus, Nagpur. 4–5th Oct 2010, pp 140–156

Malik SK, Chaudhury R, Kalia RK, Dulloo E (2011b) Seed storage characteristics and cryopreservation of tropical underutilized fruits in India. Acta Hortic 918:189–198

Mango Resources Information System. Department of Biotechnology and Indian Council of Agricultural Research, New Delhi. www.mangifera.org

Mcphee J (1967) Oranges. Farrar, Straus and Giroux, New York

Mohd Shukor N, Zulhairil A, Rosliza J, Wan Darman WA, Mohammad Ghaddafi D (2007) Country report on the state of plant genetic resources for food and agriculture (PGRFA) in Malaysia (1997–2007). MOA/MARDI/FAO, 101 p

Mohd Shukri MA, Salma I, Mirfat AHS, Mohd Shukor N (2009) Rare and underutised fruits and ulam species: New sources for high antioxidant and potential health benefits. Proceedings national conference on new crops and bioresources. MARDI, DOA, FRIM, TFNet. Negeri sembilan, Malaysia 15–17 December 2009, pp 73–86

Mukherjee SK (1949) A monograph on the genus *Mongifera* L. Lloydia 12:73–136

Mukherjee SK (1951) Origin of mango. Indian J Gen Plant Breed 11:49–56

Mukherjee SK (1985) Systematic and ecogeographic studies on crop genepools 1. *Mangifera* L. IBPGR, Rome

Mukherjee SK (1997) Introduction: botany and importance. In: Litz R (ed) The mango. CAB International, UK

Nares D, Ramanatha Rao V, Jarvis D (2001) In situ conservation of tropical fruit trees in Southeastern Asia. In: Mal B, Ramamani YS, Ramanatha Rao V (eds) Proceedings of the first annual meeting of the project, 'conservation and use of native tropical fruit species biodiversity in Asia, 6–9 Feb 2001. IPGRI South Asia Office, New Delhi, pp 51–60

Ngobese NZ, Sershen N, Pammeter NM, Berjak P (2010) Cryopreservation of the embryogenic axes of Phoenix reclinata, a representative of the intermediate seed category. Seed Sci Technol 38:704–716

Nik Masdek NH, Salma I, Rusli P, Masrom H (2008) Resistance to Phytophthora palmivora in *Durio lowianus*. Poster presented at the biodiversity and national development: achievements, opportunities and challenges conference. The Legend Hotel, Kuala Lumpur, 28–30 May 2008

Normah MN, Choo WK, Yap LV, Zeti Azura MH (2011) In vitro conservation of Malaysian biodiversity – achievements, challenges and future directions. In Vitro Cell Dev Biol Plant 47: 26–36

Normah MN, Clyde MM, Cho EG, Rao VR (2002) Ex situ conservation of tropical fruit species. Acta Hortic 575:221–230

Oguntola S (2011) Scientists corroborate additional uses of bush mango. www.Tribune.com.ng/index.php/natural-health/20745-scientists-corroborate-additional uses-of bush-mango. Accessed on 15 February 2012

Oldfield SF (2009) Botanic gardens and the conservation of tree species. Trends Plant Sci 14(11):581–583

Popenoe W (1920) Manual of tropical and subtropical fruits. Macmillan, New York

Raebild A, Larsen AS, Jensen JS, Ouedraogo M, Degroote S, Van Damme P, Bayala J, Diallo BO, Sanou H, Kalinganire A, Kjaer ED (2011) Advances in domestication of indigenous fruit trees in the West African Sahel. New Forests 41:297–315

Rajan S, Negi SS, Kumar R (1999) Catalogue on mango (*Mangifera indica*) germplasm. Central Institute for Subtropical Horticulture, Lucknow

Rajwana IA, Tabassam N, Malik AU, Malik SA, Rehman M, Zafar Y (2008) Assessment of genetic diversity among mango (*Mangifera indica* L.) genotypes using RAPD markers. Sci Hortic 117:297–301

Ramanatha Rao V (1998) Complementary conservation strategy. In: Arora RK, Rao VR (eds) Tropical fruits in Asia: diversity, maintenance, conservation and use. Proceedings of the IPGRI-ICAR-UTFANET regional training course on the conservation and use of germplasm of tropical fruits in Asia held at Indian Institute of Horticultural Research, Bangalore, 18–31 May 1997, pp 142–151

Ramanatha Rao V, Mal B (2002) Tropical fruit species in Asia: diversity and conservation strategies. Acta Hortic 575:179–190

Reed BM, Engelmann F, Dulloo E, Engels J (eds) (2005) Technical guidelines for the management of field and in vitro germplasm collections. IPGRI/FAO/SGRP, Rome

Saad MS, Ramanatha Rao V (2001) Establishment and management of field genebank: a training manual. IPGRI-APO, Serdang

Salma I (2005) Report for project monitoring. Strategic Resource Research Centre. MARDI, Melaka

Salma I, Masrom H, Mohd Nor A (2005) A catalogue of germplasm collections of tropical fruits in MARDI (unpublished)

Salma I, Mohd Yusoff A, Mohd Shukor N, Tan SL, Abd Rahman M (2010) Status of genebank and plant breeding activities in Malaysia. Paper presented at 6th national seed symposium. IOI Palm Garden Resort, Putrajaya, 23–24 Mar 2010

Schreckenberg K, Awono A, Degrande A, Mbosso C, Ndoye O, Tchoundjeu Z (2006) Domesticating indigenous fruit trees as a contribution to poverty reduction. Forests Trees Livelihoods 16:35–51

Simmonds NW (1962) The evolution of the bananas. Longman, London

Simmonds NW (1994) Bananas. In: Smartt J, Simmond NW (eds) Evolution of crop plants. Longman, UK

Singh RB (1993) Fruit production in Asia-Pacific region. In: Singh RB (ed) Research and development of fruits in Asia-Pacific region. RAPA/FAO, Bangkok, p 207

Singh IP, Jagtap DD, Patel RK (2010) Exploiting citrus genetic diversity for improvement and livelihood. In: Souvenir and abstracts, national seminar on 'citrus biodiversity for livelihood and nutritional security', NRC Citrus, Nagpur, pp 35–56

Song Beng Kah, Clyde MM, Ratnam W, Normah MN (2000) Genetic relatedness among *Lansium domesticum* accessions using RAPD markers. Ann Bot 86(2):299–307

Solley PG (1997) Citrus- lemons, limes and oranges, tangerines, grapefruits and pummelos (1999-02-07) http://www.sonprong.com/citrus.html

Swingle WT, Reece PC (1967) The botany of citrus and its wild relatives. In: Reuther W, Batchelor LD, Webber HJ (eds) The citrus industry. University of California Press, Berkeley, pp 190–340

Thammasiri K (1999) Cryopreservation of embryonic axes of jackfruit. Cryo Lett 20:21–28

Tirtawinata MR (2002) Mekarsari fruit garden: tropical fruit germplasm conservation in Indonesia. Acta Hortic (ISHS) 575:191–197

USDA, ARS (2011a) National genetic resources program. Germplasm Resources Information Network – (GRIN). (Online database) National Germplasm Resources Laboratory, Beltsville. Available: http://www.ars-grin.gov/cgi-bin/npgs/html/site_holding.pl?HILO. Accessed on 8 August 2011

USDA, ARS (2011b) http://www.ars.usda.gov/research/projects. Accessed on 8 August 2011

Vavilov NI (1951) The origin, variation, immunity and breeding of cultivated plants. Chron Bot 13:1–346

Verheij EWM, Coronel RE (1991) Plant resources of South-East Asia no. 2. Edible fruits and nuts. Pudoc, Wageningen

Wan Darman WA (2008) Conserving fruits, vegetables and medicinal plants of genetic resources. In: Mohd Shukor N, Salma I, Mohd Said S (eds) Agrobiodiversity in Malaysia. MARDI, Kuala Lumpur, pp 59–65

Wang ZC, Deng XX (2004) Cryopreservation of shoot-tips of citrus using vitrification: effect of reduced form of glutathione. Cryo Lett 25(1):43–50

Wen B, Cai C, Wang R, Tan Y, Lan Q (2010) Critical moisture content windows differ for the cryopreservation of pomelo (*Citrus grandis*) seeds and embryonic axes. Cryo Lett 31:29–39

Wong WWW, Chong TC, Tananak J (2008) Conservation and sustainable utilization of fruit species in Sabah. In: Mohd Shukor N, Salma I, Mohd Said S (eds) Agrobiodiversity in Malaysia. MARDI, Kuala Lumpur, pp 30–46

Wu YJ, Huang XL, Chen QZ, Li XJ, Engelmann F (2007) Induction and cryopreservation of embryogenic cultures from nucelli and immature cotyledon cuts of mango (*Mangifera indica* L. var. Zihua). Plant Cell Rep 26:161–168

Yadav IS (1999) Occurrence of races *Mangifera* species. Indian J Hortic 56:96–103

Yadav IS, Rajan S (1993) Genetic resources of *Mangifera*. In: Chadha KL, Pareek OP (eds) Advances in horticulture, vol 1. Malhotra Publishing House, New Delhi

Chapter 9
Conservation of Orchids the Gems of the Tropics

Irawati

9.1 Introduction

The orchid family is the most diverse family within the plant kingdom with an estimated 25,000 species known to science. As a result, any policy on plant management would have a considerable impact on orchids. Orchids often receive greater attention in conservation activities because of their beauty, specific appearance, and relative fragility.

In the wild, orchids depend on mycorrhiza and a suitable environment for germination. The germinating seeds would only grow further when their environment is suitable, and their leaves assimilate CO_2 to support their own growing requirements. When the orchid plant reaches maturity, another specific environment condition will stimulate the development of its flowers. Since most modern orchids have a waxy mass of pollen grains supported by a caudicle with sticky viscidium, the pollinia would be deposited to the stigma only by suitable pollinators; this specific relationship was first postulated by Charles Darwin in his famous paper, which "predicted" moths as the pollinator of *Angraecum sesquipedale*. Almost all orchid pollinators are insect but recently sunbirds was proved acting as pollinator of *Disa chrysostachya* and *Disa satyriopsis,* therefore the existence of sunbird is important in *Disa* conservation (Johnson and Brown 2004).

Following maturation, orchid fruit capsules open and the dry tiny seeds would spread, mostly by wind or water currents, to start a new life cycle. When the suitable habitat is undisturbed, with constant stimuli inducing flowers and the presence of insect pollinators are preserved, new growths have a better chance of survival when they mature scattered around the mother plant, where the roots are inhabited by suitable mycorrhiza. Scientists have demonstrated the need for specific mycorrhiza; for this reason, although the number of seeds in a capsule exceeds 100,000; only a

Irawati (✉)
Bogor Botanic Gardens, Ir. H. Juanda 13, Bogor 16003, Indonesia
e-mail: irawati@indosat.net.id

M.N. Normah et al. (eds.), *Conservation of Tropical Plant Species*,
DOI 10.1007/978-1-4614-3776-5_9,

fraction of them would germinate and even less would grow into mature plants. This already low probability of orchid propagation is worsened by recent climate changes, which result in unfavorable conditions for orchids (Keel 2005).

Due to human intervention, precious orchids with the most beautiful flowers vanish from their habitat faster than less valuable ones. People would spend the time and money required to hunt these valuable orchids, even when they have to be collected in very remote areas. Once obtained, the collected orchid doesn't always survive for a long time in the new habitat or personal collection. This practice is the most damaging on the survival of endemic species, especially the ones with a narrow distribution area in nature.

There is a constant search for new orchid species; this is in part due to the endless breeding of orchid hybrids to fulfill the demand of the market for new appearances, shapes, colors, arrangements, and sizes, as well as other criteria such as resistance to diseases, high productivity or ease of maintenance. Over harvesting of some endemic orchid species create a great concern to conservationist particularly in places where human activities drastically diminishes the natural habitat of orchids. Moreover the species is also facing a great pressure to extinction caused by various disturbance and habitat encroachment. There is an urgent need for conservation, as many orchids are now already or nearly extinct.

9.2 Geographical Distribution of Tropical Orchids

Although the geographical distributions of orchids are almost around the world, the greatest diversity is found in the tropics. Each continent has different orchids and they show disjunction or transoceanic distribution in related to the dispersal, and Dressler (1981) postulated that with the geographical pattern three ages of dispersal were observed: "Old dispersal" for a transoceanic tropicals such as *Corymborchis, Vanilla, Palmorchis-Diceratostele, Epistephium-Clematepistephium* and *Tropidia*; "Middle-aged hoppers" probably represent long distance dispersal in the mid-Tertiary period such as *Bulbophyllum, Calanthe,* and *Nervillea*; finally, "Young hoppers" represent relatively recent dispersal such as *Acampe, Angraecum, Taeniophyllum.*

In Mount Kinabalu, where previous orchid explorations have been conducted and nearly 5,000 specimens were observed, the elevation distribution of orchids showed that the highest diversity of orchid (epiphytic and terrestrial) occur at around 1,500 m above sea level (Wood et al. 1993).

Orchid seeds distribute by air current; on Krakatau island, where a total eruption in 1883 wiped out all living organisms, orchids were found after 3 years even though the nearest distance to the coastline is 40 km (Arditti 1992). Both terrestrial and epiphytic orchids are later found in the island, demonstrating the ability of orchid seeds with air pockets on their surface to float a long distance.

To develop management schemes for orchid conservation, population studies are necessary. Some orchids have dormant, vegetative or generative periods, therefore the right time for these studies should be taken into consideration. A population study of *Ophrys sphegodes* Mill., a subtropical species, showed that this orchid has complex population dynamics: without recruitment of new plants, the population declines rapidly. Mortality occurred in vegetative, dormant and generative stages; therefore the population should be counted at regular intervals (Waite 1989).

Many tropical orchid species also have dormant periods; however, there has been little work on the ecology of natural population of tropical orchids as background information to be able to manage orchid populations effectively either in the wild or in captivity (Koopowitz 2001). Some information on orchids often rely on a single specimen at the herbarium; when no new specimens are found, the information is stagnate. If the original location of the orchid has changed, the orchid is recorded as lost or extinct; on occasion, the orchid is later rediscovered elsewhere (Table 9.1).

9.3 Threats to the Survival of Orchids

The orchids' specific habitat and pollinator requirements, as well as their need for specific mycorrhiza/s for the germination process, cause their existence to decline faster compared to other plants naturally. Even the effect of climate change has been shown to cause a considerable impact on the survival of orchids but each species of orchid has a strong adaptation character; they adapt to the extreme environment very well by having pseudobulb for water storage, velamen of the roots, thick leaves cuticles, and pathways of carbon fixation (Arditti 1992). However, the epiphytic orchids totally depend on their host, and many terrestrials would not survive if their host or shade plants disappear.

In general Papua New Guinea is in a better position than most other tropical Asian countries, because large-scale logging and agriculture started relatively late so Papua New Guinea still has vast amounts of untouched rainforest (as of 1991, and the low land areas are always more vulnerable than the higher elevations). The species known to be most at risk are: *Paphiopedilum violascens, Paph. glanduliferum* (uncommon and overcollection), *Dendrobium cuthbertsonii* (the prime target for commercial collectors and exported to other countries), *Dendrobium tangerinum* (populations severely depleted by collectors), *Dendrobium lasianthera* (heavily collected in a large numbers), *Dendrobium* sections Latourea, Oxyglossum and Pedilonum (high demand), *Bulbophyllum fletcherianum* and *Bulb. macrobulbon* (vulnerable due to their small scattered populations) (O'Byrne 1994). In some cases, an orchid could already be threatened even before it is widely known, as in the case of *Bulbophyllum kubahense* J.J. Verm. and A.L. Lamb, an endangered species from Borneo (Vermeulen and Lamb 2011).

A study on the dispersion of epiphytes in Mexico showed that the diversity of orchids is negatively affected by forest disturbances or conversion to secondary vegetation, and *Lycaste aromatic* completely disappeared after forest disturbances; the epiphytes tend to attached at lower positions on the trees (Hietz et al. 2006). For terrestrial orchids in subtropical Calestienne region, Belgium, population studies were conducted to learn the effect of forest management on genetic diversity and

Table 9.1 Status of some orchid species

Species	Distribution	Status	Reference
Arachnorchis pumila	Australia	Extinct	http://www.orchidconservationcoalition.org/pr/apumila.html (accessed: 18 June 2011)
Arethusa bulbosa	New York	Rare	http://nyflora.wordpress.com/2010/06/10/dragons-mouth-orchid-rediscovered-on-long-island/
Bulbophyllum singaporeanum	Singapore	Extinct after more than 100 years in Singapore and rediscovered on 17 February 2009	Yam et al. 2010b
Caladenia branchysca	Tasmania, Australia	Extinct	Sullivan 2011
Caladenia pumila	Victoria, Australia	Two individuals in 2009, recently rediscovered after not been seen since 1926	Sullivan 2011
Coelogyne rochussenii	Himalaya to SE Asia	Extinct in Singapore	Lok et al. 2011
Corunastylis superb	Australia	40 plants left in single site by the side of a road	Sullivan 2011
Cymbidium bicolor spp. *pubescens*	Singapore	Only one specimen known naturally, rediscovered 10 years ago, thought to be extinct	Yam et al. 2010a
Cymbidium hartinahianum		Last location burned, only few seedlings left in laboratory	Unpublished report
Cymbidium rectum	Sarawak?	Extinct	http://www.ionopsis.com/fromseeds.htm (accessed, 18 June 2011)
Dendrobium bioculare	(Papua +) New Guinea	Discovered in 1904 by J. J. Smith in Papua New Guinea	http://www.oocities.org/~marylois/arch209.html#1 (accessed, 18 June 2011)
Galeola nudifolia	Singapore	Believed to be extinct, rediscovered in May 2007	Yam et al. 2010a

structure of tuberous orchid, *Orchis mascula*. Genetic erosion was observed and the research shows that populations that have recently gone through a demographic bottleneck may harbor lower genetic diversity. The study suggested regular coppicing of forest to increase the flowering and fruit and seed set (Jacquemyn et al. 2009). A similar study in South Korea of *Amitostigma gracile* showed genetic drift as a consequence of a small effective population size, coupled with a limited gene flow, are major factors leading to extremely low levels of genetic variation (Chung and Park 2008).

A Chinese study showed that habitat fragmentation of *Paphiopedilum micranthum*, whose distribution is restricted to the karst limestone hills of southwestern China, caused genetic drift and limited gene flow among populations of this predominantly insect-pollinated species (Li et al. 2002). Similarly, the epiphytic *Dendrobium officinale* also has a low genetic diversity at population level but high at species level, demonstrating the need for habitat protection (Li et al. 2008); limited migration between the three subpopulations tested and a significant degree of genetic differentiation between populations suggested that gene flow has been disturbed as a result of habitat fragmentation (Rodrigues and Kumar 2009). For ex situ conservation of these species, diversity of the species should be taken into consideration. For endemic orchid species, complete biological assessment is important in establishing in situ conservation area (Funk et al. 1999)

Nervilia nipponica is a terrestrial orchid which is critically endangered because its solitaire flower does not attract pollinators. Pollination biology reveal that position of the stigma overlaps the clinandrium, friable pollinia and the absence of a rostelum ensure efficient no-mechanical autopollination, but inbreeding depression may be limiting fitness at subsequent stages (Gale 2007). Another pollination study on *Disa pulchra* proved that by adding artificial nectar to the spur significantly increased the number (2.6-fold) of flowers probed by flies, the time spent on a flower (5.4-fold). The number of pollinia removed per inflorescence (4.8-fold) and the proportion of removed pollen involved in self pollination (3.5-fold). This is another way to reduce inbreeding depression during seed production by artificial cross-pollination (Jersáková and Johnson 2006). Pollen ecological study on *Spiranthes romanzoffiana* an endangered species and *Spiranthes spiralis*, a cosmopolitan species, demonstrated that increases both conspecific and heterospecific coflowering density may ameliorate the negative effects of rarity on pollination and consequently to the overall reproductive success. The role of insect pollinator on conspecific and heterospecific orchids were important, especially for orchid growing in fragmented habitat was shown here (Duffy & Stout 2011).

An unusual study on endangered edible orchids were conducted in Tanzania where wild orchid biodiversity hotspots were found. The study found 42 vernacular names of gathered orchid species which correspond to seven botanical species belonging to genera *Disa, Satyrium, Habenara, Eulophia and Roeperocharis*. Additionally, 97% of HIV/AIDS affected households stated that orchid gathering is their primary economic activity, compared to non-HIV/AIDS affected households at 9.7%. This phenomenon shows that orchid gathering is an easy source of food and income for HIV/

Table 9.2 In vitro culture for propagation of endangered orchid species

Species	Distribution	Medium		Initial culture	References
		Basal	Additives		
Anoectochilus formosanus	Vietnam	MS, KC, Hyponex	BA, TDZ, Kinetin	Shoot tip	Ket et al. (2004)
Calanthe sieboldii	Korea	MS; MSH (MS inorganic + Schenk & Hildebrant organic salts) Hyponex	Putrescine and Adenine sulfate	Seed and encapsulation seed	Park et al. (2000)
Calanthe tricarinata	India, Nepal China, Japan	New Dogashima and ½ MS, MS	BA, NAA	Seed	Godo et al. (2010)
Dendrobium candidum	SE Asia, New Guinea, Australia	MS	BA + NAA	Mature capsule	Yin and Hong (2009)
Encyclia mariae	Mexico	MS	IAA, IBA, NAA	Seed, leaves, protocorm	Díaz and Álvarez (2009)
Geodorum densiflorum	India	MS	BA	Rhizome from germinated seeds	Sheelavantmath et al. (2000)
Habenaria macroceratitis	Florida, Mexico, the West Indies, Central America	OMA	Fungal inoculum	Seed from mature capsule	Stewart and Kane (2007)
Ipsea malabarica	India and Sri Lanka	½ MS	BA, Kinetin, IBA, IAA, NAA	Rhizome	Martin (2003)
Laelia albida	Mexico	KC	BAP, NAA	Seed	Santoz-Hernandez et al. (2005)
Paphiopedilum armeniacum	Yunnan, China	MS	NAA + BA	Seed	Long et al. (2010)
Paphiopedilum bellatulum	Yunnan, China	MS	NAA + BA	Seed	Long et al. (2010)
Paphiopedilum insigne	Yunnan, China	MS	NAA + BA	Seed	Long et al. (2010)
Paphiopedilum villosum var. *densisimum*	Yunnan, China	MS	NAA + BA	Seed	Long et al. (2010)
Phalaenopsis celebensis	Celebes, Indonesia	MS and Hyponex	BA	Flower stalk	Handini and Mursidawati (2008)
Vanda spathulata	India and Sri Lanka	Mitra	BA + IAA	Node of mature plant	Decruse et al. (2003)
V. aphylla, V. pilifera, V. andamanica and *V. wightiana*	SE Asia and India	MS	BA + IBA	Encapsulating regenerated shoot buds, slow growth and micropropagation	Divakaran et al. (2006)

Key: MS Murashige and Skoog, MSH Murashige and Skoog inorganic + Schenk & Hildebrandt organic salts, OMA oat meal agar

AIDS affected households and this shows that harvesting some orchid species as food sources is another threats to orchid species (Challe and Price 2009).

Another example of human intervention is found in a study on the pollinators of *Bulbophyllum patens,* the fruit-fly *Bactrocea species (B. carambolae, B. papaya and B. cucurbitae).* The study found an important problem because these pollinators, which are attracted to the floral fragrance of *Bulb. patens,* are also pests that infest tropical and subtropical fruit species. As a result, the use of attractant to manage these fruit flies could threaten the existence of *Bulb. patens* (Tan and Nishida 2000).

The increasing rate of biodiversity loss has stimulated the development of different conservation strategies, the most effective of which is to protect the natural habitats of orchids (Tsiftsis et al. 2009). However, when their habitats have deteriorated, another approach i.e. ex situ conservation has to be conducted. Many conservation methods have been developed recently, including the application of in vitro culture (Table 9.2).

Limitations in time, funds, and number of capable personnel are the main obstacles in orchid conservation, especially in the tropics where the natural forest is lost faster that the ability to save the many endangered orchid species. Therefore, priority has to be established on which of the species need to be conserved first.

A review on terrestrial orchids in West Australia which also believed occurs in the tropics suggested an integrated approaches including seed storage, propagation and translocation. The role of mycorrhiza as well as pollinators were also elaborated to support the in situ conservation program (Swarts & Dixon 2009b).

The Red Data Book for vascular plants by International Union for Conservation of Nature (IUCN) describes the present status of plant species. Generally, more plant listed in due time and the status mostly declining especially cause by habitat destruction. The conservation status of particular species often changing from time to time.

9.4 National and International Cooperation to Prevent the Loss of Species

Many aspects are involved in orchid conservation; therefore, it is impossible to act alone to save all of the orchid species worldwide. A great concern of losing plant diversity as the basis of human life, and the urgency to save the plants on our planet, encouraged the Convention on Biological Diversity as well as several other organizations to formulate a strategy on plant conservation. In the sixth Conference of the Parties of the Convention on Biological Diversity in the Hague, the Global Strategy for Plant Conservation was adopted with 16 outcome-oriented global targets (CBD 2002).

Following an evaluation of the adoption of this strategy, in 2010 the Conference of the Parties adopted the Updated Global Strategy for Plant Conservation 2011–2020. The Strategy's vision is "to stop the continuing loss of plant diversity and to secure a positive, sustainable future where human activities support the diversity of plant life, and in turn the diversity of plants support and improve human livelihoods". The global targets should be viewed as a "flexible framework" for nations and/or regions to develop their own targets based on national priorities and capacities, while also taking differences in plant diversity between countries into account.

The Convention on International Trade in Endangered Species of wild Fauna and Flora (CITES) is an international agreement between governments to ensure that international trade in specimens of wild animals and plants does not threaten their survival. With 175 party members, CITES is among the largest conservation agreement in the world.

CITES works by subjecting international trade in specimens of selected species to certain control. All import, export and re-export of species covered by the convention have to be authorized through a licensing system. The species covered by CITES are listed in three Appendices: Appendix I includes species threatened with extinction: trade of specimens of this species is permitted only in exceptional circumstances. Appendix II includes species not necessarily threatened with extinction, but in which trade must be controlled in order to avoid utilization incompatible with their survival. Appendix III contains species that are protected in at least one country, which has asked other CITES parties for assistance in controlling the trade.

Currently two genera of orchid: *Paphiopedilum* and *Phragmipedium* and six species of orchids (*Aerangis ellisii*, *Dendrobium cruentum*, *Laelia jongheana*, *Laelia lobata*, *Peristeria elata*, *Renanthera imschootiana*) are found in Appendix I of the CITES list, which represents more than 240 species and more than 180 varieties or subspecies and 30 natural hybrids. All orchids not listed in Appendix I (the rest of the Orchidaceae family, with 20,000–30,000 species) are listed in Appendix II, making orchids the largest family affected by CITES.

There is a controversy regarding the role of CITES on the conservation of orchids. CITES have set the target: "No plant or animal extinct by the international trade". The problem is that the vast Orchidaceae family, which has more than a 100,000 hybrids, largely depend on their flowers for identification, while most of the time orchid species are traded without flower. This difficulty in identifying the target means that although the target (orchid species) is actually less than 10% (not all species are on international trade), it greatly affects the orchid family. Following a long discussion and many meetings, it was decided that artificially propagated orchids (seedlings from in vitro culture as well as propagated orchids in the nursery) and horticultural commodity of orchid hybrids were exempted from CITES. Additionally, each country is allowed to apply stricter regulation of CITES, and many countries have implemented their own regulations to protect their plant species including orchids to guarantee their sustainability in the future.

Orchid societies are found around the world, and most of them are independent with membership from different backgrounds bound together by their interest in the orchid world. Many members of the orchid societies are self educated in the morphology and taxonomy of orchids as well as orchid care, and some of them have even developed their interest into businesses. These "orchid people" play important roles in the world of orchid conservation and hybridization, because they get direct benefit from the orchids; with strong networking and modern communication, orchid communities can also inform each other about new species, hybrids, cultivation and conservation efforts easily.

9.5 In Situ Conservation

The most ideal and economical way to conserve orchids is to conserve them in their original habitat. When the human population and the natural forest are in a good balance, human activities within the forest will not cause a significant effect on the existence of plant species. However, the population explosion indirectly affects the existence of plants due to increased needs for food, shelter, and jobs, which is fulfilled through the expansion of land development for agriculture, housing, roads, and industrial areas. Therefore the boundary of the conservation site need to be declared as a nature reserve or protection area. Many protection areas are animal conservation areas, especially for charismatic species; few protection areas are created for plants, especially orchids. Fortunately, orchids have a wide distribution, and many species are present within the legal boundaries of protected areas or nature reserves.

The deforestation occurring in many parts of the world partly serves the need of people for food and settlements; however, most deforestation is done to enable the harvest of timber, as well as clearing land for mining activities. These activities, which are usually done over a wide area, only targeted on timber and ore; however, the damage caused affects the whole ecosystem, including the land, water and other living organisms, including orchids. Extinction of an orchid species is hastened by changes to their environment such as the climate change, human activities, invasive alien species or competition.

Ecology and population studies are often conducted to manage a particular orchid in nature, especially for terrestrial orchids (Kindlmann et al. 2006; Wells and Willems 1991). Many orchids are endemic and unique, particularly when the growing area was isolated for long periods; these orchids are face a greater threat of extinction. For most threatened species, in situ conservation alone is not enough, and different method of ex-situ conservation is needed as backup on conservation efforts.

A study of *Lepanthes rupestris*, which has epiphytic and lithophytic habitats, revealed that in the lithophytic environment there is a significant tendency for non-reproductive stages of the orchid to increase along with the increase in the density of reproductive individuals. For the epiphytic environment, positive correlations with adult plant density were only observed for plants in the juvenile stage; this suggest that the survival rate of seedlings is much higher in the epiphytic environment than in the lithophytics (Gomez et al. 2006). The response of orchid to natural disturbance was studied in oak forest of 'El Tepozteco', Mexico showed that the dead standing trees in the forest (old forest) was positively related to abundance of genus *Malaxis* (Cruz-Fernández et al. 2010).

Population genetic study on *Dendrobium officinale* revealed that genetic population was higher among population (72.95%) and only 27.05% was between populations. The result of this study recommends that habitat protection as well as ex situ conservation should be set up to ban over exploitation (Ding et al. 2008). Another genetic diversity studies on *Changnienia amoena* and *Cypripedium macranthos* var. *rebunense* were used as a strong background in future conservation recommendation (Li and Ge 2006; Izawa et al. 2007). Similar study has been conducted on *Liparis loeselii* an endangered orchid from England, therefore populations of dune slacks and the orchid should be managed separately (Pillon et al. 2007)

The limited availability of resources (including manpower, funds, and time) especially in the tropics where the most threatened orchids are found means that the determination of the conservation area has to be conducted carefully, as to capture the rarest and most threatened species. In many tropical regions knowledge on orchid conservation is needed especially in hot spot areas that spread in all countries. According to Tsiftsis et al. (2009), multi-scale estimation of rarity to calculate the conservation values, species specialization for assessing taxa conservation, species richness or conservation value, safeguards in all regions, and richness of the rarest species are needed in the evaluation.

Reintroduction of orchid species is an effort to bring back the propagated selected species for recovery in their original habitat. The main reason for this method is to promote in situ conservation, as well as to act as a bridge between ex situ and in situ conservation. This program may seem expensive initially, but not over the long time frame. Orchid reintroduction in the tropics using seedlings from flask or nursery-established seedlings have been successfully done for *Laelia cinnabarina, L. pumila* and *L. crispa* as adult plants after recollection from the neighbouring areas in Brazil (Warren 2001). *Paphiopedilum rothschildianum* and *Paphiopedilum sanderianum*, two famous endangered slipper orchids have also been reintroduced at Mt. Kinabalu (Grell et al. 1988) and *Ipsea malabarica* an endangered species has been reintroduced back to Vellarimala, India (Martin 2003). Establishment of *Vanda spathulata* seedlings on tree trunk were considered as an efficient method for ecorestoration (Decruse et al. 2003). In vitro symbiotic seed germination has become a favored and useful methodology for orchid seed propagation and for plant reintroduction. In *Pecteilis susanae*, symbiotic seed germination was very high and at the same time also conserving the fungal isolates (Chutima et al. 2011). Singapore Botanic Gardens has been conserving their native orchids through seeding culture or seed bank and carried out experiments of reintroduction in five species of more than 3,000 plants i.e. *Grammatophyllum speciosum, Bulbophyllum vaginatum, Bulbophyllum membranaceum, Cymbidium finlaysonianum* and *Cymbidium bicolor*, after 8 years the survival percentages ranged from 10% to 95% for *Grammatophyllum speciosum.* Other species has been successfully reintroduced are *Bulbophyllum membranaceum, Cymbidium finlaysonianum, Cymbidium bicolor* spp. *pubescens, Robiquetia spathulata* (Yam et al. 2010a). Detail location of reintroduction area often facing a risk for re-exploitation similar to the detail habitat of the valuable orchids.

There are controversial opinions on artificial pollination, by assisting pollination or orchid flower in the wild as this is considered as unnatural, but others consider this activity would help when the pollinator is limited (Koopowitz 2001).

A management plan is important to sustaining the management of an in situ conservation area. As a written document, the master plan should refer to the currently existing regulations, and the involvement of stakeholders is important in the formulating this plan. According to the Strategic Plan for Biodiversity 2011–2020, in situ conservation is found in Target 7, and at least 75% of known threatened plant species should be conserved in situ.

9.6 Ex Situ Conservation

The goal of ex situ orchid conservation is to save the orchid species as it disappears in situ. As orchids become extinct in their natural habitats, ex situ takes on a greater role in orchid conservation. Ex situ conservation usually takes the form of a living collection in a collection garden such as in botanic gardens, as well as university, park, and privately owned orchid collections. A program of the National Council for the Conservation of Plants and Gardens in the United Kingdom is "National Plant Collection", which aims to make a complete representation of a genus or section of a genus (Sullivan 2011).

The introduction of asymbiotic seed germination method by Knudson (1946) and the shoot tip culture by Morel (1960) are the most important tools for orchid propagation. These methods, also known as in vitro culture, were later developed further and often specified according to the material used for initiating the culture such as embryo culture, immature embryo culture, tissue culture, organ culture, leaf culture, bud culture, axillary bud culture, flower bud culture, stem culture, rhizome culture, petiole culture, cell culture, pollen culture, callus culture, protoplast culture etc. In vitro culture as a propagation method is an ideal method for propagation of endangered species of plants, including orchids.

An orchid seed bank, as suggested a decade ago by Koopowitz (2001) and other plant conservationists became a prominent program of the Kew gardens since the year 2000, and known as Millennium Seed. Kew collaborates with orchid laboratory staff in many countries. In the "Millennium Seed" program, studies on the longevity of orchid seeds, germination media, and other information is conducted in the home country for the orchid database. The Orchid Seed Stores for Sustainable Use (OSSSU) is a Darwin Initiative project designed to establish orchid seed banks around the globe using conventional seed banking techniques. OSSSU initially aimed to collect and store seeds of at least 250 species, focusing on orchid hot spots in the Asian and the Central South American regions, representing the orchid floras in 16 participating countries: China, India, Indonesia, the Philippines, Singapore, Thailand and Vietnam, Bolivia, Brazil, Chile, Colombia, Costa Rica, Cuba, Ecuador, Guatemala and Mexico. The target is conserving 1,000 orchid species in storage by 2015 with 30 countries involved (Kew's Millennium Seed Bank, accessed 20 June, 2011).

Unfortunately, orchid seeds are non-orthodox seeds which are dependent on humidity and temperature; seeds would have their longevities doubled when the temperature is decreased to −5°C. So far for some orchids after 3 years of storage, some seeds are still viable. Many orchids produce flower and fruit so rarely that vegetative propagation is the only method to propagate them.

The CITES legislation and national laws in restricting trade of species originated from the wild has led many nursery propagating the species in vitro, such as Ecuagenera in Southern Ecuador (Christenson 2004)

There are several stages in ex situ conservation: first is to identify the species target, then to study their distribution and ecological condition of the particular species to give the best method of conservation, and finally to study the best cultivation method for propagation and reintroduction whenever possible.

Mycorrhiza, the symbiont which has a very important role in orchid seed germination in nature was neglected when the asymbiotic germination method was introduced. Later when problem in germination of some orchids were found then the role of mycorrhiza was explored again. Symbiotic germination is considered more natural than the asymbiotic ones, especially for the terrestrial species of orchids and for the establishment of seedlings.

The Kew gardens, in collaboration with English Nature and Wildlife trusts, have succeeded in planting *Liparis loeselii* and *Cypripedium calceolus* in public areas, nature reserves and protected areas as part of their Species Recovery Programme. The role of botanic gardens and private orchid collections in conserving orchid species is very important, as many orchid species are in their good care and these are sometimes propagated and distributed. There are cases where some species thought to be extinct was later rediscovered in a nursery. Botanic Gardens have an important role in orchid conservation (Swarts and Dixon 2009b) such as in Korea in establishing both in situ and ex situ conservation (Kim 2006), Bogor Botanic Gardens also had been known as a sanctuary for many Indonesian orchid species and recently a 100 species have been successfully propagated in vitro from seeds, among them are rare, endangered and protected orchid species (Mursidawati and Handini 2008). Propagation of orchid species through seeds as a result of self pollination causes a probability of inbreeding depression. For species propagation, it is suggested to pollinate from the same species, but different specimen. In *Serapias*, a Mediterranean species, different level of inbreeding depression as a result of outcrossing and selfing were observed (Bellusci et al. 2009), while self incompatibility and myophily were found in *Octomeria* from Brazil (Barbosa et al. 2009). Different strategies are required for the conservation of these orchids.

Inbreeding depression is currently considered as one of the most important threats to the persistence of rare taxa, while self incompatibility and inbreeding depression apparently are the factors responsible for maintaining high levels of genetic variability (Bellusci et al. 2009; Barbosa et al. 2009). To minimize the problem of narrow genetic diversity in ex situ conservation, Koopowitz (2001) suggested that a minimum amount of 50 randomly selected specimens is collected for each species; this is not an easy task to accomplish, especially in third world countries with a great diversity in orchids.

Efforts to conserve orchids are supported by many scientific studies describing distribution patterns, pollinator relationships, mycorrhizal fungi dependency, and even orchid taxonomy to describe the genetic relationships among the species; these studies are very useful in saving the orchid species (Brundrett 2007; Horak 2010). Applications of mycorrhizal association in ex situ conservation was proved by Rasmussen (2002) and she suggested of encapsulated orchid seeds with suitable inoculums to secure an initial compatible symbiosis for commercial production. Mycorrhizal research proved that after germination germination, orchid mycorrhizas have a very important role in carbon, water and other inorganic and organic nutrients transport and even more in adult photosynthetic orchids. Therefore orchid population and orchid mycorrhizal conservation would maintain the global biodiversity (Dearnaley 2007).

Engelmann (2010) suggested that the use of biotechnologies for conserving plant biodiversity, including orchids, in a gene bank would be very useful. The use of some techniques such as in vitro collection of unripe fruit and shoots for vegetatively propagated species are also useful, especially during long periods of travel in remote areas. Slow growth storage techniques are also being routinely used for medium-term conservation of many species. The classical cryopreservation technique for long term storage of plants involve a slow cooling down to a defined prefreezing temperature, followed by rapid immersion in liquid nitrogen; recently, new cryopreservation techniques using cryoprotective media and/or air desiccation follow by rapid cooling are also used.

In vitro cultures are the most popular methods used in orchid conservation, especially when conventional methods cannot be conducted. The in vitro methods are relatively more expensive, but they may be necessary for species under higher threat (Pence 2010). Many orchid species has been cultured in vitro for conservation purposes (Table 9.2).

Identification of all orchid species, the database of orchids, identify the true species using DNA and extensive hybridization has been done in orchids; therefore, the challenge is to conserve the genetic purity of the species. There is a great risk of "genomic extinction through hybridization" (Sullivan 2011).

Similar to in situ conservation, priority should also be set in choosing the species to be conserved as to optimize the available resources. With the immense species diversity in the tropics, many aspects should be taken into consideration to include as many threatened species as possible while maintaining a good probability of success. The Bogor Botanic Gardens has already published the priority setting for Indonesian orchids, and 44 out of an estimated 5,000 species were selected as the most important species for conservation. These species are: *Ascocentrum aureum, Arachnis hookeriana, Bulbophyllum phalaenopsis, Cymbidium hartinahianum, Dendrobium ayubii, Den. capra, Den. devosianum, Den. jacobsonii, Den. laxiflorum, Den. militare, Den. nindii, Den. pseudoconanthum, Den. taurulinum, Den tobaense, Paphiopedilum gigantifolium, Paph. glaucophyllum, Paph. kolopakingii, Paph. mastersianum, Paph. moquettianum, Paph. niveum, Paph. primulinum, Paph. sangii, Paph. schoseri, Paph. supardii, Paph. victoria-mariae, Paph. victoria-regina, Paph. violascens, Papilionanthe tricuspidata, Paraphalaenopsis denevei, Paraphal. labukensis, Paraphal. laycockii, Paraphal. serpentilingua, Phalaenopsis celebensis, Phal. floresensis, Phal. gigantea, Phal. inscriptiosinensis, Phal. javanica, Phal.*

modesta, Phal. tetraspis, Phal. venosa, Phal. viridis, Vanda devoogtii, V. jennae and *V. sumatrana* (Risna et al. 2010)

Electronic communication plays an important role in distributing information and also give substantial support in orchid conservation. Horak (2010) suggested that collaborative ex situ activities would have a greater impact in orchid conservation. The Strategic Plan for Biodiversity 2011–2020 targeted at least 75% of threatened plant species in ex situ collections, preferably in the country of origin, and at least 20% available for recovery and restoration programmes.

9.7 Present and Future Prospects of Orchid Conservation

In 1992, Koopowitz predicted that if 45% of the tropical forests were to be cut, approximately 22% of all orchid species will be lost. On the other hand, without thorough exploration it is difficult to say whether a species is extinct or endemic. Studies on the Mediterranean orchid species *Ophrys* by Devey et al. (2008) have been able to recognize the species better than the conventional method, which result in better decisions on their conservation.

Rediscovery of species of orchids is a progress on evaluation of a species therefore, the endangered or endemic status of a species has to be updated from time to time. However, precautionary measures should be done before an orchid species is really extinct.

Research is needed in many fields including orchid ecology, population studies, genetics, studies about epiphytes, studies on the life cycle to know which part of the orchid's life cycle is the most vulnerable.

To conserve orchids, different methods of conservation are needed, including living collection, seed collection, and tissue collection. Habitat conservation is important especially for pollinator dependence orchids. Conservation strategies should be taken into consideration especially for high diversity species of small populations as a result of habitat fragmentation. Recently different aspect of biotechnology was applied on many plants including orchids and became a basic recommendation on orchid conservation.

Due to limited resources, it is impossible to conserve all area (in situ) or all species threatened as a result of natural and artificial threats (caused by people) should be taken into consideration in choosing the priority of orchid conservation.

The main problem with tropical countries is that the habitat of the greater part of orchids in the world is the lack of basic information even just to recognize orchids among other plants. Information about orchid species usually only found in a very limited institutes in contrast with the developed countries with suitable support for conducting basic and applied research on orchid. Therefore, more people dedicated to orchid research and good facilities are urgently needed to overcome problems in orchid conservation in the tropics.

Acknowledgments I would like to thank Krisna and Lana for reading this article.

References

Arditti J (1992) Fundamental of orchid biology. Wiley, New York

Barbosa AR, de Melo MC, Borba EL (2009) Self-incompatibility and myophily in *Octomeria* (Orchidaceae, Pleurothallidinae) species. Plant Syst Evol 283:1–8

Bellusci F, Pellegrino G, Musacchio A (2009) Different levels of inbreeding depression between outcrossing and selfing *Serapias* species. Biol Plant 53(1):175–178

Brundrett MC (2007) Scientific approaches to Australian temperate terrestrial orchid conservation. Aust J Bot 55(3):293–307

CBD (2002) Global strategy for plant conservation. The Secretariat of the Convention on Biological Diversity, Montreal

Challe JFX, Price LL (2009) Endangered edible orchids and vulnerable gatherers in the context of HIV/AIDS in the Southern Highlands of Tanzania. J Ethnobiol Ethnomed 5:41 (Available from: http://www.ethnobiomed.com/content/5/1/41)

Christenson EA (2004) Conservation in action: Ecuagenera. Orchid Dig 68(2):114–117

Chung MY, Park CW (2008) Fixation of alleles and depleted levels of genetic variation within populations of the endangered lithophytic orchid *Amitostigma gracile* (Orchidaceae) in South Korea: implications for conservation. Plant Syst Evol 272:119–130

Chutima R, Dell B, Vessabutr S, Bussaban B, Lumyong S (2011) Endophytic fungi from *Pecteilis susannae* (L.) Rafin (Orchidaceae), a threatened terrestrial orchid in Thailand. Mycorrhiza 21:221–229

Cruz-Fernández QT, Alquicira-Artega ML, Flores-Palacios A (2010) Is orchid species richness and abundance related to the conservation status of oak forest? Plant Ecol. doi:10.1007/s11258-010-9889-4

Dearnaley JDW (2007) Further advances in orchid mycorrhizal research. Mycorrhiza 17:475–486. doi:10.1007/s00572-007-0138-1

Decruse SW, Gangaprasad A, Seeni S, Menon VS (2003) Micropropagation and ecorestoration of *Vanda spathulata*, an exquisite orchid. Plant Cell Tiss Org Cult 72:199–202

Devey DS, Bateman RM, Fay MF, Hawkins JA (2008) Friends or relatives: phylogenetics and species delimitation in the controversial European orchid genus *Ophrys*. Ann Bot 101(3):385–402

Díaz MSS, Álvarez CC (2009) Plant regeneration through direct shoot formation from leaf cultures and from protocorm-like bodies derived from callus of *Encyclia mariae* (Orchidaceae), a threatened Mexican orchid. In Vitro Cell Dev Biol Plant 45:162–170. doi:10.1007/s1627-009-9201-2

Ding G, Zhang D, Ding X, Zou Q, Zhang W, Li X (2008) Genetic variation and conservation of the endangered Chinese endemic herb *Dendrobium officinale* based on SRAP analysis. Plant Syst Evol 276:149–156. doi:10.1007/s00606-008-0068-1

Divakaran M, Babu KN, Peter KV (2006) Conservation of *Vanilla* species, in vitro. Sci Hortic 110(2):175–180

Dressler RL (1981) The orchids natural history and classification. Harvard University Press, Cambridge

Duffy KJ, Stout JC (2011) Effects of conspecific and heterospecific floral density on the pollination of two related rewarding orchids. Plant Ecol. doi:10.1007/s11258-011-9915-1

Engelmann F (2010) Use of biotechnologies for the conservation of plant biodiversity. In Vitro Cell Dev Biol Plant. doi:10.1007/s11627-010-9327-2

Funk VA, Zermoglio MF, Nasir N (1999) Testing the use of specimen collection data and GIS in biodiversity exploration and conservation decision making in Guyana. Biodivers Conserv 8:727–751

Gale S (2007) Autogamous seed set in a critically endangered orchid in Japan: pollination studies for the conservation of *Nervilia nipponica*. Plant Syst Evol 268:59–73

Godo T, Komori M, Nakaoki E, Yukawa T, Miyoshi K (2010) Germination of mature seeds of *Calanthe tricarinata* Lindl., an endangered terrestrial orchid, by asymbiotic culture in vitro. In Vitro Cel Dev Biol Plant. doi:10.1007/s11627-009-9271-1

Goméz NR, Tremblay RL, Meléndez-Ackerman E (2006) Distribution of life cycle stages in a lithophytic and epiphytic orchid. Folia Geobot 41:107–120

Grell E, Haas-von Schmude NF, Lamb A, Bacon A (1988) Reintroducing *Paphiopedilum rotschildianum* to Sabah, North Borneo. Amer Orchid Soc Bull 57:1238–1246

Handini E, Mursidawati S (2008) Inisiasi kultur tangkai bunga anggrek langka *Phalaenopsis celebensis* Sweet (Orchidaceae) secara in vitro. Warta Kebun Raya 8(1):46–51

Hietz P, Buchberger G, Winkler M (2006) Effect of forest disturbance on abundance and distribution of epiphytic bromeliads and orchids. Ecotropica 12:103–112

Horak D (2010) Orchid conservation, creating effective strategies in modern times. Orchids 79(1):38–43

Izawa T, Kawahara T, Takahashi H (2007) Genetic diversity of an endangered plant, *Cypripedium macranthos* var. *rebunense* (Orchidaceae): background genetic research for future conservation. Conserv Genet 8:1369–1376

Jacquemyn H, Brys R, Adriaens D, Honnay O, Rodán-Ruiz I (2009) Effects of population size and forest management on genetic diversity and structure of the tuberous orchid *Orchis mascula*. Conserv. Genet. 10:161–168

Jersáková J, Johnson SD (2006) Lack of floral nectar reduces self-pollination in a fly-pollinated orchid. Oecologia 147:60–68

Johnson SD, Brown M (2004) Transfer of pollinaria on birds' feet: a new pollination system in orchids. Plant Syst Evol 244:181–188. doi:10.1007/s00606-003-0106-y

Keel BG (2005) Climate change and assisted migration of at-risk orchids. Orchid Conserv News 6:9–10

Ket NV, Hahn EJ, Park SY, Chakrabarty D, Paek KY (2004) Micropropagation of an endangered orchid *Anoectochilus formosanus*. Biol Plant 48(3):339–344

Kim YS (2006) Conservation of plant diversity in Korea. Landscape Ecol Eng 2:163–170. doi:10.1007/s11355-006-0004-x

Kindlmann P, Kull T, Whigham D, Willems J (2006) Ecology and population dynamics of terrestrial orchids: an introduction. Folia Geobot 41(1):1–2

Knudson L (1946) A nutrient for the germination of orchid seeds. Amer Orchid Soc Bull 15:214–217

Koopowitz H (2001) Orchids and their conservation. Timber Press, Portland

Kew's Millennium Seed Bank – Orchid Seed Stores Project (2011) http://www.kew.org/science-conservation/save-seed-prosper/millennium-seed-bank/projects-partners/more-seed-projects/orchid-seed-stores/index.htm. Accessed 20 Jun 2011

Li A, Ge S (2006) Genetic variation and conservation of *Changnienia amoena*, an endangered orchid endemic to China. Plant Syst Evol 258:251–260. doi:10.1007/s00606-006-0410-04

Li A, Lou Y, Ge S (2002) A preliminary study on conservation genetics of an endangered orchid *(Paphiopedilum micranthum)* from southwestern China. Biochem Genet 40(5/6):195–201

Li X, Ding X, Chu B, Zhou Q, Ding G, Gu S (2008) Genetic diversity analysis and conservation of the endangered Chinese endemic herb *Dendrobium officinale* Kimura *et* Migo (Orchidaceae) based on AFLP. Genetica 133:159–166

Lok AFSL, Ang WF, Chong KY, Yeo CK, Tan HTW (2011) Rediscovery in Singapore of *Coelogyne rochussenii de* Vriese (Orchidaceae) Nat Singapore 4:49–53

Long B, Niemiera AX, Cheng Z (2010) In vitro propagation of four threatened *Paphiopedilum* species (Orchidaceae). Plant Cell Tiss Org Cult 101:151–162. doi:10.1007/s11240-010-9672-1

Martin KP (2003) Clonal propagation, encapsulation and reintroduction of *Ipsea malabarica* (Reichb.f.) J.D. Hook., an endangered orchid. In vitro Cell Dev Biol Plant 39:322–326. doi:10-1079/IVP1002399

Morel GM (1960) Producing virus free *Cymbidium*. Amer Orchid Soc Bull 29:495–497

Mursidawati S, Handini E (2008) Perkecambahan seratus jenis anggrek alam koleksi Kebun Raya Bogor secara in vitro. Warta Kebun Raya 8(1):40–45

O'Byrne P (1994) Lowland orchids of Papua New Guinea. National Parks Board Singapore Botanic Gardens, Singapore

Park SY, Murthy HN, Paek KY (2000) In vitro seed germination of *Calanthe sieboldii*, an endangered orchid species. J Plant Biol 43(3):158–161

Pence VC (2010) Evaluating costs for the in vitro propagation and preservation of endangered plants. In Vitro Cell Dev Biol Plant (Published online: 25 Nov 2010)

Pillon Y, Qamaruz-Zaman F, Fay MF, Hendoux F, Piquot Y (2007) Genetic diversity and ecological differentiation in endangered fen orchid (*Liparis loeselii*). Conserv Genet 8:177–184. doi:10.1007/s10592-006-9160-7

Rasmussen HN (2002) Recent development in the study of orchid mycorrhiza. Plant Soil 244:149–163

Risna RA, Kusuma YWC, Widyatmoko D, Hendrian R, Pribadi DO (2010) Spesies Prioritas Untuk Konservasi Tumbuhan Indonesia, vol I, Arecaceae, Cyatheaceae, Nepenthaceae and Orchidaceae. PKT Kebun Raya Bogor – LIPI, Bogor

Rodrigues KF, Kumar SV (2009) Isolation and characterization of 24 microsatellite loci in *Paphiopedilum rothschildianum*, an endangered sloipper orchid. Conserve Genet 10:127–130

Santos-Hernandez L, Martinez-Garcia M, Campos JE Aguirre-Leon E (2005) In vitro propagation of *Laelia albida* (Orchidaceae) for conservation and ornamental purposes in Mexico. HortScience 40(2):439–442

Sheelavantmath SS, Murthy HN, Pyati AN, Kumar HGA, Ravishankar BV (2000) In vitro propagation of the endangered orchid, *Geodorum densiflorum* (Lam.) Schltr. through rhizome section culture. Plant Cell Tiss Org Cult 60:151–154

Stewart SL, Kane ME (2007) Symbiotic seed germination and evidence for in vitro mycobiont specificity in Spiranthes brevilabris (Orchidaceae) and its implications for species-level conservation. In Vitro Cell Dev Biol 43:178–186

Sullivan M (2011) The challenges of *ex situ* orchid conservation. http://www.orchidconservation-coalition.org/pr/exsitucon.html. Accessed 10 Jun 2011

Swarts ND, Dixon KW (2009a) Terrestrial orchid conservation in the age of extinction. Ann Bot 104:543–556. doi:10.1093//aob/mcp025

Swarts ND, Dixon KW (2009b) Perspectives on orchid conservation in botanic gardens. Trends Plant Sci 14(11):590–598. doi:10.1016/j.tplants.2009.07.008

Tan KH, Nishida R (2000) Mutual reproductive benefits between a wild orchid, *Bulbophyllum patens*, and *Bactrocera* fruit flies via a floral synomone. J Chem Ecol 26(2):533–546

Tsiftsis S, Tsiripidis I, Karagiannakidou V (2009) Identifying areas of high importance for orchid conservation in east Macedonia (NE Greece). Biodivers Conserv 18:1765–1780

Vermeulen JJ, Lamb A (2011) Endangered even before formally described: *Bulbophyllum kubahense* n.sp., a beautiful and assumedly narrowly endemic orchid from Borneo. Plant Syst Evol 292:51–53

Waite S (1989) Prediction population trends in *Ophrys sphegodes* Mill. In: Pritchard HW (ed) Modern methods in orchid conservation, the role of physiology, ecology and management. Cambridge University Press, Cambridge

Warren RC (2001) A private conservation project in the coastal rainforest in Brazil: the first ten years. In: Pricard H W (ed.) Modern Methods in Orchid Conservation, the Role of Physiology, Ecology and Management. 153–158

Wells TCE, Willems JH (1991) Population ecology of terrestrial orchids. SPB Academic Publishing BV, The Hague

Wood JJ, Beaman RS, Beaman JH, (1993) The plants of mount kinabalu, 2. Orchids. RBG. Kew. England

Yam TW, Chua J, Tay F, Ang P (2010a) Conservation of the native orchids through seedling culture and reintroduction – a Singapore experience. Bot Rev 76:263–274. doi:10.1007/s12229-010-9050-z

Yam TW, Leong PKF, Chew PT, Liew D, Huat WNK (2010b) The re-discovery & conservation of *Bulbophyllum singaporeanum*. GardenWise 35(July):14–17

Yin M, Hong S (2009) Cryopreservation of *Dendrobium candidum* Wall. ex Lindl. Protocorm-like bodies by encapsulation–vitrification. Plant Cell Tiss Org Cult 98:179–185

Chapter 10
Conservation of Oil Palm and Coconut Genetic Resources

N. Rajanaidu and M.M. Ainul

10.1 Introduction

The world population is expected to reach eight billion by the year 2020 and food grain production will have to be doubled from the current level of about 5 billion tonnes per year. To meet the need for more food, it will be necessary to make better use of a broader range of plant genetic diversity. Yet, genetic resources are disappearing at unprecedented rates. The reasons for this loss are many including deforestation, development and agricultural activities and introduction of new and uniform varieties (Rao 2004; Borokini et al. 2010; Jain 2011). Therefore, conservation is needed to diversify important crops for world food production.

Conservation is defined as the management of human use of the biosphere so that it may yield the greatest sustainable benefit to present generations while maintaining its potential to meet needs and aspirations of future generations. Thus, conservation embraces preservation, maintenance, sustainable utilization and restoration, and enhancement of the natural environment (Borokini et al. 2010). For conservation to succeed there is an urgent need for multidisciplinary approach which links the molecular biologist, plant physiologist, biochemist, plant pathologist, crop modelling expert, breeder and tissue culturist together in a team, working in a concerted manner towards improving multiple aspects of the crop.

Oil palm and coconut are the two major oleaginous crops cultivated in the pantropical area. Their breeding improvement by conventional methods is limited. Firstly, they are perennial crops and a long period is required in order to assess the value of a progeny. Secondly, they are allogamous and large heterogeneity is observed in hybrid planting material. Lastly, coconut is only seed propagated and there is no classical vegetative propagation technique available. For these reasons,

N. Rajanaidu (✉) • M.M. Ainul
Advanced Biotechnology and Breeding Centre, Malaysian Palm Oil Board,
6 Persiaran Institusi, Bandar Baru Bangi, Kajang 43000, Selangor, Malaysia
e-mail: rnaidu@mpob.gov.my

M.N. Normah et al. (eds.), *Conservation of Tropical Plant Species*,
DOI 10.1007/978-1-4614-3776-5_10, © Springer Science+Business Media New York 2013

in vitro propagation has been investigated for more than 30 years with the aim of propagating the best hybrid palms on a large scale (Duval et al. 1998).

10.1.1 Oil Palm

Wherever it grows naturally, oil palm has for centuries provided local communities with a large number of benefits such as palm oil, sauces, soap, wine, fertilizer using ashes, roofing with leaves, the trunk as building material and woven baskets, medicines using roots and feeding livestock with palm kernel cake (Komolafe and Joy 1990; Noll 2008; Bergert 2010). All of these traditional uses are until today very much part of the African culture in oil palm countries.

It is extremely difficult to find reliable figures on the area covered by oil palms in Africa due to a number of issues, among which the difficulty of separating forest areas containing oil palm trees as one of their components from natural palm groves where oil palms constitute the sole or main tree species; the difficulty of distinguishing between wild palms and palm groves that have been part of local communities' agricultural practices for centuries; the difficulty of classifying palm stands as family plantations or as out-grower plantations contractually linked to an industrial plantation unit; the existence of abandoned industrial plantations that are being used by local communities as if they were natural palm stands and the lack of updated inventories of natural palm stands, small scale plantations and industrial plantations. Table 10.1 provides a very broad idea on the area covered by oil palms in the 23 countries of Africa identified by World Rainforest Movement (WRM) as palm oil producers in Africa (Carrere 2010).

10.1.2 Coconut

Another important tree crop in the world is the coconut (*Cocos nucifera* L.). Coconut plays an important role in the life and welfare of millions of people in tropical countries. The world coconut production in 2007 was 61 million tonnes, with the three main producing countries, Indonesia, the Philippines and India, accounting for over 75% of the total production (Engelmann et al. 2011). In 2008, the production of coconut oil (crude and refined) in Malaysia is 42.3 thousand tonne metric [Jabatan Perangkaan Malaysia (Malaysian Department of Statistics) 2009].

The country of origin of the coconut is unknown and various regions have been indicated as such by various scientists from South America and Melanesia. Because of competition from other plants or predation by animals, coconut could not reach or survive inland on larger islands or on continents before it was domesticated. The continental coast and larger islands of Malaysia would be the obvious site for such domestication. Wild forms of coconut have been found in various countries, such as on the island of Java, Indonesia, Queensland, Australia and in the Philippines (Ohler 1999). Current theories mainly suggest that it must have originated in the Indonesian Islands

Table 10.1 Area covered by oil palms in 23 countries in Africa (Carrere 2010)

Country	Natural palm stands (ha)	Industrial plantation (ha)
Angola	–	–
Benin	300,000	20,000
Berundi	–	10,000
Cameroon	25,000	76,500
Central African Republic	18,000	1,000
Congo, R.	–	10,000
Congo, R.D.	1,000,000	147,000
Cote d'Ivoire	140,000	88,000
Equatorial Guinea	–	7,000
Gabon	–	10,000
Gambia	–	–
Ghana	–	300,000
Guinea	2,000,000	9,000
Guinea Bissau	–	–
Liberia	–	70,000
Madagascar	–	–
Nigeria	2,500,000	36,000
Sao Tome and Principe	–	–
Senegal	50,000	–
Sierra Leone	32,000	18,000
Tanzania	–	–
Togo	600,000	2,000
Uganda	–	10,000

Table 10.2 Malaysia – agricultural acreage 1970–2000, 1,000 ha (Arif and Ariff 2001)

Crops	1970	1985	1990	1995	2000
Oil palm	320	1,482	2,030	2,540	3,338
Rubber	2,182	1,949	1,837	1,679	1,590
Rice	533	655	681	673	692
Coconut	349	334	316	249	116

and later spread to become pantropical, though the date of its spread has been under considerable debate (Rajanaidu and Ramanatha Rao 2002). Wild coconut palms are not necessarily primitive or small fruited. It usually shows visible signs of growth occurring 3 months after germination (Thampan 1982). Normal coconuts have a hard and crisp endosperm at maturity but some coconuts have a soft, jelly-like endosperm, which fills the nut cavity. Such coconuts are highly priced in the ice cream and pastry industries as well as for preparing sweetened preserves (Engelmann 2011).

In Malaysia, government efforts to diversify exports to reduce the negative effects of poor terms of trade in rubber and tin focused on oil palm (Rasiah and Shahrin 2005). As a consequence, rubber plantations gave way to oil palm plantations. While agricultural land use has gradually expanded, rubber acreage has declined in absolute terms (Table 10.2). Oil palm acreage grew from 320,000 ha in 1970 to 3.3 million ha in 2000.

10.2 Collections

10.2.1 *Elaeis guineensis* Collection

10.2.1.1 *Elaeis guineensis* Collection by Malaysian Palm Oil Board (MPOB)

Oil palm commercial cultivation began in 1917 in Malaysia, but initial growth of the industry was slow. It was only during the last 50 years that plantation development was accelerated through large-scale investments in the cultivation of oil palm. It is one of the approved crops for diversifying the country's agricultural development. Malaysia is also known as a major producer of rubber, cocoa and to some extent, coconuts. Preference for oil palm has led to a rapid expansion of its planted areas at the expense of rubber and other crops over the last four decades (Yusof 2007).

Current area and production of palm in the country has increased. The planting material is derived from narrow genetic base. To broaden the genetic base, MPOB collected *Elaeis guineensis* germplasm from centres of origin of African oil palm as listed in Table 10.3. In addition, the location of all collected materials is mapped in Fig. 10.1.

10.2.1.2 *Elaeis guineensis* Collection by Agricultural Services Department (ASD), Costa Rica

The first oil palm plants in Central America were planted by ASD (formerly United Fruit Company) in 1926 and 1929, using seeds brought from Malaysia, Indonesia and Sierra Leone (Richardson 1995). From 1967 onwards, through international germplasm exchange programs, ASD consolidated one of the broadest collections of *Elaeis guineensis* in the world.

10.2.1.3 *Elaeis guineensis* Collection by Centre de Coopération Internationale en Recherche Agronomique pour le Développement (CIRAD), France

Meunier (1969) carried out a survey of 11 natural oil palm groves in Ivory Coast. The population from Yocoboué region was selected for utilization. Fourteen parents were introduced in the form of selfs and sib crosses.

The second survey conducted by CIRAD in the wild palm groves of Cameroon (Chaillard 1977). Seven populations were observed for their agronomic and morphological traits. The Widikum population was chosen and introduced into the CIRAD oil palm breeding programme.

CIRAD was able to acquire a sample of germplasm collected by MPOB and Nigerian Institute for Oil Palm Research (NIFOR) in Nigeria in 1973 (Rajanaidu

Table 10.3 List of *E. guineensis* germplasm collection available in MPOB (Kushairi et al. 2011)

Species	Country of origin	Year of collection
Elaeis guineensis	Nigeria	1973
	Cameroon	1984
	Zaire	1984
	Tanzania	1986
	Madagascar	1986
	Senegal	1993
	Gambia	1993
	Sierra Leone	1994
	Guinea	1994
	Ghana	1996
	Angola	1991, 2010

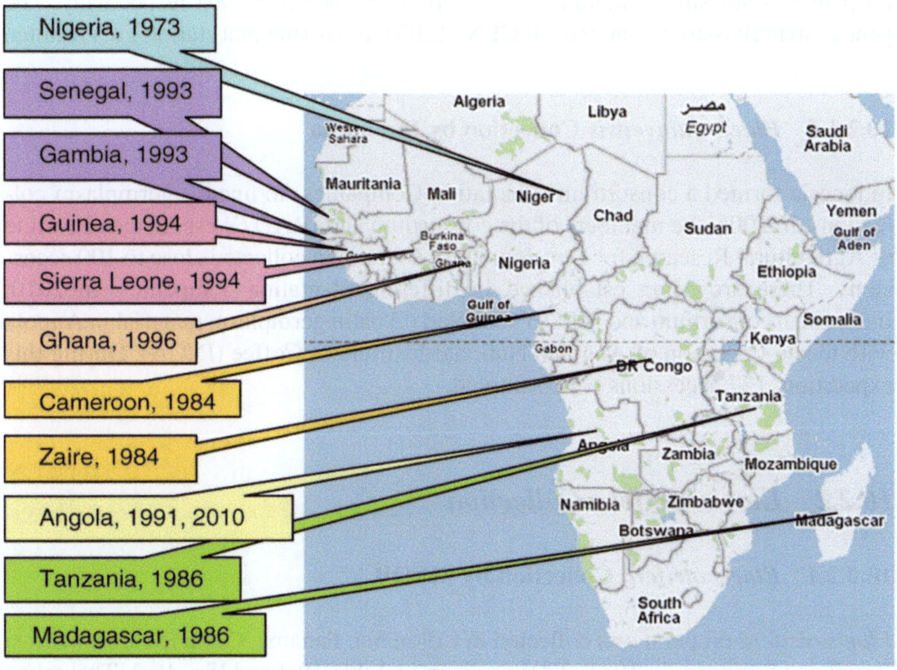

Fig. 10.1 Collection of *E. guineensis* germplasm in Africa (Kushairi et al. 2011)

1985). Out of 919 accessions, 21 accessions were introduced in the form of open-pollinated families and established in Ivory Coast.

The populations acquired through surveys, Yocoboué, Widikum and Nigeria are represented by 2,214 palms covering an area of 14.1 ha at La Mé.

Table 10.4 List of *E. oleifera* germplasm collection available in MPOB (Kushairi et al. 2011)

Species	Country of origin	Year of collection
Elaeis oleifera	Nicaragua	1982
	Honduras	1982
	Costa Rica	1982
	Colombia	1982, 2004
	Suriname	1982
	Brazil	1982, 2004
	Panama	1982
	Peru	2004
	Ecuador	2004

10.2.1.4 *Elaeis guineensis* Collection in Colombia

Corporación Centro de Investigación en Palma de Aceite (CENIPALMA) collected oil palm germplasm in Angola and Cameroon in 2002 and 2004 respectively. The genetic materials are maintained at CENIPALMA experimental station La Vizcaina.

10.2.1.5 *Elaeis guineensis* Collection by Indonesia

Indonesia formed a consortium (Plantation Companies) to finance germplasm collections. In 2006, the members of the consortium and IRAD (West African Institute of Agricultural Research for Development), Cameroon collected close to 100 accessions. These are being established in different plantation companies. In 2010, Indonesian consortium and MPOB collected oil palm germplasm material in Angola with the co-operation of Angolan National Institute of Coffee (INCA). During this expedition, 127 accessions were sampled.

10.2.2 *Elaeis oleifera Collection*

10.2.2.1 *Elaeis oleifera* Collection by MPOB

Elaeis oleifera oil palm was collected in Colombia, Panama, Costa Rica, Honduras, Brazil and Suriname in 1981–2004 as given in Table 10.4 and Fig. 10.2. The palms were screened for fatty acid composition (FAC). The collections from Colombia, Panama and Costa Rica had an IV (iodine value) of more than 90. The C18:1 level ranged from 52% to 66% and C18:2 levels varied between 15% and 23%. The IV in the Brazilian *E. oleifera* ranged from 76 to 81 and the level of C18:1 was lower than the accessions from Colombia, Panama, Costa Rica and Honduras.

When MPOB started their collection on *Elaeis oleifera*, other oil palm breeders put in extra work into the species' collection and many useful collections have been established in Malaysia, Ivory Coast, Costa Rica, Brazil and others. Despite this collection, many hybrid in Malaysia derived from a single uncertain palm known as

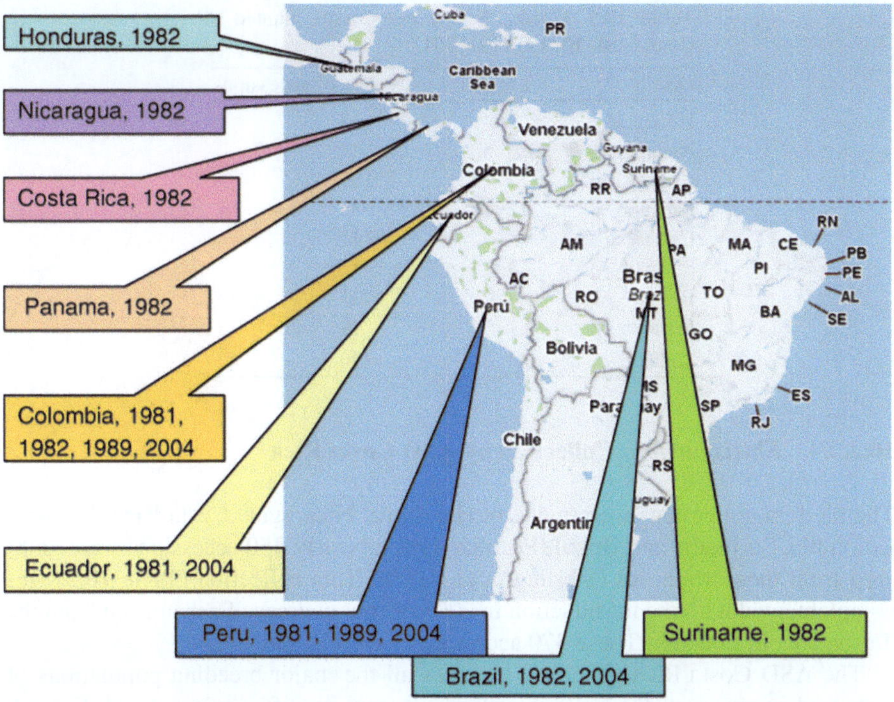

Fig. 10.2 Collection of *E. oleifera* germplasm in South America (Kushairi et al. 2011)

Kuala Lumpur Melanococca (KLM) (Corley and Tinker 2003). Four geographically distinct population that is Brazil, Peru, Central America and Suriname have been identified using molecular markers (Barcelos et al. 2002)

10.2.2.2 *Elaeis oleifera* Collection by Brazil and CIRAD

Empresa Brasileira de Pesquisa Agropecuária (EMBRAPA) and CIRAD mounted an extensive expedition in Brazilian Amazon Basin in 1980. A sample of 226 accessions comprising of 326 individual palms was obtained along the Amazon River Basin. The accessions were established as a field genebank at Rio Urubu research station, Manaus.

10.2.2.3 *Elaeis oleifera* Collection by Ecuador

A private company, Palma Del Rio collected *E. oleifera* at Taisha in Ecuador. This population has unique fruit characteristics with a thick mesocarp. The Taisha *oleiferas* are being progeny tested with advanced *pisiferas* such as AVROS (Verneiging Rubber Planters Oostkust, Sumatra) and La Mé.

Table 10.5 *Elaeis oleifera* germplasm planted at Coto, Costa Rica (1970–2004)

Origin	Localities	No. of accessions
Honduras	3	43
Nicaragua	6	43
Costa Rica	7	107
Panamá	12	88
Colombia	5	41
Surinam	3	13
Brazil	7	31
Ecuador	1	4
Total	44	370

10.2.2.4 *Elaeis oleifera* Collection by ASD Costa Rica

The oleifera collections were made in Honduras, Nicaragua, Costa Rica, Panama, Colombia, Suriname and Brazil (Escobar 1981). Nearly 350 accessions were gathered from these countries (Sterling et al. 1999). This collection (Table 10.5) was complemented with the introduction in 2003 of four sources of germplasm from the Taisha area in Ecuador. These 370 accessions were planted at Coto.

The ASD Costa Rica genebank includes all the major breeding populations of restricted origin, as well as wild introductions from specific environments. *E. oleifera* lines from Brazil, Suriname, Colombia, Panama, Costa Rica and Nicaragua are also present in the collection (Sterling and Alvarado 2002).

10.2.3 Other Economically Important Palms Collection

Palm species (*Bactris gasipaes, Jessenia bataua, Oenocarpus mapora, Orbingya martiana,* and *Euterpe precatoria*) of economic importance were collected and conserved in MPOB. The interest in theses palms in Malaysia is quite limited compared to the planting and utilization of the genus *Elaeis.* These exotic palms are most valuable as a food source. At present, these exotic palms except for *Bactris gasipaes* have not been exploited fully (Pesce 1985).

10.2.4 Coconut Collection

Coconut germplasm is largely maintained in the form of field genebank. A number of field collections are found in coconut research institutes conserving 1,426 accessions worldwide. To date, 28 countries have national collections to conserve important coconut germplasm. Coconut Genetic Resources Network (COGENT) is making efforts to establish multisite international coconut genebank (Rajanaidu and Ramanath Rao 2004).

10.3 Establishment and Evaluation

10.3.1 *Elaeis guineensis*

The oil palm genetic materials collected in the wild were planted in the form of open-pollinated' families at MPOB Research Station, Kluang, Johor. Experimental designs such as the cubic lattice, randomised complete block design (RCBD) and completely randomised design (CRD) were used to study the performance of the material.

It takes more than 20 years from the collection of germplasm in the wild to the release of oil palm planting materials. During this time detailed data are collected on an individual palm basis. They comprise:

(a) Yield (FFB)
(b) Bunch number (BNO)
(c) Average bunch weight (ABW)
(d) Mean nut weight (MNW)
(e) Fruit to bunch (FTB)
(f) Mesocarp to fruit (MTB)
(g) Oil to dry mesocarp (OTDM)
(h) Oil to wet mesocarp (OTWM)
(i) Oil to bunch (OTB)
(j) Kernel to fruit (KTF)
(k) Shell to fruit (STF)
(l) Frond production (FP)
(m) Ranchis length (RL)
(n) Leaflet number (LN)
(o) Height (H)
(p) Physiological parameters.
(q) Fatty acid composition (FAC)
(r) Oil/palm/year
(s) Kernel/palm/year
(t) Total economic product (= total oil + total kernel X 0.6)

10.3.2 *Elaeis oleifera*

The FAC of Surinam *E. oleifera* is rather unique. It has the highest level of C18:1 and the lowest C18:2 when compared to other populations. The mean iodine value of the Suriname population is the lowest found i.e. 67.5. The lowest level of C16:0 in the *E. oleifera* collections is 13% (Rajanaidu et al. 1994).

The mean fruit weight varied from 3.79 (Colombia) to 4.69 (Panama) and the highest value (6.5 g) was found in the Panama population. For mesocarp, the mean values ranged from 28.6% (Honduras) to 36.0% (Colombia) and the highest value

(53.4%) was found in fruits collected from Costa Rica. Escobar (1981) has given values for single fruit weight for Costa Rica as 3.2 g, for Panama, 3.3 g and for Colombia, 3.59 g. In the case of mesocarp the same author gave values for Costa Rica as 37.2%, Panama, 36.4% and Colombia 35.6%.

The percentage value of the total variance for each trait contributed by the components, that is, country, site and palm are summarized and can be noticed that σ_0^2 (palm) and σ_1^2 (site) components accounted for 40% each and σ_2^2 (country) about 20%. In a similar study on the *E. guineensis* Nigerian population (Rajanaidu et al. 1979) the σ_0^2 (palm) alone contributed nearly 80% to the total variance. It shows that *E. oleifera* palm/site is much more uniform in comparison with *E. guineensis* found in Nigeria. This has important implications on the sampling strategy. It may be possible, in the case of *E. oleifera,* to reduce the number of palms sampled per site and increase the number of site in order to capture the maximum genetic variability.

Since the analysis of variance had shown significant differences between sites, a closer examination of the population was carried out. In general, fruit sampled along the Atlantic Coast, i.e. at Limon, Puntarenas, Panama Atlantic and Guabala are bigger in size than those along the Pacific Coast. In the case of mesocarp content the Colombian material found at Chimichagua, Crete and Caucasia is superior to others.

The iodine value *E. oleifera* is much higher than *E. guineensis* oil. However, the oil yield of pure *E. oleifera* is much lower with an oil to bunch ratio of 5% as compared to >25 with *E. guineensis* (*tenera*). One way to exploit the *E. oleifera* oil is through the medicinal/pharmaceutical industry rather that catering for culinary purposes (Choo and Yusof 1996).

10.3.3 *Cocos nucifera*

Coconut germplasm is being evaluated beginning at the nursery stage (number of days to germination, height, leaf length, leaf width, girth at collar and leaf showing first splitting).

Juvenile stage evaluation include leaf production, leaf length, increase in girth and other growth parameters recorded before flowering (5–6 years). Agronomic evaluation determines the duration of the male and the female phase, female flower distribution and length of peduncles, number of inflorescences, aborted bunches, mean number of female flowers, number of mature nuts and setting percentage. The nut character includes the evaluation analysis of data on fruit shape, nut shape and weight of fruit, nut and nut without water, meat and copra (Ratnambal and Nair 1994).

The main objective in most plantations is to produce adult coconut having high yield copra with early bearing of fruit. In addition, some coconut breeders are looking into high content of copra, resistance to bud rut and nut fall diseases, high yield of copra per unit area with medium input, good root to shoot ratio, high harvest index, long economic life-span, resistance to insect pest, tolerant to tidal swampy area, tolerant to drought, number of bunches per palm, number of fruits per palm, high lauric acid content in oil and high protein content in meat (Duhamel 1994; Henky et al. 1994; Mkumbo and Kullaya 1994).

10.4 Conservation Methods

Two basic techniques are employed to conserve genetic diversity of oil palm. The two strategies are ex situ and in situ. Ex situ conservation means the conservation of components of biological diversity outside their natural habitat (Engelmann 1991). In contrast, in situ conservation is conserving germplasm in the natural habitat where the target species is found and in habitats such as farms and home gardens, where the species have developed their distinctive properties as a result of long term selection by humans and the environment (Dulloo 2005). Approaches to ex situ conservation include methods like seed storage, field genebanks and botanical gardens. DNA and pollen storage also contribute indirectly to ex situ conservation of plant genetic resources (Rao 2004).

10.4.1 Field Genebank

Generally, field genebank have been used for perennial species producing little or no seed, have a long life cycle and species producing recalcitrant seeds (Jain 2011). For example, the amount of land utilised for field genebank for MPOB's germplasm collection is given in Table 10.6.

Some major disadvantages of field genebank are labour and time intensive, extreme care in labelling and managing fields, many biotic and abiotic factors affect the safe conservation of germplasm, inconvenient for transportation and restriction in germplasm exchange internationally (Rajanaidu and Ramanatha Rao 2002).

For coconut, field genebank presently is the most feasible ex situ conservation method that can be used and it plays a major role in conserving coconut genetic diversity. There are many field collections connected with coconut research institutes conserving 1,426 accessions worldwide. To date, 28 countries have established national collections to conserve important coconut germplasm (Rajanaidu and Ramanath Rao 2004). One great advantage of a living collection of palms is that the material is readily available for utilization. Seed or pollen and tissue culture techniques may provide cheaper way of storage of genetic resources but the material may not be available immediately for use by the plant breeders; more than 6 years are required to regenerate, evaluate and utilize the material starting from seed and callus stages.

10.4.2 Seed Storage

Conservation in the form of seed storage is an efficient and reproducible technique. It is the most common method used for the conservation of plant genetic resources (Jain 2011). A large number of agricultural crops have seeds which are termed as orthodox. However, there are a number of species whose seeds cannot be dried to

Table 10.6 Number of palms and area occupied by field genebank in MPOB (Rajanaidu et al. 1994)

Origin	Types of oil palm	No. of palms planted in the field	Area (ha)
Nigeria	*Elaeis guineensis*	31,434	212
Cameroon	*Elaeis guineensis*	3,590	24
Zaire	*Elaeis guineensis*	13,750	93
Tanzania	*Elaeis guineensis*	3,104	21
Madagascar	*Elaeis guineensis*	38	1
Senegal	*Elaeis guineensis*	815	6
Gambia	*Elaeis guineensis*	212	1
Sierra Leone	*Elaeis guineensis*	986	9
Guinea	*Elaeis guineensis*	1,344	14
Ghana	*Elaeis guineensis*	2,800	21
Angola	*Elaeis guineensis*	2,507	17
Nicaragua	*Elaeis oleifera*	4,248	29
Honduras	*Elaeis oleifera*		
Costa Rica	*Elaeis oleifera*		
Colombia	*Elaeis oleifera*		
Suriname	*Elaeis oleifera*		
Brazil	*Elaeis oleifera*		
Panama	*Elaeis oleifera*		
Peru	*Elaeis oleifera*		
Ecuador	*Elaeis oleifera*		
Colombia, Peru	*Bactris gasipaes*	30	1
Colombia, Peru	*Jessenia bataua*	639	5
Colombia, Peru	*Oenocarpus mapora*	265	2
Colombia, Peru	*Euterpe precatoria*	20	1
Colombia, Peru	*Babasu*	6	1
Taisha	*Elaeis oleifera*	8	2
Total		65,796	460

low levels of moisture for optimum storage under cool conditions. Such seeds have been referred to as recalcitrant.

Earlier desiccation experiments have shown the oil palm seed to be orthodox, not recalcitrant, in character (Engelmann 1991). There is a significant difference in water content between the whole seed and the embryo which is maintained despite desiccation, and results in the failure of sub-zero storage of whole seeds. But according to current reports, African oil palm seeds have been classified as displaying intermediate storage behaviour since their viability can be maintained only up to 12–15 months when stored with 10–15% moisture content at 15°C (Engelmann et al. 1995; Rajanaidu and Ramanatha Rao 2002).

The main problem with this technique is its inability of seeds to withstand desiccation at lower temperatures and that may hamper seed germination rate or survival of seeds may seriously effect and ultimately die (Jain 2011). Oil palm seeds germination relies strongly on the genotypes of different lineages and the seed storing period (Martine et al. 2009). Therefore, preservation of collections through storage of seeds is not practical as the seeds can be stored for only 2 years (Rajanaidu 1980).

Coconut seeds however are highly recalcitrant (Engelmann et al. 2011). In coconut, there are also problems using this technique due to the length of growth phase of up to 8–10 years and sometimes 12–14 years elapse between generations; the lack of vegetative reproduction system and the low ability to reproduce new seedlings from seeds (150–200 times lower than that of oil palm); predominantly but not strictly allogamy; large, heavy seeds without dormancy period which considerably limits germplasm exchange and the need for large experimental areas (Lamothe 1991).

10.4.3 In Vitro Technique

A particular problem with the breeding of tree crops is their long life span compared to annual crops. In vitro methods provide tools that can be used in a variety of ways, depending on the need of the species. Collecting cuttings of plants and seeds is generally the most cost-effective method for providing material for ex situ conservation. However, there are many instances in which seeds are sterile or not available, or so few plants remain that collecting whole plants would negatively impact the population (Reed et al. 2011). For example, an oil palm tree will not bear a useful crop of fruit until it is about 3 years old. After that, the trees have a commercial lifetime averaging about 25 years, although an oil palm tree can still produce fruit for over 50 years (Murphy 2007; Kushairi et al. 2011).

These factors have made it impractical to set up classical breeding programmes for trees, but mass clonal propagation can be used as a much faster and cheaper alternative to multiplying up the best genetic stock. In this way, a single elite tree can give rise to an entire plantation, or even a whole series of plantations, in a very short period. In vitro collecting of tissues is also known to be less invasive than removing whole plants and allows for an efficient sampling of a large number of plants when seeds are not available (Engelmann 1991). It will also help alleviate the current shortage of clones at a marginally higher price as compared to the current conventional seed and is expected to contribute significantly towards improving the oil yield and therefore profitability for the oil palm industry into the medium term. In this respect, there are some elite DXP combinations which give fresh fruit bunch (FFB) yields of more than 30 tonnes per hectare per year and oil extraction rate (OER) of 27% or more (Sharma 2007).

With coconut, in vitro culture protocols have been developed with two main objectives, viz. the large scale production of particular types of coconuts and the international exchange and conservation of coconut germplasm (Engelmann et al. 2011). Blake and Eeuwens (1981) reported that the route of clonal propagation through callus and embryogenesis seems the most hopeful for coconut, especially since oil palm and date palm have been successfully propagated by this method Moreover, exchanging coconut germplasm in the form of embryos would allow both avoiding the phytosanitary problems and reducing costs linked with transportation of whole nuts. Using in vitro techniques for collecting, exchanging and conserving coconut germplasm requires efficient protocols for in vitro germination of embryos, development of embryos into whole plantlets, their acclimatization to in vivo conditions and further development into seedlings, which can be transferred to the field (Engelmann et al. 2011).

10.4.4 Cryopreservation

Cryopreservation refers to the storage of a living organism at ultra- low temperature such that it can be revived and restored to the same living state as before it was stored. Indefinitely long storage times require that the organism be maintained below the glass transformation temperature (temperature at which a supercooled liquid becomes a glass) of aqueous solutions, approximately −130°C, the temperature at which frozen water no longer sublimes and recrystallizes (Panis and Lambardi 2005).

Although ultra-cold freezers may stabilize some living cells for weeks or even years, liquid nitrogen is required for longer storage times. Cryopreservation is also known to be the current method for long-term storage without subculture (Hao et al. 2001). The choice of material includes cells, protoplasts, shoot apices, somatic embryos, seeds or excised zygotic embryos. Cryopreservation requires limited space, protects materials from contamination, involves very little maintenance and is considered to be a cost-effective option (Rao 2004).

Cryopreservation of oil palm and coconut can be used as a strategy to back up the establishment of living collections, which are expensive to maintain, and are under constant threat from biotic and abiotic factors.

Oil palm has been extensively cryopreserved but more experiments are needed to determine the desiccation method and critical moisture content for different type of tissues. In the case of coconut, only a limited amount of research has been conducted towards the development of cryopreservation protocols, involving research teams in Malaysia, Côte D'Ivoire, France and the United Kingdom. Cryopreservation has been performed with zygotic embryos, plumules and pollens (Engelmann 2011).

10.4.4.1 Zygotic Embryo

The successful recovery of viable, excised embryos from liquid nitrogen, and their subsequent regrowth in vitro, suggests a practical technique for the long term conservation of the genetic resources of the oil palm species (Panis and Lambardi 2005). MPOB has successfully developed a technique which stores zygotic embryos for a long period of time. The moisture contents of the embryos were initially lowered by exposing them under laminar air flow. The moisture content of some of the embryo samples from Guinea and Senegal could be brought down to 12% while

Table 10.7 Samples of zygotic embryos kept in the cryotank in MPOB's genebank, Kluang

Samples from	No. of accessions
Tanzania	1,050
Guinea	9,300
Madagascar	5,400
Cameroon	4,300
Angola	1,350
Sierra Leone	3,900
Senegal	5,550
Nigeria	1,950
Zaire	450
Total	33,250

samples from Angola, Cameroon and Ghana could reach moisture content as low as 7%. The amount of zygotic embryo samples currently stored in the cryotank in MPOB is summarized in Table 10.7.

A single coconut embryo survived 15 months after freezing using a classical protocol of cryoprotectant with dimethyl sulfoxide (DMSO) and slow freezing (Chin et al. 1989). Assy-Bah and Engelmann (1992) have developed an efficient cryopreservation protocol based on pre-treatment of coconut embryos with high sugar medium, partial desiccation and rapid freezing in liquid nitrogen. This protocol has been applied to ten different varieties, with 44–100% cryopreserved embryos giving rise to whole in vitro plantlets (Assy-Bah and Engelmann 1993). Sajini et al. (2006) also reported successful cryopreservation of coconut.

10.4.4.2 Pollen

Cryopreservation of oil palm pollen had been utilized in numerous laboratories. However, no published studies of cryopreservation of these techniques have been reported (Duval et al. 2006). Pollens have been stored at sub-zero temperature or lower for a limited duration in MPOB as a routine in many breeding programmes, giving high viability. Therefore, cryopreservation of pollen is seen as a very good prospect for oil palm.

In the coconut experiment, it was recognized that the pollen could be stored at room temperature following freeze-drying. These pollens are seen viable for at least 1 year. The optimum residual moisture content for successful storage at room temperature was found to be in the range 3.5–10.0%. (Whitehead 1965).

10.4.4.3 Somatic Embryos

It was established for more than 30 years that polyembryonic tissues could be cultured from torpedo-shaped embryos through somatic embryogenesis (Engelmann and Dereuddre 1988; Dumet et al. 2000). Mass production is ensured by the proliferation of fast growing calluses on media with auxin content, followed by the differentiation of meristematic clumps into somatic embryos after the addition of cytokinins to the medium (Duval et al. 1998). Dumet et al. (1993) also reported cryopreservation of oil palm (*Elaeis guineensis*) somatic embryos involving a desiccation step.

10.4.4.4 Apical Meristems

To date, samples of apical meristems of oil palm have found to have 45% positive growth after cryostorage of up to 24 h (Ainul et al. 2009). Successful cryopreservation requires the optimization of numerous variables including the size of specimen, the correct type and concentration of cryoprotectant, sample water content and rate of freezing and thawing (Bekheet et al. 2007). In the near future, other experiments will be tested for viability after cryopreservation.

10.4.5 DNA Genebank

For many species that are difficult to conserve by conventional means (either as seeds or vegetatively) or that are highly threatened in the wild, DNA storage may provide the ultimate way to conserve the genetic diversity of these species and their populations in the short term, until effective methods can be developed. After all, genomic DNA samples represent the entire genetic component of the target organism (Hodkinson et al. 2007). DNA of oil palm germplasm have been collected and stored at MPOB's Research Station, Kluang for the past years. These collections include samples from various countries such as Nigeria, Tanzania, Angola and some samples from *E. oleifera* palms.

The DNA genebank is extremely useful to verify legitimacy of controlled crosses, genetic finger-printing and to study genetic diversity.

10.4.6 In Situ Conservation

In contrast to ex situ conservation, in situ methods offer the best opportunities for the conservation of multiple species, particularly recalcitrant species, while allowing the dynamic evolutionary processes to continue, especially in producing new resistance to pests and diseases. Provided in situ reserves are designed optimally, they may conserve a broader range of variability of the target plant than is possible in ex situ collections, particularly for recalcitrant seed species. However, the difficulties are that the materials are not easily accessible for use and they may be vulnerable to natural and man-made calamities and other biotic interferences such as invasive alien plants. In fact, for both plants and animals, loss of habitat and altered environmental conditions has led to the decline and disappearance of some species and the endangerment of others. It is a fact that no in situ conservation programme would ever be able to protect all the populations and the full range of genetic variability, except for a limited number of highly threatened species (Dulloo et al. 2006).

In the case of oil palm, African countries are urged to preserve a cross-section of natural oil palm groves for the posterity and conserve a sample of traditional varieties.

10.5 Utilization of Germplasm Collections/Sampling

Oil palm will continue as one of the main crops for the future in Africa and other countries because it is used as versatile food product; it contains minor components such as carotenoids, tocopherols, tocotrienols, sterols, phosphatides, triterpenic and aliphatic alcohols; it is used widely in the non-food sector such as soaps, cosmetics and pharmaceutical products; it is an environmental friendly crop because it provide forest cover for at least 25 years and oil palm is a green fuel alternative (Yusof 2007).

The objective of germplasm collection is to gather the maximum amount of genetic variability in minimum number of samples (Marshall and Brown 1975). The basic sampling strategy to attain this objective is to collect a few individuals per site from as many sites as possible. In case of oil palm, 2–32 palms per site were collected and the number of sites visited in a country depends on the distribution and density of natural oil palm groves. For each bunch sampled, data on bunch weight, bunch length, bunch breadth, bunch depth, fruit diameter, nut diameter, kernel diameter, mesocarp to fruit (%), shell thickness, fruit weight and nut weight were recorded in situ (Rajanaidu et al. 1979).

In addition, MPOB also provided advance breeding populations for utilization by the industry such as parental materials for setting up of seed gardens and also lease of mother palms for commercial seed production. Being the custodian of the oil palm industry, MPOB is being funded by the industry in the form of cess from crude palm oil (CPO) production to do research and development.

Some interesting attributes of the germplasm are dwarf, high bunch number, high iodine value, high kernel such as in Nigerian and high vitamin E in Cameroon. The Cameroon X Zaire showed tolerance to ganoderma. High palmitic and linoleic acids, low stearic and oleic acids was found in Madagascar collection. Some palms from Tanzania are pronounced for thin-shelled *tenera* and high bunch index. The Angolan palms exhibited large fruit *dura* with bunch and fruit characteristics similar to Deli. Individuals from both the Tanzanian and Angolan collections are used for long-stalk breeding programme. Some palms collected from Guinea are found to be high in vitamin *E. oleifera* is an important source of high iodine value and high carotene content. Palms that showed interesting characters were selected and introduced into the existing breeding populations.

10.6 PS Series

Oil palm planting materials based on MPOB-Nigerian *dura x* AVROS *pisifera* – MPOB Series 1 (PS1) and MPOB Series 2 (PS2) which are formerly known as PORIM Series 1 and PORIM Series 2 – were introduced to the industry with the objectives of reducing the palm heights and increasing the iodine value of the palm oil. Progeny testing of the PS-type planting materials over 4 years on inland soil indicated that palm height increment of PS1 was 40 cm/year while iodine value of PS2 was 56. MPOB have the largest oil palm germplasm collection in the world covered by extensive oil palm germplasm collections from Africa and Latin America.

To date, MPOB had developed and transferred a total of 13 oil palm breeding and planting materials as new technologies to the industry. These technologies were announced to the industry during MPOB annual Transfer of Technology (TOT) Seminars. These technologies cover traits for high yielding dwarf (PS1), high iodine value (PS2), large kernel (PS3), high carotene *E. oleifera* (PS4), thin-shelled *teneras* (PS5), large fruit *duras* (PS6), high bunch index (PS7), high vitamin E (PS8), exotic palm – *Bactris gasipaes* (PS9), long stalk (PS10), high carotene *E. guineensis* (PS11), high oleic acid (PS12) and low lipase (PS13). Various technologies had captured keen interest from the industry for uptake into their oil palm breeding and improvement programmes.

Extensive evaluation for yield, bunch traits, fatty acid composition, physiological parameters and vegetative characters was carried out in 1982–1987. Elite Nigerian palms for high yield, high iodine value (≥60) and low stature were distributed to the members of the industry for progeny testing, introgression into the current breeding materials and to initiate new breeding lines for future seed production. These palms are popularly known as PS1.

More than 3,000 palms of the MPOB-Nigerian germplasm have been screened for high IV with prospects of reducing the palmitic and increasing the oleic acid content for PS2. Many individual palms from the collection have IV in excess of 60, which is higher than the current palm oil with IV values of 50–53. Selected *dura* and *tenera* palms have been used to produce planting material (unsaturated oil) with high IV.

The oil palm planting materials, DXP (*tenera*) are produced by crossing *dura* with *pisifera*. The *dura* mother palms with more than 20% kernel to fruit ratio and 13% kernel to bunch ratio are crossed with *pisiferas* which are known for their general combining ability, such as AVROS *pisifera*. A number of studies have shown that the heritability estimates for kernel to bunch and kernel to fruit are high in oil palm and these traits are maternally inherited. The selfs and sibs of promising high kernel *dura* mother palms have been distributed to the industry known as PS3. These *duras* have been progeny-tested with AVROS and other sources of *pisiferas* which are known for their general combining abilities.

MPOB's *E. oleifera* germplasm collections have been screened for high IV and high carotene content for PS4. Almost all palms screened have high IV in excess of

80, which is higher than the current commercial DXP with IV of 50–53. As for carotene, palms with values in excess of 3,000 ppm have been identified. The carotene of current commercial DXP is between 500 and 700 ppm.

In improvement programmes, heritability estimates and correlation among traits emphasized. Generally, the heritability of bunch yield and its components is low, and higher estimates are obtained for bunch quality traits, such as mesocarp, shell and oil contents of the fruit. Since shell to fruit ratio is selected against, negative correlations with the oil-related-traits are advantageous in selection for oil yield improvements. PS5 has 2.80% and 7.40% of oil yield as compared with that of a typical Deli X AVROS, which is about 12%. Subsequently, PS5 is high in mesocarp to fruit, above 80% and oil to bunch ratio ranging from 26% to 29%.

Mean fruit weight is positively correlated with the mesocarp to fruit, oil to dry mesocarp, oil to wet mesocarp, oil to bunch and oil yield and negatively related with kernel yields and fruit to bunch. The oil to bunch is the product of, and strongly associated with fruit to bunch, mesocarp to fruit and oil to wet mesocarp. The mean fruit weight of PS6 is between 24 and 34 g as compared with that of current *tenera* from *dura* x *pisifera* (DXP) planting materials, which is about 10 g.

PS7 Materials selected for BI are high yielding palm with low value of vegetative dry matter. In practice, it will often be simpler to select for BI directly; the heritability of BI is similar to vegetative dry matter. However, BI appears to be less sensitive to soil fertility. Correlation analysis showed that BI is highly correlated with fresh fruit bunch, bunch number and oil yield.

The oil palm germplasm in MPOB Genebank at Kluang Research Station was screened for vitamin E (tocopherols and tocotrienols) using high performance liquid chromatography (HPLC). Evaluation and selection for bunch yield, oil yield, growth and physiological parameters were also carried out. Within the *E. guineensis* germplasm, the *tenera* has higher level of total vitamin E compared with *dura*. Some 35 palms with vitamin E content of 1,300–2,496.57 ppm were identified. However, only *dura* palms with oil yields >2 tonnes/ha/year and those of the *tenera* palms with more than 4.5 tonnes/ha/year were selected as PS8 breeding population. Among the selected palms is *tenera* 0.150/500 with 2,496.57 ppm of total vitamin E. This palm comes from Population 12, which is known for the dwarf characteristic. The *tenera* palm 0.150/338 with 1,364.67 ppm of vitamin E had oil yield of 11.1 tonnes/ha/year.

Bactris gasipaes is considered for palm heart (PS9) because it produces multiple stems and grows rapidly. The palm heart has excellent taste and texture, and low level of calcium oxalate which accounts for less browning of shoot.

The MPOB oil palm germplasm collection has been screened for long stalk for PS10. The stalk length is measured from the stalk ring to the lowest spikelet. Ten palms (eight *dura*s and two *tenera*s) with stalk lengths between 28 and 36 cm are currently being used in the breeding programme. These palms showed mean FFB from 171.9 to 221.3 kg/p/year and oil yield between 28.2 and 52.9 kg/p/year. Biparental crossing programme was used for progeny testing and production of long stalk breeding material.

Carotene is much used in pharmaceuticals. In 2002, MPOB offered *E. oleifera* (PS4) with a carotene of >3,000 ppm and iodine value >80 to the oil palm industry as planting material. Even though it had extremely low oil yield (*ca.* 0.5 tonnes/ha/year), the oil can be directly used as carotene. In the current DXP (*E. guineensis*) oil palm planted, carotene content is only 500–700 ppm. However, as its oil yield is much higher than that of *E. oleifera*, carotene can be extracted from the oil. Thus, it would be worthwhile to screen for high carotene *E. guineensis* palm to raise the carotene content in normal palm oil (PS11).

Screening of MPOB oil palm germplasm collection for fatty acid composition (FAC) using gas chromatography (GC) has revealed a wide variation. The current oil palm planting materials have an oleic acid (C18:1) content of between 37% and 40%. Fifteen palms with oleic acid content exceeding 48% were selected as PS12 breeding population. These palms have the potential to increase the oleic acid and IV contents in the current commercial materials. With fractionation, the IV of the olein should reach 70. This population would be useful for developing planting materials with more liquid oil.

A wide variation was found in the levels of free fatty acid (FFA) in the fruits. Some palms from Cameroons, Guinea, Sierra Leone, Senegal and Tanzania had <10% at 5°C, considerably lower than that in the current planting materials. Among them were four *teneras* and four *duras*, which produced a mean FFB yield of more than 144.08 kg/p/year and oil to bunch of more than 11.57%. They will be used in the breeding programme to produce planting materials for quality oil.

10.7 Regeneration of Genetic Resources

Oil palm genetic resources are conserved in the form of field genebanks for the past 20 years. When the palms are too tall and more than 25 years, the accessions in the genebank are re-established by making random pairwise crosses within each population. This methodology enables the preservation of co-adapted gene complexes. In coconut, assisted pollination is carried out to produce seed nut for next planting and when this becomes too expensive; seed nut are collected from plants located in the centre of the plot of the accession (Rajanaidu and Ramanath Rao 2004).

Regeneration of germplasm is a critical activity to preserve for posterity. To date, very limited research has been carried out on adequate sampling method for regeneration of germplasm of perennial tree crops with long life-cycle.

10.8 Conclusion

The primary conservation method for oil palm and coconut genetic resources is field genebank. The materials are readily available for utilization but expensive to maintain and always at risk of being lost due to natural or human-driven calamities

(Duval et al. 2006). However, other tools of conservation such as cryopreservation of zygotic embryos, polyembryogenic cultures, somatic tissues and pollen could complement field genebanks.

In future, estimation of heterozygosity level in populations using molecular markers such as SSR (simple sequence repeat), RFLP (restriction fragment length polymorphism), AFLP (amplified fragment length polymorphism) and RAPD (random amplification of polymorphic DNA) can be used as a yardstick for optimum long-term conservation and regeneration.

Acknowledgement The authors wish to thank the Director General of Malaysian Palm Oil Board for permission to publish this paper.

References

Ainul MM, Tarmizi AH, Kushairi A (2009) Isolation and characterization of the promoter sequences of genes involved in fatty acid biosynthesis from oil palm cryopreservation of oil palm (*Elaeis guineensis)* apical meristem on tissue culture clonal materials. In: Proceedings of the 8th Malaysia congress on genetics, 4–6 Aug 2009, Genting Highlands, Malaysia, 478 pp

Arif S, Ariff MA (2001) The case study on Malayisan palm oil. Papers presented in Regional Workshop on Commodity Export Diversification and Poverty Reduction in South East Asia, Bangkok, Thailand, 3–5

Assy-Bah B, Engelmann F (1992) Cryopreservation of mature embryos of coconut (*Cocos nucifera* L.) and subsequent regeneration of plantlets. CryoLetters 13:117–126

Assy-Bah B, Engelmann F (1993) Medium term conservation of mature embryos of coconut. Plant Cell Tiss Org Cult 33:19–24

Barcelos E, Amblard P, Berthaud J, Seguin M (2002) Genetic diversity and relationship in American and African oil palm as revealed by RFLP and AFLP molecular markers. Pesquisa Agropecuária Brasileira 37(8):1105–1114

Bekheet SA, Taha HS, Saker MM, Solliman ME (2007) Application of cryopreservation technique for in vitro grown date palm (*Phoenix dactylifera* L.) cultures. J Appl Sci Res 3(9):859–866

Bergert DL (2010) Management strategies of *Elaeis guineensis* in response to localized markets in South Eastern Ghana, West Africa. Master's thesis. Michigan Technological University

Blake J, Eeuwens CJ (1981) Culture of coconut palm tissue with a view to vegetative propagation. In: Rao AN (ed) Proceedings of international symposium, 28–30 Apr 1981. National University of Singapore, Singapore, pp 145–148

Borokini TI, Okere AU, Giwa AO, Daramola BO, Odofin WT (2010) Biodiversity and conservation of plant genetic resources in field genebank of the National Centre for Genetic Resources and Biotechnology, Ibadan, Nigeria. Int J Biodivers Conserv 2(3):37–50

Carrere R (2010) Oil palm in Africa: past, present and future scenarios. Paper presented in World Rainforest Movement, Congo Basin, Democratic Republic of Congo

Chaillard H (1977) Prospection des palmeraies naturalles des provinces du sudouest et du nordouest su Cameroun. IRHO

Chin HF, Krishnapillay B, Hor YL (1989) A note on the cryopreservation of embryos of coconut (*Cocos nucifera* L. var. *Mawa*). Pertanika 12:183–186

Choo YM, Yusof B (1996) *Elaeis oleifera* palm for the pharmaceutical industry. PORIM information series no. 22, pp 1–4

Corley RHV, Tinker PB (2003) The oil palm. Blackwell Science, Oxford, pp 133–200

Duhamel G (1994) Vanuatu national coconut breeding programme. In: Batugal PA, Ramanatha RV (eds) Coconut breeding. IPGRI, Serdang, pp 92–97

Dulloo A (2005) Energy balance and body weight hemeostasis. In: Kopelman PG, Caterson ID, Dietz WH (eds) Clinical obesity, 2nd edn. Blackwell, Oxford, pp 67–80

Dulloo E, Nagamura Y, Ryder O (2006) DNA storage as a complementary conservation strategy. In: de Vicente MC, Andersson MS (eds) IPGRI: DNA banks – providing novel options for genebanks? IPGRI, Rome, pp 11–24

Dumet D, Engelmann F, Chabrillange N, Duval Y (1993) Cryopreservation of oil palm (*Elaeis guineensis* Jacq.) somatic embryos involving a desiccation step. Plant Cell Rep 12:352–355

Dumet D, Engelmann F, Chabrillange N, Dussert S, Duval Y (2000) Cryopreservation of oil-palm polyembryonic cultures. In: Engelmann F, Takagi H (eds) Cryopreservation of tropical plant germplasm: current research progress and application. Proceedings of an international workshop, Tsukuba, October 1998. International Plant Genetic Resources Institute, Rome

Duval Y, Rival A, Verdeil JL, Buffard-Morel J (1998) Advances in oil palm and coconut micropropagation. In: Proceedings of the Southeast Asian regional workshop on propagation techniques for commercial crops of the tropics, Ho Chi Minh City, 7–12 Feb 1993

Duval Y, Chabrillange N, Dumet D, Engelmann F (2006) Ex situ conservation of oil palm (*Elaies guineensis* Jacq.) genetic resources using biotechnology. In: Rajanaidu N, Hensen IE, Ariffin D (eds) Proceedings of the international symposium on oil palm genetic resources and their utilization. MPOB, Bangi, pp 471–479

Engelmann F (1991) Current development of cryopreservation for oil palm somatic embryos. In: Proceedings of the XVIIIth international refrigeration congress, vol IV, 10–17 Aug 1991, Montreal, pp 1676–1680

Engelmann F (2011) Cryopreservation of embryos: an overview. Methods Mol Biol 710:155–184. doi:10.1007/978-1-61737-988-8_13

Engelmann F, Dereuddre J (1988) Effects du mileu de culture sur la production d'embryons des destines a la cryoconservation chez le palmier a huile (*Elaeis guineensis* Jacq). CR Academia de Science Paris 30:515–520

Engelmann F, Dumet D, ChabriIlangel N, Abdelnour-Esquivel A, Assy-Bah B, Dereuddre J, Duval Y (1995) Factors affecting the cryopreservation of coffee, coconut and oil palm embryos. Plant Genet Res Newsl 103:27–31

Engelmann F, Malaurie B, N'Nan O (2011) In vitro culture of coconut (*Cocos nucifera* L.) zygotic embryos. In: Thorpe TA, Yeung EC (eds) Plant embryo culture: methods and protocols, methods in molecular biology, vol 710. Springer, London

Escobar R (1981) Preliminary results of the collection and evaluation of the American oil palm (*Elaeis oleifera* HBK Cortes) in Costa Rica. In: Proceedings of the international conference on oil palm in agriculture in the eighties, Kuala Lumpur, 17–20 Jun 1981. The Incorporated Society of Planters, Kuala Lumpur, pp 79–97

Hao Y-J, Liu Q-L, Deng X-X (2001) Effect of cryopreservation on apple genetic resources at morphological, chromosomal and molecular levels. Cryobiology 43:46–53

Henky N, Rompas T, Darwis SN (1994) Coconut breeding programme in Indonesia. In: Batugal PA, Ramanatha RV (eds) Coconut breeding. IPGRI, Serdang, pp 28–41

Hodkinson TR, Wldren ES, Pamell JAN, Salamin N (2007) DNA banking for plant breeding, biotechnology and biodiversity evaluation. J Plant Resour 120:17–29

Jabatan Perangkaan Malaysia (Malaysian Department of Statistics) (2009) Buku maklumat perangkaan Malaysia 2009. Percetakan Nasional Malaysia Berhad, Putrajaya, 18 pp

Jain SM (2011) Prospects of in vitro conservation of date palm genetic diversity for sustainable production. Emirate J Food Agric 23(2):110–119

Komolafe MF, Joy DC (1990) Agricultural science for senior secondary schools, book 1. University Press, Ibadan

Kushairi A, Mohd Din A, Rajanaidu N (2011) Oil palm breeding and seed production. In: Mohd Basri W, Choo YM, Chan KW (eds) Further advances in oil palm research (2000–2010) MPOB, Bangi, pp 47–101

Lamothe MN (1991) Coconut improvement – needs and opportunity. Papers presented on International Board for Plant Genetic Resources (IBPGR) workshop, Cipanas, 8–11 Oct 1991

Marshall DR, Brown AHD (1975) Optimum sampling strategies in genetic conservation. In: Frankel OH, Hawkes JG (eds) Crop genetic resources today and tomorrow. Cambridge University Press, Cambridge, UK, pp 53–80

Martine BM, Laurent KK, Pierre BJ, Eugene KK, Hilaire KT, Justin KY (2009) Effect of storage and heat treatments on the germination of oil palm (*Elaeis guineensis* Jacq.) seed. Afr J Agric Res 4(10):931–937

Meunier J (1969) Etude de populations naturelles d'Elaeis guineensis en Côte d'Ivoire. Oléagineux 24:195–201

Mkumbo KE, Kullaya A (1994) Coconut breeding in Tanzania. In: Batugal PA, Ramanatha RV (eds) Coconut breeding. IPGRI, Serdang, pp 15–27

Murphy DJ (2007) Future prospects for oil palm in the 21st century: biological and related challenges. Eur J Lipid Sci 109:296–306

Noll RG (2008) The wines of West Africa: history, technology and tasting notes. J Wine Econ 3(1):85–94

Ohler JG (1999) The coconut palm and its environment: historical background. In: Ohler JG (ed) Modern coconut management: palm cultivation and production. SRP, Exeter, pp 3–10

Panis B, Lambardi M (2005) Status of cryopreservation technologies in plants (crops and forest trees). In: The role of biotechnology seminar, Turin, 5–7 Mar 2005

Pesce C (1985) Oil palms and other oil seeds of the Amazon. Reference Publication, Algonac, pp 51–90

Rajanaidu N (1980) Oil palm genetic resources: current methods of conservation. Paper presented at the international symposium on conservation inputs from life sciences. MPOB, Bangi, pp 25–30

Rajanaidu N (1985) *Elaeis oleifera* collection in Central and South America. In: Proceedings of the international workshop on oil palm germplasm and utilization, Selangor, Malaysia, pp 84–94

Rajanaidu N, Ramanath Rao V (2004) Ex situ conservation of plantation crop genetic resources. In: Encyclopaedia of Plant and Crop Science 1(1):1–9

Rajanaidu N, Ramanatha Rao V (2002) Managing plant genetic resources and the role of private and public sectors: oil palm as model. In: Engels JMM, Ramanath Rao V, Brown AHD, Jackson MT (eds) Managing plant genetic diversity. CAB International/IPGRI, Wallington/Rome, pp 425–436

Rajanaidu N, Arasu NT, Obasola CO (1979) Collection of oil palm (*Elaeis guineensis*) genetic material in Nigeria II. Phenotypic variation of natural populations. MARDI Res Bull 7: 1–27

Rajanaidu N, Kushairi A, Jalani BS, Tang SC (1994) Novel oil palm from exotic palms. Malays Oil Sci Technol J 3(2):22–28

Rao NK (2004) Plant genetic resources: advancing conservation and use through biotechnology. Afr J Biotechnol 3(2):136–145

Rasiah R, Shahrin A (2005) Development of palm oil and related products in Malaysia and Indonesia. Universiti Malaya, Kuala Lumpur, pp 1–54

Ratnambal MJ, Nair MK (1994) National coconut breeding programme in India. In: Batugal PA, Ramanatha RV (eds) Coconut breeding. IPGRI, Rome, pp 1–14

Reed BM, Sarasan V, Kane M, Bunn E, Pence VC (2011) Biodiversity conservation and conservation biotechnology tools. In Vitro Cell Dev Biol Plant 47:1–4

Richardson DL (1995) The history of oil palm breeding in the United Fruit Company. ASD Oil Palm Paper 11:1–23

Sajini KK, Karun A, Kumaran PM (2006) Cryopreservation of coconut (*Cocos nucifera* L.) zygotic embryos after pre-growth desiccation. J Plant Crops 34:576–581, 131

Sharma M (2007) Research and development: the role of oil palm planting materials. Planter 83(973):227–230

Sterling F, Alvarado A (2002) Historical account of ASD's oil palm germplasm collections. ASD Oil Palm Paper 24:1–16

Sterling F, Richardson DL, Alvarado A, Montoya C, Chaves C (1999) Performance of OxG E. oleifera Central American and Colombian biotype x *E. guineensis* interspecific hybrids. In: Rajanaidu N, Jalani BS (eds) Proceedings of the seminar on worldwide performance of DXP oil palm planting materials. Clones and interspecific hybrids. Palm Oil Research Institute of Malaysia, Kuala Lumpur, pp 114–127

Thampan PK (1982) Handbook on coconut palm. Oxford & IBH Publishing, New Delhi, pp 1–25

Whitehead RA (1965) Freeze-drying and room temperature storage of coconut pollen. Econ Bot 19(3):267–275

Yusof B (2007) Palm oil production through sustainable plantations. Eur J Lipid Sci Technol 109:289–295

Chapter 11
Conserving Tropical Leguminous Food Crops

N. Quat Ng

11.1 Introduction

Species of leguminous food crops are the second most important plant species, after cereal (wheat, rice and maize), used by man for food purposes. In addition to their use as human food, many leguminous species are also used as animal feeds, medicines, and cover crops for protection of soil from erosion and improving soil fertility, green manures for enhancing condition/restoring soil fertility, for pharmaceutical and industrial uses.

Tropical food legumes refer to leguminous plant species cultivated or grown in the tropics and subtropics, for human foods. These include species used for extraction of oils, and production of pulses (dry bean seeds), fruits, green beans and pods, or bean sprouts, leaves and, tubers for uses as foods and vegetables. Among some 600 genera of the Leguminosae family with approximately 18,000 species distributed all over the world, less than 15 species are of major importance used extensively today for human food (NAS 1979; Duke 1981). This chapter is intended to provide some basic information on the uses of the leguminous food species, their origin and, distribution, and the results of an assessment on the current status of genetic resources of the food legumes being conserved ex-situ.

N.Q. Ng (✉)
Seri Alam Plantation SDN. BHD, Lot 57897, Jln Bulit Kemuning, Batu 6,
Shah Alam, Selangor 40470, Malaysia
e-mail: quatng@hotmail.com

M.N. Normah et al. (eds.), *Conservation of Tropical Plant Species*,
DOI 10.1007/978-1-4614-3776-5_11,

11.2 Coverage of Crops Under This Chapter

Leguminous food species to be introduced and covered for discussion under this chapter include 11 species of major economic importance which are widely spread and cultivated. In addition, 16 species of minor importance, presently limited in area of cultivation, neglected and/or under-utilized species, are also included.

The species considered as major importance are soybean, peanuts, common bean, lima bean, cowpea, mung bean, chickpea, pea, pigeonpea, broadbean, and lentil. The less important ones include lablab bean, Bambara groundnut, winged bean, yard-long bean, catjan bean, creole bean, scaret runner bean, runner bean, tepary bean, yam bean, African yam bean, Jack bean, sword bean and tamarind.

11.3 Uses, Origin and Distribution of the Species Under Discussion

The uses, origin, and distribution of each of the species are briefly described below to give some knowledge about the crops, using crop name as a subheading. Areas of origin and centers of diversity of crops are where one can find gene pools that are more diverse.

11.3.1 Soybean (*Glycine max* (L.) Merr.)

Uses: Soybean is the most important among all food legumes, because its seeds contain up to 40% protein (dry weight) and are high in all the eight essential amino acids except methionine. The oil content of the seeds are also very high (>17% in dry weight). Soybean oil extracted from seeds is widely used for cooking, in salad, for manufacture of margarine and shortening, paints, soap, insecticides and many other industrial uses as well as for bio-fuel.

Dried seeds are processed to give a soy-milk high in protein (as good if not better than cow milk or other animal milks) low in fat content with no cholesterol, a very healthy high protein drink for human consumption. They are processed to make bean curds, cheese, high protein vegetarian diets, and soy sauce. Unripe seeds are eaten as vegetable and dried seeds eaten whole, split, or sprouted. Soybean sprouts are highly nutritious, which are widely consumed in Asia, and is also popular in other part of the world.

Origin and Distribution: Two closely related annual wild species of soybean, *Glycine soja* (presumed ancestor of the cultivated soybean) and *G. gracilis*, are native to China, where the centers of diversity of soybean and the two closed relatives are found. Soybean is a very ancient crop, with written records of its cultivation in China for over 3,000 years. It is believed that soybean originated in China.

On the other hand, nine related wild species within the same *Glycine* subgenus of soybean (*G. argyrea, G. canescens, G. clandestina, G. cyrtoloba, G. falcata, G. latifolia, G. latrobeana, G. tabacina* and *G. tomentella*) are native to Australia (Hymowitz and Newell 1981). Soybean grows well in subtropical region, now widely spread out and cultivated through the world, extending from the tropics to 52°N. Varieties suitable for lowland tropics are also available.

11.3.2 Peanut, or Groundnut (*Arachis hypogaea* L.)

Uses: The high oil and protein contents of the peanut seeds serve important needs for human food and energy. Oil extracted from dried seeds are used as cooking oil, in salads, canning and for shortening in pastry and bread, in pharmaceutical industry and for making margarines, peanuts butter, soaps, lubricants, diesel fuel and in emulsions for insect control. The cake is used as feed and may be used as flour, for human. Shelled or unshelled seeds are used whole or split, roasted, salted or unsalted as snack. Young pods and seeds may be used as vegetables and snack.

Origin and Distribution: Peanut is native to South America, the crop is now cultivated throughout tropical, subtropical and warm temperate regions of the world extending from 40°N to 40°S. It has large diversity with numerous cultivars grown in different parts of the world with its center of diversity found in South America (Krapovickas 1969; Gregory and Gregory 1976; Singh and Simpson 1994; Singh and Nigram 1997). Wild forms of peanut and wild *Arachis* species are found throughout South America.

11.3.3 *Phaseolus* Beans

Species of the *Phaseolus* genus originated from Mesoamerica and Andes (Kaplan 1965). There are 55 species, with 5 species domesticated and cultivated for human foods. The cultivated species are *P. vulgaris* L, *P. lunatus* L., *P. acutifolius* Asa Gray, *P. coccineus* L. and *P. polyanthus* Greenman. The first two species are of major economic important which are widely grown throughout the world; while the others are minor important, with their cultivation limited and confined to small geographical distribution. The uses and origin of the five species are given below.

11.3.3.1 Common Bean (*Phaseolus vulgaris* L.)

Uses: Common bean is most cultivated of all food legumes in the temperate regions, and widely cultivated in the tropical regions in all continents, especially in Latin America and Africa (Pachico 1989; Duke 1981). It is cultivated chiefly for the green immature pods in most parts of the world. The green immature pods are cooked and eaten as vegetable. Immature pods are marketed fresh, frozen or canned, whole or cut. Dried mature seeds are also widely consumed. In some part of the tropics, leaves are used as pot-herb and young leaves as salad and the green-shelled seeds are also eaten.

Origin and Distribution: Of all the cultivated *Phaseolus* species, the common bean is the most widely spread and cultivated species throughout the world, with the greatest diversity and broadest range of cultivars. Precise location of the domestication is unclear, probably is in Central Mexico and Guatemala. Very old archaeological remains of the common bean discovered in Tehuacan, Mexico, and in Ancash, Peru, dated back to 7,000 years BP and 8,000 years BP respectively. Diverse forms of the ancestor wild species and cultivars of common bean evolved sympatrically in the Middle American and Andean South American regions. Nine distinctive races of the cultivated species are recognizable in the region (Singh et al. 1991). Six races commonly cultivated are Mesoamerica, Durango and Jalisco from the Middle America and, Nueva Granada, Chile and Peru from Andean South America (Hidalgo and Beebe 1997).

11.3.3.2 Lima Bean (*Phaseolus lunatus* L.)

Uses: Lima beans are marketed green or dry, canned or frozen. Green and dried seeds are eaten cooked and seasoned or mixed with other vegetable or foods. Green immature pods of some cultivars are sometimes eaten cooked as vegetable.

Origin and Distribution: Remains of the cultivated lima beans found in Tehuacan, Mexico and Chilca, Peruvian Andes dated back to 1,400 years BP and 5,300 years BP respectively. This indicates the crop had been brought into cultivation long time ago. Both the cultivated and wild forms are widely distributed with their primary center of diversity found in the Mesoamerican region and in the North and South Andean mountain ranges (Debouck 1986; Hidalgo and Beebe 1997). Lima beans are well adapted to the lowland tropics, is one of the most widely cultivated pulse crops, both in the temperate and subtropical regions. Some cultivars are also doing well in lowland tropical rainforest zone in Africa.

11.3.3.3 Tepary Bean (*Phaseolus acutifolius* Asa Gray)

Uses: The dry shelled beans of tepary are marketed and consumed. They are eaten like other dry bean, first soaked and then boiled or baked. Some American Indians also parch the beans and grind them to a meal that can be added to boiling water as instant food. The beans are also popular as a base soups and stews in northern Mexico. In Africa, the beans are boiled and ground, and then added to soups.

Origin and Distribution: Tepary bean is native to north Mexico and southwestern United State, might have been brought into cultivation in the region at least 5,000 years ago. It is a drought-tolerant crop, thrives in arid and semiarid region, can withstand heat. Both the domesticated and the wild ones still found throughout much of North America, stressing from north Mexico to southwestern United State, the region thought to be the center of origin and diversity of the crop and its wild forms. From this region, the crop was introduced to other parts of the world. It is grown year run in tropical Africa as a food crop.

11.3.3.4 Scaret Runner Bean (*Phaseolus coccineus* L.)

Uses: The crop is cultivated chiefly for its green pods. The young, green tender pods boiled, steamed and sautéed are eaten; the flowers used in salads and the young leaves are eaten as a potherb. Green and dry seeds are also used like other *Phaseolus* beans and the tubers are sometimes boiled and eaten as well in Central America (Duke 1981).

Origin and Distribution: Its origin and center of diversity are thought to be in Mesoamerica and northern Andean. Remains of the crop excavated in Ocampo, Mexico dated back to about 7,500 year BP. This crop has been introduced to Asia, Australia and, Africa and is cultivated in limited areas.

11.3.3.5 Runner Bean (*P. polyanthus* Greenman; Often Times Its Synonym *P. dumosus* Macfad. Is Used)

Uses: Its uses are similar to *P. coccineus*, the scaret runner bean described above.

Origin and Distribution: Archaeological remains of the runner bean found in Ocampo, Mexico dated back to about 7,500 years BP. Centers of diversity of both the cultivated and wild forms of the runner bean are found in Mesoamerican and north Andes (Schmit and Debouck 1991). It has been suggested that Guatemala is the center of origin and domestication of the species. The crop is more adapted to tropical mountain rain forest than the scaret runner bean (Hidalgo and Beebe 1997).

11.3.4 Cultivated Species of the African *Vigna*

Two species (*V. unguiculata* and *V. subterranea*) were domesticated from among some 50 wild *Vinga* species found in Africa, and cultivated for human foods. The crops commonly called cowpea (or black eye pea), yard-long bean and catjan bean belong to *V. unguiculata*. Bambara groundnut belongs to *V. subterranea*.

Cowpea, yard-long bean and catjan bean had previously been classified as three distinct botanical species *V. unguiculata* (L.), *V. sesquipedalis* (L.) Fruhw., and *V. cylindrical* (L) Skeels respectively. These have been widely recognized and classified under a single species *V. unguiculata* subsp. *unguiculata* (L.) (Walp.) Verdc., and use Cultigroup Unguiculata for cowpea, Cultigroup Sesquipedalis for yard-long bean, and Cultigroup Biflora for catjan bean to distinguish them from one and another (Marechal et al. 1978; Ng and Marechal 1985; Ng 1990, 1995;

Ng and Singh 1997). Over 50 species of wild *Vinga* and the closely related species and weedy forms of the cultigens are abundantly found throughout Africa south of Sahara and Ethiopia. Some wild *Vigna* species are pan-tropical distribution.

11.3.4.1 Cowpea {*V. unguiculata* (L) subsp. *unguiculata* (L.) (Walp.) Verdc. Cultigroup Unguiculata}

Uses: Cowpea, or black eye pea, is the most important and widely spread crop among all the cultivar-groups of *V. unguiculata*. Cowpea is cultivated mainly for its dry seeds in the tropics throughout the world, especially in Africa and Brazil. Cowpea seeds can be shelled green or dried, but they are mostly marketed and consumed as dried seeds. Protein content of dried seeds is high; varying from 24–30% and up to 35%. The seeds are used in plain cooking (soups, stews, boiled or roasted), mixed with other foods (such as rice and sorghum), or meats and in processed dishes (boiled, steamed, dried or baked). Young leaves, tender green pods and, immature seeds are also used as vegetable.

Origin and Distribution: Cowpea and its closely related wild species originated from Africa. Great diversity of the crop and its close relative and, many wild *Vigna* species are found in Sub-Sahara Africa and Ethiopia. The center of diversity of the crop appears to be in West Africa, where the crop is believed to have first been domesticated (Ng and Marechal 1985; Ng and Padulosi 1991), whereas more wild forms of the close relatives are found in the south eastern part of Africa. The crop was brought to Indian sub-continent, Sri Lanka and South East Asia and cultivated there in ancient time. Two new forms, Cultigroup Biflora and Cultigroup Sesquipedalis, aroused in these new found lands, through intensive cultivation and selection by local farmers after cowpea was brought there (Steele and Mehra 1980).

11.3.4.2 Yard-Long Bean {*V. unguiculata* (L) subsp. *unguiculata* (L.) (Walp.) Verdc. Cultigroup Sesquipedalis}

Uses: The yard-long bean is mostly climbing annual herb, but semi-erect to erect types are also available. It is cultivated mainly for its long pods (30–100 cm long), which are succulent when young and used as vegetable. Only the whole immature young green pods are marketed. Young leaves, shelled immature and mature (dried) seeds are also used as vegetable. Yields of green pods are very high, may up to 10 MT/ha. Raw green pods are nutritious; contain 3% protein, 8% carbohydrate, 1.6% fiber, vitamins and minerals.

Origin and Distribution: It is commonly cultivated on trellis or stakes in vegetable garden and sometimes inter-cropped with other crops that provide support for the climbing vines in many parts of Asia, South East Asia and the pacific regions. It is also found growing in small patches in Africa and other parts of the world, as in the Mediterranean region Caribbean and Latin America. Much diversity is found in

India and South East Asia. As already mentioned in the above section, the crop evolved in India and South East Asia, after the cultigen of *V. unguiculata* was brought there for cultivation.

11.3.4.3 Catjan Bean {*V. unguiculata* (L) subsp. *unguiculata* (L.) (Walp.) Verdc. Cultigroup Biflora}

Uses: The catjan bean is cultivated, mainly for its dried seeds for use as pulse. The dried grains are rounded and cylindrical in shape with hard seed coat, generally smaller than the seeds of Cultigroup Unguiculata. The dried seeds are used whole or split in making dishes similar to those of cowpea meals. The young leaves, tender green pods, and immature seeds are also used as vegetable.

Origin and Distribution: Center of diversity of the catjan bean is in India and Sri Lanka and many varieties are also found growing in East Africa.

11.3.4.4 Bambara Groundnut {*Vigna subterranea* (L.) Verdc.; Synonym = *Voandzeia subterranea* (L.) Thours)}

Uses: The crop is grown like peanut; it forms pods and seeds on or just beneath the ground. Like peanut, immature and mature (dried) seeds are consumed as food. Seeds may be consumed fresh, grilled, boiled, or made into flour to form cakes. Seeds are also canned in gravy and marketed. Immature shelled or unshelled nuts are pounded and boiled to a stiff porridge used as food. In much of Africa, Bambara groundnut is the third most important legume crop after peanut and cowpea (Howell 1994). It is a preferred food crop of many local people in Africa, and it provides a good supplement to a cereal-based diet there (Azam-Ali et al. 2001).

Origin and Distribution: Bambara groundnut is of African Origin with its greatest diversity found in West Africa (Olukolu et al. 2011; Goli et al. 1997). It is cultivated extensively by smallholding farmers of the semi-arid regions throughout Africa and is widely intercropped with cereals, root and tuber crops. The crop is relatively unknown outside Africa, except grown in small scale in India, Indonesia, Malaysia, the Philippines, Sri Lanka and Thailand, New Caledonia and South America, particularly Brazil.

11.3.5 Cultivated Species of the Asiatic *Vigna*

Five domesticated species (*V. radiata, V. mungo, V. aconitifolia, V. umbellata* and *V. reflexo-pilosa*) in the Asiatic genus *Vigna* subgenus *Ceratotropis* are cultivated in the tropical and sub-tropical regions for use as human foods (Tomooka et al. 2002). *V. anguilaris*, another domesticated species of the same subgenus *Ceratotropis,* is a temperate crop cultivated mostly in East Asia, the status of the conservation of its generic resources is not included in the assessment in this report. Among the six cultivated species, mung bean *V. radiata*, is most important, which is widely spread, cultivated, and used in many parts of the world. The other species are cultivated in a relatively limited geographical region. Some 20 wild *Vinga* species and the close relatives and weedy forms of the Asiatic domesticated species are found throughout most of the Asia continent, parts of South East Asia and the Pacific Islands.

11.3.5.1 Mung Bean (*Vigna radiata* (L.) Wilczek)

Uses: Mung bean is cultivated mainly for its dried seeds for making various foods and vegetable dishes. In South Asia, whole or split dried seeds are made into dhal soup with spices. In China, Japan and South East Asia, the bean is used to make various kinds of sweets, bean jams, sweetened bean soup, cakes, buns, vermicelli, bean sprouts or eaten mixed with stem rice. Bean sprouts are widely used as vegetables all over the world.

Origin and Distribution: Mung bean is believed to have been domesticated in India (Vavilov 1926; Singh et al. 1974). Some studies showed the greatest diversity of the land races of Mung bean is found in West Asia (Afghanism-Iran-Iraq) (Tomooka et al. 1992). It had also been suggested that the crop might have moved from West Asia eastward, through India, to China and South East Asia. The wild form of Mung bean *V. radiata* var. *sublobata* (Rohb.) Verdcourt, is widely distributed in Asia, parts of Africa (especially in the eastern region) and Australia. The crop is widely cultivated throughout the tropic and sub-tropic around the world.

11.3.5.2 Black Gram (*Vigna mungo* (L.) Hepper)

Uses: It is cultivated for its matured dried seeds, used whole or split and cooked to make dhal soup in South Asia. The seeds are also ground into flour and used to make cakes, breads, and porridge, or used for production of bean sprouts for used as vegetable. It is relatively important pulse in India.

Origin and Distribution: Black gram is native to South Asia. The crop was probably domesticated in India, where the center of diversity of the species is found. The cultivation of the bean for food is relatively restricted to India and the surrounding areas and recently spread to other tropical regions where Indians have migrated.

11.3.5.3 Rice Bean (*Vigna umbellata* (Thunb.) Ohwi and Ohashi)

Uses: Rice bean seeds are used as pulse for human food in India, Myanmar, China, South East Asia, Fuji and Mauritius. Dried seeds are usually boiled and used as vegetable. Immature pods are briefly boiled with salt and eaten with rice among the hill tribe in Northern Thailand. Young leave are also used as vegetable.

Origin and Distribution: The crop is widely grown in small scale across tropical Asia, and to a limited extent in South East Asia, Fiji, Australia, Mauritius, Africa and the United State. Wild forms were found from southern Himalayas across Burma, Indo-China to south China in the East, and to Malaysia and Indonesia in the South. The crop probably originated in Southeast Asia and, high diversity is found among cultivars from across the tropical Asia (Duke 1981; Tomooka et al. 2002).

11.3.5.4 Creole Bean {(*V. reflexo-pilosa* Hayata var. *glabra*) (Marechal, Mascherpa and Stainier) N. Tomooka and Maxted}

Uses: This little known cultigen is cultivated for use as a forage and food crop. The use of the bean for human food is much the same way like mung bean (Tomooka et al. 2002)

Origin and Distribution: Reports indicate that the crop has been cultivated in India, Vietnam, Philippines, Mauritius, and Tanzania. Its wild form is widely distributed in the Pacific islands and, from across South Asia to South East and East Asia, Papua New Guinea and Northern Australia. The diversity of the species is not well understood.

11.3.5.5 Moth Bean (*V. aconitifolia* (Jacq.) Marechal)

Uses: Green pods, and ripe seeds, whole or split are coked as vegetable. Sometimes seeds are sprouted and eaten with or without salt, or fried and salted (Duke 1981; Jain and Mehra 1980).

Origin and Distribution: It is believed that Moth bean was domesticated over a wide geographical region in the Indian subcontinent, including Myanmar and Sri Lanka (Purseglove 1974; Marechal et al. 1978). The centers of diversity of both the cultivated and wild forms are found in this region. The cultivation of the crop has spread to China, South East Asia, Africa, and Southern United State (Duke 1981).

11.3.6 Broadbean, or Faba Bean (*Vicia faba* L.)

Uses: Several species of the genus *Vicia* are economically important forage legumes. Broadbean is the only cultivated species in the genus used for human food as well as forage for animal feed. The bean of the crop is used green or dried, fresh, cooked or canned and processed for human consumption. Green pods are sliced and used as green bean vegetable. Green seeds are eaten cooked, canned, and frozen as vegetable and used in salad. Dried seeds, whole, ground into flour or processed are used for making various foods (like medames, cous-cous, canned bean, fried snacks, fafafel, and soup, soya sauce substitute, bean paste, noodles and vermicelli). Dried seeds are high in protein content (18–39%), has been considered suitable for making a meat or a skim milk substitutes. A chief limiting factor for the utilization of the broadbean as a human food is because of some antinutritional substances, such as the glucosides, convicine and vicine, contain in the plant and seeds which can cause haemolytic anaemia in human (Marquardt 1982; Robertson 1997). The levels of the glucosides are higher in the green seeds.

Origin and Distribution: The species is only known in cultivation and is native to the Middle East. It is believed to have been domesticated in areas between Afghanistan and the eastern Mediterranean around 8,500 years BP (Hanelt 1972). The crop spread from this region to the other part of the world. It is grown and used quite extensively in the Middle East, many parts of Pakistan and the Mediterranean countries, China, Egypt, Ethiopia, and Australia. It is also cultivated in small scale in many parts of the world in other countries including the Andes in South America for use as human food (Duke 1981; Maxted 1995; Robertson 1997). China and Ethiopia are the main growers and the major producers of the crop.

11.3.7 Chickpea (*Cicer arietinum* L.)

Uses: Chickpea is cultivated for its nutritive seeds with protein content ranges from 13% to 30% dried weight basis. Seeds are eaten fresh or as dry pulse, parched, boiled, fried, or cooked in various dishes. Dried seeds are ground into flour for making many dishes and snacks, popular in the Indian subcontinent and the Middle East. Sprouted seeds are eaten as vegetable. Young plants and green pods are also eaten like spinach. Unripe seeds are eaten as snack

Origin and Distribution: Chickpea originated in the areas encompassing southeast of Turkey, northern Iran and the Eastern Mediterranean region, where the center of diversity of the crop and its wild relatives are found (Vavilov 1926; Singh et al. 1997). Wild species are most abundant in Turkey, Iran, Afghanistan, and central Asia. Seed remnants of the cultivated species found in Hacillar near Burder dated back to 6,450 years BP indicating the crop had been domesticated in the region since antiquity. From this region, the crop moved west- and northwest-ward to the Mediterranean basin and the Europe, north to Central Europe, south to Africa, south-east to the Indian subcontinent and east to China. Chickpea reached India at least 4,000 years BP.

Two main types of chickpea are grown: desi, with angular and colored seeds, primarily grown in South Asia, Iran, Ethiopia; and kabuli, with large, owal-head shape and beige-colored seeds, grown in the Mediterranean region, Afghanistan and Pakistan. The crop has now been widely cultivated in many parts of the world, including the America and Australia.

11.3.8 Pea (*Pisum sativum* L.)

Uses: The crop is used as a vegetable, marketed fresh, frozen or canned. Green seeds and tender green pod are eaten as vegetable. It is also grown to produce dry peas (seeds), which are used whole, split or ground into flour for making many different dishes or, cooked into soup. Pea sprouts are also used as vegetable. The crop grow mostly in temperate countries, is included in this chapter because of its economic important, as well as potential for the higher altitudes region in the tropics.

Origin and Distribution: Pea is a cool season crop grown in many temperate countries in the Mediterranean, Europe, North America, Australia, China, Japan and in many other parts of the world. It thrives well in a cool and relatively humid climate, not suitable for the lowland tropic, but can grow well in the higher altitudes in the tropics. The distribution of wild pea species is more restricted to the Mediterranean basin and the Near East. The earliest archaeological remain of pea was found in Syria, Turkey and Jordon. The findings in Egypt date back to over 6,000 years BP. The crop probably originated in Europe and Western Asia, and hundreds of cultivars have been widely used (Duke 1981).

11.3.9 Lentil (*Lens culinaris* Medikus)

Uses: Lentil is cultivated for its nutritious seeds, which are high in protein content (20–35% of dry matter). It is a dietary mainstay and one of the most important pulses in the drier regions of the Middle East, North Africa, and Indian subcontinent. Lentil seeds should not be eaten raw, due to the presence of anti-nutrient substances such as phytic acid and tannin. Green seeds and young pods are used as green vegetable. Dried mature seeds, whole or split, hulled or dehulled, or ground into flour, are used for preparing soup, fried lentil and many different dishes. Lentils are frequently used for cooking mixed with rice (in some popular dishes like kushari, mujaddara and khichdi) and sometime mixed with other cereals.

Origin and Distribution: Wild progenitors and relatives are primarily found in the West Asian region (Cubero 1981). The crop probably was domesticated in the Middle East, with archaeological records showed that lentil seeds had been disseminated in the region some 10,000 years BP, and much diversity is found among the cultivars from this region (Cubero 1981; Zohary 1972; Ladizinsky 1979). The crop spread to Cyprus and southern Europe about 8,000 years BP, Egypt 7,000 years BP and India around 4,000 BP. Now the crop is cultivated in most of subtropical and warm temperate regions of the world, and high altitudes of the tropics.

11.3.10 Pigeonpea (*Cajanus cajan* (L.) Millsp.)

Uses: Pigeonpea is cultivated for human foods, animal feeds and cover crop. The seeds are nutritious, with a protein content ranging between 19% and 26%, and starch content between 51% and 59%. In India subcontinent, it is used mainly as human food. The seeds are used whole, or split (dhal), for cooking in many different dishes and soup. In Africa, it is cooked together with vegetable or meat.

Green seeds are consumed as a vegetable in eastern part of Africa, the Caribbean Islands, South/Central America, Indonesia, the Philippines and part of India. Fermented pigeonpea is made into sauce, as a substitute for soya sauce in Indonesia.

Origin and Distribution: Pigeonpea originated in India, probably in the central and eastern parts of the country (Remanandan and Singh 1997). The diversity of the cultivars found in this region is remarkable and the wild species, *Cajanus cajanifolius* (Haines) van der Maesen comb.nov., the supposed progenitor of the cultivated species is also found in this region. Six other wild species of *Cajanus* closely related to pigeonpea are found in India. The crop was thought to have been cultivated more than 3,000 years ago and spread to Africa over 1,000 years ago, where a secondary center of diversity established (van der Maesen 1990). It was later introduced and spread to other parts of the world where Indian people have migrated. It is now cultivated in the tropical and subtropical areas between 30°N and 30°S latitudes.

11.3.11 Winged Bean (*Psophocarpus tetragonolobus* (L.) DC.)

Uses: Winged bean is a nutritious plant. All parts of the plant, young leave, flowers, young pods, immature and mature seeds, and tubers are edible. Per 100 g edible portions of the green pods contain 2–4 g protein, 8 g carbohydrate and 1.6 g fiber, of mature seeds 30–39 g protein and 24–42 g carbohydrate, of immature seeds 5–10 g protein and 6–42 g carbohydrate, of raw roots/tubers 3–15 g protein, 27–30 g carbohydrate. However, winged bean is primarily cultivated for its immature edible pods, and cooked as a vegetable; young pods are also eaten raw. The immature pods are marketed and popularly used in South East Asian countries, Papua New Guinea, Myanmar, Sri Lanka and parts of India. They are also sold as a specialty vegetable in markets in North America and Western Europe, mostly for the Asian community. Yong leaves and flowers are also used as vegetables and in soups. Dry seeds are often made into a fermented food product "tampeh" in Indonesia. Fresh and roasted tubers are eaten and sometime sold in markets in Papua New Guinea and Myanmar.

Origin and Distribution: Winged bean is undoubtedly a crop of South East Asia, and the Islands of the Western Pacific region, Burma and India, even though its origin is still unclear. The crop is widely cultivated in this part of the world, and where two centers of the diversity are found; one in Papua New Guinea and another one around Thailand, Myanmar and Indo-china region (Khan 1982; Harder and Smartt 1992). Uncertainty about the origin of the winged bean is streaming from the facts that only the cultigen, with no wild form or any of its wild relatives, is found in the Asian and Pacific region. On the other hand, winged bean cultigen, along with eight wild *Psophocarpus* species, including its presumed progenitor species *P. grandiflorus*, are found in Africa, where the cultivation of winged bean and the diversity of the crop varieties found there are negligible.

11.3.12 Jack Bean (*Canavalia ensiformis* (L.) DC.)

Uses: Jack bean is mainly cultivated for green manure, as soil cover for erosion control, and for forage. Yong pods and immature seeds are used as a vegetable. Flowers and young leaves are steamed as a condiment in Indonesia. Roasted beans are sometimes used as a substitute of coffee. Ripe dry beans may be eaten after long cooking, but contain a mild poison. Immature pods and beans contain per 100 g; 6.9% protein, 0.5% fat, 13.3% carbohydrate, 3.3% fiber.

Origin and Distribution: Jack bean is a prehistoric American Indian domesticated crop in southwest United States. It is native to the region from Mexico south to Brazil and Peru, and the West Indies (Sauer and Kaplan 1969). The crop is now cultivated in many parts of the tropics, including India, Indonesia, China, Tanzania, Kenya and Hawaii.

11.3.13 Sword Bean (*Canavalia gladiata* (Jacq.) DC.)

Uses: Sword bean is used as a vegetable, forage and cover crop. Yong green pods are used quite extensively as a vegetable in India, Myanmar, Sri Lanka, and other Asian countries. Fully grown seeds of white seeded varieties may be cooked and eaten as substitute for broadbeans. Seeds may be soaked overnight, boiled until soft in water with sodium bicarbonate, then rinsed, boiled in new water and finally pounded for use in curries. Dried seeds contain 27% protein and 54% carbohydrate; green seeds 2.7% protein and 6.4% carbohydrate; raw green pods 2.8% protein and 7.3% carbohydrate.

Origin and Distribution: Sword bean is originated from the Old World and, probably derived from *C. virosa.* The centers of diversity of the sword bean cultigen are in the Indochina – Indonesian center, and the Chinese–Japanese center. It is extensively cultivated in India, Myanmar, Sri Lanka, Indochina and East Asian countries, and has spread throughout the tropics (Sauer 1964)

11.3.14 Lablab Bean (*Lalab purpureus* (L.) Sweet)

Uses: Lablab bean is used as a human food, green manure and cover crop, forage and ornamental plant, but is cultivated mainly for human food. Yong pods make an excellent vegetable; dried seeds are either cooked and eaten directly, or processed to bean cake/paste. Its leaves and flowers are cooked and eaten like spinach; sprouts are comparable to soybean or mung bean sprouts. Perennial form produces edible root tubers.

Origin and Distribution: Lablab bean has been cultivated in Asia since ancient time and, is now widespread throughout the tropics with wide adaptability. Some believe it is native to India and South East Asia, others believe it is of African origin (Purseglove 1974; Zeven and Zhukovsky 1975).

11.3.15 Yam Bean (*Pachyrhizus* DC.)

Uses: The three yam bean species (*Pachyrhizus erosus* (L.) Urban. *P. tuberosus* (Lam.) Spreng., *P. aphipa* (Wedd.) Farodi) are cultivated mainly for their edible tuberous roots. Their seeds unlike other food legumes are poisonous, which are not consumed. The texture of the tuber is succulent, juicy, crunchy and slightly sweetish in taste, appeal to most plates. The tubers are eaten raw as salad or fruit, whole or sliced, and cooked in different dishes on its own or mixed with other vegetables or meats for use as vegetable, in soups and wraps. They are also sliced and made into chips. Yong pods of *P. erosus* are also eaten as a vegetable, but the mature pods are poisonous. The tubers of *P. aphipa* and *P. tuberosus* may be squeezed to make a juicy drink in South America.

Origin and Distribution: The three cultivated species in the *Pachyrhizus* genus, *P. erosus, P. tuberosus* and *P. aphipa* are Neotropic origin (NAS 1979; Sorensen 1996). *P. erosus* and *P. tuberosus* are the two main cultivated ones, whereas *P. aphipa* is only cultivated in a very limited scale in its native land in the subtropical east Andean valley of Bolivia and northern Argentina. *P. erosus* originated in Mexico and Central America, and was introduced to Asia by the Spanish in the sixteenth century. It is now widely cultivated in the tropical America and many other parts of the world especially in Asia, South East Asia and the Islands of the Western Pacific. *P. tuberosus* is native to Amazon headwater region of South America to parts of Caribbean. Presently, it is cultivated chiefly in the tropical region of South America.

11.3.16 African Yam Bean {*Sphenostylis stenocarpa* (Hochst. ex A. Rich.) Harms}

Uses: African yam bean is cultivated in West Africa for both seeds and tubers for human foods, which are an important source of protein and starch for the local people. The tubers produced underground look like elongated sweet potatoes, smaller in size, taste like Irish potato. They contain about 12% protein (dry weight), twice the amount in sweet potatoes or Irish potato. The dried seeds are high in protein content, varying from 21% to 28%. They are cooked and eaten plain, with spices or mixed with other foods.

Origin and Distribution: The cultivated African yam species *S. stenocarpa,* along with six other species in the genus *Sphenostylis* are native to Africa. The crop is found growing wild throughout much of the tropical Africa, and is cultivated in central and Western Africa, especially in Nigeria, Cote d'Ivoire, Ghana, Togo, Gabon, Zaire and Central Africa. The crop is relatively unknown outside the African continent. Diversity of the crop and its wild relatives are in Africa.

11.3.17 Tamarind (*Tamarinus indica* L.)

Uses: Tamarind trees are cultivated mainly for the pulp in the fruits, used in making beverages, flavoring dishes and confections, curries and sauces. Tamarind is cultivated large scale in India and Thailand for market purposes. The fruits are usually gathered from its natural distribution, trees growing in the wild vegetation, planted around the edges of farmlands or home garden plots, along the roads and around villages and dwellings in many countries in the tropics.

The fruit pulp extracted from mature fruits is made into preserves, pastes, or syrups and marketed. The pulp usually taste sour, but sweet varieties are available. The mature pods of sweet varieties, unbroken entire fruits, are seen marketed in India and Thailand and sold directly to consumer for use as fruits and deserts. The pulp is rich in organic acids as source of vitamins and mineral. Yong leaves and pods are also used. It is of considerable importance in parts of the world.

Origin and distribution: Tamarind is indigenous to Africa, where it continues to grow wild, with the greatest diversity of the species found there. It is cultivated in Part of Africa, especially Nigeria, Cameroon and Tanzania, and in India and Thailand and widely planted in the dry zones and areas with well drained soils of many other tropical countries throughout the world (El-Siddig et al. 2006).

11.4 Status of Conserving Food Legume Genetic Resources

In order to assemble information on existing germplasm collections of the tropical food legumes referred to in this chapter presently being held, or preserved, at different genebanks and institutions around the world, the author studied and reviewed many reports and publication, including the SoWPGR2 and country reports for the SoWPGR2 (FAO 2010). He accessed and reviewed PGR information available onlines Internet, especially those of the CGIAR centers (CIAT, ICARDA, ICRISAT, and SINGER), AUSPGRI, EURISCO, GRIN-CA, NISM-GPA, USDA-GRIN, and WIEWS. He also contacted many researchers in national and international institutions to help providing information, through e-mail communication. He noted some differences in the number of accessions of a particular collection, documented between databases, and some reports. The information on existing collections presented in this report is based mainly on information retrieved from the WIEWS PGR databases, supplemented by information from other databases, reports, and some limited information obtained through personal contacts (WIEWS n.d.). Acronyms for the genebanks and institutions, organizations and common terms used in this report are given in the Appendix 1.

Seeds of all the food legume species and their wild relatives store well under dry conditions (typically at a seed moisture content varying between 4% and 7%) and, cold storage environments (typically at a temperature below 0°C to −20°C) (see Chap. 3 , this volume). This is the traditional and universal method for ex-situ conservation of the genetic resources of seed crops. Food legume seeds remain viable after long period of storage under these conditions. Some legume seeds can also be stored under liquid nitrogen (at −196°C); theoretically for infinite time. Certain forms or genotypes especially of some wild relatives, or types like tamarind trees (although their seeds store well using the traditional seed bank method) with special features, may require a combination of methods, including the use of field/orchard collection, maintaining live plants in pots or other containers or structure under contained conditions, on-farm or in-situ, and in vitro cultures for the conservation of the genetic resources.

However and with exception in several major genebanks, detailed information on the storage conditions of the germplasm collections held at most genebanks/institutions is unclear. Information on the standards of the seed storage conditions, health status, and quality of the seeds under storage are not easily accessible, or available. Thus, the present assessment on the status of the genetic resources conservation is based on how well the germplasm of a crop have been collected and preserved in ex-situ genebanks, in term of the number of accessions and their geographical coverage.

The number of accessions of the different food legume crops and their wild relatives held at the genebanks around the globe is summarized in Tables 11.1, 11.2, 11.3, and 11.4. Only the major genebanks, with the top rankings in number of accessions of each crop are listed in the tables, with a summary of the total number of accessions held in each genebank, and the sum total of the accessions in the rest of the other genebanks around the world, and the grand total accessions of the crop are presented. A very impressive total number of germplasm accessions of the food legume crops and their wild relatives, exceeding one million (>1,100,000), are being maintained or preserved in many institutions/genebanks around the world.

The information shows the *Phaseolus* collection with a total number of about 250,000 accessions is the largest, followed by soybean (230,026), *Vigna* (>135,409), peanut or groundnut (128,570), chickpea (98,313), pea (94,001), lentil (58,405), broadbean (43,695) and pigeonpea (40,509). Four CGIAR centers (CIAT, ICARDA, ICRISAT, IITA) and AVRDC, together with a small number of national genebanks as listed in Tables 11.1, 11.2, 11.3, and 11.4 accounted for over half the total number of accessions of all crops.

Among the *Phaseolus* cultivated species, the common bean is the most important which is widely cultivated and spread throughout the world. It has the largest number of accessions (over 184,814) held in 192 institutions worldwide. The next important is lima bean, with a total of 13,991 accessions in 75 institutions; followed by the three minor crops in the genus the scaret runner bean, runner bean and tepary bean each with a total number of 4,813, 934 and 865 accessions respectively. A total of about 6,000 accessions of wild and weedy species of the genus *Phaseolus* are recorded in some 33 genebanks worldwide, with the majority (>4,100) belonging to weedy or wild forms of the common bean.

Table 11.1 Germplasm collections of soybean, groundnut, chickpea, pea, lentil and broadbean in genebanks/institutions around the world

Institution acronym, country	No. acc.	% ws	Institution acronym, country	No. acc.	%ws
Soybean (*Glycine max*)			**Groundnut (*Arachis hypogaea*)**		
ICGR-CAAS, China	32,021	21	ICRISAT, India	15,419	3
SOY, USA	21,075	10	NBPGR_Delhi, India	13,144	7
RDAGR-GRD, Rep. Korea	17,644	<1	S9/USDA, USA	9,964	2
AVRDC, Taiwan	15,314		BBC-INTA, Argentina	8,347	4
CNPSO, Brazil	11,800		ICGR-CAAS, China	6,565	
NIAS, Japan	11,473	5	CENARGEN_Green, Brazil	2,042	
VIR, Russia	6,439		FCRI_DA/TH, Thailand	2,030	
NBPGR_Delhi, India	3,690		ICABIOGRAD, Indonesia	17,30	
TARI, Taiwan	2,745		VIR, Russia	1,667	
IPK, Germany	2,661	1	MRS, Zambia	1,500	
ATCFC, Australia	2,121	3	UzRIPI, Uzbekistan	1,438	
IITA, Nigeria	1,909		NPGRL/IPB-UPLB, Philippines	1,250	
AMFO, France	1,582		ATCFA, Australia	1,196	5
FCRI-DA/TH, Thailand	1,510		NIAS, Japan	1,181	1
INIA-Iguala, Mexico	1,500		CIFP, Bolivia	1,040	2
NPGRL/PBI-UPLB, Philippines	1,444		IPGR, Bulgaria	887	
ICA/REGION 1, Colombia	1,235		Others (133)	59,170	3
Others (173)	93,863	7	***Arachis* global total**	**128,570**	**3**
***Glycine* global total**	**230,026**	**6**			
Chickpea (*Cicer arietinum*)			**Pea (*Pisum sativum*)**		
ICRISAT, India	20,140	1	ATFCC, Australia	7,230	1
NBPGR_Delhi, India	14,704	2	VIR, Russia	6,653	
ICARDA, Syria	13,219	2	ICARDA, Syria	6,129	4
ATFCC, Australia	8,655	3	IPK, Germany	5,508	1
W6/USDA, USA	6,195	3	W6/USDA, USA	5,399	3
NPGBI-SPII, Iran	5,700		IGV, Italy	4,090	
PGRI, Pakistan	2,146	1	ICGR-CAAS, China	3,825	
VIR, Russia	2,091		SASA, U.K.	3,302	3
AARI, Turkey	2,075	1	NBPGR_Delhi, India	3,070	<1
INIA-Iguala, Mexico	1,600		SHRWIAT, Poland	2,960	<1
IBC, Ethiopia	1,173		NORDGEN, Sweden	2,821	2
RCA, Hungary	1,170	<1	CNPH, Brazil	1,958	
UzRIPI, Uzbekistan	1,055		IBC, Ethiopia	1,768	
IR, Ukraine	1,021		IR, Ukraine	1,671	<1
Others (104)	17,369	1	Others (156)	37,617	
***Cicer* global total**	**98,313**	**1**	***Pisum* global total**	**94,001**	2
Lentil (*Lens culinaris*)			**Broadbean (*Vicia faba*)**		
ICARDA, Syria	10,864	5	ICARDA, Syria	9,186	
NBPGR_Delhi, India	9.989	<1	ICGR-CAAS, China	4,207	
ATFCC, Australia	5,251	4	ATFCC, Australia	2,565	<1
NPGBI-SPII, Iran, Islamic Rep.	3,011	11	IPK, Germany	1,921	<1

(continued)

Table 11.1 (continued)

Institution acronym, country	No. acc.	% ws	Institution acronym, country	No. acc.	%ws
Lentil *(Lens culinaris)*			**Broadbean (*Vicia faba*)**		
W6/USDA USA	2,874	5	INRA-RENNES, France	1,700	
VIR, Russia	2,375		UC-ICN, Ecuador	1,650	
INIA CARI, Chile	1,345		IGV, Italy	1,420	
PGRC, Canada	1,171	1	VIR, Russia	1,259	
RCA, Hungary	1,074		INIACRF, Spain	1,252	
AARI, Turkey	1,073	1	IBC, Ethiopia	1,143	
SCARPP, Armenia	1,001		Others (122)	17,392	2
Others (97)	18,377	2	***Vicia* global total**	**43,695**	**1**
***Lens* global total**	**58,405**	**3**			

Table 11.2 *Phaseolus* germplasm collections in genebanks/institutions around the world

Institution acronym, country	No. acc.	Institution acronym, country	No. acc.
Common bean (*Phaseolus vulgaris*)		**Tepary (*Phaseolus acutifolius*)**	
CIAT, Colombia	29,766	CIAT, Colombia	*170*
W6/USDA, USA	12,057	W6/USDA, USA	115
CENAGREN_Em, Brazil	12,475	CENARGEN_Em, Brazil	100
IPK, Germany	7,920	UACH, Mexico	88
ICGR-CAAS, China	7,216	ATCFA, Australia	63
VIR, Russia	5,972	VIR, Russia	42
BCA, Malawi	6,000	INIFAP, Mexico	40
IREGEP, Mexico	5,210	ISRA, Senegal	39
RCA, Hungary	4,047	IR, Ukraine	36
KARI-NGBK, Kenya	3,514	IPGR, Bulgaria	31
NBPGR_Delhi, India	2,878	CUNSUROC-USAC, Guatemela	31
DENAREF, Ecuador	2,441	Others (21)	110
Others (180)	85,318	**Tepary global total**	**865**
Common bean global total	**184,814**		
Lima bean (*Phaseolus lunatus*)		**Scarlet runner bean (*Phaseolus coccineus*)**	
LBN, Indonesia	3,846	AISLJ, Slovenia	995
CIAT, Colombia	2,740	CIAT, Colombia	766
UFCG, Brazil	2,500	W6/USDA USA	374
CENARGEN_Green, Brazil	1,225	IPK, Germany	439
W6/USDA, USA	1,047	RCA, Hungary	215
NPGRL/IPB-UPLB, Philippines	689	UACH, Mexico	209
NIAS, Japan	554	IPGR, Bulgaria	175
UNALM, Peru	200	DENAREF, Ecuador	168
DENAREF, Ecuador	149	CENARGEN_Em, Brazil	147
INIA-CENIP, Venezuela	114	SKV, Poland	137
Others (65)	927	Other (43)	1,188
Lima bean global total	**13,991**	**Scarlet runner bean global total**	**4,813**
Runner bean (*P. polyanthus*)		**Wild *Phaseolus* species**	
CIAT, Colombia	475	INIFAP, Mexico	2,112

(continued)

Table 11.2 (continued)

Institution acronym, country	No. acc.	Institution acronym, country	No. acc.
Runner bean (*P. polyanthus*)		**Wild *Phaseolus* species**	
UACH, Mexico	104	CIAT, Colombia	1,934
UNALM, Peru	100	W6/USDA, USA	862
W6/USDA, USA	89	HBBRMAI, Belgium	406
CORPOICA, Colombia	51	ILRI, Ethiopia	272
ICA/REGION 1, Colombia	32	IPK, Germany	72
HBBRMAI, Belgium	25	DENAREF, Ecuador	57
CATIE, Costa Rica	24	CENARGEN_Em, Brazil	47
UNA, Peru	18	AMGRC, Australia	35
Others (4)	16	ATCFC, Australia	33
Runner bean global total	**934**	DLEG, USA	18
		UPM-BGV, Spain	16
		RBG, U.K.	16
		Others (20)	171
		Wild *Phaseolus* global total	**6,051**

Note: About 46,000 accessions of unknown taxa and status, designated as *Phaseolus* spp. or sp., are listed in the collections of some 40 institutions are not included. Major ones are 10,000 accessions with INIA-Iguala, 10,600 accessions with INIFP, 7,460 with IAPAR, and 1,712 with NARC

Table 11.3 *Vigna* germplasm collections in genebanks/institutions around the world

Institution acronym, country	No. acc.	Institution acronym, country	No. acc.
Cowpea (*V. unguiculata* Cultigroup Uguiculata)		**Mung bean (*V. radiata*)**	
IITA, Nigeria	15,019	NPGRL/IPB-UPLB, Philippines	6,978
S9/USDA, USA	7,593	AVRDC, Taiwan	5,897
CENARGEN, Brazil	5,426	ICGR-CAAS, China	5,158
LBN, Indonesia	3,930	AICRP-Mullarp, India	4,432
NBPGR_Delhi, India	3,307	S9/USDA, USA	3,981
ICGR-CASS, China	2,818	NBPGR_Delhi, India	3,123
NIAS, Japan	2,229	NBPGR_Jodhpur, India	2,466
DAR, Botswana	1,432	FCRI-DA/TH, Thailand	2,250
VIR, Russia	1,334	NIAS, Japan	1,579
NPGRL/IBP-UPLB, Philippines	1,307	ATCFC, Australia	667
AVRDC, Taiwan	674	PGRI, Pakistan	643
ATCFC, Australia	697		
Others (112)	15,574	Others (48)	4,770
Cowpea global total	**61,340**	**Mung bean global total**	**41,944**
Yard-long bean (*V. unguiculata* Cultigroup Sesquipedalis)		**Black gram (*V. mungo*)**	
NPGRL/IPB-UPLB, Philippines	573	AICRP-Mullarp, India	2,137
AVRDC, Taiwan	478	NBPGR_Delhi, India	1,503
NIAS, Japan	202	AVRDC, Taiwan	762
S9/USDA, USA	194	PGRI, Pakistan	799
PGRC, Viet Nam	157	NIAS, Japan	404
BARI, Bangladesh	147	S9/USDA, USA	302

(continued)

Table 11.3 (continued)

Institution acronym, country	No. acc.	Institution acronym, country	No. acc.
Yard-long bean (*V. unguiculata* Cultigroup Sesquipedalis)		**Black gram (*V. mungo*)**	
IITA, Nigeria	100	VIR, Russia	210
Others(13)	247	Others (31)	1,318
Yard-long bean global total	**2,098**	**Black gram global total**	**7,435**
Bambara groundnut (*V. subterranea*)		**Rice bean (*V. umbellata*)**	
IITA, Nigeria	1,703	NBPGR_Delhi, India	1,770
ORSTOM-MONT, France	1,416	ICGR-CAAS, China	1,415
DAR, Botswana	338	NIAS, Japan	476
PGRRI, Ghana	296	AVRDC, Taiwan	266
NPGRC, Tanzania	283	CPBBD, Nepal	157
SGRC, Zambia	232	IPB-UPLB, Philippines	150
RAS PGRC, South Africa	140	PGRC, Viet Nam	117
CRAF, Burkina Faso	126	LBN, Indonesia	100
24 others	1,271	Others (18)	364
Bambara groundnut global total	**5,805**	**Rice bean global total**	**4,815**
Wild *Vigna* and weedy forms of cultigents		**Moth bean (*V. aconitifolia*)**	
IITA, Nigeria	1,686	NBPGR_Jodhpur, India	1,720
ILRI, Ethiopia	1,142	NBPGR_Delhi, India	1,546
NIAS, Japan	1,088	IGFRI, India	242
CIAT, Colombia	1,050	PAK001, PGRI, Pakistan	66
ATCFC, Australia	801	S9/USDA, USA	56
HBBRMAI, Belgium	782	KARI-NGBK, Kenya	50
NBPGR_Delhi, India	399	VIR, Russia	49
S9/USDA, USA	352	ATCFC, Australia	37
NBPGR_Thrissur, India	292	9 others	45
KARI-NGBK, Kenya	185	**Moth bean global total**	**3,811**
Others (26)	384		
Wild *Vigna* global total	**8,161**		

Note: 3,017 accessions of unknown taxa and status, designated as *Vigna* spp. or sp., are not included

Within the *Vigna* genus, cowpea and mung bean, are the two most important crops each has a total of 61,340 and 41,944 accessions respectively being conserved globally. The sizes of the germplasm of the five minor *Vigna* crop species, balck-gram (7,435), Bambara groundnut (5,609), rice bean (4,815) and moth bean (3,811) are relatively small. The collections of yard-long bean (2,098), catjan bean (<100) and creole bean (<50) are even much smaller. A considerable high number of accessions of wild *Vigna* of about 8,000 comprises mainly the weedy and wild forms of cowpea and mung bean, and the wild *Vigna* species are recorded in 36 genebanks worldwide (Table 11.3). IITA holds the largest number of accessions covering 50

Table 11.4 Germplasm collections of pigeonpea, lablab bean, winged bean, yam bean, African yam bean, Jack bean, sword bean and tamarind in genebanks/institutions around the world

Institution acronym, country	No. acc.	ws %	Institution acronym, country	No. acc.
Pigeonpea (*Cajanus cajan*)			**Lablab bean (*Lalab purpureus*)**	
ICRISAT, India	13,289	1.6	NBPGR_Dehi, India	1,672
NBPGR_Delhi, India	12,859	4.1	BARI, Bangladesh	551
AICRP, India	5,195		AVRDC, Taiwan	413
KARI-NGBK, Kenya	1,288		ICGR-CASS, China	373
ATCFC, Australia	406	51.5	KARI-NGBK, Kenya	355
NPGRL/IPB-UPLB, Philippines	318		ILRI, Ethiopia	347
CENARGEN_Em, Brazil	278		ATCFC, Australia	207
CPBBD, Nepal	228		NPGRL/IPB-UPLB, Philippines	203
TISTR, Thailand	201		CIAT, Colombia	155
LBN, Indonesia	200		S9/USDA, USA	139
SAARI, Uganda	200		Others (54)	1,105
Others (78)	6,047	5.6	**Lablab bean global total**	**5,520**
Pigeonpea global total	**40,509**	**3.19**		
Winged bean (*Psophocarpus tetragonolobus*)			**Jack bean (*Canavalia ensiformis*)**	
NPGRL/IPB-UPLB, Philippines	652		CIBA UCV-FAGRO, Venezuela	519
DOA, Papua New Guinea	455		NBPGR_Delhi, India	38
DGCB-UM, Malaysia	435		ICGR-CAAS, China	31
TROPIC, Czech Republic	413		ATCFC, Australia	30
IDI, Sri Lanka	400		S9/USDA, USA	22
LBN, Indonesia	380		PGRRI, Ghana	16
Puslitkaret, Indonesia	234		CATIE, Costa Rica	15
NBPGR_Delhi, India	245		CENARGEN_Em, Brazil	14
AVRDC, Taiwan	245		ECPFSS, Cuba	14
S9/USDA, USA	177		Other s (27)	101
Others (30)	1,183		**Jack bean global total**	**800**
Winged bean global total	**4,696**			
Tamarind (*Tamarindus indica*)			**Sword bean (*Canavalia gladiata*)**	
CTPRSC-DA/TH, Thailand	166		NPGRL/IPB-UPLB, Philippines	40
NKTPRS-DA/TH, Thailand	149		NBPGR_Thrissur, India	37
PTPRSC-DA/TH, Thailand	125		NBPGR_Delhi, India	27
NBPGR_Delhi, India	43		AVRDC, Taiwan	16
S9/USDA, USA	35		PGRRI, Ghana	13
UACH, Mexico	15		NIAS, Japan	8
SOFRI, Viet Nam	13		BARI, Bangladesh	5
NPGRL/IPB-UPLB, Philippines	12		IITA, Nigeria	4
CNSF, Burkina Faso	9		Others (11)	21
ILRI, Ethiopia	5		**Sword bean global total**	**171**
Others (34)	70			
Tamarind global total	**642**			
Yam bean (*Pachyrhizus*)			**African yam bean (*Sphenostylis stenocarpa*)**	
CATIE, Costa Rica	181		IITA, Nigeria	140
DENAREF, Ecuador	72		PGRRI, Ghana	38

(continued)

Table 11.4 (continued)

Institution acronym, country	No. acc.	ws %	Institution acronym, country	No. acc.
Yam bean (*Pachyrhizus*)			**African yam bean (*S.stenocarpa*)**	
INIFAP, Mexico	49		S9/USDA, USA	12
AVRDC, Taiwan	49		RSA PGRC, South Africa	4
NIAS, Japan	40		LBEV, Togo	3
BNGTRA-PROINPA, Bolivia	40		RBG, U.K.	3
PRC, Viet Nam	39		NIAS, Japan	3
CIAT, Colombia	15		PREPSC, South Africa	3
IPK, Germany	14		HBBRMAI, Belgium	1
FCA-UNSM, Peru	11		**African yam bean global total**	**207**
Others (15)	189			
Yam bean global total	**699**			

species, mostly of African origin. Accessions of the wild and weedy forms of the minor economic important *Vigna* species are very minimal.

The sizes of the accessions of the seven minor food legumes shown in the Table 11.4 are very small. Among these, lablab bean has the highest number of 5,520 accessions, followed by winged bean (4,696), Jack bean (800), yam bean (699) and, tamarind (642). African yam bean and sword bean each has about 200 accessions.

11.5 Observations and Discussions

It should be pointed out that the information on number(s) of accessions of an individual crop held in genebanks across the globe alone cannot be used to indicate the amount of diversity of the crop being conserved at an individual or in all the genebanks. A higher number of accessions of a particular crop collection in one genebank do not necessarily indicate that collection contains a greater diversity than a collection of another genebank with a lower number of accessions. The extent of coverage of the total diversity of different crops in existing ex-situ collections is difficult if not impossible to estimate with real precision. Nonetheless, the sizes of a collection, together with other available information at the accession level such as their passport data, and diversity as revealed by phenotypic descriptors and agronomic traits or molecular markers can be used to speculate the extent of a relative amount of diversity of the species in the collection. On the other hand, if only a very small number of accessions of a particular crop or its wild relatives are collected and preserved, this does suggest that only a very small portion of the diversity of the crop is being preserved ex-situ.

Judging from the information available and knowing the histories of past practices used by many researchers and plant explorers in collecting and acquiring genetic resources, the diversity of the major food legume species have been very well collected, assembled and preserved in CIAT, ICARDA, ICRISAT, IITA, AVRDC, and in many national institutions around the globe. It is not uncommon that duplicates of the same materials were collected or acquired, and preserved in the same genebank unintentionally. There are many instances that the same plant materials in the fields might have been collected by different individuals at different time, or same germplasm accessions from one genebank being transferred around and ended up being preserved in one or several genebanks, resulted in multiple duplicates of the same germplasm materials being held and preserved in one or several genebanks. Storing duplicates, for safety back-up is essential, as is being practice by all the CGIAR centres, and many national genebanks, by storing a subset of their genebank materials in other genebanks with high standard, such as those in USDA, where several CGIAR centres deposited parts of their duplicate germplasm accessions for safety back up. In a few cases, CGIAR centers' genebanks and a few large national genebanks also offer services for back-up storage of national materials. On the other hand, preserving multiple duplicates of the same genetic materials in multiple locations and genebanks, not for security or practical purposes, is wasteful of resources.

In connection with the subject on safety storage of germplasm, it is good that the Svalbard Global Seed Vault has finally been built on a Svalbard islands in the Arctic Ocean to provide security back up of crop diversity. The idea of having such a system was first floated by IBPGR/FAO around 1978. This Seed Vault opened in 2008 for genebanks that wish to deposit duplicate of their accessions there. Since its opening, there has been a concerted effort at depositing duplicate accessions from the CGIAR, some national and regional germplasm collections (GCDT 2011).

The large numbers of accessions of the collections of the major crops that have been assembled and held by many genebanks is correlated with the economic importance of the crops which are widely spread and cultivated throughout the world, and amount of research activities related to crop improvement, conservation, and biosystematic studies being conducted on those crops and their wild relatives. Many national and international institutions, notably the CGIAR centers and AVRDC, devoted considerable resources to collecting and acquiring the genetic resources of the major legume crops under present discussion for conservation and use, since their establishment in late 1960s and early 1970s. These centers and IBPGR in collaboration with national programs and their research partners had systematically collected germplasm from the fields for conservation and use in crop improvement related activities. For instance, IITA researchers had acquired and collected over 4,000 accessions of cowpea from existing collections held in many institutions around the world, as well as from farmer fields in parts of West Africa, when they began breeding activities in the early 1970s. Between 1976 and 1990, IITA in collaboration with its research partners, IBPGR, and many national programs had mounted numerous plant exploration missions throughout Africa. This resulted in collecting over 10,000 accessions of cowpea and Bambara groundnut and over 1,500 accessions of wild cowpea relatives and wild *Vigna* (Ng 1990; Ng and Padulosi 1991). Similarly, other CGIAR centers in collaboration with IBPGR and national researcher partners explored and collected large number of germplasm accessions of their mandate crops during this period.

The existing collections of the crop germplasm preserved at CIAT, ICARDA, ICRISAT, IITA and AVRDC constituted a big proportion to the total accessions held by all the genebanks around the world. The common bean is accounted for 16% of the world total number of accessions, lima bean 20%, soybean 7%, groundnut 12%, pea 7%, chickpea 34%, lentils 19%, broadbean 21%, pigeonpea 33%, cowpea 26%, and mung bean 16%. Majority of the collections at the four centers were collected over a very wide geographical distribution in the centers of diversity of those crops. Thus, the gene pools of the crop collection(s) available at the centers are very diverse. The values of the individual crop collections at these centers are not easy to measure; certainly, their relative worth in comparison to other genebank's collection, in terms of the total diversity of an individual crop collection contains, should be far greater than the percentages to the global total number of accessions they accounted for.

As earlier indicated, duplicate(s) of the same germplasm accessions might be held in different genebanks, and even within the same genebank at the CGIAR centers and national genebanks. Based on on-line WIEWS or other PGR database information, it is impossible to distinguish which accessions are duplicates, or unique. For instance, it is unclear if the reported 7,262 accessions of groundnut land races held in Niger-ICRISAT are indeed unique accessions totally different from those genebank accessions (15,419) already held at India-ICRISAT. Similarly, it is not possible at this stage to tell from existing information if the collection of 2,466 accessions of mung bean reported in NBPGR_Jodhpur are indeed different from, or part or all of these are already included in the 3,147 genebank accessions of the same species held at the Headquarters NBPGR_Delhi.

Despite of the very large number of accessions of the 11 major food legumes have been maintained and preserved globally, there are still some pocket areas where land races might not have been collected and preserved, as has been reported by several national reports and in SoWPGR2 (2010). Further analyses of the information available, some carefully planned and well executed collecting expedition trips to be carried out in those pocket areas would yield additional diversity not presently represented in genebanks.

It is highly alarming to note that only a very small number of germplasm accessions, hence the diversity, of the minor food legumes species has been preserved ex-situ. With the exception of Bambara groundnut and winged bean, it seems past efforts in promoting the awareness of the importance of plant biodiversity and the conservation of the natural resources over the last two decades by many organizations have yet to impact on the conservation of the minor food legumes in ex-situ genebanks.

Genetic erosion in most food crops occurs in the fields every day and is serious. Several country reports and the SoWPGR2 documented cases of genetic erosion have been taken places in many places. The situation face by many of the neglected minor legume crops is very critical. This is certainly the case with African yam bean and Kersting's groundnut (*Macrotyloma geocarpum* (Hams) Marechal and Baudet. synonym *Kerstingiella geocarpa* Harms). Kersting's groundnut is another minor food legume indigenous to Africa not mentioned earlier. It faces the danger of being extinct from farmer's fields, as it is being replaced by other crops (personal observation). Presently the combined total number of no more than 40 accessions of the Kersting's groundnut is held at IITA in Nigeria, LBEV in Togo, PGRRI in Ghana and RBG in U.K.

There is urgency for a more concerted and effective efforts to be taken in the future by both the national and international agencies to ensure the genetic resources of those minor crops are not left out in their current and future emphasis in preserving biodiversity.

The diversity of wild relatives of all the food legume species have not been adequately collected and conserved ex-situ. The situation in soybean appears to be better than other crops. With the exception of soybean, the existing collections of the wild relatives of food legumes are conserved mainly in the four CGIAR centers (i.e. CIAT, ICARDA, ICRISAT and IITA), and a few major national genebanks in Australia, China, India, Japan, Europe, U.S.A., and Latin America.

Majority of the germplasm accessions of wild *Glycine* and wild soy bean are preserved in ICGR-CAAS (China), CSIRO (Australia) and Soy (U.S.A.), and several other national genebanks in the three countries and NIAS in Japan. The diversity in the collection of 2,089 accessions of wild *Glycine* comprising of 21 species held by SOY, perhaps is the greatest among all the wild soybean collections in genebanks around the world. The accessions in Soy came from the centers of diversity and origin of the species. There is also an exceedingly valuable collection of about 2,110 accessions of 16 wild *Glycine* (non-*G.max* or *G.soja*), presently held in CSIRO, Australia. This collection might contain the highest diversity of those species among all genebanks. A collection of about 6,700 accessions of wild *Glycine* is reported in ICGR-CASS genebank, however the species coverage and the geographical distribution of the collection are unclear.

It has been noted that past efforts in collecting wild crop relatives for biosystematics related research by researchers in several institutions had contributed to the collection and conservation of large number of accessions of the very valuable wild gene pools of the crop relatives. Examples are the *Phaseolus* and *Vigna* collections in HBBRMAI (Belgium) and, *Glycine* in CSIRO (Australia) and the University of Illinois (U.S.A.).

More efforts are needed to collect wild crop relatives for ex-situ conservation, for utilization and research, as well as develop effective conservation strategies and national policies for conserving wild relatives in situ. It is encouraged to note that IITA recently revamps its effort in collecting wild cowpea and wild *Vigna* in Africa in collaboration with research partners. A wild cowpea collecting trip carried out in Nigeria in 2010 with support from Global Crop Diversity Trust netted some 261 samples of wild and weedy species of *V. unguiculata.* IITA is also pursuing some strategic work to further enhance the conservation and documentation of cowpea germplasm (Dominique 2011, personal communication).

11.6 Conclusion

The food legumes under discussion are immensely important, second only to cereals, as a source of human food and animal feed. They offer a variety of edible products that include dry seeds as sources for protein and carbohydrate, immature seeds and pods, young leaves, bean sprouts and roots (of a few species) are succulent, and/or crunchy, are used as green vegetable and salad. These vegetables are high in vitamins and soluble carbohydrate and fiber. Dry legume seeds are 2–3 times richer in protein than cereal grains, some varieties with their protein contents as high as between 40% and 60% on dry weight basis. They provide a more economical source of protein than from animal especially in the tropics, and are healthier foods than animal products as they also contain high fiber and carbohydrate with no cholesterol. Some species like soybean and groundnut are also very rich in oil. Unimproved tamarind trees have been widely exploited in many parts of the tropics since long ago, for its fruits with pleasant taste, and leaves used in many ways. Legume plants can also help maintaining soil fertility and preventing soil erosion.

Multiple uses of legumes offer important advantages to farmers in the developing countries. There are species and varieties suitable for sole-crop field cultivation, as well as for inter-crop cultivation with cereals and other crops in many subsistence farming systems throughout the world. Legume vegetables also constitute part of the component crops in the peri-urban agriculture and home gardens that they provide food security and opportunity for income generation to farmers. Thus the use of legumes should be encouraged and promoted, hence research for sustainable or increasing the production, improving nutritive quality and resistant to pests and diseases (be less dependence on the use of pesticides which are harmful to human health) in these crops for specific agro-ecological conditions and niches should be continued and enhanced. This requires the availability of genetic resources.

The study revealed very large collections of over one million accessions of the genetic resource of the 11 major and 16 minor food legumes and their wild relatives have been collected and preserved in hundreds of genebanks worldwide. The largest collection is soybean (230,026), followed by common bean (184,814), groundnut (128,570), chickpea (98,313), pea (94,001), cowpea (61,340), lentil (58,405), broadbean (43,695), mung bean (41,944), pigeonpea (40,509) and lima bean (13,991). Among the minor food legumes the largest one being black gram (7,435), followed by Bambara groundnut (5,609), lablab bean (5,520), rice bean (4,815), scarlet runner bean (4,813), winged bean (4,696), moth bean (3,811) and yard-long bean (2,098). Collections of others, the runner bean (934), tepary bean (865), Jack bean (800) and, tamarind (642), African yam bean (207), sword bean (171), catjan bean (<80), creole bean (<40) and kersting's groundnut (<40) are very small. Fairly large number of accessions of wild relatives of the food legumes have also been collected and preserved. The highest number of wild germplasm accessions is found in soybean (13,800), followed by *Vigna* (8,161), *Phaseolus* (6,051), groundnut (3,734), pea (1,880), lentil (1,752), pigeonpea (1,291), chickpea (983), and broadbean (437).

The most significant collections are those held at the five internal centres, AVRDC, CIAT, ICARDA, ICRISAT and IITA, and several large and well equipped national genebanks such as those in Australia, Brazil, China, Germany, India, Japan, Mexico and the USA (not listed in any order of ranking, nor the list is exhausted; see Tables 11.1, 11.2, 11.3, and 11.4 for further details).

Special recognition should be given to the international centres for their outstanding effort in collecting and preserving very large and diverse germplasm accessions of the valuable leguminous crops at their centre (15,314 soybean, 5,897 mung bean, 762 black gram, 478 yard-long bean and 266 rice bean in AVRDC; 29,766 common bean, 2,740 lima bean, 766 scarlet runner bean, 475 runner bean, 170 tepary bean, 1,934 wild *Phaseolus* species, and 1,050 wild *Vigna* species in CIAT; 13,219 chickpea, 187 wild *Cicer*, 10,864 lentil, 88 wild Lens, 9,186 broadbean, pea 6,129, 376 wild *Pisum* in ICARDA; 20,140 chickpea, 98 wild *Cicer*, 15,419 groundnut, 116 wild *Arachis* species, 13,286 pigeonpea in ICRISAT, and 15,019 cowpea, 100 yard-long bean, 1,703 Bambara groundnut, 140 African yam bean, 1,686 wild *Vigna* in IITA). These collections are well preserved, characterized, and their information properly documented. Many of the germplasm materials are fairly well evaluated for sources of resistance to prevailing pests and diseases, and other useful agronomic traits for use in crop improvement at their centre and research sites in many countries. The germplasm materials and information available in these centres are shared freely with all interested users. In recent years, these centers use a germplasm transfer agreement in order to continue supplying germplasm materials freely to bona fide users, by recognizing the source origin of the materials, and fair and equitable sharing of benefits arising from the use of the materials.

The existing germplasm collections of most minor food legumes are very small, they are held mostly by national genebanks. The more significant ones are those in India (>3,500 black gram, >3,200 moth bean, 1,672 lalab bean), China (1,415 rice bean), France (1,416 Bambara groundnut), Philippines (652 winged bean, 40 sword bean), Papua New Guinea (455 winged bean), Thailand (440 Tamarind), CATIE in Costa Rica (181 yam bean). Wild crop relatives of all food legumes are also not adequately collected and preserved.

It is noted that the Convention on Biological Diversity (CBD) and the International Treaty on Plant Genetic Resource for Food and Agriculture (ITPGRFA) are two most significant International Treaties promoted the conservation of plant biodiversity in recent years. The provision on recognition of sovereignty right of nation over its genetic resources of the CBD, and the sharing of benefits arising from the use of plant genetic resources of the ITPGRI have had great impact on conservation and use of plant genetic resources for food and agriculture. They promoted political support for the conservation of biological resources. The mushrooming of many national plant genetic resources programs and genebanks, resulted in collecting and preservation of very huge number of germplasm accessions of many nationally important crops (not only legumes) over the last one to two decades are testimony to the positive impact of the treaties.

However, it has also been noted that many nations, their national programmes and authorities responsible for the conservation or use for the germplasm become overly jealous on genetic materials perceived to have originated from their territories, and are very reluctant, at times even refuse to share materials or information with others, for fear of bio-piracy and other reasons. Often times, it becomes very difficult to acquire germplasm materials, or get accurate information on existence of the germplasm collections.

A spillover effect from the implementation of the CBD and ITPGRFA has resulted in enhancing the public awareness of the importance of plant genetic resources, and acquiring and preserving lot of plant genetic materials in many national genebanks worldwide. Same genetic materials could have been maintained and preserved in different genebanks in many countries, which would make them more securely preserved. On the other hand, this might lead to multiple duplicates of the same genetic materials being preserved in many different genebanks, and overly duplication of efforts in preserving the same genetic materials by many genebanks. This is wasteful of financial and human resource, which could be used for collecting and preserving other genetic materials not presently represented in ex situ genebanks, and doing other important tasks.

The diversity of all the minor food legumes species are poorly represented in existing genebanks. Genetic erosion in these crops occurring in the fields is serious. A few species like Kersting's groundnut is in danger of facing extinction. There is an urgent need for a more concerted and effective efforts be taken in the future by both the national and international agencies to ensure the genetic resources of the minor crops are not left out in their current and future emphasis in preserving biodiversity. A greater effort should also be placed on collecting and preserving wild relatives, for ex-situ conservation and use. It is also necessary to develop effective strategies and national policies for conserving wild relatives in situ.

Even though the diversity of the major cultivated food legume species have been relatively well collected and preserved ex-situ, there are some pocket areas where local land races have not been collected and preserved in genebanks. Further analyses of all information available, together with reports made by several countries and the SoWPGR2 on specific areas where further collection might be needed, some carefully planned collecting expedition trips be filled out in those pocket areas would yield additional diversity not presently represented in the genebanks.

There is need for improving documentation of germplasm both in the software development for enhancing retrieval of information, and quality of information, with better passport data and georeference information and other germplasm data. Regular updating is important to ensure the information is current and up-to-date.

Acknowledgement Many individuals whose names are too numerous to acknowledge had responded positively to the request by the author for information and assistance in clarifying matters related to genebank holdings and others. Specific thank goes to the staff of the IITA Genebank, the national focal points and stakeholders of the NISM-GPA project in many Asian countries for their support, enthusiasm in conserving genetic resource, and willingness in sharing information

Appendix 1. Acronyms for Names of Institutes, Genebanks and Terms Used in the Food Legume Chapter

Acronym	Name, country
AARI	Plant Genetic Resources Department, Izmir, Turkey
AICRP-Mullarp	All India Coordinated Research Project on Mullarp, India
AISLJ	Crops and Seed Production Department, Agricultural Institute of Slovenia, Slovenia
AMFO	G.I.E. Amelioration Fourragere, France
AMGRC	Australian Medicago Genetic Resources Centre, South Australian Research and Development Institute, Australia
ATCFC	Australian Tropical Crops and Forage Genetic Resources Centre, Australia
ATFCC	Australian Template Filed Crops Collection, Horsham Victoria, Australia
AUSPGRIS	Australian Plant Genetic Resource Information Service
AVRDC	Asian Vegetable Research and Development Center (also known as World Vegetable Center), Shanhua, Taiwan
BARI	Plant Genetic Resources Centre, BARI, Bangladesh
BBC-INTA	Banco Base de Germoplasm, Instituto de Recursos Biologicos, Instituto Nacional de Tecnologia Agropecuria, Hurlingham, Argentina
BCA	Bunda College of Agriculture, Lilongwe, Malawi
BNGTRA-PROINPA	Fundación para la Promoción e Investigación de Productos Andinos, Bolivia
CATIE	Centro Agronómico Tropical de Investigación y Enseñanza, Costa Rica
CENARGEN_Em	Embrapa Recursos Geneticos e Biotecnologia, Brazil
CENARGEN_Green	Greenhouse Collection, Embrapa Recursos Genéticos e Biotecnologia, Brazil
CGIAR	Consultative Group on International Agricultural Research, Washington, DC, USA
CGN	Centre for Genetic Resources, the Netherlands Plant Research International, Wageningen, Netherlands
CIAT	Centro Internacional de Agricultura Tropical, Cali, Colombia
CIFP	Centro de Investigaciones Fitoecogeneticas de Pairumani, Cochbamba, Bolivia
CNPH	Embrapa Hortalicas, Brasilia, Brazil
CNPSO	Embrapa Soja, Londrina, Brazil
CNSF	Centre National de Semences Forestieres, Burkina Faso
CORPOICA	Centro de Investigación La Selva, Corporación Colombiana de Investigación Agropecuaria, Colombia
CPBBD	Central Plant Breeding and Biotechnology Division, Nepal Agricultural Research Council, Nepal
CRAF	Centre de Recherches Environnementale, Agricole et de Formation de Kamboinse, Burkina Faso
CSIRO	Indigenous Crop Relatives Collection, Commonwealth Scientific and Industrial Research Organization, Australia
CTPRSC-DA/Th	Chaiyapoom Technical and Production Resources Service Center, Thailand
CUNSUROC-USAC	Centro Universitario de Sur Occidente, Universidad de San Carlos, Guatemela
DAR	Department of Agricultural Research, Ministry of Agriculture, Botawana

(continued)

Appendix 1. (continued)

Acronym	Name, country
DENAREF	Departmento Nacional de Recursos Fitogeneticos y Biotecnologia, Ecuador
DGCB-UM	Department of Genetics and Cellular Biology, University Malaysia, Malaysia
DLEG	Desert Legume Program, Tucson, Arizona, USA
DOA	Department of Agriculture, Papua New Guinea University of Technology, Papua New Guinea
ECPFSS	Estación Central de Pastos y Forrajes de Sancti Spiritus, Cuba
EEF-UCR	Estación Experimental Fraijanes – Universidad de Costa Rica, Costa Rica
EURISCO	European Network of Ex-Situ National Inventories, Rome, Italy
FCA-UNSM	Facultad de Ciencias Agrarias, Universidad Nacional de San Martín, Peru
FCRI-DA/TH	Field Crop Research Institute, Department of Agriculture, Thailand
GPA	Global Plan of Action
GRIN-CA	Germplasm Resources Information Network-Canadian Version, Canada
HBBRMAI	National Botanical Garden of Belgium, Belgium
IAPAR	Area de Documentacao, Instituto Agronomico do Parana, Brazil
IBC	Institute of Biodiversity Conservation, Adis Ababa, Ethiopia
IBPGR	International Board for Plant Genetic Resources, FAO (the predecessor of IPGRI and Bioversity International)
ICA/REGION 1	Corporación Colombiana de Investigación Agropecuaria Tibaitata, CORPOICA, Colombia
ICABIOGRAD	Indonesian Center for Agricultural Biotechnology and Genetic Resources Research and Development, Bogor, Indonesia
ICARDA	International Centre for Agricultural Research in Dry Areas (CGIAR), Aleppo, Syrian Arab Republic
ICGR-CAAS	Institute of Crop Germplasm Resources, Chinese Academy of Agricultural Sciences, Beijing, China
ICRISAT	International Crop Research Institute for the Semi-Arid Tropics (CGIAR), Patancheru, India
IDI	The International Dambala (Winged Bean) Institute, Sri Lanka
IFVCNS	Institute for Field and Vegetable Crops, Novi Sad, Serbia
IGFRI	Indian Grassland and Fodder Research Institute, India
IGV	Istituto di Genetica Vegetale, Consiglio Nazionale delle Richerche, Italy
IITA	International Institute of Tropical Agriculture (CGIAR), Ibadan, Nigeria
ILRI	International Livestock Centre for Africa (CGIAR), Addis Ababa, Ethiopia
INIA CARI	Centro Regional de Investigacion INIA Carillanca, Temuco, Chile
INIA-CENIP	INIA – Centro Nacional de Investigaciones Agropecuarias, Venezuela
INIA-Iguala	Estacion de Iguala, Instituto Nacional de Investigaciones Agricolas, Iguala, Mexico
INIFAP	Instituto Nacional de Investigacioes Forestales, Agricolas y Pecuarias, Mexico
INRA-MONTPEL	Genetics and Plant Breeding Station, ESRA-INAR SGAP, France
INRA-RENNES	Station d'Amelioration des Plantes, INRA, France

(continued)

Appendix 1. (continued)

Acronym	Name, country
INRA-VERSAIL	Station de Genetique/Amelioration des Plantes, INRA, Versailles Cedex, France
IPGR	Institute for Plant Genetic Resources "K. Malkov", Sadovo, Bulgaria
IPGRI	International Plant Genetic Resource Institute (predecessor of Bioversity International), Rome, Italy
IPK	Genebank, Leibniz Institute of Plant Genetics and Crop Plant Research, Germany
IR	Institute of Plant Production n.a. V.Y. Yurjev of UAAS, Ukraine
IREGEP	Instituto de Recursos Genéticos y Productividad, Colegio de Postgraduados, Mexico
ISRA	Institut Sénégalais de Recherches Agricoles, Senegal
IGV	Istituto di Genetic Vegetale, Consiglio Nazionale delle Richerche, Bari, Italy
KARI-NGBK	National Genebank of Kenya, Crop Plant Genetic Resources Centre – Muguga Kenya
LBEV	Laboratoire de Botanique et Ecologie Végétale, Togo
LBN	National Biological Institute, Bogor, Indonesia
MRS	Msekera Research Station, Chipata, Zambia
NARC	National Horticultural Research Centre, Thika, Kenya
NBPGR_Delhi	National Bureau of Plant Genetic Resources. New Delhi, India
NBPGR_Jodhpur	Regional Station Jodhpur, NBPGR, India
NBPGR_Thrissur	Regional Station Thrissur, NBPGR, India
NIAS	National Institute of Agrobiological Sciences, Tsukuba-shi, Japan
NISM-GPA	National Information Sharing Mechanism on the Implementation of the Global Plan of Action for the Conservation and Sustainable Use of PGRFA, FAO, Rome, Italy
NKTPRS-DA/TH	Nong Kai Technical and Production Service Center, Thailand
NORDGEN	Nordic Genetic Resource Center, Alnarp, Sweden
NPGBI-SPII	National Plant Genebank of Iran, Seed and Plant Improvement Institute, Iran, Islamic Rep. of
NPGRC	National Plant Genetic Resources Centre, Tanzania
NPGRL/IBP-UPLB	National Plant Genetic Resources Laboratory, Institute of Plant Breeding, University of the Philippines Los Banos, Philippines
ORSTOM-MONT	Laboratoire des Ressources Génétiques et Amélioration des Plantes Tropicales, ORSTOM, France
PGRC	Plant Gene Resources of Canada, Saskatoon Research Centre, Agriculture and Agri-Food Canada, Canada
PGRFA	Plant Genetic Resource for Food and Agriculture
PGRI	Plant Genetic Resources Institute, Islamabad, Pakistan
PGRRI	Plant Genetic Resources Research Institute, Ghana
PRC	Plant Resource Centre, Vietnam
PREPSC	Division of Plant and Seed Control, Department of Agriculture, Technical Service, South Africa
PTPRSC-DA/TH	Phetchaburi Technical and Production Resources Service Center, Thailand
Puslitkaret	Indonesian Rubber Research Institute, Indonesia
RAS PGRC	RSA, Plant Genetic Resources Centre, South Africa

(continued)

Appendix 1. (continued)

Acronym	Name, country
RBG	Millennium Seed Bank, Royal Botanical Gardens, Kew, Wakehurst, U.K.
RCA	Institute for Agrobotany, Tapioszele, Hungary
RDAGB-GRD	Genetic Resources Division, National Institute of Agricultural Biotechnology, Rural Development Administration, Suweon, Republic of Korea
S9/USDA	Southern Regional Plant Introduction Station, U. Georgia, USDA-ARS, Griffin, USA
SAARI	Serere Agriculture and Animal Production Research Institute, Uganda
SASA	Science and Advice for Scottish Agriculture, Scottish Government, U.K.
SCARPP	Scientific Center of Agriculture and Plant Protection, Armenia
SGRC	SADC Plant Genetic Resources Centre, Zambia
SHRWIAT, Poland	Plant Breeding Station, Wiatrowo, Poland
SINGER	System-wide Information Network on Genetic Resources, CGIAR, Rome, Italy
SKV	Plant Genetic Resources Laboratory, Research Institute of Vegetable Crops, Poland
SOFRI	Southern Fruit Research Institute, Vietnam
SOY	Soybean Germplasm Collection, USDA-ARS, Urbana, USA
SoWPGR2	The second report on the state of the world's PGRFA, FAO, Rome, Italy
SUMPERK	Agritec, Research, Breeding and Services Ltd., Sumperk, Czech Republic
TARI	Taiwan Agricultural Research Institute, Taichung, Taiwan
TISTR	Thailand Institute of Scientific and Technological Research, Thailand
TPOPIC	Institute of Tropical and Subtropical Agriculture, Czech University of Agriculture, Czech Republic
UACH	Banco Nacional de Germoplasma Vegetal, Departamento de Fitotecnia, Universidad Autónoma de Chapingo, Mexico
UC-ICN	Instituto de Ciencias Naturales, Ecuador
UFCG	Universidade Federal de Campina Grande, Brazil
UNA	Programa de Investigación en Hortalizas, Universidad Nacional Agraria La Molina, Peru
UNALM	Universidad Nacional Agraria La Molina, Peru
UNHEVAL	Universidad Nacional Hermilio Abad del Cusco, Centro K'Ayra, Peru
UPM-BGV	Comunidad de Madrid, Universidad Politecnica de Madrid, Escuela Techica Superior de Ingenieros Agronomos, Banco de Germoplasma, Spain
USDA-GRIN	Germplasm Resources Information Network, United States Department of Agriculture, USA
UzRIPI	Uzbek Research Institute of Plant Industry, Uzbekistan
VIR	N.I. Vavilov All-Russian Scientific Research Institute of Plant Industry, St. Petersburg, Russia
W6/USDA	Western Regional Plant Introduction Station, USDA-ARS Washington State University, USA
WIEWS	FAO World Information and Early Warning System on Plant Genetic Resources, FAO, Rome, Italy
ZARI	Zambia Agriculture Research Institute, Zambia

References

Azam-Ali SN, Sesay A, Karikari SK, Massawe FJ, Aguilar-Manjarrez J, Bannayan M, Hampson KJ (2001) Assessing the potential of an underutilized crop: a case study using Bambara groundnut. Exp Agric 37:433–472

Cubero JL (1981) Origin, taxonomy and domestication. In: Webb C, Hawtin G (eds) Lentils. CAB, Farnham, pp 15–38

Debouck DG (1986) Primary diversification of *Phaseolus* in the Americas: three centres? Plant Genet Resour Newsl 67:2–8

Duke JA (1981) Handbook of legumes of world economic importance. Plenum Press, New York

El-Siddig K, Gunasena HPM, Prasad BA, Pushpakumara DKNG, Ramana KVR, Vijayanand P, Williams JT (2006) Tamarind, *Tamaridus indica*. Southampton Centre for Underutilized Crops, Southampton.

FAO (2010) The second report on the state of the world's plant genetic resources for food and agriculture. FAO, Rome

GCDT (2011) Global crop diversity trust. Retrieved from http://www.croptrust.org, June 2011

Goli AE, Begemann F, Ng NQ (1997) Characterization and evaluation of IITA's Bambara groundnut collection. In: Heller J, Begemann F, Mushonga J (eds) Bambara groundnut (*Vigna subterranean* (L.) Verdc.). Promoting the conservation and use of underutilized and neglected crops, 9. Proceedings of the workshop on conservation and improvement of Bambara groundnut (*Vigna subterranean* (L.) Verdc.), 14–16 Nov 1995, Harare, pp 101–118. Institute of Plant Genetics and Crop Plant Research/Department of Research & Specialist Services/International Plant Genetic Resources Institute, Gaterleben/Harare/Rome

Gregory WC, Gregory MP (1976) Groundnut. In: Simmonds NW (ed) Evolution of crop plants. Longman, London, pp 151–154

Hanelt P (1972) Zur Geschichte des Anbaues von *Vicia faba* L. und ihrer verschiedenen Formen. Kulturpflanze 20:209–223

Harder DK, Smartt J (1992) Further evidence on the origin of the cultivated winged bean, *Psophocarpus tetragonolubus* (L.). DC. (Fabaceae): chromosome numbers and the presence of a host-specific fungus. Econ Bot 46:187–191

Hidalgo R, Beebe S (1997) *Phaseolus* beans. In: Fuccillo D, Sears L, Stapleton P (eds) Biodiversity in trust. Cambridge University Press, Cambridge, UK, pp 139–155

Howell JA (1994) Common names given to Bambara groundnut (*Vigna subterranea*: Fabaceae) in central Madagascar. Econ Bot 48:217–221

Hymowitz T, Newell CA (1981) Taxonomy of the genus *Glycine*, domestication and uses of soybeans. Econ Bot 35:272–288

Jain HK, Mehra KL (1980) Evolution, adaptation, relationships and uses of species *Vigna* cultivated in India. In: Summerfield RJ, Bunting AH (eds) Advances in legume science. Volume 1 of the proceedings of the international legume conference, Kew, pp 459–468

Kaplan L (1965) Archeology and domestication in American *Phaseolus*. Econ Bot 19:358–368

Khan TN (1982) Winged bean production in the tropics. Food and Agriculture Organization Plant Production Paper 38, Rome

Krapovickas A (1969) The origin, variability, and spread of the groundnut (*Arachis hypogea*). In: Dimbleby CW, Ucko RJ (eds) The domestication and exploration of plant and animals. Duckworth, London, pp 427–440

Ladizinsky G (1979) The origin of lentil and its wild genepool. Euphytica 28:179–187

Marechal R, Mascherpa JM, Stainier F (1978) Etude tax-onomique dún groupe complexe déspecies des genres *Phaseolus* et *Vigna* (Papilionaceae) sur la base de donnees morphologiques et polliniques, traitees par lánayse informatique. Booissiera 28:1–273

Marquardt RR (1982) Favism. In: Hawtin GC, Webb C (eds) Faba bean improvement: world crops: production, utilization, description, vol 6. Martinus Nijhoff, The Habgue, pp 343–353

Maxted N (1995) An ecogeographic study of *Vicia* subgenus *Vicia*. Systematic and ecogeographic studies in crop genepools 8. IBPGR, Rome, p 184

NAS (1979) Tropical legumes: resources for the future, 4th edn. National Academy of Sciences, Washington, DC

Ng NQ (1990) Recent development on cowpea germplasm collection, conservation, evaluation and research. In: Ng NQ, Monti LM (eds) Cowpea genetic resources. IITA, Ibadan, pp 13–28

Ng NQ (1995) Cowpea *Vigna unguiculata* (Leguminosae – Papilionoideas). In: Smartt J, Swimmonds NW (eds) Evolution of crop plants, 2nd edn. Longman, Essex, pp 326–332

Ng NQ, Marechal R (1985) Cowpea taxonomy, origin and germplasm. In: Singh SR, Rachie KO (eds) Cowpea research, production and utilization. Wiley, Chichester, pp 11–21

Ng NQ, Padulosi S (1991) Cowpea genepool distribution and crop improvement. In: Ng NQ, Perrino P, Attere F, Zedan H (eds) Crop genetic resources of Africa, vol 2. IITA/CNR/IBPGR/UNEP, Ibadan, pp 161–174

Ng NQ, Singh BB (1997) Cowpea. In: Fuccillo D, Sears L, Stapleton P (eds) Biodiversity in trust. Cambridge University Press, Cambridge, UK, pp 82–99

Olukolu BA, Mayers S, Stadler F, Ng NQ, Fawole I, Dominique D, Azam-Ali SN, Abbott AG, Kole C (2011) Genetic diversity in Bambara groundnut (*Vigna subterranea* (L.) Verdc.) as revealed by phenotypic descriptors and DArT marker analysis. Genet Resour Crop Evol. doi:10.1007/s, 1O722-011-9686-5

Pachico D (1989) Trends in world common bean production. In: Schwartz HF, Pastor-Corrales MA (eds) Bean production problems in the tropics. CIAT, Cali, pp 1–8

Purseglove JW (1974) Tropical crops: dicotyledons. Longman, London

Remanandan P, Singh L (1997) Pigeonpea. In: Fuccillo D, Sears L, Stapleton P (eds) Biodiversity in trust. Cambridge University Press, Cambridge, UK, pp 156–167

Robertson LD (1997) Faba bean. In: Fuccillo D, Sears L, Stapleton P (eds) Biodiversity in trust. Cambridge University Press, Cambridge, UK, pp 168–180

Sauer J (1964) Revision of *Canavalia*. Brittonia 16:106–181

Sauer J, Kaplan L (1969) *Canavalia* beans in American prehistory. Am Antiq 34:417–424

Schmit V, Debouck DG (1991) Observations on the origin of *Phaseolus polyanthus* Greeman. Econ Bot 45(3):345–364

Singh AK, Nigram SN (1997) Groundnut. In: Fuccillo D, Sears L, Stapleton P (eds) Biodiversity in trust. Cambridge University Press, Cambridge, UK, pp 114–127

Singh AK, Simpson CE (1994) Biosystematics and genetic resources. In: Smartt J (ed) The groundnut crop: a scientific basis for improvement. Chapman & Hall, London, pp 96–137

Singh HB, Joshi BS, Chandel KPS, Pant KC, Saxena RK (1974) Genetic diversity in some Asiatic *Phaseolus* species and its conservation. Indian J Genet 34A:52–57

Singh SP, Gepts P, Debouck DG (1991) Races of common bean (*Phaseolus vulgaris*, Fabaceae). Econ Bot 45(3):379–396

Singh KB, Pundir RPS, Robertson LD, Rheenen HA, Singh U, Kelley TJ, Rao PP, Johanses C, Saxena NP (1997) Chickpea. In: Fuccillo D, Sears L, Stapleton P (eds) Biodiversity in trust. Cambridge University Press, Cambridge, UK, pp 100–113

Sorensen M (1996) Yam bean (*Pachyrhizus* DC.). Promoting the conservation and use of underutilized and neglected crops 2. Institute of Plant Genetics and Crop Plant Research/International Plant Genetic Resources Institute, Gatersleben/Rome

Steele WM, Mehra KL (1980) Structure, evolution and adaptation to farming system and environment in *Vigna*. In: Summerfield RJ, Bungting AH (eds) Advances in legume science. HMSO, London, pp 393–404

Tomooka N, Lairungreang C, Nakeeraks P, Egawa Y, Thavarasook C (1992) Center of genetic diversity and dissemination pathways in mung bean deduced from seed protein electrophoresis. Theor Appl Genet 83:289–293

Tomooka N, Vaughan DA, Moss H, Maxted N (2002) The Asian *Vigna*: genus *Vigna* subgenus *Ceratotropis* genetic resources. Kluwer, London

van der Maesen LJG (1990) Pigeonpea: origin, history, evolution, and taxonomy. In: Nene YL, Hall SD, Sheila VK (eds) The pigeonpea. CAB International, Wallingford, pp 15–46

Vavilov NI (1926) Studies on the origin of cultivated plants. Institute of Applied Botany and Plant Breeding, Leningrad

WIEWS (n.d.) World information and early warning system on PGRFA, FAO, Rome. Retrieved from http://apps3.fao.org/wiews/wiews.jsp?i_l=EN, between April and June 2011

Zeven A, Zhukovsky PM (1975) Dictionary of cultivated plants and their centres of diversity: excluding ornamentals, forest trees and lower plant. Centre for Agricultural Publishing and Documentation, Wageningen

Zohary D (1972) The wild progenitor and the place of origin of the cultivated lentil: *Lens culinaris*. Econ Bot 26:326–332

Chapter 12
Tropical and Subtropical Root and Tuber Crops

David Tay

12.1 Introduction

Almost all root and tuber crops (RTC) in the world are of tropical and subtropical origin and they include potato (*Solanum tuberosum* L. and other diploid, triploid and pentaploid spp.), cassava (*Manihot esculenta* Crantz), sweet potato (*Ipomoea batatas* (L.) Lam.), yams (*Dioscorea* spp.), the aroids which include taro (*Colocasia* spp.), cocoyam (*Xanthosoma* spp.), and genera including *Alocasia*, *Amorphophallus* and *Cyrtosperma* (giant swamp taro), and a group of nine minor Andean RTC consisting of achira (*Canna indica* L.), ahipa (*Pachyrhizus ahipa* (Weddell) Parodi and two other species including the Mexican jicama (*Pachyrhizus erosus* (L.) Urb. a commonly grown species in Mexico and Southeast Asia), arracacha (*Arracacia xanthorrhiza* Bancroft), maca (*Lepidium meyenii* Walpers), mashua (*Tropaeolum tuberosum* Ruiz and Pavón), mauka (*Mirabilis expansa* Ruiz and Pavón), oca (*Oxalis tuberose* Mol.), ulluco (*Ullucus tuberosus* Caldas) and yacon (*Smallanthus sonchifolius)* (Poeppig and Endlicher Robinson). This group of crops is usually not traded widely as international commodities and thus is not subjected to wide price increases and fluctuation as that for grain crops such as rice, wheat and maize during international market crises in recent years. Additionally, their tropical origin allows them to be grown anytime of the year when there are available water and temperature for growth and they complement the growing of grain cereals as rotational crops and during the off-season. Some are also grown in marginal lands with limitations like drought, water-logging, salinity and poor soil conditions. Once they start to set storage roots and tubers, harvesting can be commenced periodically without waiting for a one-time maturation harvest as in cereal crops. Thus, they play an important role in stabilizing local and regional food security and prices. The current

D. Tay (✉)
Genetic Resources Conservation and Characterization Division,
International Potato Center (CIP), Apartado 1558, Lima 12, Peru
e-mail: dt26012012@gmail.com

M.N. Normah et al. (eds.), *Conservation of Tropical Plant Species*,
DOI 10.1007/978-1-4614-3776-5_12,

low yield realized in most farmer fields in developing countries offers enormous, untapped potential for yield improvement, combating malnutrition and vitamin A deficiency specifically the orange-flesh cultivars with high beta-carotene, and improving livelihoods. They can produce better yields in poor conditions with fewer inputs making them particularly suitable for households threatened by displacement, civil disorders or diseases such as AIDS. They therefore play an important role toward food security in developing countries.

Potato, the fourth most important food crop in the world ranked by production at 330 million T, is eaten by millions of people in Asia, Africa, and Americas (FAOSTAT 2009). Potato, commonly related as staple of temperate countries, in fact, has its origin in the Andean highlands in South America. It is still the main stable in the isolated and deprived regions of the Andes. In term of productivity, potato harvest index is up to 85% of the plant (Reynaldo et al. 1986). Upon cooking, potato diet provides significant daily needs of energy, protein, vitamins C, B6, and B1, folate, iron, potassium, phosphorus, calcium, zinc and other essential trace elements such as manganese, chromium, selenium, and molybdenum. It has high dietary fiber and is rich in antioxidants, including polyphenols and tocopherols. In most developing countries it is grown as a high value income generating crop considering that French fries has become one of the most popular fast food in the world. China, largest potato producing country in the world, is planning the development of more potato production to sustain its future food need because it is the only staple crop that new production areas are available for expansion (CCCAP on CIP website – www.cipapa.org).

Cassava is predominately cultivated in developing countries of Asia, Africa, and Latin America. It ranks fifth in term of total production in the world at 234 million T (FAOSTAT 2009). It is thus an important energy source in the tropics and subtropics. The vitamin and mineral rich young leaves are used as vegetable green in Southeast Asia and Africa. The short shelf-life after harvesting means that it is mainly used and traded locally. In fact, the bulk of Asia production is processed into starch.

Sweet potato is the seventh most important food crop in the developing countries with a total production of 102 million T (FAOSTAT 2009). It is one of the main staples in the Pacific including Papua New Guinea and Indonesia Irian Jaya. Sweet potato originated in tropical Americas and was taken across the Pacific as far as New Guinea before Columbus by ancient civilizations. In fact, the sweet potato culture is more elaborated in the Pacific than in the Americas as the 'kumara' in the Maori culture. The orange-flesh cultivars (OFSP) have very high levels of pro-vitamin A and are used to reduce vitamin A deficiency in the tropics and subtropics. As for cassava, the nutritious young leaves are eaten in Southeast Asia and the Far East as green vegetable, and the whole vine for animal feed.

Yam, a multispecies crop with origins in Africa and Asia rank eleventh in global food crops with a total production of 49 million T (FAOSTAT 2009). Production is mainly in tropical Africa (96%) by smallholders consisting mainly of Guinea yams, *D. rotundata* (white yam) and *D. cayenensis* (yellow yam) of African origin. They are preferred for the organoleptic properties of the tubers. The water yam, *D. alata*, originated in Asia, is widely grown in the world as minor crop because of its wide adaptation.

12.2 Genetic Diversity and Genebank Collections

The genetic resources of a crop (germplasm collections) are the 'building blocks' consisting of genetic linkage groups available for plant breeders to put them together in the best combining ways to create the 'winning varieties'. The gene linkage groups are the results of 1,000 years of selection by generations of ancient farmers to present days to adapt to the ever changing biotic and abiotic environments and the changes in the anthropological and aesthetic needs through time. The domestication of a crop could be said as the first big leap forward of the crop genepool from the wild ancestors. As the crop spreads and migrates to new territories it faces new selection pressure and thus new genetic linkage groups are created. When it comes into contact with new related species (both wild and cultivated) they intercross, introgression takes place and new genetic linkage groups are formed. When it reaches edges of near extreme environments founder effect due to chance and genetic drift happens and it adapts by polyploidization and often becomes self-pollinated to survive and new linkage groups are stabilized. Through time mutation accumulated, new alleles and genes are created adding another dimension to the total genetic diversity.

A quality crop germplasm collection is a collection with a good representation of the total genepool of the crop where most of the genetic diversity specifically in term of the genetic linkage groups, genes and alleles is represented. Harlan and de Wet (1971) proposed a genecological approach to define the different segments of a crop genepool based on their ability to exchange genes where the primary genepool (GP1) contains relatives that readily intercross with the crop; secondary genepool (GP2) contains relatives that hybridize with the crop but show sterility problems in the progenies; and tertiary genepool (GP3) contains relatives that can be crossed but with difficulty. However, in most crops crossing information between a crop and its relatives is unavailable and thus this concept can be applied only to a certain degree. An alternative concept was thus proposed by Maxted et al. (2008) based on the taxonomy of a crop which to some degree reflects their genetic relationships and crossability. The classification is based on Taxon groups where Taxon Group 1a is the crop; Taxon Group 1b is the same species as the crop e.g. wild or semi-wild subspecies or varieties; Taxon Group 2 is of species in the same series or section as the crop; Taxon Group 3 is of species in the same subgenus; Taxon Group 4 is of species in the same genus; and Taxon Group 5 is of species in different genera. Information on taxonomic treatments and domestication of a crop is thus essential in managing the genetic quality of a crop collection in term of genepool representation.

Currently, most root and tuber crop collections have poor GP2 and GP3, or poor Taxon Group 2–5 representations. The following crop summaries will provide some insight on the state of the collections for different root and tuber crops:

Potato (tuber): Potato is in the family Solanaceae, genus *Solanum*, subgenus *Potatoe*, section *Petota* (formerly *Tuberarium*), series *Tuberosa* (Hawkes 1990). The local name in the Andes is *papa*. It is an ancient crop and was found in archeological remains in the coastal Peruvian pre-Inca civilizations as far back as 7,000 years ago (Ugent and Peterson 1988). Spooner et al. (2005) using AFLP data postulated a single domestication and origin for cultivated potato from *S. bukasovii* in the broad area of southern Peru. It then spread to the whole Andes from western Venezuela in the north to Argentina and Chiloe Islands in the south and evolved into multispecies with ploidy levels from diploid (2n = 2x = 24) to pentaploid (2n = 5x = 60). Through time introgression with wild potatoes, selection by the Andean ancient farmers, mutation, genetic drift and intercrossing between cultivated forms have created a wide diversity of many thousands of native Andean potato cultivars (not subjected to modern breeding). Ugent et al. (1987) reported wild potato remains in archeological sites in southern Chile as far back 13,000 years ago.

Based on Ochoa (1999, 2003) cultivated potato can be classified into nine species, namely, *Solanum stenotomum* (a 2× species believed to be the first potato derived from *S. bukasovii* (Spooner et al. 2005)) and possibly other related wild species), *S. goniocalyx* (a 2× species with lesser distribution, generally yellow flesh, high dry matter and good eating quality, and classified as a subspecies of *S. stenotomum* by Hawkes 1990), *S. phureja* (a 2× species with early maturing, no tuber dormancy in the lower altitude down to 1,700 m asl in the eastern slope of the Andes toward Amazon from Colombia to Bolivia), *S.* x *ajanhuiri* (a 2× natural hybrid species between *S. stenotomum* and a wild species, *S. megistacrolobum*) (Huaman et al. 1980) with only two main forms – ajanhuiri (non-bitter) and yari (bitter with frost tolerance), *S.* x *chaucha* (a very variable 3× species resulted from the natural crosses between diploid species and tetraploid *S. tuberosum* ssp. *andigenum* or unreduced and normal gamete crosses of diploid species), *S.* x *juzepczukii* (a bitter 3× natural hybrid species between *S. stenotomum* and a wild species, *S. acuale*) (Schmiediche et al. 1982), *S. tuberosum* (a 4× species with two subspecies – ssp. *andigenum* and ssp. *tuberosum* which represents 78% of the diversity in the Andes and almost all the potato growing in the world, respectively), *S. hygrothermicum* (a 4× species from the lower altitude of Andes toward the Amazon (Ochoa 1984) and Hawkes (1990) classified it as a form of *S. phureja*) and *S.* x *curtilobum* (a bitter 5× natural hybrid species between ssp. *andigenum* and *S.* x *juzepczukii* with very narrow diversity) (Schmiediche et al. 1982). Recently, Spooner et al. (2007) using SSR markers and T chloroplast deletion data reclassified these cultivated species into four species – the three natural F1 hybrid species (*S.* x *ajanhuiri, S.* x *juzepczukii* and *S.* x *curtilobum*) accounting for only 1.3% of the known diversity in cultivated native potatoes in CIP Genebank, and lumping all the remaining diploids, triploids and tetraploids together as *S. tuberosum* in two groups, the Andigenum group (accounting for 74.3% of the diversity) and Chilotanum group from Chiloe region (3.8% of the diversity). This was concluded despite reasonable clear groupings were obtained in their analysis between the diploids and tetraploids. This lumping of 2×, 3× and 4× together as a species (98.7% of diversity) is not practical as it would create serious communication difficulty between researchers. These entities although have overlapping morphological characteristics are able to be distinguished to large extend in the field by field researchers (based on 70 man-year experience of CIP potato curators).

Table 12.1 The global in trust potato collection held under the International Treaty on Plant Genetic Resources for Food and Agriculture at CIP in 2011 (CIP data)

Species	Acc	Distribution
Cultivated sp. (Ochoa 1999, 2003)		
Native potato		
2n = 2x = 24		
S. stenotomum	299	Argentina (2), Bolivia (88),Colombia (2), Ecuador (2), Perú (205)
S. goniocalyx	98	Bolivia (2), Chile (1), Peru (95)
S. phureja	204	Colombia (99), Ecuador (85), Peru (20)
S. x *ajanhuiri*	14	Bolivia (13), Peru (1)
2n = 3x = 36		
S. x *chaucha*	116	Bolivia (24), Ecuador (7), Peru (85)
S. x *juzepczukii*	36	Argentina (1), Bolivia (20), Peru (15)
2n = 4x = 48		
S. tuberosum ssp. *andigenum*	3,148	Argentina (183), Bolivia (375), Colombia (147), Ecuador (254), Perú (2,153), Venezuela (36)
S. tuberosum ssp. *tuberosum*	163	Argentina (13), Chile (142), Colombia (2), Ecuador (1), Perú (4), Venezuela (1)
2n = 5x = 60		
S. x *curtilobum*	6	Argentina (1), Bolivia (1), Peru (4)
To be classified	151	Argentina (6), Bolivia (18), Colombia (3), Ecuador (12), Perú (112)
Landraces[a]	119	Bangladesh (18), Bhutan (5), Costa Rica (1), Guatemala (32), India (2), Mexico (32), New Zealand (5), Philippines (3), Russia (11), Switzerland (8), Unknown (3)
Improved material	92	–
Wild potato sp. (section *Petota*)		
141 species	2,174	Argentina (80), Bolivia (451), Chile (9), Colombia (48), Costa Rica (7), Ecuador (88), Guatemala (3), Mexico (106), Panama (2), Paraguay (17), Peru (1,281), Uruguay (47), USA (26), Venezuela (9)
Total	**6,620**	

[a]Native potato types from outside the Andes

The diploids are self-incompatible and thus outcross, the triploids are mostly sterile, the tetraploids are highly self-compatible and the pentaploids are crossable with tetraploids. The species intercross to produce a wider segregation of new forms. There are some 30 collections of potato worldwide conserving about 65,000 accessions. The major ones represent some 59,000 accessions are listed in the Global Strategy for the Ex situ Conservation of Potato, 2006 (http://www.croptrust.org/documents/web/Potato-Strategy-FINAL-30Jan07.pdf). The collection at CIP includes 4,235 accessions of cultivated native potatoes, 2,174 accessions of wild potatoes (*Solanum* section *Petota*), 119 landraces, 92 bred cultivars, and 3,752 genetic stocks and advanced breeding lines (Table 12.1). These collections are underutilized in breeding work. For example, only 246 of the 4,235 native potato accessions representing about 5.6% were used in the breeding program at CIP in the last 40 years and they were represented in only 30% of the crosses, i.e. 6,324 out of 20,922 total crosses (based on 2010 data). The key causes contributing to this low level of utilization are the lack of evaluation information of the collection for important agronomic traits; the inadequate documentation on and ease of access of this information on the collection; and the need of a specific germplasm enhancement pre-breeding program to isolate useful genes and gene linkage groups of native potatoes. The classical example is the use of *S. tuberosum* ssp. *tuberosum* (female parents) and ssp. *andigenum* (male parents) crosses to exploit the resulting adaptation and heterosis in the Andes. Using this concept Carlos Ochoa bred more than ten released cultivars in Peru in the 1960s and 1970s (Ochoa 2008), and several of them have made their ways into the traditional potato production system and thus continue to be important cultivars to date, e.g. cv. *Yungay* is still one of the most important cultivars in cultivation in Peru.

Similarly, the neo-tuberosum program to select for long photoperiod genotypes from new genepool of ssp. *andigenum* in the 1960s and 1970s in Scotland (Simmonds 1964; Glendinning 1975a, b, c, d) and in USA (Plaisted 1971) is another important event to widen the genetic base of the ssp. *tuberosum* genepool in the world. There is a lot of genetic diversity in the native cultivated potato of the Andes available for this purpose without having to resort to the use of wild related species which tend to introduce glycol-alkaloids (bitterness) into the tubers and many other unwanted agronomic characteristics that are difficult to breed out.

There are 187 species of wild potato in the world (based on CIP revision of *Petota* under preparation) with a polyploidy series from diploids to hexaploids and they are distributed from southwest USA in the north continuously along the cordillera of the two Americas to Argentina to Pacific coast of Chile around Chiloe Islands in the south (Fig. 12.1). The differentiation between wild potatoes and other *Solanum* species is tuber bearing and this characteristic forms the section *Petota*. They are endemic from the high Andes of up to 4,300 m above sea level, to the west the edge of Pacific Ocean and to the east the foothills of the Andes and cloud forests toward the Amazon Basin. Along this distribution there are two centers of diversity with the South American center concentrates in Peru with 83 endemic species and the North American center in Mexico with 29 endemic species. The genetic diversity is many times richer than the cultivated native potatoes and thus a tremendous genetic resource for breeding because of their wide range of ecological adaptation. The useful traits include both biotic and abiotic stresses such as pest and disease resistance, and frost, drought, salinity and heat tolerance, respectively. Many of species are inter-specific cross-compatible and also with the cultivated species. In fact, the three natural hybrid native potatoes species, *S.* x *ajanhuiri* is the result of nature crosses of cultivated native species directly with *S. megistacrolobum* ($2n=2x=24$), and *S.* x *juzepczukii* and *S.* x *curtilobum* with *S. acuale* ($2n=4x=48$), meaning that *ajanhuiri* has a half of wild genome, *juzepczukii* has two thirds wild genome and *curtilobum* has two fifth wild genome. This is the reason for their bitter tubers.

Since the formation of CIP in 1971 emphasis has been to collect, conserve, evaluate and use the wild species. Currently CIP genebank holds in trust the most

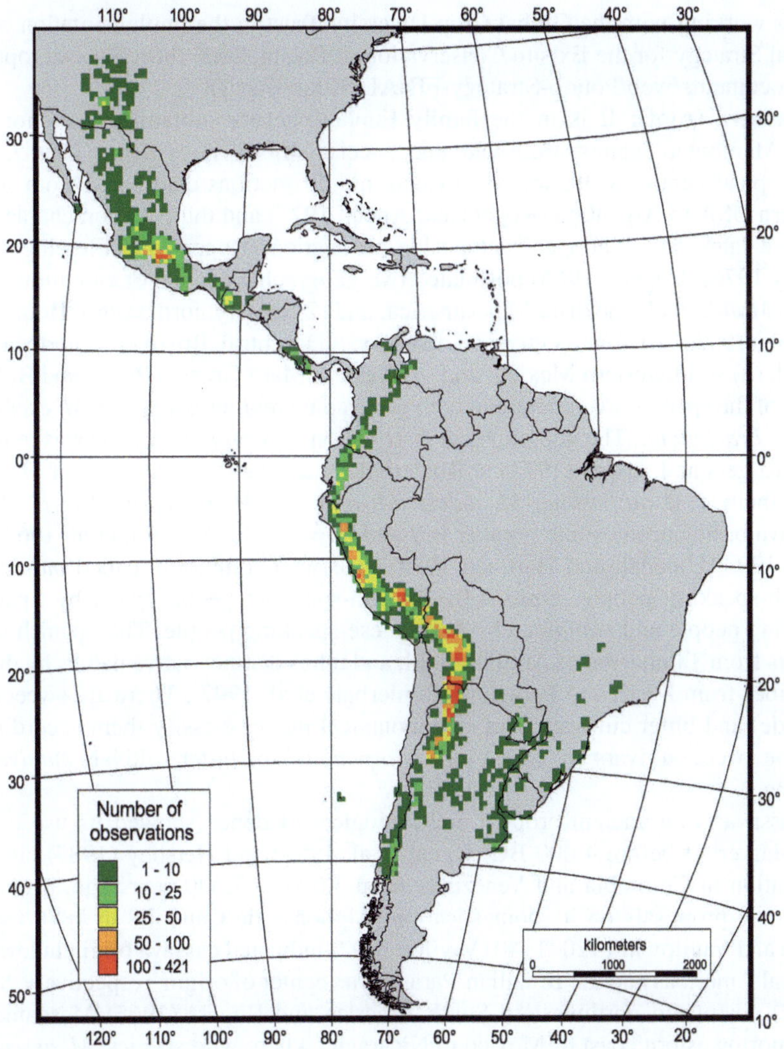

Fig. 12.1 Geographical distribution of wild potatoes (*Solanum* section *Petota*) in their natural endemic habitats

diverse collection of this germplasm with 2,164 accessions of 141 species (Table 12.1). This represents one third of the in trust Global Potato Collection maintained at CIP – the highest % holding of CWR germplasm in all the major crops. However, the use of these genetic resources in plant breeding is always faced with the occurrence of bitter tuber, day-length sensitivity, long stolon, small multiple tuber formation per stolon in tandem and other poor agronomic traits. The recommendation is that when a trait is present in the native potato germplasm breeding should focus on them first rather than going directly at the CWR. Currently,

CIP is working with the Global Crop Diversity Trust in the implementation of the Global Strategy for the Ex situ Conservation of Potato, 2006 (http://www.croptrust.org/documents/web/Potato-Strategy-FINAL-30Jan07.pdf).

Cassava (root): It is in the family Euphorbiaceae, subfamily Crotonoideae, tribe Manihotae, genus *Manihota* and species *Manihot esculenta* Crantz. The Neotropical genus has 98 species divided into 19 sections distributed from southwestern USA to Argentina (Rogers and Appan 1973) and the common characteristics are latex and cyanogenic glucoside production (Rogers and Fleming 1973; Bailey 1976). Rogers (1965) postulated two geographic centers of evolution in the genus *Manihot*: (1) the drier Mesoamerica, and (2) the dry northeastern Brazil; and Nassar (1978a, b) four centers of diversity: (1) central Brazil, (2) northeastern Brazil, (3) southwestern Mexico, and (4) western Mato Grosso, Brazil and Bolivia. None of the species are endemic in both geographic regions except for *M. esculenta* and *M. brachyloba*. The geographical distribution of *Manihot* sections was redrawn from Rogers and Appan (1973) in Bonierbale et al. (1997). Synonyms of *M. esculenta* include *M. utilissima, M. dulcis, M. aipi* and *M. palmata* (Rogers 1965). Cassava basic chromosome number is 9 and 2n=4x=36 (Magoon et al. 1969; Bai et al. 1993; Umanah and Hartman 1973). It is also commonly called *manioc* by French-speaking people, *tapioca* by English-speaking people, *yuca* by Spanish-speaking people and *mandioca* by Portuguese-speaking people. The Spanish name *yuca* is from Taino tribe of Antillas/Haiti, and it has distinct native names by different tribes from Mexico to Paraguay (Bonierbale et al. 1997). There are sweet (low cyanide) and bitter cultivars, and ethnobotanical names classify them accordingly, e.g. the sweet cultivars are called *aipi* or *aipim* and the bitter cultivars *maniyua* or *maniva*.

Cassava is an ancient crop as archaeological evidence showed its use on the Peruvian coast before 4,000 BC (Ugent et al. 1986) and Hershey (1987) cited its cultivation in Colombia and Venezuela from 3,000 to 7,000 years ago. There are numerous proposals on its domestication. Alphonse de Candolle in 1886 (Smith 1968) and Vavilov in 1920–1940 (Vavilov 1992) indicated cassava origin in lowland tropical Americas and the Brazilian-Paraguayan center of origin, respectively. Sauer (1952) postulated northwestern South America and Rogers (1965) Mesoamerica from northwestern coast of Mexico to Nicaragua where wild species, *M. aesculifolia, M. pringlei* and *M. isoloba* could have involved in the forming of cassava. Rogers and Fleming (1973) indicated that *M. esculenta* is a complex species with multiple domestication sites. Allem (1987, 1994) indicated that cassava derived from two primitive forms where cassava is *M. esculenta* subsp. *esculenta* and the two wild primitive forms, *M. esculenta* subsp. *peruviana* and *M. esculenta* subsp. *flabellifolia* which is morphologically similar to cassava. AFLP studies (Roa et al. 1997) supported the close relationship with *flabellifolia*. Similarly, study on the single-copy nuclear gene coding glyceraldehyde 3-phosphate dehydrogenase (Olsen and Schaal 1999) and SSR markers (Olsen and Schaal 2001) also pointed to *flabellifolia* and that domestication occurred along the rim of southern Amazon Basin in the Brazilian states of Acre, Rondonia and Mato Grossa and probably extending south to Bolivia. Brucher (1989) indicated that the Arawak tribes of

Central Brazil took cassava to the Caribbean and Central America in the eleventh century. After Columbus discovery, the Portuguese took it to west coast of Africa in the sixteenth century (Jones 1959), to east coast of Africa in eighteenth century (Barnes 1975; Jennings 1976), to India beginning of nineteenth century, and then to the Pacific by the Spanish (Jennings 1976). Since then cassava has developed wide diversity in different parts of the world especially in Africa (Gulick et al. 1983).

There are more than 50 collections of cassava worldwide conserving about 10,000 accessions ex situ of the estimated 27,000 distinct landraces found in situ worldwide (Table 12.2). To represent the complete diversity of cassava 15,000 landraces have to be conserved with about half of them from the Americas. Two international centers (CIAT and IITA) conserve cassava germplasm where CIAT targets on Americas and Asia diversity and IITA on Africa. The collection at CIAT includes 5,301 accessions of landraces, 883 accessions of wild related species and 408 accessions of breeding material (2011 data provided by Daniel Debouck and Ericson Aranzales, CIAT genebank). The collection at IITA consists of 2,556 accessions (2011 data provided by Dominique Dumet, IITA genebank).

Sweet potato (root): All the diversity of sweet potato belongs to one species, *Ipomoea batatas* ($2n=6x=90$), in the family *Convolvulaceae*, genus *Ipomoea*, subgenus *Eriospermum*, section *Eriospermum*, series *Batatas* (Austin and Huaman 1996). Austin (1977) studied a range of morphological $2n=4x=60$ forms and considered all of them as synonyms to *I. batatas*. It is an ancient crop of American origin. Carbon-dated sweet potato remains in the Chilca canyon in Peru were estimated to be from 8,000 to 10,000 BP (Engel 1970; Yen 1974). However, Austin (1988) postulated its origin was between the Yucatan peninsula of Mexico and the mouth of Orinoco River in Venezuela and it was widespread by 2,500 BC from southern Mexico to southern Peru (O'Brien 1972). Austin (1988) described two main groups – the *aje* (an Arawakan word) group with starchy less sweet tuber and the *batata* group with starchy sweeter tuber. To date two distinct groupings between the Central American and the South American cultivars can be detected based on chloroplast and nuclear SSR marker studies and this has led to the suggestion of duo independent domestications, in Central/Caribbean America and in the north-western part of South America (Roullier et al. 2011). There is no wild *I. batatas* but it can survive in abandoned cultivated fields either directly from vegetative growth or seedlings of seed naturally produced. This could be the case as for accession, K123 with $2n=6x=90$, that Nishiyama (1963) collected in Mexico as *I. trifida* and Jones (1967) concluded that it could be *I. batatas* derivative from morphological, crossing and chromosome studies. The hexaploid nature of sweet potato indicates its complex genetic origin. According to the Global Strategy for Ex Situ Conservation of Sweet potato Genetic Resources (http://www.croptrust.org/main/identifyingneed.php?itemid=513), the series *Batatas* in addition to sweet potato (*I. batata*) has 13 wild species of which all are endemic to the New World except *I. littoralis* from Australia and Asia. These species (Taxon Group 2 genepool) are therefore closely related to sweet potato and thus could all contribute to the genome of modern sweet potato. postulated that The origin of sweet potato was postulated by Nishiyama (1971) and Nishiyama et al. (1975) to come from $6\times$ *I. trifida* which came from

Table 12.2 Status of ex situ cassava collections, estimates of unique local landraces, estimate of accessions not in CGIAR centers and proposed minimum number of accessions to be conserved in the world (based on Table 12.3 of the 'Global Conservation Strategy for Cassava and Wild *Manihot* Species'; accessions for the two CGIAR centers (CIAT in light blue and IITA in purple) were based on 2011 data provided by Daniel Debouck and Ericson Aranzales, CIAT genebank, and Dominique Dumet, IITA genebank)

	Status of ex situ collections		Estimates of unique local landraces (excluding duplicates and breeding/experimental material)[c]			
Region/country	GCDT survey[a] (2008)	In CGIAR centers[b]	Ex situ	In situ	Estimate in situ accessions missing from CGIAR centers[d]	Proposed minimum ex situ no. of accessions[e]
Americas						
Argentina		122 (2.3%)	160	250	128	200
Bolivia	30	7 (0.1%)	18	300	293	200
Brazil	3,075	1,281 (24.7%)	1,600	8,000	6,719	4,000
		25 (1.0%)				
Colombia	2,000	2,000 (38.5%)	1,800	3,000	1,000	2,500
Costa Rica	72	81 (1.5%)	70	100	19	100
Cuba		82 (1.5%)	75	100	18	100
Dominican Rep		5 (0.1%)	25	50	45	25
Ecuador	93	116 (2.2%)	80	250	134	200
El Salvador		10 (0.2%)	8	25	25	20
French Guiana			0	50	50	50
Guatemala		92 (1.7%)	50	75	0	100
Guyana	29		25	50	50	50
Haiti			0	100	100	75
Honduras		27 (0.5%)	20	50	23	25
Jamaica		20 (0.4%)	0	50	50	25
Mexico		106 (2.0%)	75	200	94	100
Nicaragua		3 (0.1%)	10	100	97	75
Panama	2	47 (0.9%)	40	75	28	50
Paraguay		208 (4.0%)	300	500	292	400

Peru	639	421 (8.1%)	550	1,500	1,079	1,000
Puerto Rico		17 (0.3%)	17	25	8	15
Suriname			0	75	75	25
USA		10 (0.2%)				
Venezuela		253 (4.8%)	225	1,000	747	500
Sub-total	5,940	4,933	5,148	15,925	11,074	9,835
Africa						
Angola		2 (0.1%)	10	300	297	100
Benin		329 (12.9%)	300	400	0	100
Botswana			10	20	20	20
Burkina Faso		6 (0.2%)	10	15	9	20
Burundi				50	50	25
Cameroon		179 (7.0%)	200	300	81	250
Cape Verde		17 (0.7%)	10	20	7	
Central African Republic		2 (0.1%)	2	200	198	100
Chad	45	4 (0.2%)	40	50	47	25
Congo, Republic of		50 (2.0%)	150	200	178	50
Cote d'Ivoire	170	25 (1.0%)	250	300	277	100
D.R.Congo	140	27 (1.1%)	300	1,000	978	500
Gabon			40	75	75	50
Gambia		6 (0.2%)	5	25	20	10
Ghana	36	264 (10.3%)	300	400	62	100
Guinea Bissau/Conakry	50	133 (5.2%)	120	175	63	75
IITA (Unknown)		24 (0.1%)				
Kenya		12 (0.5%)	150	200	190	50
Liberia		8 (0.3%)	75	100	94	50
Madagascar		3 (0.1%)	4	200	196	100
Malawi	192	6 (0.2%)	150	175	170	100

(continued)

Table 12.2 (continued)

Region/country	Status of ex situ collections: GCDT survey[a] (2008)	Status of ex situ collections: In CGIAR centers[b]	Estimates of unique local landraces (excluding duplicates and breeding/experimental material)[c]: Ex situ	Estimates of unique local landraces (excluding duplicates and breeding/experimental material)[c]: In situ	Estimate in situ accessions missing from CGIAR centers[d]	Proposed minimum ex situ no. of accessions[e]
Mali		1 (0.1%)	1	25	24	20
Mozambique	25		75	250	250	150
Niger	124	10 (0.4%)	50	75	65	20
Nigeria	40	1,207 (47.0%)	500	800	253	200
		19 (0.4%)				
Rwanda		1 (0.1%)	125	150	148	75
Senegal			10	50	50	25
Sierra Leone	118	69 (2.7%)	100	200	90	100
South Africa	0		5	25	25	10
Sudan	10		10	10	10	10
Swaziland	10		10	20	20	10
Togo	209	128 (5.0%)	100	200	24	100
Uganda		14 (0.5%)	250	1,000	986	500
U.R.Tanzania		3 (0.1%)	250	1,000	997	500
Zambia	103		75	200	200	150
Zimbabwe			6	25	25	10
Sub-total	1,272	2,549	3,743	7,480	5,368	3,675
Asia–Oceania						
Australia		1 (0.1%)				
Cambodia				25	25	10
China	4	2 (0.1%)	10	15	13	20
Fiji Islands		6 (0.1%)	5	25	19	20
India			600	750	750	200
Indonesia	130	253 (2.6%)	150	1,000	864	500

Malaysia	52	61 (1.1%)	50	100	39	50
Micronesia				25	25	10
Myanmar			7	50	50	20
Papua New Guinea	95		75	100	100	50
Philippines		6 (0.1%)	130	500	494	250
Polynesia			0	25	25	10
Sri Lanka			50	100	100	50
Thailand	11	37 (0.7%)	10	11	6	11
Vanuatu	150		120	50[f]	150	25
Vietnam	31	9 (0.2%)	20	50	41	25
Sub-total	473	375	1,132	2,965	2,708	1,170
Total	**7,685**	**7,857**	**10,068**	**26,986**	**19,954**	**14,791**

[a]Information provided by survey respondents

[b]Accessions for the two CGIAR centers (CIAT in light blue and IITA in purple) were based on 2011 data provided by Daniel Debouck and Ericson Aranzales, CIAT genebank, and Dominique Dumet, IITA genebank

[c]These are approximations based on region (primary or secondary center of diversity), area planted, ex situ accessions reported, and personal knowledge of the author about diversity in individual countries

[d]Estimates of in situ unique landrace varieties minus number of accessions in CGIAR centers

[e]Estimated number of accessions that would be required to fully represent a country's cassava genetic diversity. More accurate estimates will be possible as more molecular information becomes available on genetic variation

[f]There appear to be a large number of landrace varieties lost from farmers' fields and home gardens in the past two decades due to declining cassava production

3× *I. trifida* and in turn came from the cross of *I. littoralis* (4×) and *I.* x *leucantha* (2×) where the former derived from the latter. However, most of *I. trifida* are diploid. Austin (1988) postulated that *I. trifida, I. triloba* and *I. tiliacea* all contributed to the genome of sweet potato.

There are some 36 collections of sweet potato worldwide and conserving about 29,000 accessions (Table 12.3). The related wild species are poorly represented in

Table 12.3 Overall composition and size of the sweet potato collections included in this assessment (Source: the Global Strategy for Ex Situ Conservation of Sweet potato Genetic Resources – http://www.croptrust.org/main/identifyingneed.php?itemid=513)

	No. of accessions		
Region/collection/country[a]	Wild	Cultivated	Total
Latin America and Caribbean			
1. International Potato Center – PER	1,160	6,360	7,520
2. Instituto Nacional de Tecnología Agropecuaria Castela – ARG	122	362	484
3. Empresa Brasilera de Pesquiza Agropecuaria – BRA		1,024	1,024
4. Instituto Nacional de Investigaciones de Viandas Tropicales – CUB	95	535	630
Sub-total	**1,377**	**8,281**	**9,658**
North America			
5. United States Department of Agriculture's Agricultural Research Service – USA	447	755	1,202
Sub-total	**447**	**755**	**1,202**
Asia			
6. CIP-East South East Asia and Pacific – IDN		1,366	1,366
7. Indonesian Agriculture Biotech and Genetic Resources Institute – IDN		1,520	1,520
8. The Philippines Root Crop Research and Training Center – PHL		801	801
9. Xuzhou Sweetpotato Research Center – CHN	40	1,044	1,084
10. MOKPO Experiment Station – PRK		497	497
11. Plant Genetic Resources Center – VNM		480	480
12. The National Plant Genetic Resources Laboratory – PHL		183	183
13. National Institute of Agrobiological Sciences – JPN		1,600	1,600
14. Central Tuber Crop Research Institute – IND	84	3,778	3,862
15. Northern Philippines Root Crops Research and Training Center – PHL		180	180
16. Malaysian Agricultural Research and Development Institute MYS		72	72
17. PHRC – THA		236	236
18. Central Agricultural Research Institute – LKA		131	131
19. South Korea – KOR		430	430
Sub-total	**124**	**12,318**	**12,442**
Africa			
20. Centre de Developpement Rural et de Recherche Appliquee – MAG		98	98

(continued)

Table 12.3 (continued)

Region/collection/country[a]	No. of accessions Wild	Cultivated	Total
21. National Crops Resources Research Institute – UGA		1,808	1,808
22. CIP-South Saharan Africa – UGA		141	141
23. Mulungu Research Center – COD		120	120
24. Kenya Agricultural Research Institute – GHA		167	167
25. University of Ibadan – NGR		90	90
26. Mpnza Research Station – ZMB		258	258
27. Agricultural Research Institute – MWI		139	139
28. Instit. Investigacao Agrarian de Mozambique – MOZ		102	102
29. Vegetable and Ornamental Plant Institute – ZAF		444	444
30. Agricultural Research Institute – AGO		34	34
31. EARI, Awasa, Ethiopia – ETH		319	319
32. Kenya Agricultural Research Institute – KEN		120	120
33. Horticulture Research Institute – TNZ		584	584
34. Rwanda Agricultural Research Institute – RWA		159	159
35. INIDA, S.J. Orgaos – CPV		11	11
Sub-total	**0**	**4,594**	**4,594**
Melanesia			
36. National Agricultural Research Institute – PNG		1,120	1,120
Sub-total	**0**	**1,120**	**1,120**
Total	**1,948**	**27,068**	**29,016**

[a]Data of 18 collections: 1–5, 6–15, 20–21 and 36 were obtained from the survey's questionnaires and make a total of 25,459 accessions; the remaining 18 collections are from the Manila workshop and make a total of 3,557 accessions

these collections. The collection at CIP includes 7,777 accessions of sweet potato from 59 countries and 1,178 accessions of 67 wild species including 183 accessions of *I. trifida*. It consists of 2,089 Peruvian landraces of which about 50% are of duplicates, 1,568 Latin American excluding Peruvian landraces of which an estimated 50% are duplicates, the AVRDC collections with Asia, Oceania and to great extent worldwide representations of landraces and improved germplasm, and the IITA breeding lines. These collections are underutilized in breeding work due to the lack of evaluation information for important agronomic traits. Major disease of sweet potato is sweet potato virus disease complex (SPVD) which consists of the combination of two viruses – sweet potato chlorotic stunt virus and sweet potato feathery mottle virus and the main pests are sweet potato weevils (*Cyclas* spp.). The crop has large genetic variation for crop duration, adaptation, nutrient composition, etc. Currently, CIP is working with the Global Crop Diversity Trust to introduce germplasm from Southeast Asia, Oceania and Africa in the implementation of the Global Strategy for Ex Situ Conservation of Sweet Potato Genetic Resources (http://www.croptrust.org/main/identifyingneed.php?itemid=513).

Yams (corm): They are in the family Dioscoreaceae and genus *Dioscorea* with over 600 species. The cultivated species are in the following sections from both the Old and New World: (1) Section *Enantiophyllum – Dioscorea alata* L. (water yam, greater yam, white yam), *D. glabra* Roxb., *D. nummularia* Lam., *D. transversa*

Br. of Asia and Oceania origin; *D. japonica* Thumb. (Chinese yam, igname de Chine), *D. opposita* Thumb. of Sino-Japanese origin; and *D. cayenensis* Lam. (yellow yam), *D. rotundata* Poir. (white Guinea yam, white yam) of Africa; (2) Section *Lasiophyton* – *D. pentaphylla* L., *D. hispida* Dennsdest, *D. dumetorum* (Knuth) Pax (bitter yam); (3) Section *Opsophyton* – *D. bulbifera* L. (aerial yam); (4) Section *Combilium* – *D. esculenta* (Lour.) Burk. (Chinese yam, lesser yam); and (5) Section *Macrogynodium* – *D. trifida* L. (cush-cush yam) of tropical America (Martin 1974a, b, 1976; Martin and Degras 1978a, b; Martin and Sadik 1977). Hanson (1985) named in total 13 food and seven medicinal species with different ploidies. Dumont et al. (1994) described 16 wild and 7 cultivated species in the domestication of yams in Cameroon. They are mainly dioecious. *D. rotundata* (2n=40) and *D.* cayenensis (2n=60 and 80), the two major species cultivated in Africa because of their preferred tuber organoleptic properties represent 95% of yam production in the world (FAOSTAT 2009). However, Martin (1976) considered *D. alata* (2n=40, 60 and 80) from Southeast Asia the most important species because of its wide introduction to both Africa and America for its agronomic and nutritive value, and wide acceptance.

D. rotundata and *D. cayenensis*, originated in West Africa, are related (Martin and Rhodes 1978; Akoroda and Chheda 1983; Terauchi et al. 1992) and some taxonomic treatments put them together as a single species complex. Terauchi et al. (1992) using RFLP analysis on chloroplast and ribosomal DNA postulated that *D. rotundata* was domesticated from either *D. abyssinica*, *D. liebrechtsiana* or *D. praehensilis* or their hybrids and *D. cayenensis* from hybrids of *D. burkilliana*, *D. minutiflora* or *D. smilacifolia* as father and *D. rotundata*, *D. abyssinica*, *D. liebrechtsiana* or *D. praehensilis* as mother. Hamon et al. (1995) using morphological studies also proposed the involvement of *D. burkilliana*, *D. abyssinica* and *D. praehensilis* in their domestication.

There are 21 yam collections in the world which include the IITA collection of 3,166 accessions (Table 12.4). The global conservation strategy for yams is under development to identify the status of the major collections globally and the missing gaps, to prioritize genepools for collecting and safe duplication at another site, and to formulate a global conservation strategy. Conservation of yams in field genebank requires a period of storing some huge tubers between annual harvesting and planting time and at IITA controlled environment room of 18°C and 50–60% RH is used (Ng 1993).

Edible Aroids (corm): The family Araceae has the following food genera: *Alocasia*, *Amorphophallus*, *Colocasia*, *Cyrtosperma* and *Xanthosoma*. Among them, *Colocasia* (taro) and *Xanthosoma* (cocoyam, tannier) are the most commonly grown in Southeast Asia to China, Korea and Japan, Melanesia, the Pacific Islands, New Zealand, Mediterranean and tropical Africa and America. Taro is an important staple in South Pacific islands such as Tonga and Samoa, and parts of Papua New Guinea. The minor aroids, *Alocasia*, *Amorphophallus*, *Cyrtosperma* (giant swamp taro) are used only in specific regions and countries. For example, the giant swamp taro is mainly grown in the atoll countries of the Pacific and the *Amorphophallus* in Japan and Taiwan as a functional food. There are no organized

Table 12.4 Number of yam accessions and percentage of the collection maintained in ex situ conditions per species (Source: draft of the global strategy for the conservation and use of edible yam including 2010 report on the state of the world's plant genetic resources from 57 countries (FAO 2010) and the yam global strategy survey)

	FAO data 2010		GCDT survey data	
Species	No. of accessions	Percentage of the collection	No. of accessions	Percentage of the collection
D. alata	3,904	24.5	2,763	35.9
D. bulbifera	334	2.1	164	2.1
D. cayenensis	843	5.3	393	5.1
D. esculenta	662	4.2	215	2.8
D. nummularia	43	0.3	81	1.1
D. opposita-japonica/ japonica	118	0.7	20	0.3
D. pentaphylla	68	0.4	65	0.8
D. rotundata/rotundata-cayenensis	4,208	26.5	3,631	47.2
D. transversa	2	0.0	10	0.1
D. trifida	154	1.0	35	0.5
Others/unknown	5,567	35.0	320	4.2
Total	15,903	100.0	7697	100.0

collections. However, many of their species are valued as ornamentals and therefore some unique collections are found in botanic gardens and arboreta. Effort should be made to collect and conserve them because they have shown to be more resistance to pests and diseases compared to taro and have been reported to be used as substitution in some countries.

The center of origin and domestication of *Colocasia esculenta* is in Southeast Asia more specifically the region from Myanmar to Bangladesh (Plucknett 1976) include northeast India where several *Colocasia* species occur (Matthews 1990; Edison et al. 2004). *C. esculenta* var. *aquatilis,* which is distributed in Indo-Malaysian region, China, Japan, Melanesia, northern Australia and Polynesia and has wide diversity across its distribution, is postulated to be the progenitor of cultivated taro (Matthews 1991, 2004) and that domestication in Southeast Asia and Melanesia was from their respective native var. *aquatilis* populations thus forming two separate genepools overlapping in Indonesia with high diversity if diploids (Matthews 1990, 1991, 2004; Irwin et al. 1998; Lebot 1992, 1999; Kreike et al. 2004; Lebot and Aradhya 1991; Yen 1993). Their overall total diversity is low in term of isozyme, RAPD, AFLP and SSR markers used on the Taro Network for South Asia and Oceania – TANSAO's collection of 2,300 accessions from Indonesia, Malaysia, the Philippines, Thailand, Vietnam and the Pacific countries of Papua New Guinea and Vanuatu (Kreike et al. 2004; Quero-Garcia et al. 2006; Noyer et al. 2004). Southeast Asia, Papua New Guinea and Solomon Island have greater diversity than in the Pacific. With its eastward migration to Polynesia its diversity progressively decreases away from Melanesia (Lebot 1992; Yen 1993). The diverse morphotypes in Polynesia are likely somaclonal variation derived from a narrow genetic base of a few mother clones (Irwin et al. 1998; Lebot and Aradhya 1991; Lebot et al. 2004). Purseglove (1972) described two cultivated forms – the dasheen (*C. esculenta* var. *esculenta*) and eddoe (*C. esculenta* var. *antiquorum*). The dasheen usually has a large central corm and higher dry matter content, and the eddoe has a small central corm with a large number of smaller cormels with lower dry matter content and slimy even after cooking (personal experience). Intermediate forms exist and isozymes, AFLP, RAPD or SSR markers have not shown consistent difference between them (Vincent Lebot of CIRAD in Edible Aroid Conservation Strategies – http://www.croptrust.org/main/identifyingneed.php?itemid=513). The trueness of the two botanical varieties was challenged (Hay 1998).

The differences between the 'wild' and cultivated forms are likely the alkaloid content which causes irritation to skin and the corm size. Wild forms are used as pig fodder and sometimes leaf petioles are eaten and their leaf blades are preferably used as food wrapper than that of the cultivated forms (personal experience in Sarawak). Other wild taros – *C. fallax* and *C. affinis* (both in the arc of Himalayan India, Nepal to Myanmar) and *C. gigantea* (eastern China, Indonesia, Myanmar, Sri Lanka, southern Japan, Thailand and Vietnam) and *C. gracilis* (Sumatra), *C. mannii* (Assam) and *C. virosa* (eastern India) are yet to be collected and studied in details. For example *C. gigantea* has closer affinity to *Alocasia* based on mitochondrial DNA analysis (Matthews 1990). A comprehensive DNA analysis of the wild and cultivated diversity is needed to clarify the taxonomy of the genus *Colocasia*, and the origin and domestication of taros.

The major collections of taro are in Asia and the Pacific. These are some of the results of TANSAO (1998–2001) and Taro Genetic Resources Network – TaroGen (1998–2003) projects (The Edible Aroid Conservation Strategies – http://www.croptrust.org/main/identifyingneed.php?itemid=513).

Under TANSAO, 2,300 accessions from Indonesia, Malaysia, the Philippines, Thailand, Vietnam and the Pacific countries of Papua New Guinea and Vanuatu were collected and 168 accessions were selected based on morphological and isozyme diversity of each country to form a core collection for network exchange and utilization. TaroGen project collected some 2,199 accessions in network countries of Pacific Islands (Cook Islands, Fiji, Niue, Papua New Guinea, Samoa, Solomon Islands, Tonga and Vanuatu) and 211 accessions were selected accordingly as regional core collection. Both the two core collections are among the 857 accessions (December 2009) that are safely conserved as in vitro collection at the Centre for Pacific Crops and Trees (CePaCT) of the South Pacific Commission in Fiji, a regional germplasm center for the conservation of core collections, back-up national holdings, and virus index, multiply and disseminate germplasm.

The genus *Xanthosoma* is originated from tropical America, probably northern South America (Clement 1994; Giacometti and Leon 1994). The edible species according to Wilson (1984) include *X. violaceum*, *X. atrovirens*, *X. caracu*, *X. jacquini*, *X. maffafa*, *X. belophyllum* and *X. brasiliense*. Brown (2000) named *X. sagittifolium* and *X. violaceum* as the main species and Reyes Castro (2006) reduced all of them to *X. sagittifolium* as the differences between them are leaf shape, pigmentation and other morphological features. *Xanthosoma* is the most grown aroid in the world (Matthews 2002). The collections in the world based on the survey in the formulation of the Edible Aroid Conservation Strategies are available at – http://www.croptrust.org/main/identifyingneed.php?itemid=514.

Minor Andean root and tuber crops (ARTC): They consist of nine underutilized crops domesticated from different botanical families to exploit all the agroecosystems of the Andes (Table 12.5). Three are tuber crops, mashua, oca and ulluco, usually grown in mixed stand with native potato in traditional cropping system for their starchy swollen stolons. Their domestication and use are parallel to that of potato. Like potato they are annual, produce seed and tubers are used in propagation. In recent years, these tubers are being developed as new crops in countries like New Zealand (King 1988). Achira is grown for its swollen starchy rhizomes in warm Andean valleys. The main use is its starch for baking local recipes but in some communities the fresh rhizomes are fermented to convert the starch into sugars for use as traditional regional snack. Ahipa, arracacha and mauka are starchy root crops grown mostly for subsistent use. However, arracacha has potential to be developed into an important specialty crop because of its easily digestible starch and this is exploited in Brazil. The Andean ahipa unlike the Asian sweet juicy type is not used as snack. Yacon is a sweet juicy root crop for use raw as refreshing snack. It is a functional food for people on diet and with diabetes because of its high contents of low calorie fructans and human unmetabolized inulin. Maca, a turnip relative, is a swollen hypocotyl crop grown at high altitude at frost zone for its nutraceutical value. ARTC are mainly cultivated as part of the food diversity and security system in the Andes from Venezuela to northern Argentina. However, in recent years maca and yacon have been developed into commercial crops because of their nutraceutical value as energy and diabetic supplement, respectively.

Table 12.5 Estimated accessions of Andean root and tuber crops maintained by Andean germplasm banks (September 2011)

	Ecuador[a]	Peru[b]	Bolivia[c]	CIP[d]	Total
Oca	130	2,961	334	492	3,917
Olluco	245	1,689	140	420	2,494
Mashua	26	828	39	54	947
Yacon	35	311	36	29	411
Achira	35	297	19	35	386
Arracacha	20	286	37	7	350
Mauka	12	103	–	4	119
Ahipa	35	–	13	54	102
Maca	–	–	–	30	30
Total	538	6,475	618	1,125	8,756

[a]INIAP
[b]INIA, UNSAAC, UNSCH, UNMSM, UNC
[c]INIAF (IBTA-PROINPA)
[d]CIP (only in trust materials under ITPGRFA)

Achira (rhizome): It belongs to the family Cannaceae, species *Canna indica* L. *C. edulis* Ker-Gawler is its synonymy (Maas and Maas 1988; Brako 1993). The name is Quechua. The genus consists of 25–60 species in America and Asia. In addition to *C. indica, C. iridiflora* from the Peruvian Andes and *C. paniculata* in Peru, Brazil and Chile also produce tuberous rhizomes, and the New World ornamental species are *C. glauca, C. iridiflora* and *C. flacida* (Maas and Maas 1988; Brako 1993). Achira comprises of both diploid ($2n=2x=18$) and triploid ($2n=3x=27$) cultivars (Darlington and Janaki-Ammal 1945; Gonzales and Arbizu 1995) and the diploids produce dark hard seed through mainly self-pollination. It is a perennial and grows from 1,000 to 2,900 m asl.

The centers of genetic diversity and production are in the upper valley of Apurimac (Arbizu 1994; Meza 1995), and low valleys of Ayacucho in Peru, in the Patate of Ecuador (Espinosa et al. 1993), and in the departments of Huila and Cundinamarca in Colombia (Morales 1969). However, it is grown from Mexico to Argentina but in small scale. Vietnam reported growing about 30,000 ha where the starch is used mainly for making noodle (Ho and Hao 1995).

There are 386 accessions in Andean genebanks (Table 12.5) which include 35 accessions maintained at CIP. The number of unique accessions among the different collections is unknown.

Ahipa (root): It is a Leguminosae, genus *Pachyrhizus* with three cultivated species, namely, *P. ahipa* (Weddell) Parodi, *P. erosus* and *P. tuberosus*, and two wild relative species – *P. panamensis* and *P. ferrugineus* (Sørensen 1988). *P. ahipa* is mainly cultivated in Bolivia and some in northern Peru (Orting et al. 1996) and in northwest Argentina (Towle 1961). *P. erosus* (known as *jicama* in Mexico) from southwest Mexico to northwest Costa Rica is widely used there and worldwide including Southeast Asia, Far East, Indian Subcontinent and west coast of Africa (Sørensen 1988). *P. tuberosus* from the Amazonia is still in cultivation in Peruvian Amazonia. *P. panamensis* is from Panama and southwestern Ecuador to 800 m asl and *P. ferrugineus* from Mexico to Colombia to 1,600 m asl (Sørensen 1988). The name ahipa is from Quechua word *aqipa* or *asipa* and its other names in Aymara are *konori* and *villu*, in Spanish *jiquima* in Peru (Yacovleff 1933) and in English yam bean (which is more referring to *P. erosus* in Asia).

The domestication of ahipa is poorly known. Brucher (1989) postulated its derivation from wild forms in the eastern slope of Andes and Rea (1995) reported wild forms in the department of La Paz, Bolivia. Archeological evidences in Peru show its cultivation in the Nazca and Mochica cultures (Yacovleff 1933; Yacovleff and Herrera 1934–1935; Brucher 1989). It was described as a 'very watery and sweet' and 'used as a fruit' by the Spanish chronicler Bernabe Cobo (Yacovleff and Herrera 1934–1935). In fact, in Asia *P. erosus* is a common vegetable and is used sometimes like a fruit in vegetable sweet salad and as refreshing snack when consumed raw after removing the peel (personal experience). This is because the dry matter is 19–25% and about half is sugar and the remaining half is 10% protein and 40% starch (Orting et al. 1996). The seed extract of *P. erosus* is used in Southeast Asia as insecticide (personal experience) due to the presence of rotenone, pachyrhizid and erosone. Propagation is by seed and as in many legumes ahipa is self-pollinated.

The Andean genebanks have 102 accessions of *Pachyrhizus* (Table 12.5). Presently, CIP has a project to collect *P. tuberosus* in the department of Junin, Peru and to date 54 accessions have been collected for research and conservation.

Arracacha (root): It is an Apiaceae (Umbelliferae), species *Arracacia xanthorrhiza*. This is the only species in the family domesticated in the New World (Leon 1967) out of 30–36 wild species of this genus in the mountain ranges of Mexico, Guatemala, Costa Rica, Panama, Peru and Bolivia (Constance 1949; Mathias and Constance 1962, 1976; Hiroe 1979). The name is from Quechua word *raqacha* and other names in Aymara is *lakachu*, Amusha (Amazonian tribe) *pueb*, other Spanish names *virraka, zanahoria blanca* and *apio criollo*, and in English white carrot, Peruvian carrot and Peruvian parsnip. The names are referring to celery and carrot because the plant is like celery and the taste of the root is that of carrot. The leaves can be eaten like celery. The root is white to yellow and purple.

It is an ancient crop (Bukasov 1930) and the ancestor is unknown. The genetic diversity is in Andean valleys of 1,500 to 3,000 m asl with warmer temperature where potato cannot be grown in Colombia, Ecuador, Peru and Bolivia. The volume of production is limited. However, starch of arracacha is easily digestible and is used in baby food production and other processed food.

There are some 350 accessions in Andean genebanks (Table 12.5) and CIP has a collection of seven accessions. The number of distinct cultivars in these collections has not been compared. Tapia et al. (1996) reported only 17 distinct morphotypes out of 93 studied in the Ecuadorian collection, and Blas and Arbizu (1995) in Peru indicated 16 out of 32 accessions studied.

Maca (hypocotyls): It is a Brassicaceae, species *Lepidium meyenii* Walper. *L. peruvianum* Chacón (Chacón 1990) is considered within *L. meyenii* in a recent revision of the genus (Al-Shehbaz 2010). The genus consists of some 150 species of annual, biennial or perennial mainly in temperate region of the world and *L. sativum* is the other cultivated species from the Old World (Bailey 1976). The name is a Quechau word, *maca.* The species is a biennial but the crop is managed as an annual in the high Andes (4,000 m) in order to harvest its swollen hypocotyls which are a traditionally functional food for impotency, stamina and female fertility. This may be attributed to its high protein content (10–14%) among tuber and root crops, high iron and calcium, high leusine and isoleucine, high palmitic, linoleic, oleic and stearic fatty acids, and high sterols (Tello et al. 1992; Dini et al. 1994). The planting propagule use is botanical seed and not vegetative sets.

The domestication of maca is postulated to be in central Andes of Peru around Lake Chinchaycocha, Junin about 2,000 year ago (Rea 1992) but its ancestor is unknown. It is an octoploid ($2n=8x=64$) and self-pollinator (Quiros et al. 1996). Wild *Lepidium* occur from Ecuador to Argentina but their relationship with maca has not been reported. The center of diversity and production is in Central Highlands of Peru in the department of Junin and Pasco at altitude of above 4,000 m asl. It is the only crop that cans tolerant frost, hail and snow at these altitudes in addition to the bitter potato. The variation in term of foliage is low but the hypocotyl skin has variable color (Tello et al. 1992). CIP holds the only maca collection with 30 accessions (Table 12.5).

Mashua (tuber): It belongs to the family Tropaeolaceae, genus *Tropaeolum*, section *Mucronata*, species *Tropaeolum tuberosum* Ruis and Pavon with two subspecies (42 chromosomes) – *tuberosum* and *silvestre*. The *silvestre* is the wild forms that do not produce tubers. In addition, there are 86 wild species endemic from Mexico to temperate South America including the ornamental nasturtium (Sparre and Anderson 1991). The name mashua is derived from Quechua names *maswa* and *mashwa* and other names are *añu* in Quechua, *isaño* in Aymara, *mashua* in Spanish in Peru and Ecuador, and *cubio* or *navo* in Colombia. The plant is similar to nasturtium but with smaller orange long peduncle flowers. The tuber shape is typically claviformis with tapered stolon end, and peel and flesh colors are diverse from whitish cream to dark purple (usually refer to as 'black' mashua) as in native potato. In Peru, the 'black' cultivar is believed to have the property against cancer (traditional knowledge) and in fact Noratto et al. (2004) reported that mashua extracts suppress tumor cell proliferation. It is also said to have anti-aphrodisiacal property (Yacovleff and Herrera 1934–1935; Johns et al. 1982) and as a diuretic for kidney ailments.

The main area of diversity and production is from central Peruvian Andes to central Bolivia and in Ecuador the provinces of Carchi and Cañar but it is grown from the Andes of Venezuela to northwestern Argentina from 2,600 to 4,000 m. It is an ancient crop based on archeological evidence in ceremonial pottery representation of Wari culture in Peru (600–1,100 AD). In traditional mixed stand Andean planting system mashua is known as the 'guardian of potato' and it is inter-planted and surrounding the potato field because it repells pests. This could be due to its isothiocyanate content (Johns et al. 1982) and for consumption the tubers have to be cured under the sun for 4–6 days before cooking to remove the chemical odor.

Some 947 accessions of mashua are conserved in genebanks in the Andean countries including 54 accessions in CIP collection (Table 12.5). These collections have not been compared to eliminate the duplicates so that the number of unique cultivars can be estimated.

Mauka (root): It is a Nyctaginaceae, species *Mirabilis expansa* Ruiz and Pavon, the only species in the genus with edible roots. The name is of Aymara origin, *mauka* and other names include *arracacha de toro, chago, miso, pega-pega, tazo* and *yucca inca*. It is a perennial and the edible part is the thickened long fusiform storage root. Propagation is by cutting from underground and basal stems (Rea and Leon 1965; Seminario 1993). It is rarely found in Peru and the Andean region and thus disappearing as a crop as compared to its ornamental relative, *Mirabilis jalapa* (The four o'clock flower or marvel of Peru) which is popular in Peru. The reasons could be that it has no reported functional food value and the root is bitter if not cured properly in the sun before use.

Its domestication and ancestor have not been studied and no archeological evidence has been reported. There are 60 species distributed in the Americas from Mexico to Chile from sea level to more than 3,000 m asl (MacBride 1937; Weberbauer 1945; Rea 1992). The Andes has the most diversity with ten species in Peru (Liesner 1993). Its cultivation is reported in Cajamarca, Amazonas and La Libertad department in northern Peru (Seminario 1993), Puno department in southern Peru (Vallenas 1995), La Paz and Cochabamba department in Bolivia (Rea 1992) and Cotopaxi and Pichincha province in Ecuador (Tapia et al. 1996) growing in altitude between 2,000 and 3,000 m asl.

The Andean genebanks have some 119 accessions (Table 12.5) and the number of unique cultivars has not been studied. CIP is conserving four accessions.

Oca (tuber): It belongs to the family Oxalidaceae, genus *Oxalis*, section *Tuberosae*, species *Oxalis tuberosa* Mol (octoploid with $2n=8x=64$). There are about 80 wild species in the Andes with many endemic in Peru (MacBride 1949; Ferreyra 1986; Pool 1993) and they have different ploidy levels diploids, tetraploids and hexaploids (de Azkue and Martinez 1990) and their relationships were illustrated by Emshwiller and Doyle (2002). The name oca is derived from Quechua names *okko, oqa* and *uqa* and it known as *apilla* in Aymara. As in potato there are sweet and bitter cultivars. The former is for direct cooking and the latter processed into *kaya*, the dehydrated storable form as for *chuño* in potato. Flores (1991) reported that during the processing of *kaya* antibiotics including penicillin, streptomycin, ampicillin and nystatin were found and postulated that the traditional consumption of *kaya* after child birth help in fast recovery is due to the antibiotics.

Oca is also an ancient crop domesticated probably 4,000 BP (Hawkes 1989). Its present cultivation ranges from the Andes of Venezuela to northern Argentina and Chile at 2,500 to 4,000 m asl. However, the main diversity and production concentrate in the central highlands of Peru to central Bolivia (Rea and Morales 1980; Arbizu and Robles 1986; King 1988). The tuber can easily be confused with that of mashua by inexperience eyes. In fact, its cylindrical shape instead of the tapering shape of mashua is quite distinctive. As in potato and mashua the variability is diverse in term of tuber skin and flesh color.

Nearly 3,917 accessions of oca are conserved in genebanks in the Andean countries including 492 accessions in CIP collection (Table 12.5). These collections have not been compared to eliminate the duplicates so that the number of unique cultivars can be estimated.

Ulluco (tuber): It belongs to the family Bassellaceae, genus *Ullucus*, species *Ullucus tuberosus* Caldas with two subspecies where subsp. *tuberosus* is the cultivated forms and subsp. *aborigineus* the wild forms (Sperling 1987). Both subspecies are diploid with $2n=2x=24$. However, cultivated triploid (Cardenas and Hawkes 1948; Gandarillas and Luizaga 1967; Larkka et al. 1992) and tetraploid (Mendez et al. 1994) forms are also found. *Ulluco* came from the Quechua word *ulluku* relating to male organ probably referring to its shape. It is called *ulluma* or *illaco* in Aymara and also called *papa lisas*, *melloco* and other names.

As for oca it is said to be cultivated some 4,000 BP in central Peruvian Andes to Bolivia (Hawkes 1989; Martins 1976) from the wild subsp. *aborigineus* (Sperling 1987). Presently, it is grown from Colombia and Venezuela in the north to northwestern Argentina in the south and in Peru it is most widely eaten among the three minor tubers. Its leaves are also eaten in some Andean communities as nutritive vegetable as in Asia for *Basella alba* and *nigra* belonging to the same family. However, the center of diversity and production is in central Peruvian Andes to central Bolivia (Rea and Morales 1980; Arbizu and Robles 1986; King 1988). In Ecuador, high diversity occurs also in the provinces of Canar, Pichincha, Imbabura and Chimborazo (Castillo et al. 1988).

There are some 2,494 accessions maintained in genebanks in Andean countries (Table 12.5) including 420 accessions in CIP genebank. However, these collections have not been compared so the number of unique cultivars is not known. Morphological study at CIP showed that about half of the collection is of duplicates with 86 morphotypes out of 160 accessions (Vivanco and Arbizu 1995) and in Ecuador 57 out of 287 accessions (Tapia et al. 1996).

Yacon (root): It is an Asteraceae (Compositae), species *Smallanthus sonchifolius* (Poeppig and Endlicher) (Robinson 1978). It was previously *Polymnia sonchifolia* (Wells 1965). The genus has 21 species endemic in the New World concentrating in Peru, Colombia and Venezuela at altitude range of 500–4,000 m asl (Wells 1965; Robinson 1978). Yacon comes from the Quechua word *yakun* referring to its sweet juicy storage roots of some 86% water content. In Aymara it is called *aricoma*, in Spanish *jicama* and *arboloco* and in English yacon. It thus makes a refreshing snack after peeling and eaten fresh like fruits. However, its value is the fructans, a group of low-calorie compounds that aid human intestinal flora and hyperpilemia (Hata et al. 1983) and inulin, a polymer of mainly fructose which is not metabolized by human (NRC 1989) and thus favor by dieters and people with diabetes. Other species with functional food value are *S. uvedalius, S. glabratus* and *S. maculatus* (Wells 1965; Uphof 1968; Robinson 1978).

Archeological evidences showed that yacon was grown during Candelaria culture (1–1,000 AD) in Argentina (Zardini 1991) and Nazca culture (100–1,000 AD) in Peru (O'Neal and Whitaker 1947). Its domestication and ancestor are not known. A species that produces tuberous roots is *S. conatus* from Argentina, Brazil, Uruguay and Paraguay. Yacon is widely grown in the Andes from Venezuela to northwest Argentina and the production is increasing because of its functional food value for fresh use and processed into medicinal products including tea making from its leaves.

The genebanks in Andean countries maintain 411 accessions (Table 12.5) and CIP conserves 29 accessions. The number of duplicates within and among the different collections has not been studied.

12.3 Conservation Strategy

The RTC are usually conserved as individual genotypes (clones) by vegetative propagation methods to ensure that they remain true to type. These clones usually can also sexually reproduce and some especially the balanced polyploids self-pollinate. However, the resulting generations segregate because of their genetic heterozygous nature. This means that most RTC genebanks use clonal conservation methods mainly as field and in vitro collection where the whole collection has to be regenerated in very short periods of a few months to a couple of years. In recently years, cryopreservation is being introduced and tested in well equipped genebanks as the long-term storage method. Because of all these the maintenance cost of RTC is several times higher as compared to seed conservation. To date, the debate continues on whether RTC should be conserved only as seed collection to reduce cost. This is a question on whether linkages of beneficial genes (genotypes) or individual genes/alleles (seed populations) should be the conservation objective. Many of the landraces might be the results of 100 years of selection and adaptation carried out by our farmer ancestors. For example, the ancient Inca artifacts of potato in pottery and stone models are very similar to some of the present day native potato cultivars and, thus, they represent the 'winning' linkages of genes bundled together. In RTC genebanks these linkages of genes are what are being conserved. On the contrary, when these clones are converted to seed populations many of the 'winning' linkages segregate and become displaced. On the other hand, the segregating seed population of an accession will provide a generation of head-start for breeders to evaluate and select the best individuals for further selection or for use as parents in breeding crosses.

The cost in conserving a RTC collection is primarily determined by the size (number of accessions) of a collection. Take the case of the native potato collection at CIP which used to be more than 17,000 accessions and after the elimination of those duplicates through both morphological, and general protein and isozyme studies only 4,235 distinct accessions are currently conserved. The cost of maintaining the collection is thus reduced by four fold. Duplicate identification and elimination are the first priority in RTC genebanks.

The RTC wild relatives (wild species) are conserved using seed conservation methods as populations to represent the genepools as being collected. Most RTC both wild and cultivated species have orthodox seed and thus suitable for long-term storage at –20°C. Similarly, the duplicates of the CIP potato collection have been converted to true seed for long term storage at –20°C. These seed collections (both CWR and cultivated duplicates) are the ones that are being put in safe duplication in the Svalbard genebank in Norway.

Ex situ Conservation: The ex situ methods include field (including in greenhouse), in vitro, cryopreservation, and seed when the focus is on living plant conservation. Additionally, a pollen genebank should be explored because pollen can be conserved well at freezing temperatures and under cryopreservation. DNA samples can be frozen at –70°C for many years and CIP has initiated a DNA bank for potato, sweet potato and ARTC. Herbarium collection is another form of preserved DNA of each genotype in a collection. CIP has established a herbarium collection of all its collections. Each of the conservation methods has its advantages and disadvantages, and thus in most genebanks a combination of methods is employed to safeguard a collection.

Field genebank: This is the most common method used especially in smaller genebanks where there is no tissue culture laboratory. Collections are either grown in the field or in containers, usually under cover in greenhouses. A new field is established when plants in the current field start to deteriorate. Similarly, new containers are planted to replace degenerated ones as required. The main advantages of this method are that it can be applied without a sophisticated laboratory and that the collection can be used for characterization, genetic identity monitoring and is readily available to provide planting materials for field trials. The disadvantages are that it is costly to plant and manage large collections yearly and the collections are subjected to the elements of natural disasters such as flood, frost, drought, and pest and disease infestation. In the latter, a collection through years of growing accessions collected and acquired from different places together in same fields allows cross-contamination of systemic diseases like viruses, phytoplasmas, bacteria, etc. In addition, adaptation is a problem where accessions collected from very different geographical regions and habitats often grow poorly at the genebank site. As the result some of the accessions become too weak to survive and this is one of the main causes of losing accessions in a collection. This also creates quarantine issues in their distribution and use. In most RTC genebanks field materials are not used for distribution. This could be minimized to certain extent by growing collections in quarantined greenhouses using stringent maintenance protocols. However, in this case the greenhouse and labor costs to manage the plants could be extremely high. At CIP the Peruvian and Latin American sweet potato collections are maintained in this manner in a slow-growth state and, similarly, the cassava collection at CIAT as 'bonsai' plants. Alternately, a cleaner growing field area has to be identified for growing the field collection with clean planting propagules. This is the case in the maintenance of the native potato collection at CIP where clean plants from tissue culture program are reconstituted into a set of clean field collection for growing in high Andes where there are fewer insect vectors to readily re-infest the clean stock. Coupled with a stringent selection of the best plants to be harvested for the next growing cycle seed tubers, this program has been very successful to reduce the recontamination of the clean stock by viruses and the resulting harvests are used for the repatriation program of cleaned accessions back to their original owner communities in CIP's 'Ruta Condor' project.

The management of field collection is easily subjected to mechanical mix-up of accessions due to two main causes: (1) handling of yearly regeneration of field collection which includes selection of suitable field plot free of pests, diseases and volunteer plants, planting, field management including accurate labeling, harvesting, storage (in potato at 4°C and 80% RH) and documentation; and (2) in vine propagated crops such as sweet potato and yams an accession with vigorous vine growth, if not managed well, can grow over the neighboring accessions and be mistakenly used for the next season planting. A protocol to compare individual accessions in the new field with that in the old field is absolutely essential. This combines with comparison of accessions in the new field with photographs or images of authentic genetics will minimize this error. The field genebank is therefore a costly conservation procedure.

In vitro genebank: This is the storage of RTC collections as meristem culture in well controlled environment. Meristem culture method is chosen because of its reduced risk of somaclonal variation (mutation) as compared to other forms of tissue culture methods such as callus culture and embryogenesis. The growth of the in vitro plantlets in storage is slowed by controlling the light, temperature and osmotic pressure in the growing medium. The specific protocols for cassava, potato, sweet potato and yam are given in Table 12.6. The most well developed protocol is

Table 12.6 Slow-growth in vitro collection protocols used in cassava at CIAT, potato and sweetpotato at CIP and yams at IITA (Source: final report on GPG2-1.2 collaborative activity: refinement and standardization of storage procedures for clonal crops – http://www.sgrp.cgiar.org/sites/default/files/1_2_FullReport_Final_OrigReply_1April.doc)

CGIAR center – crop	Slow-growth in vitro protocol
CIAT – cassava	Conservation (SN – silver nitrate) medium: MS (2% sucrose) + 0.02 mg/l BAP + 0.1 mg/l GA + 0.01 mg/l ANA + 10 mg/l silver nitrate. Agar 0.7%. pH 5.7–5.8
	Storage growing conditions: temperature 23–24°C; light 18.5 μmol. $m^{-2}.s^{-1}$; photoperiod 12 h; light quality fluorescent lamps, light day type; relative humidity 50–70%
CIP – potato	Conservation (S42) medium: MS based medium with 2 mg/l glycine, 0.5 mg/l nicotinic acid, 0.5 mg/l pyridoxine, 0.4 mg/l thiamine, 2.5% sucrose, 4% sorbitol, 7 g/l agar
	Storage growing conditions: temperature of 6–8°C, photon flux density of 5–20 μmol/(m^2.seg) with a 16 h/8 h photoperiod of light and darkness (fluorescent lamp COOL DAYLIGHT, 36°W)
CIP – sweetpotato	Conservation medium: MS based medium with 0.2 g/l ascorbic acid, 0.1 g/l calcium nitrate, 2 mg/l calcium panthotenate, 0.1 g/l L-arginine, 20 mg/l putrescine, 30 g/l sucrose, and 3 g/l phytagel
	Storage growing conditions: temperature of 19–21°C, photon flux density of 45 μmol/m^2/s with a 16 h/8 h photoperiod of light and darkness (fluorescent lamp COOL DAYLIGHT, 36 W)
IITA – yams	Conservation medium: MS based medium with 100 mg/l myo-inositol, 30 g/l sugar, 1 mg/l kinetin, 20 mg/l L-cysteine and 7 g/l purified agar
	Storage growing conditions: temperature of 18–20°C, photon flux density of 43 μmol/m^2/s with a 12 h/12 h photoperiod of light and darkness (fluorescent lamp COOL DAYLIGHT, 36 W)

for potato developed at CIP where meristem culture is stored in a chamber under controlled conditions: temperature of 6–8°C, photon flux density of 5–20 μmol/(m^2.seg) with a 16 h/8 h photoperiod of light and darkness (Fluorescent lamp Cool Daylight, 36 W) and sorbitol of 30 g/l medium for an average of 2 years in more than ten thousand clones. For cassava at CIAT silver nitrate medium is used at 20°C (day)/15°C (night) temperatures, 12-h photoperiod and 500 to 1,000 lux illumination. Similarly, yams are maintained at IITA, and sweet potato and ARTC at CIP with varying storage periods as shown in Table 12.6. This is known as the slow-growth protocol and in genebanking terminology the medium-term storage of clonally propagated crops. It is becoming the standard method in the conservation of RTC due to the fact that with the in vitro state, a collection can be cleaned of systemic diseases and contaminates such as viruses, bacteria, phytoplasmas and fungi, and then maintained in the clean state for safe international distribution to clear both international and national quarantine requirements. The combination of slow-growth protocol and safe movement status of a collection for the first time allows the implementation of safety duplication of a collection, also known as a 'blackbox' at another site of different risk factors, i.e. in another country, both from natural disaster and national politics. It is also in the in vitro state that sufficient clean meristems can be easily multiplied for cryopreservation in liquid nitrogen at −196°C for long-term conservation (see below on this method). These are some of the advantages of conservation as in vitro collection.

However, the establishment of an in vitro genebank requires a major infrastructure investment including a well equipped laboratory with the following rooms: slow-growth storage, incubation growing room, transfer room, media preparation room, washing room and backup electric generator in case of main power line cut, and virus and bacteria elimination thermo-therapy precision incubator and insect-proof greenhouse for virus diagnostic works. The laboratory and greenhouse complex should have air-locked anteroom to prevent the entry of insects and micro-organisms from outside and the air-conditioning system HEPA filtered to prevent dust, mist and spores in outside air from getting into the genebank. Ant infiltration through building gaps and conduit causes mite infestation and contamination. Maintaining stable quality technicians and equipment quality of the tissue culture laboratory are fundamental to maintain consistent standards. The capital and maintenance costs are expensive and more importantly annual budget has to be obligatory to cover staff salary, utility costs, consumable and equipment, and infrastructure maintenance and replacement costs.

Technicians have to be well trained and disciplined because the frequent subculturing in the regeneration cycle allows plenty of opportunities to commit errors, such as mixed labeling, mix-up of accessions, missed place in wrong position in the shelving system, etc. At CIP, the application of a fully wireless barcode system in the management of the in vitro genebank has prevented many of these mistakes. The CIP's in vitro genebank management software is custom-made to suit the different component protocols and all these are held together and institutionalized by the ISO 17025 accreditation system (see below).

Finally, the genetic stability of collections going through cycles of regeneration in stressful conditions has not been extensively studied. To date, decision to replace field genebank with only an in vitro genebank is not possible.

Cryopreservation genebank: This is the preservation of meristems in liquid nitrogen (LN) at −196°C. In RTC germplasm conservation it is the long-term storage system because it is generally believed that meristems frozen at this ultra low temperature will remain alive from hundreds of years or more. Potato and cassava have stable cryopreservation protocols (Source: Final report on GPG2-1.2 collaborative activity: Refinement and standardization of storage procedures for clonal crops – http://www.sgrp.cgiar.org/sites/default/files/1_2_FullReport_Final_OrigReply_1April.doc).

In potato, research commenced some 40 years ago (Grout and Henshaw 1978) and, currently, at CIP a stable method can obtain high survival and high plant recovery rate in most genotypes of native potato (90%) tried on to date. Satisfactory results were also obtained in a verification trial by cryo-laboratories of CIAT, IITA and Bioversity International.

This is a costly conservation system because it can only be established with an adjoining tissue culture laboratory. The direct costs are the technician cost, laboratory space, cryo-equipment, cryo-storage tanks and LN (which could be expensive in developing countries). The protocols are tedious and involve many treatment steps. Technicians have to be very accurate in chemical preparation, formulation, treatment and incubation time in each step, and thus the introduction of a complete collection will take many years. Presently, CIP has only more than 1,000 of its 4,235-accession native potato collection in its cryo-genebank after 7 years of work. However, once in LN the maintenance cost is low comparing with the other conservation methods.

On the whole cryo-genebank could be said in a pilot stage. The long-term survival and recovery rates have not been fully evaluated. The genetic stability of an accession subjected to intense cryo- stresses at pre-freezing, storage and thawing treatments has not been studied in details. There is no safe duplicate (blackbox) system in place as that for seed crops at Svalbard, Norway. The international clonal crop genebanking community's proposal to use cryo-genebank to keep the base collections and in vitro slow-growth method the working collections in order to reduce cost could not be implemented until the above issues are fully resolved.

Seed genebank: The seed of most RTC both cultivated and related wild species are orthodox seed which mean that they can be stored at low seed moisture content of around 5% (wet weight basis) at –20°C for long period. In the case of potato it is about 30–40 years. At CIP, all the duplicates of the native potato cultivars were converted to seed using bulked pollen pollination within an accession for the tetraploids before they were eliminated from the field and in vitro collection. All the related wild species of potato are conserved as seed populations using bulked pollen pollination in greenhouse with 20–25 plants at CIP based on findings of del Rio and Bamberg (2003). Similarly, wild relatives of sweet potato are kept this way. These are the seed lots that are being deposited at Svalbard vault for safe duplication. The seed banking management processes are similar to that for seed crops.

In situ and on-farm conservation: In recent years in situ and on-farm methods are applied for both CWR and cultivated species. The Global Plan of Action (GPA) on Plant Genetic Resources for Food and Agriculture (FAO 1996) states the need to promote in situ conservation of CWR and wild plants for food production. Recently, a manual on in situ conservation focusing on CWR was published (Hunter and Heywood 2011). Collections maintained ex situ can be said to be in a static state in evolution. In situ conservation provides the needed interaction for the changing environments and agricultural systems to act on the genetics of the accessions and the introgression of CWR to the cultivated forms. This may be the most cost effective and feasible way for managing large diversity of CWR with limited information on their geographical distribution and reproductive biology to efficiently collect and maintain them in ex situ collections. The model of Nature Conservancy (http://www.nature.org/) to procure lands with endemic targeted species has proven to be effective in some countries. Biosphere reserves and heritage sites for the conservation of traditional cultures and crops are being declared and implemented. Community-based natural resources management and community conserved areas such as the Potato Park are increasingly promoted. Public awareness in the value of these different models is an important message to get through.

At CIP, in the last 15 years, a dynamic conservation strategy linking the well established ex situ collection with in situ/on-farm conservation activities by the

Andean farming communities has been developed and the in situ-ex situ complementary has proven its benefits in the case of native potato conservation. In the project, the ex situ accessions under custodianship at CIP collected some 20–40 years ago are returned back to their respective original farming communities from where they were collected. In the recent 10 years (1998–2008) 3,608 samples of 1,250 accessions (out of 4,235 accessions) have been returned back to 41 communities using clean virus indexed tubers. In return the communities of the Potato Park in Cusco, the most progressive field site, through trust established in the engagement and collaboration, have solicited CIP to safeguard their landraces that are not yet conserved in CIP ex situ genebank. This dynamic in situ-ex-situ conservation strategy is being implemented in the last 10 years as the 'Ruta Condor' project in the repatriation of clean native potato cultivars back to their respective original owner communities and guide them in the establishment and maintenance of communal genebanks of native potato. Micro-centers of native potato diversity are established using the passport information of the 17,000 accessions collected by CIP and they are the priority sites of the project. However, in situ conservation activities should be treated as complementary and not a substitution to the well established ex situ conservation strategies and collections.

12.4 Quality of RTC Collections

The quality of a collection is equivalent to how useful it is. The factors determine this include the following:

- Good representation of the crop genepool covering primary, secondary and also to some extent tertiary genepool;
- Accurate passport and genetic identity of all accessions;
- Good characterization and evaluation information;
- Well documented with user friendly searchable database; and
- Cleaned of quarantined pathogen for safe distribution.

The RTC collections of the CGIAR centers (CIAT, CIP and IITA), the South Pacific Commission (SPC) and some major national collections are to large extent representative of the primary genepool for cassava, potato, sweet potato and yams, but lesser extent for taro and aroids, and ARTCs. Upon the completion of the implementation of the global crop conservation strategies by the Global Crop Diversity Trust, CG centers and national genetic resources programs in the coming years the primary genepool of RTC should be well in place. The recently initiated CWR collection and utilization project – 'Adapting agriculture to climate change: New global search to save endangered crop wild relatives' of the GCDT and Kew Gardens will increase the representation of cassava, potato, sweet potato and yams secondary genepool (http://www.eurekalert.org/pub_releases/2010-12/bc-aat120610.php).

The quality and accuracy of passport information are fundamental relating to genepool assessment, the management of the genetic integrity at accession level and taxonomic treatment. New acquisitions have to ensure the best accompanying information are acquired. In RTC a collection is subjected to frequent yearly regeneration both in the field and in vitro genebank and a systematic genetic identity verification program has to be in place to correct any mechanical mix-up of accessions. This program is not well developed in most RTC genebanks. Together with good characterization and evaluation information and a good collection catalog (hard copy and online) the utilization of a collection will be greatly enhanced. Proven crop descriptors are available for all the major RTC. A user-friendly computerized software package for managing a collection accession information and genebanking management information is essential. In CIP genebank, a custom-built software package is used together with a wireless barcode system for its in vitro, seed and field genebanks since 2004. Minimum uses of pens and pencils have eliminated handwriting mistakes completely. The barcode system has been introduced to other CGIAR centers through the World Bank Global Public Goods Project II (http://www.sgrp.cgiar.org/?q=gpg2).

Safe movement of clonal germplasm around the world should be a vital concern of any genebank to prevent the spread of quarantine diseases to new areas in the world. FAO and IPGRI have published guidelines for the main clonal crops including that for RTC (http://typo3.fao.org/fileadmin/templates/agphome/documents/PGR/PubPGR/FAOIPGRI_list.pdf). At CIP accessions qualified for international distribution have to have a health status (HS) level of 2, i.e. HS2. To obtain this level of cleanliness, a potato accession has to be cleaned through meristem culture and thermotherapy and indexed for some 30 known viruses and potato spindle-tuber viroid, and any detected bacteria and phytoplasma, and for a sweet potato accession for 10 known viruses and bacteria. The process involves meristem culture, micropropagation, ELISA and DAS-ELISA tests, thermotherapy and greenhouse diagnostic tests. They are all held together with the wireless barcoding system in real-time. Standardized 'best practice' protocols were developed and documented into operation manuals.

ISO accreditation: With the fundamentals mentioned above at CIP the ISO 17025 accreditation documentation was put together consisting of the following components:

- General organization and policies
- Quality system and related documents both internal and external
- Workflows (19 under ISO accreditation) and operational procedures (41 under ISO accreditation) of all the processes
- Records and forms used and auditing reports both internal and external
- Tools and views

CIP became the first genebank in the world to be accredited with this standard from February 2008 to date. The maintenance of the accreditation required programmed internal and yearly external auditing on staff competency, staff succession plan (shadow-training) and staff training plan at all levels, equipment calibration and monitoring plan and implementation documentation, equipment renewal plan, consumable procurement and proper storage procedures, suppliers plan, laboratory and greenhouse procedures, quality control and monitoring procedures, validation testing with both within and outside reputable laboratories, information documentation and backup plan and procedures, germplasm acquisition and distribution procedures, quarantine procedures, etc. All non-conformities and observations have to be corrected and a corrective procedure in place and documented to prevent a repeat occurrence. All these mean additional costs. However, the benefits are many and the single most important benefit is that all the best practices are actually implemented exactly as they should be in the documentation. If not, 'best practices' could exist only on paper and not being implemented as should be.

Decentralized conservation and distribution strategy: As mentioned above RTC (clonal) collections are expensive to conserve. Two components of a clonal genebank, the 'blackbox' and safe distribution, are particularly challenging relating to their short regeneration and replacement cycle and the quarantine related delay in shipment. A strategy to duplicate sub-collections in different regions of the world for conservation and distribution will take away these two difficulties. In a situation like this a duplicated subset will act as the 'blackbox' for that subset. When the whole collection is completely duplicated as subsets in different regions of the world there is no more need for a 'blackbox'. When a subset is in a region with similar quarantine risks the distribution of these germplasm will be facilitated in clearing the regional quarantine requirements. The selection of an appropriate subset for a particular region will enhance the effectiveness of this strategy. This strategy will call for a good global real-time database system.

Core collection: It is a subset of 10% of a collection that represents the total diversity of the whole collection (Brown 1989). A core collection could be formulated with the use of passport data in combination with molecular and morphological data, ecological adaptation data, evaluation data, nutritional chemistry data, and from breeding aspect combining ability data, and genomic and other omics data. Core collections could vary according to focus on particular traits or regions. The formulation of these fine-tune cores will need evaluation data from multi-location trials. A comprehensive core should include the wild relative species so as to include both the primary and secondary genepools. In the case of RTC, core should exclude closely related clones such as mutated variants and sister lines which could be common because many of the clones could have in cultivation for hundreds of years.

The purpose of a core collection is to provide a tool in the management of a large collection. A core is a scientific starting point for the evaluation of a specific trait because germplasm evaluation is a tedious and costly process and a core allows effort to concentrate on the most diverse set of the whole collection. Additionally, because the maintenance of RTC collections is very costly a strategy to put all the accessions of a collection in cryopreservation and only to maintain the core as active collection and this will reduce the maintenance cost significantly.

Securing the funds for the conservation of RTC: The Global Crop Diversity Trust and CGIAR Consortium in 2010 jointly did a detailed costing study on the conservation component of all CGIAR center's crop genebanks. Based on this study, the funds for RTC conservation at CIAT, CIP and IITA in 2011 have been allocated and commitment to further this funding into future years is under consideration. The present funding plan is the sharing of responsibility between the GCDT endowment long-term grant and the CGIAR allocation to each center. As the GCDT endowment grows, the GCDT long-term grant will increase and the CGIAR contribution will decrease until a stage that GCDT will fund the whole conservation operation.

12.5 Future Challenges and Possibilities

Increasing utilization of the Genetic Resources: The utilization of the genetic resources for crop improvement is lagging behind conservation effort and this is a challenge. For example, only an estimated 5.6% of CIP in trust Native Potato collection has been used in CIP breeding program in the last four decades by CIP breeders. The low usage is partly attributed to the limited ‘good quality’ evaluation information available on this collection (Tables 12.7 and 12.8) and also the bias toward using known clones already described, improved and/or proven for making crosses. The use of the CWR is even less. The publication of the CIP wild potato catalog in 2009 (Salas et al. 2009) has not seen an increase in the request for this group of germplasm. In fact there are only a few breeding programs in the world with pre-breeding activities looking at enhancing the use of these wild genepool. The recent publication of the potato genome (The Potato Genome Sequencing Consortium 2011) may provide the genetic information and knowledge to pace the utilization of this broader genepool in the native potatoes and the wild relatives. The use of the sweet potato collection is even less as compared to potato.

On the other hand, their direct use may be quite impressive. In the case of the native potato, the collection at CIP has been extensively used in the repatriation program at CIP in the ‘Ruta Condor’ project. In the recent past 10 years, 30% (1,250 accessions) of the collection of 4,235 accessions have been returned to their original communities (41 farming communities) in the Andes.

The key breeding objectives for cassava, potato, sweet potato and yams identified in the 2011 CGIAR Research Program proposal on root and tuber crops and banana (CRP3.4 or CRP-RTB) is shown in Table 12.9. This means increasing germplasm evaluation efforts both in genotyping and phenotyping the collections to look for new novel genes for abiotic and biotic traits to be bred into the existing cultivars. Equally important is the use of new genepools of both landraces and CWR to widen the gene base of RTC. For example, innovative ideas proposed include durable resistance to major diseases, earliness in tuberization in potato, role of mycorrhizal fungi in mineral nutrition in different yams accessions, etc.

The ‘chuno’ (traditional freeze-dried potato) factor – facing climate change for native potato in the Andes: Rosegrant (2009) using modeling on crop growth and other factors on climate change indicated by 2050 severe negative impacts on yield and production for all root and tuber crops and thus sharp price increases. Climate change in the Andes is real. The high mountain glaciers are shrinking (Thompson et al. 2011) and the winter-month snow caps are disappearing in many high tropical mountains. Growing seasons in the Andean highlands are experiencing more and more frequent adverse weather conditions. There are increasing

Table 12.7 Evaluation summary for the in trust potato collection maintained at CIP in 2011

Biotic/abiotic stresses and traits evaluated		Number of accessions Evaluated	With useful genes[a]
Fungi	Reaction to late blight leaf	700	142
	Reaction to late blight tuber	318	218
	Reaction to pink rot	1,334	4
	Reaction to scab	546	41
	Reaction to wart	855	62
	Reaction to smut	199	2
	Reaction to *Fusarium*	59	0
	Reaction to *Phoma* blight	440	277
	Reaction to charcoal rot	40	8
Bacteria	Reaction to potato soft rot	464	34
	Resistance to black leg	77	4
Viruses	Reaction to PVX	2,092	19
	Reaction to PVY	0	0
	Reaction to PLRV	3,196	3
	Reaction to PVS	3,068	0
Nematodes	Reaction to cyst nematode race pa2	1,339	148
	Reaction to cyst nematode race pa3	1,333	180
	Reaction to cyst nematode in Cusco	737	72
	Reaction to golden nematode	660	87
	Reaction to root-knot nematode	1,376	570
Insects	Reaction to potato tuber moth	2,011	142
	Reaction to Andean potato weevil	574	59
Environmental stress	Reaction to frost	43	9
	Reaction to hail	10	3
Other desirable traits	Dry matter content	938	408[b]
	Protein content	273	36[c]
	Tuber dormancy	1,605	167[d]
Total		**24,287**	**2,695**

Evaluations at CIP using material from native and landrace potato
[a]Includes highly resistant, resistant, moderately resistant, highly tolerant and tolerance cultivars
[b]Percentage of dry matter content in freshly harvested tuber >24%
[c]Percentage protein content >10%
[d]Tuber dormancy about 6 months

drought and hail storm frequency, raising temperature and followed by frost events in the growing season. Diseases like potato late blight and insect pests are moving up the Andean slopes and so the traditional native potato planting system. This means a reduction in arable cultivated area because of the conical effect of mountains and the infringement into the paramo and puna wetlands, the 'water-tank' of the Andes and Amazon. At the same time the soil becomes thinner and once the soil organic matter is exhausted its renewal is slow because of the low temperature and low rainfall in the winter months which are unfavorable for vegetation growth to build up new organic material. The reduced cultivated land and increasing

Table 12.8 The key breeding objectives for cassava, potato, sweet potato and yams in the coming decades as indicated in the 2011 CGIAR Research Program for root and tuber

	Number of accessions	
Traits evaluated	Evaluated	With useful genes[a]
Nematodes		
Reaction to root-knot nematode	2,761	744
Reaction to brown ring out	758	25
Insects		
Reaction to west sweet potato weevil	1,596	395
Fungi		
Reaction to java black rot	604	231
Reaction to foot rot	168	42
Viruses		
Reaction to SPFMV virus	587	58
Environmental stress		
Reaction to salinity	615	44
Hot tropical climate adaptation	302	53
Nutritive quality		
Dry matter content	1,654	(>45%) 40[b]
Storage root starch content	2,698	(>75%) 34[c]
Storage root beta carotene content	1,949	(>3) 125[d]
Protein content	848	(10%) 7[c]
Total	**14,540**	**1,855**

[a]Including highly resistant, resistant, moderately resistant, highly tolerant and tolerance cultivars
[b]Percentage of dry matter content in freshly harvested tuber
[c]Based on a dry weight basis
[d]mg/100 g fresh weight

Table 12.9 Key breeding objectives for root crops (CRP3.4 or CRP-RTB) as prepared by Bioversity International, CIAT, CIP and IITA (Source: internal document)

Crop	Key breeding objectives
Cassava	Yield, high dry matter, resistance to viruses such as cassava mosaic disease and cassava brown streak disease, compatibility with integrated pest management (IPM), tolerance to drought and low fertility, and low toxins, high-carotene, and fodder
Potato	High stable tuber yield, durable resistance to diseases late blight (LB) and bacterial wilt, resistance to multiple viruses (potato virus Y, potato leafroll virus) and pests, adaptation to heat and drought, short vegetative cycle, tuber quality, and nutritional attributes
Sweetpotato	Wide adaptation, stable tuber yield, high nutritious genotypes and high foliage production for animal feed
Yam	High stable tuber yields, resistance nematodes, viruses, anthracnose, and scale insects, tuber quality, ease of harvest and long storage, suitability to cropping systems and markets, tolerance to abiotic stresses, and textural and nutritional attributes

population mean that the traditional fallowing period is shortening and as the result reducing yield and depleting soil fertility.

Many of the native potato are adapted to the long growing season of up to six months from October–November planting to March–May harvesting. With climate change the growing season could be reduced to less than four months in the future because of late rain and early frost events. This means that many of the traditional late varieties will not be able to tuberize or yield well and thus will be selected against and disappear. The early tuberizing cultivars will persist and at the same time those that can tolerate frost will be selected for.

The most frost tolerance cultivars among the native potato are the bitter potato of *S.* x *ajanhuiri*, *S.* x *juzepczukii* and *S.* x *curtilobum*. When the effect of climate change becomes extreme serious at one stage only bitter potato will be able to survive the severe weather. All the 'sweet' potato will be swept off in situ in the Andes and only bitter potato will remain. This is the 'chuno' factor. Ex situ conservation will be the only home of most native potato.

Potato breeding has to concentrate on abiotic stress tolerance especially against frost in combination with drought and heat adaptation and at the same time be able to resist the increasing occurrence of late blight, viruses and pests. Breeding at triploid and pentaploid level based on *S.* x *juzepczukii* and *S.* x *curtilobum* model (Schmiediche et al. 1982) should be revisited. *S.* x *curtilobum* has limited cultivar forms but they perform well in the altiplano. The increasing use of wild species with frost tolerance such as *S. acaule, S. megistracolobum, S. commersonii* and others as parents means that an understanding on the genetic of bitterness and a high throughput evaluation method should be in place so that frost tolerance non-bitter cultivars can be selected for. Late blight resistance is also important. Many of the native potato cultivars may have been in cultivation for hundreds of years. They should have some minor genes in order to endure the disease all these years. Breeding to accumulate for minor genes should be emphasized.

An alternative model is the breeding of potato that can be irrigated with saline water. Accession CIP 703254 (OCH 2699), known as *Darwin Potato*, an escaped *S. tuberosum*, is said to grow right to the ocean in Low Bay on Guaitecas Island, Chile (Ochoa 1975) and probably survived in brackish water. The genetics is thus available.

Similarly, specific scenarios on other RTC in Africa, Asia and Oceania, where most RTC systems exist, have to be identified, analyzed and mitigation measures proposed and tested.

The '42-day super-quick' potato – the 'poultry farm' potato concept: Growing enough food under decreasing arable land and climate change for the estimated 9.2 billion people in 2050 is the ultimate challenge. Poultry chicken is marketed in 42 days in the Peruvian poultry industry. This is the result of good poultry genetics and poultry farm management. At CIP there are elite bred lines that can be harvested in 70 days from planting to harvesting. The author believes that potato has the genetics for the development of a '42-day' cultivar (the 'super quick' potato). Potato harvesting index could be as high as 85% (Reynaldo et al. 1986), i.e. excellent sink-source photosynthetic assimilates translocation system. Preliminary screening of some native potato accessions at CIP in 2011 showed the presence of very early tuberization in 45 days after planting and day-neutral for long photoperiodic growth. The large sinks for assimilates makes potato a crop for photosynthetic enhancement in response to enriched CO_2 growing environment. Studies have shown that RTC are more responsive to elevated CO_2 than other C3 crops (Miglietta et al. 1998). The 'super quick' potato will be growing in a 'potato' factory, a computerized environmental controlled greenhouse, using the accompanying plant management technologies including temperature and humidity controlled, hydroponic, quarantine controlled, 24-h day length with combination light waves to promote vegetative growth, tuber initiation and tuber bulking, CO_2 enrichment management, etc. The assembling of all the required genetics into a '42-day super-quick' potato is a multidisciplinary project. Germplasm evaluation to identify, understand and isolate all the required genetic components is the first step and then the assembling of these components through breeding and biotechnology. In parallel greenhouse technologies have to be designed and tested according to the 'super quick' potato.

Sweet potato – the space age staple, vegetable and 'food for the mind': Sweet potato is a survival crop. It travelled across the vast Pacific from Latin America to New Guinea before Columbus. Many lives were sparse from starving to death during the Second War World in Far East and Southeast Asia. The orange-flesh cultivars have high beta-carotene and in Asia, the leaves are a high nutritious vegetable of equivalent to spinach. The author was the first to use the yellow and purple leaf accessions as ornamental plants at AVRDC – The World Vegetable Center's genebank in 1985 and this concept was then taken to the world. Currently, sweet potato is an ornamental species in its own right. These three qualities of sweet potato have prompted the author's recommendation of sweet potato as a crop for the International Space Station (http://www.nasa.gov/mission_pages/station/main/) when Texas A&M University was looking for crops that can be grown at low pressure in the space station for NASA. The tuber will be the 'space' staple, the young leaves the 'space' vegetable and the growing vines the garden of the space station providing nature greenness and morning glory purple flowers in the harsh metallic environment, the 'food for the mind'. Sweet potato, a humble vine with a proven past, will have a great future – the 'space age' crop.

12.6 Conclusion

RTC have had proven to be important survival crops in history, e.g. many lives survived on sweet potato in Asia and the Pacific during the Second World War and in recent years they are proving to be important local staple food not suffering to the fluctuation in prices as the grain commodity crops. They are the food security of many people in the tropics and subtropics. The genetic resources conservation of the major RTC is progressing as for other important crops. However, RTC being a vegetatively propagated group the conservation strategies and methodologies are different from seed crops and as the result more research investment and experimentation have to be done. The conservation of the wild crop relatives is weak and has to be intensified in the near future because of continuous degradation of their natural habitats and climate change effect. The breeding of RTC also presents special issues and requirements and the immediate need is the evaluation of the available germplasm for the required agronomic traits. The potential of RTC in contributing to the global food security is certain and they have specific traits that allow them to be exploited as that in potato to factory produce them as the '42-day super-quick' potato concept and the space-age sweet potato described in the chapter.

References

Akoroda MO, Chheda HR (1983) Agro-botanical and species relationships of Guinea yams. Trop Agric 60:242–248

Allem AC (1987) *Manihot esculenta* is a native of the neotropics. FAO/IBPGR Plant Genet Resour Newsl 71:22–24

Allem AC (1994) The origin of *Manihot esculenta* Crantz (Euphorbiaceae). Genet Resour Crop Evol 41(3):133–150

Al-Shehbaz IA (2010) A synopsis of the South American *Lepidium* (Brassicaceae). Darwiniana 48(2):141–167

Arbizu C (1994) The agroecology of achira in Peru. CIP Circular 20(3):12–13

Arbizu C, Robles E (1986) Catalogo de los recursos geneticos de raices y tuberculos andinos. Universidad Nacional de San Cristobal de Huamanga, Facultad de Ciencias Agrarias, Prog. De Investigaciones en Cultivos Andinos, Ayacucho

Austin DF (1977) Hybrid polyploids in *Ipomoea* section *Batatas*. J Hered 68:259–260

Austin DF (1988) The taxonomy, evolution and genetic diversity of sweet potatoes and related wild species. In: Exploration, maintenance and utilization of sweet potato genetic resources. Report of the first sweet potato planning conference 1987. International Potato Center, Lima, pp 27–59

Austin DF, Huaman Z (1996) A synopsis of *Ipomoea* (Convolvulaceae) in the Americas. Taxon 45(1):3–38

Bai KV, Asiedu R, Dixon AGO (1993) Cytogenetics of *Manihot* species and interspecific hybrids. In: Proceedings of the first international meeting of the cassava biotechnology network. CIAT, Cartagena

Bailey H (1976) Hortus third. A concise dictionary of plants cultivated in the United States and Canada. McMillan, New York

Barnes H (1975) The diffusion of the manioc plant from South America to Africa: an assay in ethnobotanical culture history. Dissertation, Columbia University

Blas R, Arbizu C (1995) Estudios preliminaries sobre la variacion de la arracacha (*Arracacia xanthorrhiza* Bancroft). In: Resumenes del Primer Congreso Peruano de Cultivos Andinos 'Oscar Blanco Galdos', Universidad Nacional de San Cristobal de Huamanga, Facultad de Ciencias Agrarias, Programa de Investigacion en Cultivos Andinos, Ayacucho, Peru, 11–16 setiembre 1995, Cultivos Andinos 5(1):17

Bonierbale M, Guevara C, Dixon AGO, Ng NQ, Asiedu R, Ng SYC (1997) Cassava. In: Fuccillo D, Sears L, Stapleton P (eds) Biodiversity in trust – conservation and use of plant genetic resources, CGIAR centres. Cambridge University Press, Cambridge, pp 1–20

Brako L (1993) Cannaceae. In: Brako L, Zaruchi JL (eds) Catalogue of the flowering plants and gymnosperms of Peru. Missouri Botanical Garden, St. Louis, p 326

Brown AHD (1989) Core collections: a practical approach to genetic resources management. Genome 31:818–824

Brown AHD (2000) Aroids. Plants of the Arum family, 2nd edn. Timber Press, Portland, 392 pp

Brucher H (1989) Useful plants of neotropical origin and their wild relatives. Springer, Berlin

Bukasov SM (1930) The cultivated plants of Mexico, Guatamala and Colombia. Bull Appl Bot Genet Plant Breed (Leningrad) Suppl 47:191–226, 513–525

Cardenas M, Hawkes JG (1948) Numero de cromosomas de algunas plantas nativas cultivades por los indios en los Andes. Revista de Agricutura, Universidad Mayor de San Simon, Cochabamba 5(4):30–32

Castillo R, Nieto C, Peralta E (1988) El germoplasma de cultivos andinos en Ecuador. In: Memorias del VI Congreso Internacional sobre Cultivos Andinos, Quito, Ecuador, 30 mayo–2 junio 1988. Instituto Nacional de Investigaciones Agropecuarias (INIAP), pp 323–331

Clement CR (1994) Crops of the Amazon and Orinoco regions. Their origin, decline and future. In: Hernández-Bermejo JE, León J (eds) Neglected crops: 1492 from a different perspective, FAO plant production and protection series. FAO, Rome, pp 195–203

Constance L (1949) The South American species of *Arracacia* (Umbelliferae) and some related genera. Bull Torrey Bot Club 76(1):39–52`

Chacón G (1990) La maca (Lepidium peruvianurn) Chacón sp. nov.) y su habitat. Rev. Peruana de Biologia 3:171–272

Darlington CD, Janaki-Ammal EK (1945) Chromosome atlas of cultivated plants. G. Allen, London

de Azkue D, Martinez A (1990) Chromosome number of *Oxalis tuberose* alliance (Oxalidaceae). Plant Syst Evol 169:25–29

del Rio AH, Bamberg JB (2003) The effect of genebank seed increase on the genetics of recently collected potato (*Solanum*) germplasm. Am J Potato Res 80:215–218

Dini A, Migiliuolo G, Rastrelli L, Saturnino P, Schettino O (1994) Chemical composition of *Lepidium meyenii*. Food Chem 49:347–349

Dumont R, Hamon P, Seignobos S (1994) Les ignames au Cameroun. Reperes, Cultures annuelles. CIRAD-CA, Montpellier

Edison S, Sreekumari SK, Pillai SV, Sheela MN (2004) Diversity and genetic resources of taro in India. In: Guarino L, Taylor M, Osborn T (eds) Third taro symposium. Secretariat of the Pacific Community, Fiji, 21–23 May 2003, pp 85–88

Emshwiller E, Doyle JJ (2002) Origins of domestication and polyploidy in oca (*Oxalis tuberosa*: Oxalidaceae). 2. Chloroplast-expressed glutamine synthetase data. Am J Bot 89(7): 1042–1056

Engel E (1970) Exploration of the Chilca canyon. Curr Anthropol 11:55–58

Espinosa P, Vaca R, Abad J (1993) Informe sobre la producion de archira en Patate: limitantes y posibilidades. Equipo de Ciencias Sociales: 1–22. CIP, Quito

FAO (1996) 'Global plan of action for the conservation and sustainable utilization of plant genetic resources for food and agriculture and the Leipzig declaration', adopted by the international technical conference on plant genetic resources, Leipzig, 17–23 Jun 1996, FAO. www.fao.org/ag/AGP/agps/GpaEN/leipzig.htm

FAO (2010) The second report on the state of the world's plant genetic resources for food and agriculture. Commission on genetic resources for food and agriculture. FAO, Rome

FAOSTAT (2009) Summary of world food and agriculture statistics. www.fao.org/faostat. Access 20 February 2012

Ferreyra R (1986) Flora y vegetación del Perú. Gran Geografía del Perú. Coedit. Manfer Mejia Baca, Barcelona España. Tomo II: pp 11–13

Flores I (1991) Estudio del processo de elaboracion de khaya. In Generacion de tecnología para procesamiento de cultivos andinos. Informe Tecnico Final. INIAA-FUNDEAGRO, Huancayo, pp 34–53

Gandarillas H, Luizaga J (1967) Numero de cromosomas de la papalisa (*Ullucus tuberosus* Caldas). Sayana Revista Boliviana de Agricultura 5(2):8–9

Giacometti DC, Leon J (1994) Tannia, yautia (*Xanthosoma sagittifolium*). In: Hernaldo JE, Leon J (eds) Neglected crops: 1492 fauna different perspective, vol 26, Plant production and protection series. FAO, Rome, pp 253–260

Glendinning DR (1975a) Neo-tuberosum: new potato breeding material. 1. The origin, composition, and development of the tuberosum and neo-tuberosum gene pools. Potato Res 18:256–261

Glendinning DR (1975b) Neo-tuberosum: new potato breeding material. 2. A comparison of neo-tuberosum with unselected Andigena and with Tuberosum. Potato Res 18:343–350

Glendinning DR (1975c) Neo-tuberosum: new potato breeding material. 3. Characteristics and variability of neo-tuberosum, and its potential value in breeding. Potato Res 18:351–362

Glendinning DR (1975d) Chilean potatoes: an appraisal. Potato Res 18:306–307

Gonzales R, Arbizu C (1995) Niveles de ploidia de las archiras cultivadas en el Peru. In: Resumenes del Primer Congreso Peruano de Cultivos Andinos 'Oscar Blanco Galdos'. Universidad Nacional de San Cristobal de Huamanga, Facultad de Ciencias Agrarias, Programa de Investigacion en Cultivos Andinos, Ayacucho, 11–16 setiembre 1995, Cultivos Andinos 5(1):17

Grout BWW, Henshaw GG (1978) Freeze preservation of potato shoot-tip cultures. Ann Bot 42:1227–1229

Gulick P, Henshey C, Eswuinas-Alcazar J (1983) Genetic resources of cassava and wild relatives. International Board for Plant Genetic Resources, Rome

Hamon P, Dumont R, Zoundjihekpon J, Tio-Toure B, Hamon S (1995) Wild yams in West Africa: morphological characteristics. ORSTOM, Paris

Hanson J (1985) Methods for storing tropical root crop germplasm with special reference to yam. Plant Genet Resour Newsl 64:24–32

Harlan JR, de Wet JMJ (1971) Towards a rational classification of cultivated plants. Taxon 20:509–517

Hata Y, Hara T, Oikawa T, Yamamoto M, Hirose N, Nagashima T, Torihama N, Watabe A, Yamashita M (1983) The effect of oligofructans (neosugar) on hyperpilemia. Geriatr Med 21:156–167

Hawkes JG (1989) The domestication of roots and tubers in the American tropics. In: Hillman BC, Harris DR (eds) Foraging and farming: the evolution of plant exploitation. Unwin Hyman, London, pp 481–503

Hawkes JG (1990) The potato – evolution, biodiversity and genetic resources. Belhaven Press, London

Hay A (1998) Botanical varieties in taro, *Colocasia esculenta*: leaving old baggage behind. A report on taro consultancy no. CO2C. IPGRI, Rome, 13 pp

Hershey C (1987) Cassava germplasm resources. In: Proceedings of the workshop on cassava breeding: a multidisciplinary review, Philippines, 4–7 Mar 1985

Hiroe M (1979) Umbelliferae of the world. Anake Book, Tokyo

Ho TV, Hao BT (1995) Studies on edible *Canna* in Vietnam. In: Chujoy E (ed) Root crops germplasm research in Vietnam. National Institute of Agricultural Sciences (INSA)/International Development Research Center (IDRC)/International Potato Center (CIP), Hanoi/Tanglin/Manila

Huaman Z, Hawkes JG, Rowe PR (1980) A biosystematic study of the origin of the diploid potato, *Solanum ajanhuiri*. Euphytica 31:665–675

Hunter D, Heywood V (2011) Crop wild relatives: a manual of in situ conservation. Earthscan, London

Irwin SV, Kaufusi P, Banks K, de la Peña R, Cho JJ (1998) Molecular characterization of taro (*Colocasia esculenta*) using RAPD markers. Euphytica 99:183–189

Jennings DL (1976) Cassava, *Manihot esculenta* (Euphorbiaceae). In: Simmonds N (ed) Evolution of crop plants. Longman, London, pp 81–84

Johns T, Kitts WD, Newsome F, Towers GHN (1982) Anti-reproductive and other medicinal effects of *Tropaeolum tuberosum*. J Ethnopharmacol 5:149–161

Jones WO (1959) Manioc in Africa. Stanford University Press, Stanford

Jones A (1967) Should Nishiyama's K123 (*Ipomoea trifida*) be designated *I. batatas*? Econ Bot 21:163–166

King SR (1988) Economic botany of the Andean tuber crop complex: *Lepidium meyenii, Oxalis tuberose, Tropaeolum tuberosum* and *Ullucus tuberosus*. Ph.D. thesis, The City University of New York, New York

Kreike CM, van Eck HJ, Lebot V (2004) Genetic diversity of taro, *Colocasia esculenta* (L.) Schott, in Southeast Asia and the Pacific. Theor Appl Genet 109:761–768

Larkka J, Jokela P, Pietila L, Viinikka Y (1992) Karyotypes and meiosis of cultivated and wild ulluco. Caryologia 45(3–4):229–235

Lebot V (1992) Genetic vulnerability of Oceania's traditional crops. Exp Agric 28:309–323

Lebot V (1999) Biomolecular evidence for plant domestication in Sahul. Genet Resour Crop Evol 46:619–628

Lebot V, Aradhya KM (1991) Isozyme variation in taro (*Colocasia esculenta* (L.) Schott) from Asia and the Pacific. Euphytica 56:55–66

Lebot V, Prana M, Kreike N, van Heck H, Pardales J, Okpul T, Gendua T, Thongjiem M, Hue H, Viet N, Yap TC (2004) Characterisation of taro (*Colocasia esculenta* (L.) Schott) genetic resources in Southeast Asia and Oceania. Genet Resour Crop Evol 51:381–392

Leon J (1967) Andean tuber and root crops: origin and variability. In: Proceedings of the international symposium on tropical root crops, University of West Indies, St. Augustine, 2–8 Apr 1967, pp 118–123

Liesner RL (1993) Nictaginaceae. In: Brako L, Zarucchi JL (eds) Catalogue of flowering plants and gymnosperms of Peru. Missouri Botanical Garden, St. Louis, pp 750–754

Maas PJM, Maas H (1988) Cannaceae. In: Harling G, Anderson L (eds) Flora of Ecuador, vol 32. Swedish Research Council, Stockholm, pp 1–9

MacBride JF (1937) Mirabilis L. In: Flora of Peru, Chicago, Field Mus Nat Hist Bot 13(2): 539–546

Magoon M, Krishnan R, Bai K (1969) Morphology of the pachytene chromosomes and meiosis in *Manihot esculenta* Crantz. Cytologia 34:612–626

Martin FW (1974a) Tropical yams and their potential. Series – part 1. *Dioscorea esculenta*. USDA agriculture handbook no. 457. U.S. Department of Agriculture, Agricultural Research Service, Washington, DC

Martin FW (1974b) Tropical yams and their potential. Series – part 2. *Dioscorea bulbifera*. USDA agriculture handbook no. 466. U.S. Department of Agriculture, Washington, DC

Martin FW (1976) Tropical yams and their potential. Series – part 3. *Dioscorea alata*. USDA agriculture handbook no. 495. U.S. Department of Agriculture, Washington, DC

Martin FW, Degras L (1978a) Tropical yams and their potential. Series – part 5. *Dioscorea trifada*. USDA agriculture handbook no. 522. U.S. Department of Agriculture, Washington, DC

Martin FW, Degras L (1978b) Tropical yams and their potential. Series – part 6. Minor cultivated *Dioscorea* species. USDA agriculture handbook no. 538. U.S. Department of Agriculture, Washington, DC

Martin FW, Rhodes AM (1978) The relationship of *Dioscorea cayenensis* and *D. rotundata*. Trop Agric 55:195–206

Martin FW, Sadik S (1977) Tropical yams and their potential. Series – part 4. *Dioscorea rotundata* and *Dioscorea cayenensis*. USDA agriculture handbook no. 502. U.S. Department of Agriculture, Washington, DC

Martins R (1976) New archeological techniques for the study of ancient root crops in Peru. Ph.D. thesis, University of Birmingham, England

Mathias ME, Constance L (1962) *Arracacia* Bancroft. In: Mathias ME, Constance L (eds) Flora of Peru vol XIII(1), part V. Field Museum of Natural History, Chicago, pp 13–19

Mathias ME, Constance L (1976) The genus *Niphogeton* (Umbelliferae) – a second encore. Bot J Linn Soc 72(4):311–324

Macbride JF (1949) Oxalidaceae. In: Flora of Peru, Field Mus. Nat. Hist. Bot. 13(2): 544–602

Matthews PJ (1990) The origins, dispersal and domestication of taro. Ph.D. thesis. Australian National University, Canberra

Matthews PJ (1991) A possible tropical wildtype taro: *Colocasia esculenta* var. *aquatilis*. Indo Pac Prehist Assoc Bull 11:69–81

Matthews PJ (2002) Potential of root crops for food and industrial resources. In: Potential of root crops for food and industrial resources. Twelfth symposium of the International Society for Tropical Root Crops (ISTRC), 10–16 Sept 2000, Tsukuba, pp 524–533

Matthews PJ (2004) Genetic diversity of taro, and the preservation of culinary knowledge. Ethnobot Res Appl 2:55–71

Maxted N, Dulloo ME, Ford-Lloyd BV, Iriondo J, Jarvis A (2008) Gap analysis: a tool for complementary genetic conservation assessment. Divers Distrib 14:1018–1030

Mendez M, Arbizu C, Orrillo M (1994) Niveles de ploidia de los ullucos cultivados y silvestres. En Resumenes de trabajos presentados al VIII Congreso Internacional de Sistemas Agropecuarios Andinos y su proyección al tercer milenio, Universidad Austral de Chile, Valdivia, 21–26 marzo 1994, Agro Sur 22:12

Meza G (1995) Variedades nativas de archira (*Canna edulis* Ker Cawler) en la Valle del Apurimac. Centro de Investigacion en Cultivos Andinos, Facultad de Agronomia y Zootecnia, Universidad Nacional de San Antonio Abad del Cusco, Cusco

Miglietta F, Magliulo V, Bindi M, Cerio L, Vaccari FP, Loduca V, Peressotti A (1998) Free air CO_2 enrichment of potato (*Solanum tuberosum* L.): development, growth and yield. Glob Chang Biol 4:163–172
Morales R (1969) Caracteristicas físicas, químicas y organolépticas del almidon de 'achira' (*Canna edulis* Ker var.). Revista de la Academia Colombiana de Ciencias Exactas, Fisicas y Naturales XIII(51):357–369
Nassar N (1978a) Conservation of the genetic resources of cassava (*Manihot esculenta*) – determination of wild species localities with emphasis on probable origin. Econ Bot 32(3):311–320
Nassar N (1978b) Microcenters of wild cassava, *Manihot* spp. Diversity in central Brazil. Turrialba 28(4):345–347
Ng NQ (1993) Annual Report 1992, Genetic Resources Unit, Crop Improvement Division, International Institute of Tropical Agriculture. IITA, Ibadan, Nigeria
Nishiyama I (1963) The origin of sweet potato plant. In: Barrau J (ed) Plants and the migrations of Pacific peoples. Bishop Museum Press, Honolulu, pp 119–128
Nishiyama I (1971) Evolution and domestication of sweet potato. Bot Mag Tokyo 84:377–387
Nishiyama I, Miyasaki T, Sakamoto S (1975) Evolutionary autoploidy in the sweet potato (*Ipomoea batatas* (L.) Lam.) and its progenitors. Euphytica 24:197–208
Noratto G, Cisneros-Zevallos L, Mo H (2004) *Tropaeolum tuberosum* (mashua) extracts suppress tumor cell proliferation. FASEB J. 18(5):A886–A886 (Suppl.).
Noyer JL, Billot C, Weber P, Brottier P, Quero-Garcia J, Lebot V (2004) Genetic diversity of taro (*Colocasia esculenta* (L.) Schott) assessed by SSR markers. In: Guarino L, Taylor M, Osborn T (eds) Third taro symposium. 21–23 May 2003, Secretariat of the Pacific Community, Fiji, pp 174–180
NRC (National Research Council) (1989) Lost crops of the Incas. Little-known plants of the Andes with promise for worldwide cultivation. National Academic Press, Washington, DC
O'Brien JP (1972) The sweet potato: its origin and dispersal. Am Anthropol 74:343–365
O'Neal LM, Whitaker TW (1947) Embroideries of the early Nazca period and the crops depicted on them. Southw J Anthropol 3(4):294–321
Ochoa CM (1975) Potato collecting expedition in Chile, Bolivia and Peru, and the genetic erosion of indigenous cultivars. In: Frankel OH, Hawkes JG (eds) Crop genetic resources for today and tomorrow. Cambridge University Press, Cambridge
Ochoa CM (1984) *Solanum hygrothermicum*, new potato species cultivated in the lowlands of Peru. Econ Bot 38:128–133
Ochoa CM (1999) Las papas de Sudamerica: Peru. Centro International de La Papa (CIP), Lima
Ochoa CM (2003) Las Papas del Peru. CIP, UNALM, CSUDE, Lima
Ochoa CM (2008) Homenaje a la trayectoria cientifica del Dr. Carlos M. Ochoa Nieves. Memoria de actividades y participantes. 13 Congreso Latinoamericano de Genetica y VI Congreso Peruano de Genetica – "Recursos Genetica Latinamericanos: Vida para la vida", Lima
Olsen KM, Schaal BA (1999) Evidence on the origin of cassava: phylogeography of *Manihot esculenta*. Proc Natl Acad Sci U S A 96:5586–5591
Olsen KM, Schaal BA (2001) Microsatellite variation in cassava (*Manihot esculenta*, Euphorbiaceae) and its wild relatives: further evidence for a southern Amazonian origin of domestication. Am J Bot 88:131–142
Orting B, Gruneberg WJ, Sørensen M (1996) Ahipa (*Pachyrhizus ahipa* (Wedd.) Parodi) in Bolivia. Genet Res Crop Evol 43:435–446
Plaisted RL (1971) A project to duplicate 400 years of potato evolution. N Y Food Life Sci 4:24–26
Plucknett DL (1976) Edible aroids. In: Simmonds NW (ed.) *Evolution of crop plants*. Longman, London. pp. 10–12
Pool A (1993) Oxalidaceae. In: Brako L, Zarucchi JL (eds) Cataloque of the flowering plants and gymnosperms of Peru. Missouri Botanical Garden, St. Louis, pp 867–875
Purseglove JK (1972) Tropical crops. Monocotyledons I. Longman, London
Quiros C, Epperson A, Hu J, Holle M (1996) Physiological studies and determination of chromosome number in maca, *Lepidium meyenii* (Brassicaceae). Econ Bot 50(2):216–223
Quero-Garcı´a J, Noyer JL, Weber A, Perrier X, McKey D, Lebot V (2006) Recombination and clonality in taro (*Colocasia esculenta* (L.) Schott): implications for the evolution of cultivar

diversity. Paper presented at the 14th Triennial Symposium of the International Society for Tropical Root Crops (ISTRC), Thiruvananthapuram, India, 21–26 November 2006

Rea J (1992) Raices andinas. In: Hernandez JE, Bermejo, Leon J (eds). Cultivos marginados, otro perspectiva de 1492 Colección FAO: Produccion y protección vegetal n. 26, Roma

Rea J (1995) Informe técnico sobre conservación in situ de raíces y tuberculos andinos. Programa Colaborativo Biodiversidad de Raices y Tuberculos Andinos, Centro Internacional de la Papa-Cooperacion Tecnica Suiza, Lima

Rea J, Leon J (1965) La mauka (*Mirabilis expansa* Ruiz & Pavon), un aporte de la agricultura andina prehispanica de Bolivia. Anales Cientificos (Universidad Agraria, la Molina, Peru) 3(1):38–41

Rea J, Morales D (1980) Catalogo de tuberculos andinos. Ministerio de Asuntos Campesinos y Agropecuarios, Instituto Boliviano de Tecnologia Agropecuaria, Programa de Cultivos Andinos, La Paz

Reyes Castro G (2006) Studies on cocoyam *(Xanthosoma spp.)* in Nicaragua, with emphasis on Dasheen mosaic virus. Diss. (sammanfattning/summary) Uppsala: Sveriges lantbruksuniv., Acta Universitatis agriculturae Sueciae, 1652–6880; 2006:7

Reynaldo GV, Moreno U, Black CC (1986) Growth, partitioning, and harvest index of tuber-bearing *Solanum* genotypes grown in two contrasting Peruvian environments. Plant Physiol 82:103–108

Roa AC, Maya MM, Duque MC, Tohme J, Allem AC, Bonierbale MW (1997) AFLP analysis of relationships among cassava and other *Manihot* species. Theor Appl Genet 95:741–750

Robinson H (1978) Studies in the Heliantheae (Asteraceae) XII. Re-establishment of the genus Smallanthus. Phytologia 39(1):47–53

Rogers DJ (1965) Some botanical and ethnological considerations of *Manihot esculenta*. Econ Bot 19(4):369–377

Rogers DJ, Appan SG (1973) Flora neotropica monograph no. 13 *Manihot* Manihotoides (Euphorbiaceae). Hafner Press, New York, pp 1–272

Rogers DJ, Fleming HS (1973) A monograph of *Manihot esculenta* with an explanation of the taximetrics methods used. Econ Bot 27:1–113

Rosegrant MW (2009) Roots and tubers: opportunities and challenges under growing resource scarcity. Presentation at ISTRC conference on roots and tubers: the overlooked opportunities, CIP, Lima, 2 Nov 2009

Roullier C, Rossel G, Tay D, Mckey D, Lebot V (2011) Combining chloroplast and nuclear microsatellites to investigate origin and dispersal of New World sweet potato landraces. Mol Ecol 20:3963–3977

Salas A, Tay D, Centeno R (2009) Catalogue of the Global FAO-International Treaty 'in trust' wild potato collection at the International Potato Center (CIP). In: Pieterse L, Hils U (eds) World catalogue of potato varieties 2009/10. Agrimedia, Clenze. ISBN 3-86037-984-4

Sauer J (1952) Agricultural origins and dispersals. American Geographical Society, New York

Schmiediche PE, Hawkes JG, Ochoa CM (1982) The breeding of the cultivated potato species *Solanum juzepczukii* Buk. and *S. curtilobum* Juz. Et Buk. II. Euphytica 31:695–707

Seminario J (1993) Aspectos etnobotanicos del chago, miso o mauka (*Mirabilis expansa* R. y P.) en el Peru. Boletin de Lima 86:71–79

Simmonds NW (1964) Studies of the tetraploid potatoes. II. Factor in the evolution of the tuberosum group. J Linn Soc (Bot) 59:43–56

Smith CE (1968) The New World centers of origin of cultivated plants and the archaeological evidence. Econ Bot 22(3):253–266

Sparre B, Anderson L (1991) A taxonomic revision of the Tropaeolaceae. Opera Bot 108:1–140

Sperling C (1987) Systematics of the Basellaceae. Ph.D. dissertation. Harvard University, Cambridge, MA

Spooner DM, McLean K, Ramsay G, Waugh R, Bryan GJ (2005) A single domestication for potato based on multilocus amplified fragment length polymorphism genotyping. Proc Natl Acad Sci USA 102:14694–14699

Spooner DM, Núñez J, Trujillo G, del Rosario HM, Guzmán F, Ghislain M (2007) Extensive simple sequence repeat genotyping of potato landraces supports a major reevaluation of their gene pool structure and classification. Proc Natl Acad Sci USA 104:19398–19403

Sørensen M (1988) A taxonomic revision of the genus Pachyrhizus Rich. ex DC. nom. cons. Nord J Bot 8(2):167–192
Tapia C, Castillo R, Mazon N (1996) Catalogo de recursos geneticos de raices y tuberculos andinos en Ecuador. Instituto Nacional Autonomo de Investigaciones Agropecuarias, Departamento Nacional de Recursos Fitogeneticos y Biotecnologia, Quito
Tello J, Hermann M, Calderon YA (1992) La maca (*Lepidium meyenii* Walp): cultivo alimenticio potencial para las zonas altoandinas. Boletin de Lima 81:59–66
Terauchi R, Chikaleke VA, Thottapilly GS, Hahn SK (1992) Origin and phylogeny of Guinea yams as revealed by RFLP analysis of chloroplast DNA and nuclear ribosomal DNA. Theor Appl Genet 83:743–751
The Potato Genome Sequencing Consortium (2011) Genome sequence and analysis of the tuber crop potato. Nature 475, 189–195
Thompson LG, Mosley-Thompson E, Davis M, Brecher H (2011) Tropical glaciers, recorders and indicators of climate change are disappearing globally. Ann Glaciol 52(59):23–34
Towle MA (1961) The ethnobotany of pre-Columbian Peru. Aldine, Chicago
Ugent D, Peterson W (1988) Archeological remains of potato and sweet potato in Peru. CIP Circular 16:1–10
Ugent D, Pozorski S, Pozorski T (1986) Archeological manioc (Manihot) from coastal Peru. Econ Bot 40:78–102
Ugent D, Dillehay T, Ramirez C (1987) Potato remains from a late Pleistocene settlement in south central Chile. Econ Bot 4:17–27
Umanah E, Hartman R (1973) Chromosome numbers and karyotypes of some *Manihot* species. J Am Soc Hortic Sci 98:272–274
Uphof JC (1968) Dictionary of economic plants, 2nd edn. Verlag von J Cramer, New York
Vallenas M (1995) Vigencia del cultivo de Mauka (Mirabilis expansa) en Puno, Perú. En: Resúmenes del Primer Congreso Peruano de Cultivos Andinos "Oscar Blanco Galdós", 11–16 de setiembre, Huamanga, pp 72–73
Vavilov NI (1992) Origin and geography of cultivated plants: (collected works 1920–1940). In: Dorfeyev V (ed) Cambridge University Press, Cambridge
Vivanco F, Arbizu C (1995) Variacion morfológica del ulluco (*Ullucus tuberosus* Caldas), oca (*Oxalis tuberosa* Mol.) y mashua (*Tropaeolum tuberosum* R.&P.). In: Resumenes del Primer Congreso Peruano de Cultivos Andinos 'Oscar Blanco Galdos', Universidad Nacional de San Cristobal de Huamanga, Facultad de Ciencias Agrarias, Programa de Investigacion en Cultivos Andinos, Ayacucho, 11–16 setiembre 1995, Cultivos Andinos 5(1):18
Weberbauer A (1945) El mundo vegetal de los andes peruanos. Estacion Experimental Agricola La Molina, Ministerio de Agricultura, Lima
Wells JR (1965) A taxonomic study of Polymnia (Compositae). Brittonia 17:144–159
Wilson JE (1984) Cocoyam. In: Goldsworthy PR, Fisher NM (eds) The physiology of tropical field crop. Wiley, New York/London, pp 589–605
Yacovleff E (1933) La jíquima, raíz comestible extinguida en el Peru. Rev Mus Nac (Lima) 2(1):51–66
Yacovleff E, Herrera FL (1934–1935) El mundo vegetal de los antigous peruanos. Rev Mus Hist Nat (Lima) 3(3):243–322, 4(1):31–100
Yen DE (1974) The sweet potato and Oceania. Bishop Mus Bull (Honolulu) 236:1–389
Yen DE (1993) The origins of subsistence agriculture in Oceania and the potential for future tropical food crops. Econ Bot 47:3–14
Zardini E (1991) Ethnobotanical notes on 'yacon', *Polymnia sonchifolia* (Asteraceae). Econ Bot 45(1):72–85

Chapter 13
Cereals

Fiona R. Hay, N. Ruaraidh Sackville Hamilton, Bonnie J. Furman, Hari D. Upadhyaya, K.N. Reddy, and S.K. Singh

13.1 Rice

13.1.1 Why Conserve Rice Germplasm

Rice is the staple food for more than half of the world's human population. Average global consumption is 52.9 kg/capita/year. However, in countries such as Bangladesh, Cambodia, Lao PDR, Myanmar, and Viet Nam, average consumption is more than 150 kg/capita/year (FAO 2011). Rice is grown in more than 100 countries, with a global total of more than 150 million hectares planted with rice, much of which is in the tropics or sub-tropics. The top five producers in 2007 were China, India, Indonesia, Bangladesh, and Viet Nam (FAO 2011). A significant proportion of the rice crop is grown by millions of small farmers and agricultural workers who are dependent on rice for their livelihood.

It is predicted that the human population will reach more than nine billion by 2050 (UN 2011) and estimated that rice production needs to increase by 1.5% per year in order to meet people's needs and keep rice affordable. In order to meet increasing demand, it is essential that rice scientists and breeders have access to the entire rice genepool, so that they can incorporate desirable genes into varieties that are evermore productive and tolerant of abiotic and biotic stresses.

F.R. Hay (✉) • N.R.S. Hamilton
T.T. Chang Genetic Resources Center, International Rice Research Institute (IRRI), Los Baños, Laguna, Philippines
e-mail: F.Hay@cgiar.org

B.J. Furman
Wellhausen Anderson Plant Genetic Resource Building, International Maize and Wheat Improvement Center (CIMMYT), Apartado Postal 6-641, Mexico, DF 06600, Mexico

H.D. Upadhyaya • K.N. Reddy • S.K. Singh
Genebank, International Crops Research Institute for the Semi-Arid Tropics (ICRISAT), Patancheru, Andhra Pradesh 502 324, India

M.N. Normah et al. (eds.), *Conservation of Tropical Plant Species*,
DOI 10.1007/978-1-4614-3776-5_13, © Springer Science+Business Media New York 2013

13.1.2 The Rice Genepool

The genus *Oryza* consists of 25 species (USDA-ARS-NGRP 2011). The two cultivated species, *O. sativa* L. and *O. glaberrima* Steud., together with six other species are grouped within the *O. sativa* complex and have the AA-genome. Asian rice, *O. sativa* appears to have been domesticated approximately 9,000 years BC in the Yangtze River basin in China, whilst domestication of African rice occurred later, ~3500 BC (Vaughan et al. 2008a, b; Molina et al. 2011). Key factors in the domestication of Asian rice (African rice is less well studied) were the loss of shattering and loss of strong secondary dormancy. The commonality of the mutant for non-shattering grains suggests that domestication arose as a single event, from the wild rice *O. rufipogon* Griff. Early domestication and extensive geographical spread of *O. sativa* has resulted in a huge amount of genetic diversity. There are two main variety groups: indica rice which is mainly grown in tropical and subtropical regions, and japonica rice which is grown in more temperate climates. The use of isozyme markers has identified four additional variety groups, aus, ashwina, rayada, and aromatic; SSR markers and SNPs have further suggested distinction between tropical and temperate japonica varieties, the tropical japonica group corresponding approximately to a group still sometimes referred to as javanica. In the late 1960s, high-yielding varieties bred by the International Rice Research Institute (IRRI) marked the Green Revolution for rice. Cultivars of *O. sativa* are now grown in many countries, whilst *O. glaberrima* is still largely confined to West Africa. Since the late 1990s, breeders have started to produce varieties known as NERICA (New Rice for Africa) by crossing *O. sativa* and *O. glaberrima*. These inter-specific varieties are higher yielding than *O. glaberrima*, yet tolerate the drier, less fertile African soils.

Within the *Oryza* genus there are three other species complexes (Vaughan and Morishima 2003): the *officinalis* complex consists of 12 species, one with each of BB, EE, or FF genomes, three with BBCC, and three with each of CC and CCDD genomes; *ridleyi* comprises *O. ridleyi* Hook.f. and *O. longiglumis* Jansen with the HHJJ genome; *O. meyeriana* Baill. and *O. neocaledonica* Morat have the GG genome and *O. schlechteri* Pilg. has an HHKK genome. The three species with the CCDD genome, *O. latifolia* Desv., *O. grandiglumis* (Döll.) Prodoehl, and *O. alta* Swallen and *O. glumaepatula* Steud. (AA) occur in Latin America. Five wild species occur in Africa (two with AA genome, one with CC, and one with BBCC); one of these, *O. eichingeri* Peter (CC) is also found in Sri Lanka. *O. australiensis* Domin (EE) is restricted to tropical Australia; *O. schlechteri* and *O. longiglumis* Jansen (HHJJ) are only found in New Guinea; *O. rhizomatis* D.A. Vaughan (CC) is restricted to Sri Lanka. Other species are generally more widely distributed across southeast and south Asia. Many of these crop wild relatives (CWR) can be found growing in damp or wet habitats close to fields of cultivated rice and may be classed as agricultural weeds. Some wild *Oryza* species with the AA genome readily hybridize with cultivated rice; the resultant offspring are described as weedy rice, although this description may also include cultivar hybrids or derivatives. Weedy rice also represents a potentially significant problem in rice cultivation. Conversely, there are instances where genes from rice CWR have been used in breeding programs to improve cultivated rice. For example *O. australiensis*, *O. officinalis* Wall, and *O. latifolia* have been used as a source of genes conferring resistance to brown plant hopper and *O. australiensis*, *O. brachyantha* A. Chev. and Roehrich, and *O. minuta* J. Presl. and C. Presl. have been used to confer resistance to blast bacterial blight. Accurate identification of wild species is difficult, and complicated by the natural occurrence of inter-specific hybrid swarms. This hampers the effective conservation and use of wild rice accessions.

13.1.3 Existing Collections

There are more than 1,750 genebanks around the world, of which more than 10% hold some rice germplasm. Based on country reports submitted between 2006 and 2008 (FAO 2010) and a 2007 survey carried out for the development of the Global Strategy for the Ex Situ Conservation of Rice Genetic Resources (Global Crop Diversity Trust 2010), there are in excess of 750,000 genebank accessions of *Oryza* species. Overall there are very many more accessions of the Asian cultivated rice, *O. sativa* than of the African cultivated rice, *O. glaberrima*, a reflection of its wider cultivation and greater genetic diversity. The proportion of all collections that are wild species is less than 3%. However, gaining an understanding of the amount of genetic diversity all these collections represent, in particular those of *O. sativa*, is difficult, not least since within countries there may be multiple collections held by different institutes with a mandate for rice research. Some collections may be extensively duplicated in another genebank, and both genebanks may be maintaining and distributing that germplasm using different genebank management systems and accession numbers. Other collections may be more unique, and may not even be safety duplicated to another genebank in case of e.g., natural disasters which result in the loss of accessions. This is less likely given recent efforts, in particular by the Global Crop Diversity Trust (GCDT), to ensure that accessions are available as part of a rational system of rice conservation and use, and that there are black box samples as safety back-up in a genebank, preferably on another continent, and/or in the Svalbard Global Seed Vault (SGSV) in Norway (Global Crop Diversity Trust 2010).

The largest single collection is that held in-trust and made freely available under the International Treaty on Plant Genetic Resources for Food and Agriculture (ITPGRFA) at IRRI. In May 2011, the T.T. Chang Genetic Resources Center at IRRI had 112,884 accessions of *O. sativa*, 2,828 accessions of *O. glaberrima*, and 4,389 accessions of wild *Oryza* species (Table 13.1). The second largest collection is held by India's National Bureau of Plant Genetic Resources (NBPGR), with 86,119 accessions of which approximately 1% is wild species (FAO 2010). There are also large collections in China: the Chinese Academy of Agricultural Sciences (CAAS) holds more than 80,000 accessions (Global Crop Diversity Trust 2010)

Table 13.1 Major ex situ collections of rice and maize

Institution	Country	Number of accessions		
Rice		*Oryza sativa*	*O. glaberrima*	Other *Oryza* species (23)
International Rice Research Institute (May 2011)	International (Philippines)	112,884	2,828	4,389
National Bureau of Plant Genetic Resources (FAO 2010)	India	ca. 85,250	Unknown	ca. 860
Chinese Academy of Agricultural Sciences (Global Crop Diversity Trust 2010)	China	67,783	0	12,000
China National Rice Research Institute (December 2010)	China	75,106	17	2,053
National Institute of Agrobiological Sciences (FAO 2010)	Japan	ca. 44,000	Unknown	ca. 450
Maize		*Zea mays* subsp. *mays*	Teosintes (*Zea* sp.)	*Tripsacum* sp.
International Center for Maize and Wheat Improvement (May 2011)	International (Mexico)	27,052	227	161
National Center for Genetic Resources Preservation	USA	27,649	459	190
Banco Portugues de Germoplasma Vegetal (FAO 2010)	Portugal	24, 529	0	0
North Central Regional Plant Introduction Station (May 2011)	USA	19,582	ca. 325	ca. 50
Chinese Academy of Agricultural Sciences (FAO 2010)	China	19,088	0	0
Instituto Nacional de Investigaciones Forestales y Agropecuarias (FAO 2010)	Mexico	14,067	ca. 136	ca. 50

whilst the Chinese National Rice Research Institute (CNRRI) reported having, at the end of 2010, 75,106 accessions of *O. sativa*, 17 accessions of *O. glaberrima*, and 2,053 of 11 other *Oryza* species [pers. comm]. Collections of between 20,000 and 45,000 accessions are held by the National Institute of Agrobiological Sciences (NIAS) in Japan; the Rural Development Agency in Korea; the National Center for Genetic Resources Preservation in the USA; AfricaRice (another international agricultural research organization located in Benin); and the Biotechnology Research Development Office (BRDO) in Thailand.

Most of the accessions (>85%) held by NBPGR in India, by CAAS and CNRRI in China, and by other national genebanks, were collected in-country; the Napok Agricultural Center in Laos reported that all their accessions were collected in country (Global Crop Diversity Trust 2010). Both India and China have conducted extensive in-country collecting missions, often without collaboration, and as such, these collections are likely to contain a high proportion of unique germplasm. Collections with a national focus are also more likely to be dominated by landraces (rather than for example, breeding lines or advanced cultivars). In contrast, not surprisingly, the international collections held by IRRI and by AfricaRice have high proportions of regionally-collected germplasm (Asia and West Africa, respectively). Indeed many national collections have been duplicated to IRRI or AfricaRice, usually for maintenance and distribution as a regular accession. Genebanks in countries that are not traditional rice producing countries tend to have a larger proportion of their collection that is introduced material for breeding and other fundamental research. The Dale Bumpers National Rice Research Center in the USA for example, reported that 93% of their 23,090 accessions were breeding lines (FAO 2010). This USA collection, like the collection held by NIAS in Japan, perhaps has a more temperate focus than the Asia-based collections.

The largest ex situ collection of wild rice species is held by CAAS, China; in 2007 CAAS reported having 12,000 accessions of wild species, approximately 15% of the entire collection (Global Crop Diversity Trust 2010). The Cambodian Agricultural Research and Development Institute (CARDI), the National Agricultural Research Centre of Pakistan, and the SADC Plant Genetic Resources Centre in Zambia also a have relatively high proportion of wild species germplasm (10–30% of their total accessions), although in terms of numbers, the second largest collection of wild rice species is held by IRRI. However, most genebanks have relatively few (<100 accessions) or no wild rice germplasm.

13.1.4 Networks and Resources

The major national genebanks in Asia have been supported by the strong national programmes in each of the countries, commensurate with the high importance attached to rice in the region. However, there is a need for greater cooperation and rationalization to understand the extent of duplication and, conversely, the gaps in the rice genepool which are not in conserved ex situ, and to ensure efficiency. In regions outside of Asia, rice is a lower priority for research and conservation and hence rice collections in those regions tend to be small, scattered, and not well resourced. A partial "rice registry" has been constructed to help identify accessions with common ancestry, cross-referencing rice accessions held in the USA Department of Agriculture, CAAS China, and the IRRI and AfricaRice genebanks. The *Oryza* accessions held by IRRI and AfricaRice can also be requested via a single on-line portal, the System-wide Information Network for Genetic Resources (SINGER: http://singer.cgiar.org). SINGER has been incorporated into Genesys (http://www.genesys-pgr.org/), a more general global portal for searching germplasm from national as well as international collections based on a broader range of passport and characterization data.

In contrast to the absence of networks for the conservation of rice germplasm, the International Network for Genetic Evaluation of Rice (INGER) is a good example of an effective network for its utilization. INGER is a partnership of national agricultural research and extension systems (NARES) and international centers including IRRI, AfricaRice, and the Centro Internacional de Agricultura Tropical (CIAT), that started in 1975. The INGER partner institutions collaborate mainly through the distribution of pre-defined "nurseries", or sets of germplasm selected on the basis of their perceived potential value for particular situations or qualities. All the passport and evaluation data is also shared. The germplasm is now exchanged following the standard material transfer agreement of the ITPGRFA. Over the last 30 years or so, it has been credited with the release of 667 varieties in 62 countries. The Latin-American fund for irrigated rice (Fondo Latinoamericano para el Arroz de Riego; FLAR) has a similar purpose to INGER but is more restricted in genetic scope and in membership, geographic coverage, and germplasm sharing.

13.1.5 Future

Given the total number of rice accessions that are held in genebanks globally, it is clear that there is considerable national and international commitment to the maintenance of this resource. This commitment is doubtless promoted by the continued use of 'new' germplasm in rice breeding programs. However, there is a bottleneck in the use of a large proportion of rice collections, due to limitations in the amount of phenotypic data that has been collected. More advanced and faster phenotyping techniques and facilities, for example as planned as part of the Global Rice Science Partnership (GRiSP), together with the increasing availability of DNA sequences should facilitate better selection of accessions for use in breeding programs (conventional breeding methods and for example, through marker assisted selection or genetic modification). Generation of large data sets for numerous accessions will require sophisticated data storage and analysis. The International Rice Information System (http://iris.irri.org) is a database system for germplasm pedigree, evaluation, genomic, and environmental data (Bruskiewich et al. 2003).

The conditions under which collections are stored vary between genebanks and not all genebanks store their seed germplasm according to the international standards for genebanks (FAO/IPGRI 1994). Of particular concern is unique germplasm that is only held in genebanks that are able to store rice seeds under short-term storage conditions or whose storage conditions or pre-storage drying conditions depend on ambient temperature and humidity. Genetic erosion of such collections is inevitable without investment to upgrade drying facilities and/or temperature-controlled storage. Further, there are strong indications of gaps in coverage particularly of wild rice species outside south and east Asia. This under-representation of crop wild relatives in ex situ collections is going to be addressed by a new initiative led by the GCDT to collect and conserve the wild relatives of rice and other major food crops. Characterisation and pre-breeding activities will help identify useful traits for introduction into advanced cultivars.

13.2 Maize

13.2.1 Why Conserve Maize Germplasm

Maize (*Zea mays* L. subsp. *mays*) is currently produced on nearly 100 million hectares in 125 developing countries and is among the three most widely grown crops in 75 of those countries (FAO 2011). Although much of the world's maize production (approximately 78%) is utilized for animal feed, human consumption in many developing and developed countries is steadily increasing. For example, maize is the most important cereal crop for food in sub-Saharan Africa and Latin America. Further, with human population growth, the demand for maize in the developing world will double between now and 2050, and by 2025, maize is likely to become the crop with the greatest production globally (Rosegrant et al. 2008).

A great deal of genetic variability is present within and between the many diverse lines, varieties, and races of maize (Goodman and Brown 1988). Maize genetic resources constitute an immensurable treasure for humankind. Their conservation and the investigation of existing variation and possible current and future uses provide:

- Resources for agricultural improvement to reduce hunger and poverty.
- A solid knowledge base for future generations of researchers.

The value of germplasm within genebanks is mostly measured by the extent of utilization of their genetic variation (mostly for breeding programmes). Landraces have the value of adaptation to specific cultivated regions or ecological conditions. Wild species may contain special genes such as those conferring disease resistance, climate adaptation, or nutritional quality.

In some areas, where populations of Native Americans predominate, their landraces of maize are still being cultivated and can still be collected today. However, economic forces are seriously eroding the wellbeing of the small farmers in Mexico with a resulting serious reduction in the number of landraces being grown and a reduction of land area devoted to most races. As a result, some races of maize have been displaced by modern cultivars while others, including virtually all the teosintes, are threatened as a result of modern farming and ranching practices.

13.2.2 The Maize Genepool

Maize was domesticated in southern or southwestern Mexico, most likely from teosinte or some extinct wild maize closely related to teosinte (Wilkes 2004; Sluyter and Dominguez 2006). While maize covered much of the New World at the time of colonisation, the teosintes have a much more restricted geographic distribution; mainly central and southwestern Mexico, with limited populations in northern Mexico, Guatemala, and Nicaragua. The genetic diversity of maize, being an outcrossing crop, is extremely broad. Maize landraces exhibit significant morphological variation and genetic polymorphism (Ortiz et al. 2010). Thus, maize has gained adaptation and productivity in all continents through introductions and breeding and is widely grown between 55°N and 55°S (Guidry 1964) and from sea level to 3,800 m (Taba 1997).

Maize has been divided into about 300 races (Goodman and Brown 1988). A maize race has been defined as 'a group of related individuals with enough characteristics (genes) in common to permit their recognition as a group' (Anderson 1944). It is reported that the total number of unique New World maize germplasm accessions exceeds 27,000 [survey carried out in 2006 for the development of the *Global Strategy for the* Ex Situ *Conservation and Utilization of Maize Germplasm* (Global Crop Diversity Trust 2007]. The major types of cultivated maize (*Zea mays* L.) under conservation are characterised by their kernel types: dent, flint, floury, sugary, pop, and morocho (a soft floury texture inside the grain, surrounded by a hard flint texture). Grain may be white, yellow, purple, orange yellow, red, sun red, mottled, or brown.

The primary maize genepool consists of cultivated maize and diploid teosintes (Tallury and Goodman 2001). Teosintes comprise seven taxa divided into two sections and five species. All but one species is diploid, $2n=20$, and inter-fertile with cultivated maize. There is also one species of perennial, tetraploid teosinte (*Zea diploperennis*). Teosinte was originally divided into six races (Wilkes 1967), but since the discovery of *Z. diploperennis*, species and subspecies names have come into favour for teosinte (Wilkes 2004).

The genus *Tripsacum* is the genus most closely related to Zea, and all species within this genus constitute the secondary gene pool of maize (Tallury and Goodman 2001). *Tripsacum* species have multiples of $x=8$ chromosomes (Anderson 1944; Stebbins 1950) and are perennials. Maize and *Tripsacum* have been hybridized (e.g., de Wet and Harlan 1978), but the offspring have varying degrees of sterility and no spontaneous hybrids have been confirmed. There are at least 16 species of

Tripsacum, ranging from the USA to Bolivia (de Wet et al. 1983). A tertiary gene pool consists of the distantly related Asiatic genera *Coix*, *Selerchne*, *Pollytoca*, and *Chinache* as well as genera in the tribe Andropogoneae (Tallury and Goodman 2001). No members of the tertiary gene pool have been used for maize improvement or backcrossed to cultivated maize.

13.2.3 Existing Collections

Maize ex situ germplasm collections include landraces (maize races), improved populations (synthetics, varieties, and cycles of selection), inbred lines (early generation homozygous), reference hybrids, genetic stocks (natural genes and transgenes) and wild species housed as caryopsis and clones (Taba 1997). Not all accessions have been classified by race, and not all can be classified due to hybridization.

The two most recent surveys of global maize collections include that for the *Global Strategy for the* Ex Situ *Conservation and Utilization of Maize Germplasm* (Global Crop Diversity Trust 2007) and the Second Report on the State of the World's Plant Genetic Resources for Food and Agriculture (SOWPGR2; FAO 2010). The SOWPGR2 surveyed 281 collections and reported almost 328,000 accessions worldwide, including approximately 33% land races and 1% wild relatives. The Global Strategy reported on unique landrace collections and reported over 42,000 accessions in 34 collections worldwide (Global Crop Diversity Trust 2007).

The largest single collection in the world, with over 27,000 accessions, is housed at The International Center for Maize and Wheat Improvement (CIMMYT) in Mexico (Table 13.1). CIMMYT's maize collection is held in Trust and made freely available under the ITPGRFA. The other key international collection of the Americas is the North Central Regional Plant Introduction Station (NCRPIS) in Ames, IA, USA. The National Center for Genetic Resources Preservation (NCGRP) in Ft. Collins, CO, USA holds duplicates for a number of collections and has over 39,000 accessions overall. However, it has very few accessions that are unique to the Center. In addition the Maize Genetic Co-operation Stock Center in Urbana, IL, USA is a bank specifically for genes. The maize stock center houses virtually all of the mutants of maize, various chromosomal stocks, multiple mutant stocks, and various other stocks of interest to the maize genetics community. All centers have back-up collections either between them, at the SGSV in Norway, or both.

The national centers of major importance are in Brazil, Colombia, Mexico, and Peru. These four centers were the original stock centers of the National Research Council Rockefeller Foundation collections of the 1940s–1960s. These collections comprised the original strains available in current collections and were the basis of maize race definition and description. Other national collections of major importance include BPGV-DRAEDM, Portugal, ICGR-CAAS, China, and INIFAP, Mexico (FAO 2010).

13.2.4 Networks and Resources

While there are numerous regional germplasm networks in the Americas, Africa, Asia, and Europe, most of the networks are more concerned with general information exchange rather than regeneration and active germplasm exchange. The networks that have been the most effective for maize are LAMP (Latin American Maize Project) and the Latin America Maize Regeneration Project. Over 12,000 Latin American accessions were evaluated through LAMP, sequentially identifying the more promising ones for further breeding work, and eventually identifying an elite set of about 300 accessions. The regeneration project eventually regenerated most of the maize accessions of Latin America, including those of LAMP.

The LAMP project eventually led to the GEM (Germplasm Enhancement of Maize) project in the USA, a cooperative public/private endeavor to quickly expand the germplasm base of commercial maize. Elite germplasm accessions are crossed to private lines from US or foreign companies (or to public lines from foreign countries), families are derived by selfing, and top crosses are tested cooperatively to identify superior families. These superior families are first distributed to cooperators (US and international) and then, with a year's delay, to the general public anywhere in the world via the NCRPIS at Ames, IA, USA. Together, these three projects represent much of the current leadership for protection, promotion, and utilization of maize germplasm resources.

13.2.5 Future

The proper utilization of germplasm collections requires access to information on its morphological, agronomic, and genomic characters as well as access to the germplasm material itself. The GCDT is working with national partners worldwide to regenerate unique and imperiled collections. One important outcome of this endeavor is that duplicates of these materials will be donated to CIMMYT to become part of its collection and thus available for all.

The government of Mexico has recently announced the program "MasAgro; the Sustainable Modernization of Traditional Agriculture." As part of this 10-year initiative, smallholder farmers are working with agricultural research and development organizations to raise and stabilize their crop yields, increase their incomes, and reduce the effects of climate change on Mexico's agricultural output (http://www.cimmyt.org). A subproject, Seeds of Discovery (SeeD), will undertake the genetic characterization of each accession in the CIMMYT genebank. In addition, subsets of CIMMYT's holdings will be evaluated for traits of agronomic importance and end-use quality. These activities will be carried out in collaboration with international and national institutions.

Finally, the inception of the GRIN-Global database (http://www.grin-global.org) should make information across collections available to users worldwide. Together, these undertakings promise to augment utilization on a global level as well as add to the conservation of materials that could otherwise become lost.

13.3 Sorghum

13.3.1 Why Conserve Sorghum Germplasm

Sorghum (*Sorghum bicolor* (L.) Moench) is the fifth most important cereal in the world after rice, wheat, maize, and barley. It is mostly grown in the semi arid tropics; the countries with the large areas of sorghum cultivation are India (7.7 million ha), Sudan (5.61 million ha), Nigeria (4.7 million ha), Niger (3.3 million ha), Burkina Faso (1.98 million ha), USA (1.95 million ha), Mexico (1.77 million ha), Ethiopia (1.62 million ha), and Mali (1.22 million ha) (FAO 2010). Sorghum constitutes the staple food for over 750 million people who live in the semi-arid tropics of Africa, Asia, and Latin America. In addition, stalk and foliage are used as fodder, fuel, and for thatching and fencing. It has a wide range of industrial uses including ethanol production, sugar and concentrated sugar syrup, popping, confectionary, etc., in developed and developing countries.

13.3.2 The Sorghum Genepool

S. bicolor (2n=20) is synonymous with *Holcus bicolor* L., *Andropogon sorghum* (L.) Brot., and *Sorghum vulgare* Pers. Sorghum was probably domesticated in the north east quadrant of Africa, an area that extends from the Ethiopia-Sudan border westward to Chad (Doggett 1970; de Wet et al. 1976). From there it was probably spread to India, China, the Middle East, and Europe. Sorghum is known by a variety of names: Guinea corn in West Africa, kaffir corn in South Africa, durra in Sudan, and mtama in eastern Africa. In India, the crop is known as jowar in the North and cholam in the South.

S. bicolor is considered an extremely variable crop-weed complex. It comprises wild, weedy, and cultivated annual forms which are fully inter-fertile. The cultivated forms fall in *S. bicolor* subsp. *bicolor* and are classified, in the most widely accepted system, into five basic races (bicolor, guinea, caudatum, kafir, and durra) and ten intermediate or hybrid or half races on the basis of spikelet morphology and panicle shape [guinea-bicolor, caudatum-bicolor, kafir-bicolor, durra bicolor, guinea-caudatum, guinea-kafir, guinea-durra, kafir-caudatum, durra-caudatum, and kafir-durra (Harlan and de Wet 1972; de Wet 1978)]. Sub-race names are being added to this system (Doggett and Prasad Rao 1995; Prasad Rao et al 1989). The wild forms are classified into *S. bicolor* (L.) Moench subsp. *verticilliflorum* (Steud.) Piper (synonyms: *S. arundinaceum* (Desv.) Stapf and *S. bicolor* (L.) Moench subsp.

arundinaceum (Desv.) de Wet and Harlan). The subspecies is further divided into four overlapping races, the most widely distributed and variable of which is *verticilliflorum*, found across the African savanna and introduced into tropical Australia, parts of India, and the Americas. The weedy forms are classified into *S. bicolor* (L.) Moench subsp. *drummondii* (Steud.) de Wet (synonyms: *S. sudanense* (Piper) Stapf and *S. drummondii* (Steud.) Millsp. and Chase), which arose and probably continues to arise from crossing between cultivated grain sorghum and close wild relatives wherever in Africa, they are sympatric. The hybrids have stabilized and occur as persistent weeds in abandoned fields and field margins. A well know forage grass, Sudan grass, belong to this complex.

The primary genepool include the *S. bicolor* cultivated species and *S. propinquum* (Kunth) Hitchc., a wild diploid complex found in southeast Asia (Acheampong et al. 1984). The secondary genepool includes *S. halepense* (L.) Pers., a rhizomatous tetraploid fodder, thought to be an autotetraploid of *S. propinquum* (Acheampong et al. 1984). Commonly known as Johnson grass, this is native to southern Eurasia east to India, but has now been introduced as a weed to warm temperate regions of the world. The tertiary genepool consists of all other sections/subgenera of sorghum.

13.3.3 Existing Collections

World collections of sorghum comprise 235,688 accessions (FAO 2010). Substantial collections (about 62%) are in 21 genebanks including 16.1% in the genebank of the International Crops Research Institute for the Semi-Arid Tropics (ICRISAT). At ICRISAT, sorghum germplasm was assembled by (1) introducing material gathered at various places across the world; (2) launching germplasm collection missions in priority areas; and (3) assembled from centre's own research. At present ICRISAT is the major repository for sorghum germplasm with a total of 37,949 accessions from 92 countries (Table 13.2). ICRISAT has collected 9,011 sorghum samples from 94 collection missions in 33 countries. 121 organizations located in 52 countries donated 28,932 accessions. The collection comprises 85.3% landraces, 13.2% breeding material, 1.2% wild species and 0.3% named cultivars. The germplasm maintained at ICRISAT is predominantly represented by three races: durra (23.5%), caudatum (20.6%), and guinea (14.8%). Of the ten hybrid races, only durra-caudatum (11.5%), guinea caudatum (9.2%), and durra bicolor (7.1%) are common. To enhance use of germplasm in crop improvement, a mini core collection (242 accessions, 1% of entire collection) representing the diversity of the entire collection has been developed (Upadhyaya et al. 2009a).

In addition to the world repository of sorghum germplasm at the ICRISAT genebank, considerable collections (>10,000 accessions) have been assembled at the Plant Genetic Resources Conservation Unit, Southern Regional Plant Introduction Station, University of Georgia, USDA-ARS, USA; the Institute of Crop Germplasm Resources, CAAS, China and National Bureau of Plant genetic Resources (NBPGR), India Table 13.2.

Table 13.2 Major ex situ collections of sorghum and millets

Institution	Country	Number of accessions
Sorghum		
Plant Genetic Resources Conservation Unit, University of Georgia, USDA-ARS	USA	44,993
International Crop Research Institute for the Semi-Arid Tropics (ICRISAT)	International (India)	37,949
Institute of Crop Germplasm Resources, Chinese Academy of Agricultural Sciences (CAAS)	China	18,263
National Bureau of Plant Genetic Resources (NBPGR)	India	17,466
Pearl millet		
International Crop Research Institute for the Semi-Arid Tropics (ICRISAT)	International (India)	22,211
National Bureau of Plant Genetic Resources (NBPGR)	India	7,444
ICRISAT-Niamey	International (Niger)	5,365
Costal Plains Experiment Station USDA-ARS Tifton	USA	5,228
***Eleusine* spp.**		
National Bureau of Plant Genetic Resources (NBPGR)	India	9,522
All India Coordinated Research Project (AICRP) on Small Millets	India	6,257
International Crop Research Institute for the Semi-Arid Tropics (ICRISAT)	International (India)	5,957
***Setaria* spp.**		
Chinese Academy of Agricultural Sciences (CAAS)	China	26,233
National Bureau of Plant Genetic Resources (NBPGR)	India	4,392
Institut de Recherche pour le Développement	France	3,500
***Paspalum* spp.**		
National Bureau of Plant Genetic Resources (NBPGR)	India	2,180
Plant Genetic Resources Conservation Unit, University of Georgia, USDA-ARS	USA	1,385
All India Coordinated Research Project (AICRP) on Small Millets	India	1,111
***Echinochloa* spp.**		
Department of Genetic Resources I, National Institute of Agrobiological Sciences (NIAS)	Japan	3,671
National Bureau of Plant Genetic Resources (NBPGR)	India	1,677
***Panicum* spp.**		
N.I. Vavilov All-Russian Scientific Research Institute of Plant Industry	Russia	8,778
Chinese Academy of Agricultural Sciences (CAAS)	China	8,451
Department of Genetic Resources I, National Institute of Agrobiological Sciences (NIAS)	Japan	6,277

13.3.4 Networks and Resources

In collaboration with national agricultural research systems (NARS), universities, and Non-Governmental Organizations (NGOs), ICRISAT has launched expeditions to collect germplasm of its mandate crops. More than 95% of the sorghum collection was placed under the purview of the ITPGRFA and are available to the world community using Standard Material Transfer Agreement (SMTA). ICRISAT's agreement with FAO requires safety back-up for long-term conservation. The Memorandum of Understanding (MOU) between the Nordic genebank and ICRISAT has facilitated the safety duplication of ICRISAT collections in the SGSV; ICRISAT has already deposited about 21,000 duplicate accession samples in the SGSV.

13.3.5 Future

Sorghum is a multipurpose crop. The demand for sorghum grain as food is likely to remain high in the traditional producing and consuming countries which have little opportunity to produce alternative crops. Under these circumstances there will be continuous demand for sorghum cultivars that are high yielding, have a broad genetic base, and are stable under abiotic and biotic stresses. In future, ICRISAT needs to ensure that the assembled germplasm is conserved, maintained safely, and distributed to all bona fide users for utilization in crop improvement programs. The strategic research on core and mini core collections and identification of new diverse sources will enhance the use of germplasm in breeding programs. Molecular characterization of mini core and trait-specific subsets will further reveal genetic usefulness of the germplasm accessions in allele mining.

13.4 Pearl Millet

13.4.1 Why Conserve Pearl Millet Germplasm

Pearl millet (*Pennisetum glaucum* (L.) R. Br.) is an important food and forage crop in Africa and Asia. It is also valued for fodder (both stover and green fodder) and poultry feed in the Americas. Traditionally, pearl millet grains are used in the preparation of conventional foods such as unleavened flat breads (chapati), fermented breads (Kisra, injera, dosa, etc.), porridge, mudde or dumpling, biscuits, snacks, and malt and opaque beer. Pearl millet is mainly grown by small farmers who are dependent on pearl millet for their livelihood. Pearl millet is mainly cultivated in Niger, Nigeria, Burkina Faso, Togo, Ghana, Mali, Senegal, Central African Republic, Cameroon, Sudan, Botswana, Namibia, Zambia, Zimbabwe, and South Africa in Africa and India, Pakistan, and Yemen in Asia.

Pearl millet is probably the world's hardiest crop and has great potential because of its suitability to the extreme limits of agriculture. Its importance is expected to increase under various climate change scenarios (Lane et al. 2007). Pearl millet is endowed with enormous genetic variability for various morphological traits, yield components, adaptation and quality traits. The genetic variability accumulated over centuries is fast eroding, mainly due to the replacement of landraces by improved cultivars, natural catastrophes (droughts, floods, fire hazards, etc.), industrialization, human settlement, over-grazing, and destruction of plant habitats for irrigation projects and dams (Upadhyaya and Gowda 2009). Therefore, the conservation of pearl millet diversity is essential to combat new pests and diseases and to produce adapted varieties for the changing climatic conditions.

13.4.2 The Pearl Millet Genepool

The genus *Pennisetum* (L.) Rich., to which pearl millet belongs, is the largest in the tribe Paniceae and consists of over 140 species (Clayton 1972). *Pennisetum* is divided into five sections: *Gymnothrix*, *Eu-Pennisetum*, *Penicillaria*, *Heterostachya*, and *Brevivalvula*. Pearl millet belongs to the section *Penicillaria*. The genus *Pennisetum* is a heterogeneous assemblage of species with large variation in chromosome number, ranging from $2n=10$ to $2n=72$, in the multiples of 5, 7, 8, and 9. There are many more species in this genus, but their chromosome number and genepool relationships are yet to be established (Hanna 1987). Among those studied, the lowest chromosome number ($2n=2x=10$) occurs in *P. ramosum* (Hochst.) Schmeinf. Those with $x=7$ chromosomes include the cultivated pearl millet and its wild and weedy subspecies, and *P. schweinfurthii* Pilger ($2n=2x=14$) and *P. purpureum* Schumach. ($2n=4x=28$). *P. massaicum* Stapf ($2n=16$ and 32) is the only known species with $x=8$. All other species have the basic chromosome number of $x=9$. The genus includes annual as well as perennial and sexual as well as asexual, including apomictic species.

On the basis of crossability and following the biological concept of species, *Pennisetum* germplasm has been classified into three genepools: primary, secondary, and tertiary. The primary genepool includes all those taxa that can easily cross with the cultivated pearl millet. This includes all variability in the cultivated *P. glaucum*, and those in its wild progenitor [*P. glaucum* ssp. *violaceum* (=*monodii* Maire)] and weedy form (*P. glaucum* subsp. *stenostachyum* Kloyzcsh ex. A. Br. and Bouche). The latter results from hybridization between cultivated pearl millet and its wild progenitor. The secondary genepool includes those species that also cross easily with the cultivated types, but do not produce fertile hybrids. Elephant or Napiergrass (*P. purpureum*), a rhizomatous perennial, belongs to this genepool. It is an allotetraploid ($2n=2x=28$) with A′A′BB genomic constitution in which the A′ genome is homologous or at least homoeologous to the A genome of the primary genepool. Its hybrids with cultivated pearl millet have $n=21$ and are sterile, but become fertile after chromosome doubling ($2n=6x=42$). The tertiary genepool includes the remainder of the species which either do not cross with cultivated pearl millet or when they do, albeit in rarest instances, the fertility in hybrids or their derivatives can only be achieved through resorting to special techniques (Hanna 1987).

13.4.3 *Existing Collections*

There are 65,400 accessions of pearl millet in 70 genebanks of 46 countries (FAO 2010). The long-term objective of ICRISAT is to serve as the world repository for the germplasm of its mandate crops, including pearl millet, and their wild relatives. By the end of 2010, the genebank at ICRISAT had registered 22,211 accessions of pearl millet germplasm from 51 countries, including 750 accessions of wild relatives belonging to 24 species of genus Pennisetum (Table 13.2). This is the largest collection of pearl millet germplasm assembled at any one place in the world. Pearl millet germplasm was assembled by (1) introducing the material that was already gathered at various places across the world; (2) launching germplasm collection missions in priority areas; and (3) assembly from center's own research. A total of 65 organizations contributed 11,381 accessions including those contributed by different disciplines at ICRISAT, in different years. The major donors of pearl millet germplasm include the Institut de Recherche pour le Développement, France (2,178); Rockefeller Foundation, India (2,022); and IBPGR (now Bioversity International), Italy (974). ICRISAT in collaboration with its partners launched a total of 216 collection missions for all its mandate crops including 76 for pearl millet germplasm in 28 countries and collected 10,830 pearl millet samples. Biological status of accessions indicated 19,063 landraces, 2,269 breeding materials, 129 advanced cultivars, and 750 wild accessions in the collection. To enhance the use of germplasm in crop improvement programs, core (2,094 accessions; Upadhyaya et al. 2009a) and mini core collections (238 accessions; Upadhyaya et al. 2011b) have been developed.

In addition to the single largest collection of pearl millet at the ICRISAT genebank, there are considerable collections at NBPGR, India; ICRISAT, Niamey; Coastal Plains Experimental Station, USDA, Tifton, USA; Institut de Recherche pour le Développement, France (4,405); Plant Genetic Resources of Canada (3,840), and Serere Agricultural and Animal Production Research Institute, Uganda (2,142). Most of these pearl millet collections were collected in-country and may, to a greater or lesser extent, be duplicates of accessions at ICRISAT (Table 13.2).

13.4.4 *Networks and Resources*

In collaboration with NARS, universities, and NGOs, ICRISAT has conducted expeditions to collect germplasm of its mandate crops. More than 96% of the pearl millet collection was placed under the purview of the ITPGRFA and are distributed according to the SMTA. Efforts have been made to conserve 5,205 pearl millet accessions at ICRISAT regional genebank at Niamey, Niger. In addition, ICRISAT has already deposited 8,050 duplicate accession samples at SGSV. ICRISAT's pearl millet passport and characterization databases can be accessed through SINGER and Genesys.

13.4.5 Future

In future, ICRISAT needs to ensure that the assembled germplasm is maintained in safe, secure and cost-effective manner and distributed to all bona fide users for utilization in crop improvement programs. Identifying gaps (Upadhyaya et al. 2009c, 2010a) in the collection and exploration for wild relatives and trait-specific germplasm are essential to further increase the diversity in collection. Trait-specific genetically diverse parents for trait enhancement are the primary need of the plant breeder. Strategic research on core and mini core collections and identification of new diverse sources will enhance the use of germplasm in breeding programs. Molecular characterization of mini core and trait-specific subsets will further reveal genetic usefulness of the germplasm accessions in allele mining. The aim is to identify the trait-specific genetically diverse and agronomically better germplasm lines through characterization, evaluation, and screening for the development of high yielding cultivars with a broad genetic base.

13.5 Small Millets

13.5.1 Why Conserve Small Millet Germplasm

Small millets are small-seeded annual coarse cereals grown throughout the world, particularly in arid, semi-arid, or mountain zones, as rain-fed crops under marginal and sub-marginal conditions of soil fertility and moisture. They include finger millet (*Eleusine coracana* (L.) Gaerth.), foxtail millet (*Setaria italica* (L.) Beauv.), proso millet (*Panicum miliaceum* L.), little millet (*P. sumatrense* Roth. ex Roem. & Schult.), barnyard millet (*Echinochloa crusgalli* (L.) Beauv.), and kodo millet (*Paspalum scrobiculatum* L.). Precise estimates of area and production are not available and data is generally combined for all the millets. South and East Asia produces about 60% of the total millet harvest; Eurasia and Central Asia, 14%; Africa, 16%; and the rest of the world, 10%. India is the leading producer contributing about 38% of the production. Finger millet is the principal small millet species grown in South Asia, followed by kodo millet, foxtail millet, little millet, proso millet and barnyard millet, in that order. Foxtail millet and proso millet are important in China and the latter is grown extensively in southern USSR. Presently, small millets are cultivated in areas where they produce a more dependable harvest than other crops. This has been largely responsible for their continued presence and cultivation in many parts of the world.

13.5.2 The Genepools of Small Millets

All small millets belong to the family Poaceae (Gramineae). Of them, the genera *Setaria, Panicum, Paspalum, Echinochloa* and *Digitaria* are classified under the tribe Paniceae of sub-family Panicoideae, and the other two genera *Eleusine* and *Eragrostis* under the tribe Eragrostideae of sub-family Chloridoideae (Prasad Rao and Mengesha 1988).

13.5.2.1 Finger Millet

Finger millet was domesticated in Africa, probably in the Ethiopian region. It is also known as African millet, koracan, ragi (India), wimbi (Swahili), bulo (Uganda), and teleburn (Sudan). It was introduced to India as a crop more than 3,000 years ago (Doggett 1989). Cultivated finger millet is *Eleusine coracana* (L.) Gaertn. subsp. *coracana*. The closest wild relative is *Eleusine coracana* subsp. *africana* (Kenn-O' Byrne) Hilu and de Wet. Wild finger millet (subsp. *africana*) is native to Africa but was introduced as a weed to the warmer parts of Asia and America. Cultivated finger millets are divided into races and sub-races on the basis of inflorescence morphology. Members of the race *compacta* are commonly referred to as cockscomb finger millet in both Africa and India. The race *plana* is characterized by large spikelets (8–15 mm long) that are arranged in two, more or less even rows along the rachis, giving the inflorescence branch a flat ribbon like appearance. The race *vulgaris* is the common finger millet to Africa and Asia; four sub-races are recognized on the basis of inflorescence morphology.

13.5.2.2 Foxtail Millet

Foxtail millet was first domesticated in the highlands of central China: remains of the cultivated form are known from the Yang-Shao period dating back some 5,000 years. Comparative morphology suggested that foxtail millet spread to Europe and India as a cereal soon after its domestication (Prasad Rao et al. 1987). The possible ancestor is *Setaria viridis*. Foxtail millet is fairly tolerant of drought; it can escape, some droughts because of early maturity. On the basis of inflorescence morphology, foxtails millets are classified into species, sub-species, races, and sub-races.

13.5.2.3 Proso Millet

Domestication of proso millet probably occurred in Manchuria (de Wet 1986). It was probably introduced into Europe about 3,000 years ago. After this date, it was introduced to the Near East and India (Zohary and Hopf 1988). On the basis of cultivated morphology study, proso millet is divided into five cultivated races.

13.5.2.4 Little Millet

The possible wild progenitor of little millet is *Panicum psilopodium* which is distributed from Sri Lanka to Pakistan and eastward to Indonesia. On the basis of morphology, little millet is divided into subspecies, race, and sub-race.

13.5.2.5 Barnyard Millet

Echinochloa crusgalli was domesticated about 4,000 years ago in Japan. *E. colona*, which occurs widely in tropical and sub-tropical areas, was domesticated in India. On the basis of morphology they are similar, but hybrids between them are sterile. *E. colona* differ from the *E. crusgalli* in having smaller spiklets with membranaceous rather than chartaceous glumes. On the basis of inflorescence morphology the genus *Echinochloa* is classified into two species, three subspecies and eight races.

13.5.2.6 Kodo Millet

Kodo millet was domesticated in India about 3,000 years ago and is now produced in West Africa and India. Crossing readily occurs between cultivated and weedy races, and seed from the hybrid harvested along with those of the sown crop; racial differentiation is not distinct despite years of cultivation in India (Prasad Rao et al. 1987). On the basis of inflorescence morphology, kodo millet is classified into three races.

13.5.3 Existing Collections

The genebank at ICRISAT conserves a total of 10,235 accessions of small millets germplasm from 50 countries. In addition, several organizations in different countries are also conserving considerable collections. There may be some duplicates across the collections (Table 13.2).

13.5.4 Networks and Resources

While collecting germplasm of its mandate crops, ICRISAT has also collected small millet germplasm. Core and mini core collections of finger millet and foxtail millet have been developed. Selected germplasm sets and mini core collections have been evaluated for agronomic performance across locations in collaboration with NARS scientists in Asia and Africa. ICRISAT has already deposited 7,399 duplicate accession samples at SGSV.

Small millets are not mandate crops of ICRISAT. To enhance use of genetic resources in crop improvement, core collections in finger millet (622 accessions; Upadhyaya et al. 2006), foxtail millet (155 accessions; Upadhyaya et al. 2008), barnyard millet (89 accessions; Upadhyaya et al. 2011b), and proso millet (106 accessions; Upadhyaya et al. 2011c), and mini core collections of finger millet (80 accessions Upadhyaya et al. 2010b) and foxtail millet (35 accessions Upadhyaya et al. 2011d) have been developed.

13.5.5 Future

The most important feature of minor millets is their short biological cycle; this trait is very important for risk avoidance under rainfed cultivation. In intercropping systems, they are used as short duration millets with slower maturing crops. The grain is small and can be stored for long periods without damage, ensuring continuous food supply during adverse situations such as failure of other crops. Small millets also perform better in adverse climatic conditions like severe drought (Kodo millet has the highest drought resistance). Most of the millets mature early so they can escape drought stress and they can be grown in sub-marginal conditions of soil fertility and moisture where the major cereal crops fail to realize production satisfactorily. Small millets are ideal for organic agriculture production and are often grown by subsistence level farmers with years of experience and traditional wisdom embedded in the variety choice and system of agriculture. Minor millets are nutritionally comparable or even superior to staple cereals such as rice and wheat (Gopalan et al. 2004). Millets are known as nutri-cereals; they are rich in vitamins, sulphur-containing amino acids, minerals, and phytochemicals, and have a high proportion of non-starch polysaccharides and dietary fiber. In view of all these benefits, small millets will play an important role in agriculture in the years to come.

References

Acheampong EN, Murthi A, William JT (1984) A world survey of sorghum and millets germplasm. IPGRI, Rome

Anderson E (1944) Cytological observations on *Tripsacum dactyloides*. Ann Missouri Bot Gard 31:317–324

Bruskiewich RM, Cosico AB, Eusebio W, Portugal AM, Ramos LM, Reyes MT, Sallan MAB, Ulat VJM, Wang X, McNally KL, Hamilton RS, McLaren CG (2003) Linking genotype to phenotype: the International Rice Information System (IRIS). Bioinformatics 19(Suppl 1):i63–i65

Clayton WD (1972) Gramineae. In: Hepper FN (ed.). Flora of West Tropical Africa, 3(2):277–574

de Wet JMJ (1978) Systematics and evolution of *Sorghum* Sect. *Sorghum* (Gramineae). Am J Bot 65:477–484

de Wet JMJ (1986) Origin, evolution and systematics of minor cereals. In: Seetharama A, Riley KW, Harinarayana G (eds). Small millets in global agriculture. Proceedings of the 1st International small millets workshop, Bangalore, India
de Wet JMJ, Harlan JR (1978) *Tripsacum* and the origin of maize. In: Walden DB (ed) Maize breeding and genetics. Wiley, New York, pp 129–141
de Wet JMJ, Harlan JR, Price EG (1976) Variability in *Sorghum bicolor*. In: Harlan JR, de Wet JMJ, Stemler ABL (eds) Origins of African plant domestication. The Mountain Press, The Hague, pp 129–141
de Wet JMJ, Brink D, Cohen CE (1983) Systematics of *Tripsacum* section *Fasciculata* (Gramineae). Am J Bot 70:1139–1146
Doggett H (1970) Sorghum. London and Harlow, Longmans Green & Co. Ltd; Published by Wiley
Doggett H (1989) Small millets – a selective overview. In: Seetharam K, Riley KW, Harinarayana G (eds) Small millets in global agriculture. Oxford and IBH, New Delhi, pp 3–18
Doggett H, Prasad Rao KE (1995) Sorghum. In: Smartt J, Simmonds NW (eds) Evolution of crop plant. Longman, Harlow, pp 173–180
FAO (2010) The second report on the state of the world's plant genetic resources for food and agriculture. FAO, Rome
FAO (2010) http://faostat.fao.org
FAO/IPGRI (1994) Genebank standards. Food and Agriculture Organization of the United Nations/ International Plant Genetic Resources Institute, Rome
Global Crop Diversity Trust (2007) http://www.croptrust.org/main/articles.php?itemid=757, Accessed on Apr 2011
Global Crop Diversity Trust (2010) http://www.croptrust.org/main/articles.php?itemid=760, Accessed on Apr 2011
Goodman MM, Brown WL (1988) Races of corn. In: Sprague GF, Dudley JW (eds) Corn and corn improvement, 3rd edn, monograph 18. American Society of Agronomy, Inc./Crop Science Society of America, Inc./Soil Science Society of America, Inc., Madison
Gopalan C, Ramashastri BV, Balasubramanium SC (2004) Nutritive value of Indian foods. ICMR, New Delhi
Guidry NP (1964) A graphic summary of world agriculture. USDA Misc. Pub. 705. U.S. Govt. Print. Office, Washington, DC
Hanna WW (1987) Utilization of wild relatives of pearl millet. In: Witcombe JR, Beckerman SR (eds) Proceedings of the international pearl millet workshop, 7–11 Apr 1986. ICRISAT Center, Patancheru, pp 33–42
Harlan JR, de Wet JMJ (1972) A simplified classification of sorghum. Crop Sci 12:172–176, http://www.croptrust.org/documents/cropstrategies/maize.pdf, http://www.croptrust.org/documents/cropstrategies/Rice%20Strategy.pdf
Lane A, Jarvis A, Atta-Krah K (2007) The impact of climate change on crops and crop areas and the role of agricultural biodiversity in adaptation. International symposium on climate change, 22–24 Nov 2007, ICRISAT, Patancheru. SAT ejournal/ejournal.ICRISAT.org, vol 4(1)
Molina J, Sikora M, Garud N, Flowers FM, Rubinstein S, Reynolds A, Huang P, Jackson S, Schaal BA, Bustamante CD, Boyko AR, Purugganan MD (2011) Molecular evidence for a single evolutionary origin of domesticated rice. Proc Natl Acad Sci USA. doi:10.1073/pnas.1104686108
Ortiz R, Taba S, Chávez Tovar VH, Mezzalama M, Xu Y, Yan J, Crouch JH (2010) Conserving and enhancing maize genetic resources as global public goods – a perspective from CIMMYT. Crop Sci 50:13–28
Prasad Rao KE, Mengesha MH (1988) Minor millets germplasm resources at ICRISAT. In: Small millets – recommendation for a network. Manuscript report IDRC-MR 171e. International Development Research Centre, Ottawa, pp 38–45
Prasad Rao KE, de Wet JMJ, Brink DE, Mengesha MH (1987) Interspecific variation and systematics of cultivated *Setaria italica*, foxtail millet (Poaceae). Econ Bot 41:108–116
Prasad Rao KE, Mengesha MH, Reddy VG (1989) International use of sorghum germplasm collection. In: Brown AHD, Frankel OH, Marshall DR, Williams JT (eds) The use of plant genetic resources. Cambridge University Press, Cambridge, pp 49–67

Rosegrant MW, Msangi S, Ringler C, Sulser TB, Zhu T, Cline SA (2008) International Model for Policy Analysis of Agricultural Commodities and Trade (IMPACT): model description. International Food Policy Research Institute, Washington, DC. http://www.ifpri.org/themes/impact/impactwater.pdf

Sluyter A, Dominguez G (2006) Early maize (*Zea mays* L.) cultivation in Mexico: dating sedimentary pollen records and its implications. Proc Natl Acad Sci USA 103(4):1147–1151

Stebbins GL (1950) Variation and evolution in plants. Columbia University Press, New York

Taba S (1997) Maize. In: Fuccillo D, Sears L, Stapleton P (eds) Biodiversity in trust, conservation and use of plant genetic resources in CGIAR centers. Cambridge University Press, Cambridge

Tallury SP, Goodman MM (2001) The state of the use of maize genetic diversity in the USA and sub-Saharan Africa. In: Cooper DH, Spillane C, Hodgkin T (eds) Broadening the genetic base of crop production. Food and Agriculture Organization of the United Nations/International Plant Genetic Resources Institute/CABI, Wallingford

UN (2011) Population Division of the Department of Economic and Social Affairs of the United Nations Secretariat, World population prospects: the 2010 revision. http://esa.un.org/unpd/wpp/index.htm. Accessed 8 Dec 2011

Upadhyaya HD, Gowda CLL, Pundir RPS, Reddy VG and Sube Singh (2006) Development of core subset of finger millet germplasm using geographical origin and data on 14 quantitative traits. Genetic Res. and Crop Evolution 53:679–685

Upadhyaya HD, Gowda CLL (2009) Managing and enhancing the use of germplasm – strategies and methodologies. Technical manual no. 10. International Crops Research Institute for the Semi-Arid Tropics, Patancheru, 236 pp

Upadhyaya HD, Pundir RPS, Gowda CLL, Reddy VG, Singh S (2008) Establishing a core collection of foxtail millet to enhance utilization of germplasm of an underutilized crop. Plant Genet Resour Char Util 7:177–184

Upadhyaya HD, Gowda CLL, Reddy KN, Singh S (2009a) Augmenting the pearl millet [*Pennisetum glaucum* (L.) R. Br.)] core collection for enhancing germplasm utilization in crop improvement. Crop Sci 49:573–580

Upadhyaya HD, Pundir RPSP, Dwivedi SL, Gowda CLL, Gopal Reddy V, Singh S (2009b) Developing a mini core collection of sorghum for diversified utilization of germplasm. Crop Sci 49:1769–1780

Upadhyaya HD, Reddy KN, Irshad Ahmed M, Gowda CLL and Haussmann Bettina (2009c) Identification of geographical gaps in the pearl millet germplasm conserved at ICRISAT genebank from West and Central Africa. Plant Genetic Resources: Characterization and Utilization 8(1):45–51

Upadhyaya HD, Reddy KN, Irshad Ahmed M, Gowda CLL (2010a) Identification of gaps in pearl millet germplasm from Asia conserved at the ICRISAT genebank. Plant Genet Resour Char Util 8(3):267–276

Upadhyaya HD, Sarma NDRK, Ravishankar CR, Albrecht T, Narasimhudu Y, Singh SK, Varshney SK, Reddy VG, Singh S, Dwivedi SL, Wanyera N, Oduori COA, Mgonja MA, Kisandu DB, Parzies HK, Gowda CLL (2010b) Developing a mini-core collection in finger millet using multilocation data. Crop Sci 50:1924–1931

Upadhyaya HD, Yadav D, Reddy KN, Gowda CLL, Singh S (2011a) Development of pearl millet mini core collection for enhanced utilization of germplasm. Crop Sci 51:217–233

Upadhyaya HD, Sharma S, Gowda CLL, Gopal Reddy V, Sube Singh (2011b) Developing barnyard millet (*Echinochloa colona* L. and *Echinochloa crusgalli* L.) core collection using geographic and morpho-agronomic data. Genet Resour Crop Evol (submitted)

Upadhyaya HD, Sharma S, Gowda CLL, Gopal Reddy V, Sube Singh (2011c) Developing proso millet (*Panicum miliaceum* L.) core collection using geographic and morpho-agronomic data. Crop & Pasture Sci 62:383–389

Upadhyaya HD, Ravishankar CR, Narsimhudu Y, Sharma NDRK, Singh SK, Varshney SK, Reddy VG, Singh S, Parzies HK, Dwivedi SL, Nadaf HL, Sahrawat KL, Gowda CLL (2011d)

Identification of trait-specific germplasm and developing a mini core collection for efficient use of foxtail millet genetic resources in crop improvement. Field Crops Research 124:459–467
USDA-ARS-NGRP (2011) USDA, ARS, National Genetic Resources Program. Germplasm Resources Information Network (GRIN). http://www.ars-grin.gov/cgi-bin/npgs/html/index.pl5. Accessed25 Jan 2011
Vaughan DA, Morishima H (2003) Biosystematics of the genus *Oryza*. In: Smith CW, Dilday RH (eds) Rice: origin, history, technology, and production. Wiley, Hoboken, pp 27–65
Vaughan DA, Lu B-R, Tomooka N (2008a) The evolving story of rice evolution. Plant Sci 174:394–408
Vaughan DA, Lu B-R, Romooka N (2008b) Was Asian rice (*Oryza sativa*) domesticated more than once? Rice 1:16–24
Wilkes G (1967) Teosinte: the closest relative of maize. Bussey Institute of Harvard University Press, Boston
Wilkes G (2004) Corn, strange and marvelous: but is a definitive origin known? In: Smith CW, Betran J, Runge ECA (eds) Corn: origin, technology, and production. Wiley, New York
Zohary D, Hopf M (1988) Common millet (*Panicum miliaceum*). In: Domestication of plants in the world. Clarendon, Oxford, pp 76–77

Chapter 14
Diversity and Conservation of Tropical Forestry Species in Southeast Asia

Marzalina Mansor

14.1 Introduction on Southeast Asian (SEA) Tropical Forestry

This chapter will explore conservation efforts of tropical forestry within the SEA region, with special attention given to Malaysia. Tropical rainforests are a complex ecosystem with extremely high biodiversity; hence these areas become a natural reservoir of genetic diversity that offers a rich source of timber species, medicinal plants, high-yield foods, and a myriad of other useful forest products. Despite their numerous roles, tropical forests are contained to the small land area between the latitudes 22.5°N and 22.5°S of the equator, i.e. between the Tropic of Capricorn and the Tropic of Cancer. Since the majority of Earth's land is located north of the tropics, rainforests are naturally limited to a relatively small area. However, due to high temperatures and availability of water, this area cradles a massive organic mass with fast growth response creating an important habitat that sustains as much as 50% of the species on Earth, as well as a number of diverse and unique indigenous cultures (Butler 2006).

Tropical rainforests play an essential role in regulating global weather in addition to maintaining a regular rainfall, while buffering against floods, droughts, and erosion. They store vast quantities of carbon, while producing a significant amount of the world's oxygen. In a humid, warm environment with non-seasonal conditions, plants produce flowers and fruits all year around, while dead leaves and other plant materials are decomposed within a short time, and their nutrients again become available to the living plant mass. Furthermore the fruits of rainforest trees germinate within a few days. Meanwhile, young trees stand in a waiting position for many years until the older trees die and make room for these new trees when light is able

M. Mansor (✉)
Forest Research Institute Malaysia (FRIM), Kepong, Selangor 52109, Malaysia
e-mail: mzalina@frim.gov.my

M.N. Normah et al. (eds.), *Conservation of Tropical Plant Species*,
DOI 10.1007/978-1-4614-3776-5_14, © Springer Science+Business Media New York 2013

to penetrate the canopy. In these highly diverse systems, studies have shown hundreds of tree species can be found within a hectare. Individuals of the same species may sometimes form clusters, but others may be distributed over a larger area. This contributes to numerous interrelated biological systems in the tropical forestry.

The dominant tree family of the rainforests of SEA is Dipterocarpaceae, a family that comprises of 17 genera and 500 species (Ashton 1982). Within the Indo-Malayan region, Dipterocarpaceae has its centre of diversity in Malesia, a floristic region encompassing Thailand, Malaysia, Singapore, Indonesia, the Philippines and the island of New Guinea. Their center of the diversity is Borneo, where in Peninsular Malaysia alone there are 155 species found comprising of 165 taxa of which 34 taxa are endemic (Chua et al. 2010). Ninety-five species are found in Sumatra and beyond that area, the number of dipterocarp species has rapidly declined with Mindanao recording only 24 species, while Java and Sulawesi have only 5. With their huge, straight stems, Dipterocarpaceae constitutes with the best timber trees of the area, but many populations of dipterocarps are on the decline mainly as a result of land use changes and harvesting.

The major forest types in Malaysia are mangrove forest; peat swamp forest; lowland dipterocarp forest, hill dipterocarp forest, upper hill dipterocarp forest that covers the oak-laurel forest and montane ericaceous forest. In addition, there also smaller areas of freshwater swamp forest, heath forest, forest on limestone and forest on quartz ridges (Fig. 14.1).

14.2 The Need to Conserve

Southeast Asia continues to possess the highest biodiversity among terrestrial ecosystems known on earth (Baskaran et al. 2002). While techniques for forest plantations, restoration and rehabilitation have been successfully developed, the genepools to source quality genetic material for such projects are fast disappearing and the options for future benefits are diminishing. These forests, however, face immense threat from human activity. Increased population and pressure for social and economic development have resulted in rapid decline in the forest areas. The main causes behind the loss and/or degradation of forests are poverty, overconsumption, inequitable distribution of land ownership, misguided policies, inept public sector debt, lack of knowledge as well as trained personnel. Topped by unsustainable forestry, illegal logging, overgrazing and conversion to agricultural activities, forest regeneration is further disturbed. Besides that forest areas in SEA have faced natural disasters such as droughts and forest fires due to El Niño/Southern Oscillation effects (Baskaran et al. 2002; Primack and Corlett 2005). Studies show that during drought stress, even the rainforest ecosystems do not act as carbon sinks crater (Werner 2003). However, as reported by Theilade and Petri (2003), the major threat to forest genetic resources (FGR) as perceived by national experts is summarized in Table 14.1.

The growing threats affect the “green lung” and carbon sink abilities of tropical forests that is important to the global ecosystem and human existence in the region

Fig. 14.1 Tropical rainforest profiles in Malaysia. Source: Hamdan, O., Khali Aziz, H., Shamsuddin I. & Raja Barizan, R.S. 2012. Status of Mangroves in Peninsular Malaysia. FRIM, pp: 8.

and to the whole atmosphere. Today more than two thirds of the world's tropical rainforests exist as fragmented remnants. For example, Indonesia has been identified as the world's third-largest emitter of green-house gases (GHGs) contributing to climate change due its high deforestation rate, particularly in lowland carbon-rich peat forests (World Bank 2007). Approximately 80% of Indonesia's emissions are

Table 14.1 Threats to forest genetic resources

Country	Major present threats to FGR
Cambodia	Encroachment and shifting cultivation at former concession sites; unsustainable harvests at concessions
Indonesia	Illegal logging and illegal trade of timber; forest fires
Lao PDR	Encroachment into forest for permanent or slash-and-burn agriculture; forest fires; logging; infrastructure development
Malaysia	Plantation and infrastructure development
Myanmar	Illegal logging, shifting cultivation, etc.
Philippines	Land hunger
Thailand	Illegal logging (high wood prices)
Vietnam	Shifting cultivation; inappropriate harvesting practices

Source: Theilade and Petri (2003)

from the forestry sector. The number of species at the brink of extinction is increasing significantly. The decline in forest areas and resources has emphasized the need to utilize these resources sustainably.

To avoid dependency of timber resources on the natural forests, tropical forestry plantations were established. Whereby in SEA, during 1970s and 1980s, 90% of the plantations were planted with exotic fast growing species. This is to cater to the wood-based industrial needs. Very little is being focused on to establish plantations of indigenous species. For example, in Cambodia alone, between 1985 and 2000 about 10,000 hectares of plantations of *Acacia auriculiformis*, *Eucalyptus camaldulensis*, *Tectona grandis* and *Pinus merkusii* have been established. As such this further affects the dwindling of sites for conservation of tropical forestry species (Greathouse 1982). But towards the 1990s, Baskaran et al. (2002) reported that many donor agencies now favor social forestry/agroforestry projects over industrial plantations. These conditions have created a larger demand for seeds of indigenous species, a demand that is difficult to meet because of lack of basic seed information on these species. Studies have shown that there are many gaps in our understanding of tropical seeds, including lack of definitive information on the phenology of flowering; plus maturation of fruits and seeds since multiple stages of seed maturity can be present at any one time on the same tree (Marzalina et al. 1993, 1994). Nevertheless, phenological studies found that some of the tropical forest trees have started to continuously produce flower-fruiting at a late age between 10 and 12 years (Marzalina and Krishnapillay 2002). An example of a-5-years study of phenological observations on mangrove forest trees is presented in Figs. 14.2 and 14.3; when a huge demand of mangrove planting materials are requested by local authorities after the massive Tsunami in 2004 hit Malaysia.

Physiological effects of the environment such as a dry spell preceding leaf flushing, accompanied by a rising gradient of daily sunshine (Ng 1981; Marzalina 1995), or a continuous low night temperature over 3–4 days may induce flowering behavior of tropical timber species (Ashton et al. 1988). Besides that another unique and fascinating phenomenon is the mass flowering and mast fruiting in dipterocarp

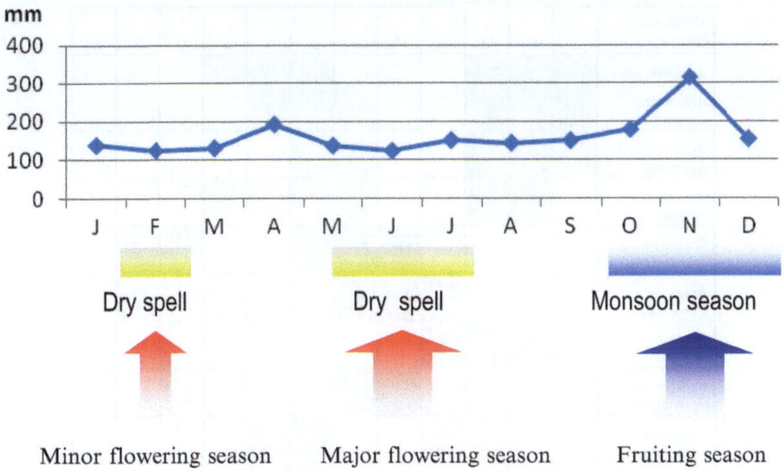

Fig. 14.2 Average monthly rainfall precipitation recorded from 2006 to 2010 by Dept. of Climatology Malaysia in coastal areas and its relation to the phenological events observed on mangrove species

forests (Yap and Marzalina 1990; Appanah 1993). Additionally, mass flowering can occur at varying intervals of 2–10 years. During this time, almost half of the mature individuals and over 80% of the canopy and emergent tree species in a forest may flower heavily. This involves over 200 tree species in the entire forest area which tend to flower over a short period of 3–4 months. The environmental cue for this irregular, but widespread mass flowering can be traced to a small dip of about 2°C below mean night-time temperature for 4 or 5 nights. The conditions for such temperature drop occur during the El Niño events. Thus, with a massive production of a majority of recalcitrant type of fruits once in such a long interval period could be a major disadvantage, thus conservation strategies must be in place to cater to the years without fruitings.

Besides that, spatial distribution or the number of trees of individual species in natural stands are commonly low, with only a few fruits are available from individual trees. This may affect collection costs since the fruit-bearing limbs of desirable trees may be as much as 35–40 m above the forest floor. Unless seeds can be collected from the ground after natural seedfall, climbing is the only practical option (Bonner 1992).

It has been estimated that more than 70% of tropical forestry seeds are recalcitrant in nature, thus they germinate promptly when dispersed (Marzalina et al. 1993; Krishnapillay et al. 1996). Seeds are described as recalcitrant if they are killed by desiccation below moisture content of 20–30%, intolerant to low moisture content and low temperature (Chin and Roberts 1980). If seeds can be dried to 10% moisture, they can be stored at subfreezing (below 0°C) temperatures. Such seeds are described as orthodox and exhibit seed dormancy.

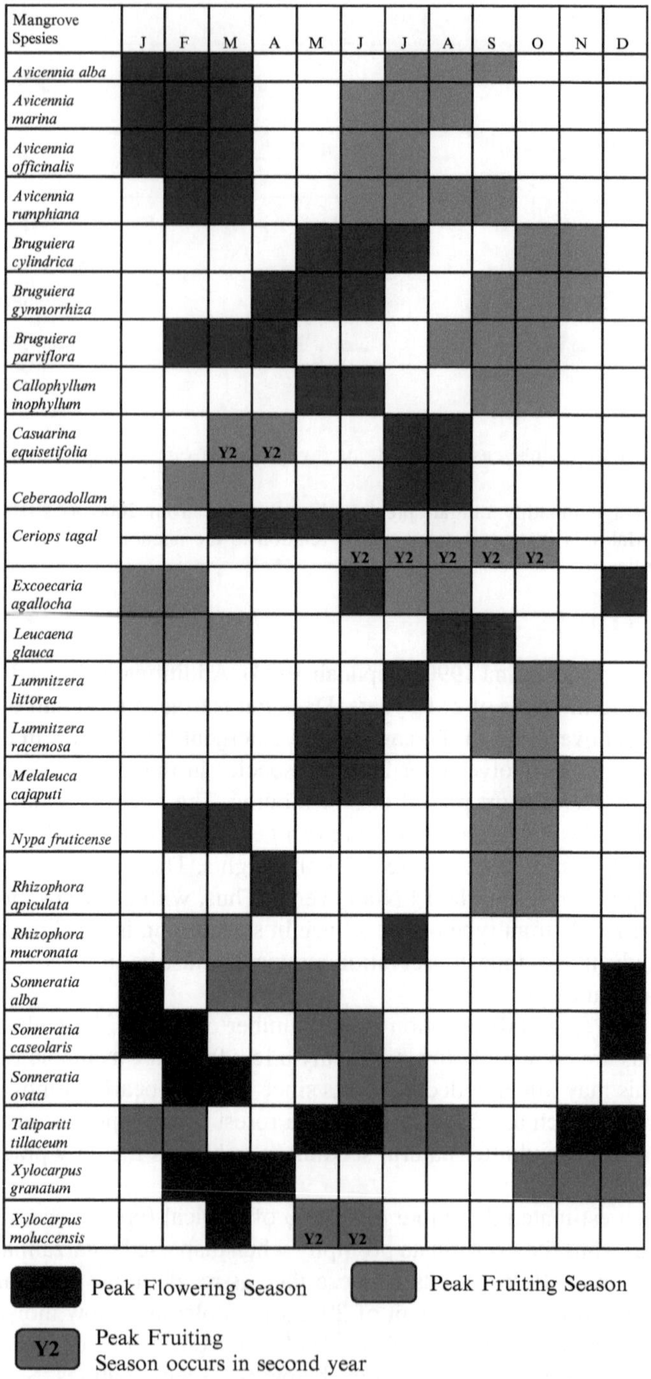

Fig. 14.3 Average monthly phenological observation on 25 mangrove species in Malaysia from 2006 to 2010 (Source: Mohd. Afendi et al. 2011) and its relation to rainfall events

14.3 Conservation Efforts

There is widespread concern about conserving diversity both among and within plant species for their actual or potential economic, social, cultural and scientific values. According to FAO report (2001), the Indo-Malayan Biogeographical Realm, which covers several countries of the Asia-Pacific region, had a total of 572 protected areas in 1985 extending over 27.5 million ha, some 10% of the world's protected areas. The report continues to mention the efforts taken in some countries to establish a more comprehensive network of national parks and other categories of reserves covering representative samples of ecosystems. Later, The World List of Threatened Trees published by World Conservation Press in 1998 was published. Later the IUCN Red List Categories and Criteria were published to provide a global listing of the conservation status (Chua et al. 2010). Several biosphere reserves have been established, in an attempt to integrate conservation and management objectives through developing models for sustainable management.

Although knowledge of tropical forest ecosystems and the genetic resources of plants and animals they contain are generally weak, the situation is better in the SEA region than in other major regions. Close cooperation among universities, research institutes and relevant stakeholders forged through many conservation projects, have resulted in an awareness of the presence of rare and threatened species in their areas of jurisdiction. However, the general objectives of these projects have focused on poverty reduction and sustainable development. In situ and ex situ conservation of trees in living stands is the most common strategies for conservation of forest genetic resources in SEA region. Jalonen et al. (2009) did report the needs for research and development in the conservation strategies and the use of forest genetic resources in seven countries within the South and SEA. Relevant national institutes and government departments in Brunei, Indonesia, Philippines, Lao PDR, Malaysia, Myanmar, Cambodia and Thailand have started several provenance trails of forestry important species, demonstration plots and pilot activities were established. Whereby majority of them are for in situ conservation of plant species. The Plant Red List project on conservation and monitoring of rare and threatened plants was initiated to assess the threats and conservation status of indigenous plant species, especially in the families of Dipterocarpaceae, Palmae and Begoniaceae (Chua et al. 2010).

Despite improvements in national legal and policy framework for conservation, such as the development of national strategies for forest genetic resources conservation in certain countries, the implementation of co-ordinated conservation programmes for valuable and potentially useful tree species calls for an increase in resources for operationalization as well as some regional collaboration. This further initiates and helps the government's ability to undertake appropriate conservation measures since forestry and forest genetic resources are important to the livelihood of mankind. As reported by Jalonen et al. (2009), several action plans have been launched and implemented to protect biodiversity, for example: Cambodia's National Protected Area Strategic Management Plan (NPASMP); Indonesian Biodiversity Strategy and Action Plan (IBSAP); and Philippine's Protected Area Management Plan (PAMP) and Strategic Environmental Plan (SEP).

14.3.1 In Situ Conservation

As stated by Palmberg-Lerche (1994), it is possible to conserve an ecosystem and still lose specific species, or to conserve a species and lose genetically distinct populations, genes or gene complexes that may be of future value. Therefore, it is important to specify clearly the level or levels targeted for conservation as well as the objectives of conservation. The basic objectives of conservation of genetic resources are the preservation of particular traits, unrecognized variation and adaptability. Many helpful practical guidelines for the conservation of forest genetic resources have been developed by Appanah and Turnbull (1998), ITTO and RCFM (2000), Thielges et al. (2001) and Theilade and Petri (2003), plus plenty more websites from FAO, DFSC, Biodiversity International (former IPGRI) that provide a recent general introduction to in situ and ex situ conservation of forest genetic resources.

For example, Brunei's undisturbed forests gazetted to 28,562 ha, amounted to 5.5% of the total land area, were set aside to preserve and conserve biodiversity for scientific, educational, and related purposes. A total of 1,100 ha within the Labi Hills Forest Reserve and 53 ha in the Badas Forest Reserve have been earmarked for seed production and conservation of genetic resources that will further go into developing of ex-situ conservation areas of important species (James 2007).

While in Indonesia, in-situ conservation of *Shorea leprosula*, *Pinus merkusii*, *Manilkara kauki*, *Loppethalum multinervium* and a number of other dipterocarp species have been initiated, including a good tree improvement programme for teak and *Pinus*. Both Indonesia and Philippines have embarked on in situ logging areas (Baskaran et al. 2002).

Malaysia continues to play an active role in the management of its genetic resources and the sustainable management of its forests. Various in-situ activities are in place like the Virgin Jungle Reserves (VJR), Genetic Resource Area (GRA) and Seed Production Areas (SPA) for important species, including those endemic such as *Hopea auriculata, H. bracteata* var. *penangiana, H. subalata, Shorea lumutensis, Vatica flavida* and *V. yeechongii.* Conservation measures are already in placed for species such as *Cotylelobium melanoxylon, Dipterocarpus sarawakensis, D. tempehes, Dryobalanops beccarii* and *Vatica yeechongii* (Chua et al. 2010).

Philippines undertakes an active programme on conservation and management of the Forest Genetic Resources that includes in situ conservation of a number of legume timber trees namely *Sindora supa*, *Albizzia akle*, *Afzelia rhomboidea* and *Intsia bijuga* and *Pterocarpus indicus,* which has been actively pursued since 1930 (Baskaran et al. 2002). It conserves plus tree selection of 23 important species and establishes a total of 38 seed production areas. A seed orchard of *Eucalyptus deglupta* was established in 1991 consisting of 500 grafted trees.

A project on the in situ conservation of the forest genetic resources in Thailand, established in 1999 by the Royal Forest Department, has emphasized the conservation of forest tree species inside their habitats. Among the 65 dipterocarp species available in Thailand, *Dipterocarpus alatus* has long been recognized as one of the most essential economic trees in Thailand. The species provides not only good quality wood for plywood and construction but also resin and oil for multiple uses. One can say that it is the most prominent species of dipterocarps planted in Thailand. The activities of in situ conservation of *D. alatus* is also far ahead of the other dipterocarps. There have been approximately 15 in situ conservation areas designated as important gene resources of the species (Boontawee 2001).

14.3.2 Ex Situ Conservation

There are some long-established botanical and zoological gardens in the region which contribute to the ex situ conservation of wild plants and animals. Ex situ conservation stands have also been established for selected provenances of some of the important plantation species of the genera *Araucaria, Pinus, Eucalyptus* and *Tectona* within the SEA region. National organizations dealing specifically with plant genetic resources have been established in a number of countries, but they focus on crop resources, meanwhile the coverage of forest species is very limited.

As explained by Thielges et al. (2001), there are two types of ex situ techniques that works for forestry species, these are:

(1) The evolutionary conservation technique that is involved in planting and domestication programmes such as the pseudo in situ plantings, field gene banks, demonstration forests; and
(2) The static conservation technique such as seed bank, slow-growth seedlings, pollen storage, tissue culture banks, macropropagation, clonal archives, in vitro gene banks, botanical gardens and arboreta.

Upon recovery from storage, the conserved population must be able to respond and adapt to changes in the environment it faces as fast as it can – otherwise the population may disappear in the long run. The choice could either be the static conservation (preservation) or evolutionary conservation (dynamic). In this chapter, several conservation techniques have been applied by SEA countries to the best of their abilities as described below.

14.4 Conservation Techniques

Several effective evolutionary conservation techniques are the least that can be applied when it comes to forest conservation.

14.4.1 Pseudo In Situ Plantings

As teak is its main commercial species in Myanmar, a well managed SPA (Seed Production Areas), Seed Orchards and Clonal orchards have been established for the species. Provenance trials have also been established. Some research activities on tissue culture of teak have also been initiated to produce elite planting materials (Baskaran et al. 2002).

14.4.2 Field Gene Banks

Malaysia also actively provides ex-situ measures including field genebanks, botanical gardens, arboreta, field trials of indigenous and exotic species, seed stand, seed banks and in-vitro gene banks. For example at Forest Research Institute of Malaysia (FRIM), its arboretum has maintained 722 species of timber trees from 80 families. Trial plots have also been regarded to serve as field gene banks. *Neobalanocarpus heimii*, locally known as chengal, an endemic species in Malaysia that possess recalcitrant seed. The Forest Department has established trial plots of this species at Bukit Sekilau, Pahang and at FRIM Kepong, Selangor. Usually, trial plots are the repositories of important genetic and clonal materials from exceptionally good quality trees that can become future seed collection areas. Besides, seed stands for *Shorea leprosula, Shorea macrophylla* and *Hopea odorata* have becomes ex situ seed production areas of these important species meant for plantations (ITTO and RFCM 2000).

Meanwhile, in Cambodia, 35 forest gene conservation stands, amounting to a total area of 691 ha, comprising 20 endangered tree species (including six priority species) have been established to date. Five species are conserved in more than one gene ecological zone, whilst other priority species are not yet conserved anywhere (Theilade et al. 2006).

Thailand did establish four small ex situ gene conservation plots at Prachuab Khiri Khan, Nakhon Ratchasima, Kanchanaburi and Chiang Rai (Boontawee 2001).

14.4.3 Demonstration Forests

Laos PDR has established both in-situ and ex-situ conservation areas of *Aquilaria crassna*, *Tectona grandis* and *Pinus merkusii* which are valuable and endangered. Meanwhile other international research institute such as CSIRO-ATSC has assisted Laos PDR on *Acacia* sp., *Eucalyptus* sp. and *Chukrasia* sp. While in the Philippines, the provenance and species trials for *Pinus*, *Acacia*, *Eucalyptus*, *Casuarina* and *Gmelina* have been established. In Philippines, some tree plantations and natural forest stands have been designated by the government as seed production areas to serve as the primary sources of quality seeds for tree plantations. The networks of protected areas serve as in situ conservation areas for many forest trees. Despite the Philippines' rapid decline of its forest habitats, through the protected areas system in situ remains as its best hope for conserving genetic resources of timber trees (Fernando 2001). To date Philippines has established eight seed storage and testing centers in strategic regions to ensure only superior planting materials came from seeds sourced from certified SPAs being utilized for industrial forest plantation especially *Paraserianthes falcataria*, *Gmelina arborea*, *Endospermum peltatum* and *Eucalyptus deglupta.*

14.4.4 Botanical Gardens and Arboreta

An arboretum is essentially a representative collection of well-grown trees maintained for the purpose of reference and as a convenient source of seed and herbarium materials for utilization and exchange (ITTO and RFCM 2000). Anticipating a worsening condition for in situ conservation in Indonesia, attempts have been made to establish botanic gardens, arboreta, or collection gardens may be the best solution for the time being for the conservation of some timber species (Thielges et al. 2001). An example is the newly established Serpong Garden, under the State Ministry of Research and Technology, has tried to accommodate collections of timber species using a representation of at least 50 individuals per species to allow a gene pool to be maintained. Again, the lack of minimum financial support prevents the garden to operate as planned. To be effective, this method of ex situ conservation should be combined with other objectives of national development such as education, research, and recreation.

Meanwhile the Government of the Philippines, in collaboration with FAO, implemented a range of conservation activities through the project "Bamboo Research and Development Project", including the establishment of the Philippine Bambusetum (Schlegel and Tangan 1990). A good collection of 68 species of which 17 species are native to the Philippines. The Bambusetum also functions as an "acclimatization area" for newly introduced bamboo species. This part of the programme is operated through an International Bamboo Species Exchange Programme with foreign research institutions.

14.4.5 Seed Storage

Timber species growing in the Malaysian aseasonal tropical forest thrive under very moist conditions. The whole process of seed storage begins with collection of good quality seeds, both physiologically and genetically. Combined with a series of seed handling procedures beginning with selection of the best quality of seed source, through collection, processing, storage, pre-treatment to germination; each link of this chain implies a potential risk of losing seed. As described by Theilade and Petri (2003) any link in the process is of equal importance (though not necessarily equally sensitive). If a seed dies due to careless handling during collection or processing, even the best storage conditions will not bring it back to life. If a seed dies during storage, the whole preceding effort is wasted.

Krishnapillay et al. (2003) reported that a majority of the recalcitrant seeds found in Malaysian forest are generally large, with the 1,000 seed weight often exceeding 500 g. About 98–99% of the seed consists of the cotyledons (4–6 g) which surrounds the small embryonic axis (7–10 mg). At shedding, the cotyledons are covered with a thin seed coat and the hard pericarp of the fruit. However, for the seeds found in seasonal climates of the tropics, the seeds are somewhat smaller in size as that compared to those found in the wet aseasonal habitat. As there is a distinct dry spell, the seeds tend to dry very swiftly after abscission and they can behave as orthodox or intermediate in their storage behaviour. As many of the species bearing recalcitrant seeds are erratic seed bearers, the reproductive biology, seed phenology and dispersal mechanisms and seed viability for such species are critical.

As mentioned earlier, it has been estimated that between 75% and 94% of the Dipterocarp seeds are recalcitrant (Tompsett 1992; Krishnapillay et al. 1996). Seeds of tropical recalcitrant species generally have the same moisture and gas-exchange requirements as seeds from recalcitrant species of the temperate regions, but they are sensitive to chilling damage at the common temperatures used to store temperate recalcitrant seeds. Even short periods below 10–15°C will cause loss of viability for many species (Chin and Roberts 1980). Therefore, tropical species are generally more sensitive to storage environment than temperate species, and viability retention is reported to be considerably less (Marzalina et al. 1996). Another confounding factor is pathogens. The relatively high moisture content and temperatures the tropical recalcitrant behaviour demands also favours profuse growth of pathogenic microorganisms. The immediate need in this problem is for safe, effective way to control pathogens in seeds during storage (Krishnapillay et al. 2003).

Seed banks are a valuable tool for short-term storage, but in most cases, the seed must sooner or later be transferred to other types of conservation (often within a year). Dipterocarpaceae is an example of a family where seed longevity may range from only a few weeks (e.g. *Shorea dasyphylla* – 14 days) up to a maximum of 1 year (e.g. *Hopea hainanensis* – 365 days) (FAO 2001). As reported by ITTO and RCFM (2000), for over the last 60 years, various methods of storage of tropical recalcitrant seeds have been practiced. These include:

- Imbibed storage in media such as sawdust, perlite and vermiculite
- Storage in airtight containers
- Storage in inflated bags with different gaseous environments
- Storage using germination inhibitors
- Fungicide treatment followed by partial desiccation and storage at controlled temperatures
- Partial desiccation without fungicide
- Storage at harvest moisture content without fungicide or partial desiccation.

As example the seed storage study on *Neobalanocarpus heimii* or chengal seeds was categorized within the recalcitrant class as the seeds' moisture content falls within the range of 42–55% on fresh weight basis. Without pre-treatment, storage of seeds will not last longer than a month. But from different storage experiments carried out by the Seed Technology Laboratory at FRIM, seeds are best stored if first treated by soaking in 1% benlate + thiram for 20 min, air-drying them on old newspapers and then placing them in polythene bags of moist vermiculite at 20 ± 2°C chamber. Polythene bags greatly restrict exchange of moisture but allow exchange of oxygen and carbon dioxide with the surrounding air. Such condition can prolong storage up to 3 months. Within the time, seeds tend to germinate and seedlings can be stored for another 6–18 months (Marzalina and Krishnapillay 2002).

14.4.6 Slow-Growth Seedlings

In situ conservation can utilize the presence of massive production of dipterocarp fruits during the gregarious period as the seedlings of some species will be able to survive for many years under the dense, shady canopy of established trees. The resources provided by a large seed are an obvious advantage here. It is well known that tropical timber seeds usually have survival and slow growth rate over a period of several months and years when grown under low light intensity. This effectively creates a "seedling bank" that can respond to opportunities created by an opening in the canopy overhead. Lee et al. (1996) and Krishnapillay et al. (1998) reported on the seedlings studies carried out from 1989 to 1999 by Seed Technology Laboratory at FRIM upon the excessive dipterocarp fruits collected elsewhere during the mast fruiting years was dumped under the dense forest canopy site nearby the lab. It was found to be a good cheap way of ex-situ conservation effort, supplement to the in-situ conservation. Thus, the field genebank has been the ex situ method of choice for those species which produce recalcitrant seeds or which are vegetatively propagated. Within the whole 10 years, 50% the seedlings survived, majority with the height of not more than 10 cm. Meanwhile in forests in Borneo, a report of the dipterocarp seedlings that last for more than 15 years on the forest floor after a single fruiting event was recorded (Guhardja et al. 2000; Primack and Corlett 2005).

However, seedlings storage can also be carried out in environmental controlled chamber as reported by several studies (Krishnapillay et al. 1998; Marzalina et al. 1998; ITTO and RCFM 2000). Freshly collected seeds are surface-sterilized with fungicide (0.1% benlate/thiram mixture) and allowed to germinate under ambient conditions in containers kept at high humidity with moistened paper. Once germinated, they are packed loosely in polythene bags or transparent boxes lined with moist paper and stored in seedling chamber with the temperature set at 16°C, relative humidity of 80% and storage photoperiod for 4 h. These conditions caused slow development of seedling growth, attaining the heights in the range between 20 and 25 cm of the storage period. Seed Technology Laboratory in FRIM conducted the following on dipterocarp species as listed in Table 14.2 (unpublished data) and found that the survival rate was within 60–80% after they were being weaned under 70% shade and gradually transferred under direct sunlight over 2–3 weeks.

Table 14.2 Storage periods of recalcitrant seeds in seedling chamber

Species	Period of storage (months)
Dipterocarpus cornatus	6
Dryobalanops aromatica	5
Dryobalanops oblongifolia	4
Hopea apiculata	6
Hopea helferi	9
Hopea mengarawan	9
Hopea odorata	9–12
Neobalanocarpus heimii	6–18
Shorea acuminata	8
Shorea bracteolata	6
Shorea curtisii	8–9
Shorea dasyphylla	6–9
Shorea hemsleyana	6
Shorea leprosula	6–9
Shorea longisperma	6
Shorea macrophylla	4
Shorea macroptera	6
Shorea maxwelliana	4
Shorea ovalis	8–9
Shorea parvifolia	8–9
Shorea pauciflora	6
Shorea platyclados	8–9
Shorea singkawang	12
Parashorea stellata	**6–9**
Vatica nitens	**6–12**

14.4.7 Macropropagation

There are several methods of vegetative forest tree propagations: grafting, air-layering and the use of cuttings. With dipterocarps, there has been considerable success using this technique as its species have been known for being difficult to root. However the production scale of vegetative propagation by cuttings is being applied by the PT Inhutani in East Kalimantan, Indonesia (Suzuki et al. 2006). Inducing roots from cuttings requires conditions to control and balance their transpiration and to promote photosynthesis simultaneously. As reported by Sakai et al. (1994), this can be achieved using fog-colling system that keeps the temperature inside the box of cutting beds below 30°C, at about 5,000 lux of sunlight, due to heat removal by evaporation of the fog, and at a relative humidity of about 95%. Via this system, root percentage of *Shorea leprosula* and *Shorea selanica* cuttings was 95% and 92% respectively. The nursery unit in FRIM has conducted plenty of research in macropropagation of important species for the past decade. Ahmad (2006) recommended the use of the simple technique of stem cutting in polyurethane box (Table 14.3). Studies indicated that not only the rooting condition, but the number of leaves per

Table 14.3 Rooting percentage and number of leaves per stem cuttings of the seven dipterocarp species tested (Source: Ahmad 2006)

Species	% rooting	Duration (weeks)	Number of leaves per cutting
Hopea odorata	90–100	3–5	One leaf
Shorea acuminata	40–50	8–12	One to two leaves
Shorea bracteolata	70–80	3–5	One third of the leaf
Shorea leprosula	40–50	6–10	One half of the leaf
Shorea ovalis	30–40	8–12	One half of the leaf
Shorea parvifolia	30–40	8–12	One leaf
Shorea roxburgii	80–90	3–5	One third of the leaf

stem cutting and juvenility may be an important factor in the rooting potential of dipterocap cuttings. Juvenile tissues of woody plant tend to have higher levels of endogenous auxins and are less differentiated (Suzuki et al. 2006). In the Philippines macropropagation work for hybrids of *Acacia* and *Eucalyptus* and also for a number of endangered species such as *Diospyros philippinensis*, *Tectona philippinensis* and *Agathis dammara* were successfully carried out.

14.4.8 Tissue Culture Banks: Micropropagation

The application of biotechnology has revolutionized forestry practices worldwide. Micropropagation or in vitro propagation is an alternative high-tech approach to the vegetative propagation of planting materials. The technique involves the excision of small parts of a plant such as embryos, shoot tips, nodal segments, young leaves, stems and roots, then placing them in an artificial media under controlled aseptic conditions to develop into little plants. New plants can be derived directly after undergoing a callus phase. The advantages of in vitro propagation are that the explants need small space to maintain and multiply into large numbers; new plants produced are free from bacteria, fungi and microorganisms due to aseptic conditions; a higher rate of multiplication is obtained compared with conventional methods; desired clones can be produced quickly in quantity; and selected clonal material can be stored using the in vitro system itself. Once the explant starts to proliferate, subculturing is done to bulk up the plant stock every 4–6 weeks (Harikrishna et al. 2002). However, this method requires advanced technical skills, expensive equipment and the methods are species specific.

Based on our experience, the attempt to get cultures started from field material of *N. heimii* did pose a big problem as microscopic bacteria and fungi covered the surfaces of leaves and buds. To date, the Seed Technology Laboratory at FRIM has been able to establish the sterilization technique of *N. heimii* seeds using chemical sterilants that can surpass the problematic convolution of the cotyledons that enable microbes to lodge within the seeds. However, the problem of phenolic compound excretion, oxidation and the production of polyphenoloxidase and peroxidaseenzymes, has often caused the discolouration of tissues and has been known to result in poor growth. The Tissue Culture Laboratory at FRIM currently attempts in vitro micropropagation of *N. heimii*. Once this species can be propagated via micropropagation, an in vitro gene bank can be established to supply the number of plantlets whenever required. Plantlets regenerated from meristem-tip cultures must be evaluated for virus indexing in order to prevent reinfection. Following virus indexing, the plantlets can serve as a source of stock plant material for provision of germplasm repositories. However, the maintenance of an in vitro gene bank can be costly as cultures are kept under laboratory conditions.

The effectiveness of tissue culture as a conservation method has been recognized and proven to be an effective technique of storing or preserving species, which are normally difficult to store, like recalcitrant seeds, using the conventional techniques. But somaclonal variation in cultured tissues has been observed to be the problem of a number of woody species and the variations can lead to changes in the species being conserved, although studies reported that the variations are considered positive and have improved the quality of some species (Lapitan 2002). Thus there is the need for proper management of in vitro gene banks.

Malaysian research in micropropagation of important species are in place with several micropropagation protocol have been developed at FRIM for teak (*Tectona grandis*), rattan (*Calamus mannan, Calamus caesius*), bamboo (*Gigantochloa and Bambusa* spp.), jelutong (*Dyera costulata*), karas (*Aquilaria malaccensis*), sesenduk (*Endospermum malaccense*), Acacia hybrids (*A. mangium* x *A. auriculiformis*), sentang (*Azadirachta excelsa*), sungkai (*Peronema palembanica*), khaya (*Khaya ivorensis*), herbal plants and horticulture plants.

In the Phillipines tissue culture protocols have also been developed for *Pterocarpus indicus*, *Acacia mangium*, *Paraserianthes falcataria*, *Endospermum peltatum*, *Eucalyptus* sp., *Pinus caribaea* and *Cratoxylon sumatranum*.

14.4.9 Cryo Storage

Cryopreservation utilizes an ultra-low temperature condition for storage. This is achieved by using liquid nitrogen as the immersion medium, which provides a storage temperature of about −196°C. At this condition, preservation can be long term. The potential for the use of cryopreservation has been proven for several timber species and has still to be realized for many others (Marzalina and Nashatul 2000).

Desiccation technique has been applied mainly to either whole seed or zygotic embryos of a large number of tropical forest trees species when the lowest safe moisture content (LSMC) was within 4–12% (Engelmann 1992; Engelmann et al. 1995; Krishnapillay et al. 1992, 1994; Chai et al. 1994; Marzalina 1995; Normah

Table 14.4 Cryo probing studies on some of tropical timber species (Source: Seed Tech Lab, FRIM)

Species	Part cryopreserved	LSMC (%)	Viability (%)
Aquilaria malaccensis	Whole seed	7–10	75
Bambusa arundinacea	Whole seed	10–12	73
C alamus manan	Embryonic axis	7–8	60
Dendrocalamus membranaceus	Whole seed	8–9	53
Dendrocalamus brandisii	Whole seed	7–8	48
Dipterocarpus alatus	Whole seed	8–12	68
Dryobalanops oblongifolia	Embryonic axis	10–12	15
Dyera costulata	Whole seed	6–10	50
Hopea odorata	Embryonic axis	6–8	15
Khaya senegalensis	Whole seed	9–10	70
Koompasia malaccensis	Whole seed	8–10	62
Pterocarpus indicus	Whole seed	7–9	80
Shorea bracteolata	Embryonic axis	6–8	10
Shorea faguetiana	Embryonic axis	8–12	15
Shorea leprosula	Embryonic axis	4–5	12
Shorea ovalis	Embryonic axis	7–10	7
Shorea parvifolia	Embryonic axis	5–7	10
Shorea macrophylla	Embryonic axis	10–12	5
Swietenia macrophylla	Embryonic axis	5–8	63
Tectona grandis	Whole seed	8–10	40
Thyrsostachys siamensis	Whole seed	7–9	86
Vatica umbonata	Whole seed	7–10	**5**

and Marzalina 1996). Seed Technology Laboratory at FRIM has conducted several studies on cryopreservation of tropical timber species as listed below (Table 14.4) and obtained survival rates ranging from as low as 5% to as high as 86% (unpublished data). The number of seeds/embryos tested was inconsistent depending on availabilities, ranging from 5 to 100 samples.

Encapsulation-dehydration technique used this far has been based on the technology developed for the production of synthetic seeds (Villalobos and Engelmann 1995). For cryopreservation purposes, apices or somatic embryos are encapsulated in a bead of alginate and pregrown for various durations in liquid medium with high sucrose concentrations. Beads are then partially dehydrated under the air current of a laminar flow cabinet or using silica gel, down to a water content of ca. 20%. Freezing is usually rapid, by direct immersion of the samples in liquid nitrogen. For recovery, samples are usually placed directly under standard culture conditions. Growth recovery of cryopreserved materials is generally rapid and direct, without callus formation. This technique has been applied to somatic embryos of carrot, apices of several temperate as well as tropical species (Dereuddre et al. 1990; Withers and Englemann 1997), and recently to zygotic embryos of two tropical forestry species, *Swietenia macrophylla* and *Shorea leprosula* (Marzalina et al. 1994, 1995; Krishnapillay and Marzalina 1995).

14.4.10 Pollen Storage

As described by Theilade and Petri (2003), pollen storage is comparable to seed storage, since most pollen can be dried to less than 5% moisture content on a dry weight basis and stored below 0°C or in cryo storage. Some species, however, produce pollen with recalcitrant type storage characteristics. There is limited experience on the survival and fertilizing capacity of cryopreserved pollen more than 5 years old (Towill 1985). However, pollen has a relatively short life compared to seeds and viability testing may be time consuming. For these reasons, despite being a useful technique for species which produce recalcitrant seed, pollen storage has been used to a limited extent in germplasm conservation (Hoekstra 1995). Other disadvantages of pollen storage are the small amount produced by many species; the lack of transmission of organelle genomes via pollen; the loss of sex-linked genes in dioecious species; the lack of plant regeneration capacity although first indications exist that pollen can be re-grown into whole plants (Hoekstra 1995). On the other hand, pollen transfer of pests and diseases is rare with the exception of some virus diseases, thus allowing the safe movement and exchange of germplasm.

14.4.11 Genetic Diversity and DNA Storage

Genetic diversity kept in forests is seen not only as a natural reserve but also as a potential resource. There is a growing interest in the conservation of the genetic resources of the much disturbed areas on account of their potential for eco-restoration and provision of multiple benefits to rural populations.

At FRIM, genetic materials or germplasm from the forests are used for selection to increase resistance to diseases, to improve the product quality, or to make them more suitable for use in agro forestry activities especially in the setting up of forest plantations.

Information on population genetics of forest plant species is essential for the successful implementation of conservation and tree improvement programmes. Several suggestions on problem related approach towards tree improvement efforts can be referred to Kjaer et al. (2001). In conservation programmes, this information would be useful in designing sampling strategies for ex situ conservation and sustainable forest management. At the same time such information can be used in tree improvement programmes to provide adequate guidelines in designing sampling strategies, clone identification and tracking desirable genes for early selection in tree improvement programme.

As reported by Norwati et al. (2010) at the Genetic Unit in FRIM, the issue of conservation of important tropical species is approached through two notable issues, which are (a) Genetic diversity evaluation of species based on their status (such as endemism, rare, threatened or endangered etc.) and; (b) effect of logging on genetic diversity. Selective logging can also increase the potential for inbreeding as logging temporarily reduces adult population densities. For the former, some of the species studied include *Aquilaria malaccensis, Dyera costulata, Shorea lumutensis, Hopea bilitonensis, H. subalata, H. odorata, Neobalanocarpus heimii.* As for the effect of logging, some of the species studied include *Shorea leprosula, S. ovalis, S. curtisii and S. macroptera.* For these purposes, the Genetic Unit has developed various types of genetic markers such as isozymes, Amplified Fragment Length Polymorphism (AFLP), Randomly Amplified Polymorphic DNA (RAPDs) and microsatellite. Research in recombinant DNA has now been established for advance research.

DNA storage method is gaining its rapid importance. DNA from the nucleus, mitochondrion and chloroplasts are now routinely extracted and immobilised into nitro-cellulose sheets where the DNA can be probed with numerous cloned genes. With the development of PCR (polymerase chain reaction) one can now routinely amplify specific oligonucleotides or genes from the entire mixture of genomic DNA. As mentioned by Theilade et al. (2006), these advances have led to the formation of an international network of DNA repositories for the storage of genomic DNA. The advantage of this technique is that it is efficient and simple and takes up little space. However, the main disadvantages, besides high standard requirements in capacity and equipment, lie in problems with subsequent gene isolation, cloning and transfer as well as that it does not allow the regeneration of entire plants.

14.5 Challenges Facing Conservation of Forest Ecosystems

While timber is economically the most important forest product, many other products are valued both on world markets and by local people. Jansen et al. (1991) found that nearly 6,000 species of rain forest plants in SEA region have many economic uses as stated in Table 14.5. The repetition of species that has multipurpose economic use may be present here. However, the efforts of conserving natural forests are not only for biodiversity but also of the important role played by in the provision of environmental and social services. As mentioned by Dennis et al. (2008), despite several decades of attempts to improve forest management in SEA, the implementation of sustainable forest management (SFM) in a significant part of the region's production forests still has a long way to go. Continued rapid forest loss from the timber estate indicates that production forestry in existing natural forests will be of major economic importance. Strengthening the role of the timber industry in biodiversity conservation will require improvements in collaborations between the timber industry and local communities, as well as the providing the alternatives to sustain the need of timber industry. The innovation of engineered lumber from waste biomass, such as the creation of MYScrim-OPT for furniture industry will indeed be an option to be considered in the era of green wood technology (Marzalina and Wan Tarmeze 2010).

Table 14.5 Selected economic uses of Southeast Asian tropical rain forest plants

Product/commodity group	Species (number)
Timber trees	1,462
Medicinal plants	1,135
Ornamental plants	520
Edible fruits and nuts	389
Fibres	227
Rattans	170
Poisonous and insecticidal plants	147
Spices and condiments	110
Others	1,790
Total	5,950

The efforts to conserve the evolutionary potential of ecosystems and species, and to ensure the enhancement and sustainable utilization of the genetic variation available, aim to meet present and future human needs. It is understandable that conservation of the genetic resources of agricultural plants is clearly oriented by species, but for forest genetic resources, the consideration must include the genetic resources of all component species in an ecosystem, not merely the scope of single species. Successful long-term conservation of forestry species even in their own natural ecosystems requires a good understanding of the ecological interactions between species, and the challenges begin with the surveying and inventorying of existing resources. Priority should be to identify species that are endangered or threatened with extinction. As described by (Bawa 1998), studies must include identifying centers of taxonomic diversity, genetic diversity, and the effects of logging. Information of breeding system and pollination mechanisms is required in maintaining genetic diversity. Dipterocarps has the potential for inbreeding as studies have shown self-incompatibility barriers of its species are not strong (Bawa 1998). Thus comparative studies at species level and infraspecific levels of reproductive output, mating patterns and regeneration processes involving trees in large contiguous forests and small fragments will pose challenges to conserve a forest ecosystem.

Research on dipterocarps is still being carried out within the confines of narrow disciplines, topped up with some reports written in local language, substantial wealth of knowledge is simply not available to the vast majority of scientists (Appanah and Turnbull 1998). Thus, much of the knowledge on tropical timber conservation appears to be accessible only to specialists. Problem oriented, multi-disciplinary approaches are indeed rare. Although numerous initiatives both national and international bodies are working together with SEA countries to address the variety of issues related to forest genetic conservation (refer to Table 14.6), there are still lack of coordination between and among external agencies and international donors for most of the initiatives. Many of the duplication of activities is seen to be occurring in countries that have substantial infrastructure and government support as compared to poor countries in SEA.

Conservation of forest ecosystems in majority of SEA countries faces several major challenges. Coverage of protected areas is not adequate, particularly in the

Table 14.6 Signatory to/party to international agreements and conventions related to biodiversity conservation and environment

Agreement/Convention	Brunei	Cambodia	Indonesia	Lao PDR	Malaysia	Myanmar	Philippines	Thailand	Vietnam
Agenda 21	☑							☑	
Convention on Biological Diversity (CBD)		☑	☑		☑	☑	☑	☑	
UN Framework Convention for Climate Change (UNCCC)					☑	☑		☑	
Convention on International Trade in Endangered species of Wild Flora and Fauna (CITES)	☑	☑			☑	☑	☑		
UN convention to combat desertification		☑				☑			
ITTO agreement		☑	☑		☑	☑			
International Centre for Integrated Mountain Development (ICIMOD)						☑			
Statement on forest principles					☑	☑			
Convention on wetlands of international important especially as waterfowl habitat (Ramsar convention)		☑	☑		☑				
Kyoto protocol					☑				
Cartegena protocol on biosafety					☑				
Montreal protocol on substances that deplete the ozone layer					☑				
Basel convention on transboundary shipment on hazardous wastes					☑				
Convention on the continental shelf					☑				
International tropical timber agreement (ITTA)					☑				
United Nations convention on the law of the sea					☑				
International donors/NGOs collaborates with									
JICA		☑	☑		☑	☑		☑	
ASEAN agreement on the conservation of nature and natural resources	☑	☑	☑	☑	☑	☑	☑	☑	☑

(continued)

Table 14.6 (continued)

Agreement/Convention	Brunei	Cambodia	Indonesia	Lao PDR	Malaysia	Myanmar	Philippines	Thailand	Vietnam
UNESCO			☑		☑			☑	
GTZ		☑	☑		☑				☑
UNDP		☑			☑		☑	☑	☑
FAO		☑	☑		☑		☑	☑	☑
USAID		☑	☑		☑				☑
Rockefeller Foundation			☑					☑	
Ford Foundation			☑		☑				☑
CIFOR			☑				☑		
ADB		☑			☑		☑		
GEF					☑		☑		
FORSPA		☑				☑			
APAFRI		☑		☑	☑	☑			
TEAKNET					☑	☑			
FRIM	☑				☑	☑			
INBAR					☑	☑			
Smithsonian Institute Foundation					☑	☑			
CAMCORE						☑			
IPGRI					☑				
UNEP			☑		☑				
IUCN		☑	☑		☑				
WWF		☑	☑		☑				
AusAID		☑	☑		☑				
SEAMEO			☑		☑				
CARE		☑							
CONCERN		☑							

EU		☑			☑				
CIDA		☑			☑				
DFID-UK		☑			☑				
UNCHR		☑			☑				
Special support to projects									
CSIRO-ASEAN Tree Seed Center	☑								
**Provenance trial		☑	☑	☑	☑	☑	☑	☑	☑
DANIDA									
**Tree Seed Project		☑	☑	☑					

Note: **only involve with this specific project

islands, and some of the major ecosystems, for example the lowland tropical rain forests of Malaysia, are almost lost due to transformation of land use (FAO/IPGRI/DFSC 2001). Protected areas often exist as islands, surrounded by densely populated areas with incompatible land uses. Absence of any direct interest to and involvement of local people makes it extremely difficult to protect these areas.

Efforts are heavily skewed toward protection from biotic interferences; active habitat manipulation is seldom undertaken. Knowledge of the dynamics of ecosystems and how they respond to interventions is inadequate. Conservation aspects in areas earmarked for other purposes are generally neglected, and very few efforts are made to arrive at acceptable levels of trade-off between competing objectives.

For many tropical tree species, there are several prerequisites which have to be fulfilled before in vitro conservation becomes possible. A minimal knowledge of the biology and physiology of the species to be conserved is needed in order to identify which type of explants can be used as a storage propagule and to define the necessary in vitro culture techniques. In many cases, simple in vitro culture procedures (embryo culture, shoot micropropagation) have to be established and/or optimized. Additional research aiming at the development of in vitro conservation techniques, including slow growth storage and cryopreservation, and at the understanding of the biological mechanisms determining recalcitrance is also required. Besides that past experiences should be analyzed and continuous effort should be given into the establishment of species trials, provenance testing, seed orchards, selection of plus trees, vegetative propagations and ex situ conservation.

14.6 Conclusions and a Strategy for Action

In situ and ex situ conservation measures are being implemented by various agencies within each country but they are fragmented and mostly focused on fauna, lacking on flora. The design of any ex situ conservation program and which technologies to use must start from considerations on the biological material in question, why do we want to conserve it and how is it going to be used in the future. The strategic action plan for plant conservation need to be crafted to suit each country's requirement. Thus, purpose must be strictly defined as a prerequisite to management planning. The result will often be a program for conservation of chosen heritable attributes rather than of species. The situation-specific nature of conservation needs makes it difficult to indicate priorities on a subregional basis. Dipterocarp is certainly among the six largest families that are exceedingly abundant in lowland forest of SEA, whereby they embrace 80% of the emergent individuals and 40% of the understory trees (Ashton 1982). Thus this need to be conserved for future generations. As suggested by Kjaer et al. (2001), it is recommended that ex situ conservation as conservation strategy for tropical trees be linked to a program for domestication and use, preferably as part of active breeding and/or research programs, to ensure long-term management and regeneration. Ex situ conservation of trees in living stands is expensive and, on a long-term basis, is likely applicable for a small number of commercially important species.

At the regional level, the emphasis in ecosystem conservation will need to be on the identification of forest ecosystems which are not afforded protection in national parks or similar reserves, with a view to promoting their protection. Many of the unique forests in the continental and insular Southeast Asia merit priority attention and require urgent conservation efforts. A priority for all of the countries in the tropical Oceania subregion should be the development of national conservation strategies too. Protection of wetland sites should also be seen as a general priority for the region. This should be implemented as part of a comprehensive approach to water resource management and land use planning.

Nationally, activities envisaged will depend on the present level of efforts. In certain countries urgent priority has to be given to national surveys to catalogue the ecosystems and species in need of protection and management. The present status reports indicate that fairly comprehensive legal and institutional frameworks for the conservation and sustainable management of tropical forest genetic resources are already widely in place in the region, and many of the countries have initiated specific national programmes and strategies related to forest genetic resources. Now the efforts should focus on their consolidation and effective management so as to further develop these initiatives and translate them to effective practical guidelines and measurable outputs. Species of local socio-economic importance are in need of attention too.

Long-term conservation of biological diversity depends on the development of location-specific techniques and systems for the management of tropical forest ecosystems and target species which are compatible with species and genetic resource conservation goals, and which at the same time ensure the sustained production of locally important goods and services. Global climatic change will influence the plant populations over time.

Meanwhile, optimizing methods are required to counter primary problems with all recalcitrant seeds related to the rapid loss of viability; seed-borne pathogens and difficulty in their appropriate time of collection. Detailed studies on the ecophysiology and storage physiology of tropical seeds and germinated seedling still warrant more attention. While we are now exploring the use of the embryonic axis of tropical seeds for storage and conservation, nevertheless, it is important that we also need to focus our efforts in trying to better understand these tropical recalcitrant whole seeds. There may be lessons that we can learn from these studies that will help us to understand better and to handle such seeds even for short term storage purposes.

There is a need for demonstration areas under management systems which are consistent with this to be established, and for information generated on approaches and techniques to be widely disseminated. This may be facilitated by the development of international networks. As mentioned by FAO/IPGRI/DFSC report in 2001, all donors, private sector, NGOs and communities should provide technical assistance and capacity building in support of sustainable resource use and best management practices, improved local governance, appropriate spatial planning, and financial incentives (e.g., Payment for Ecosystem Services, voluntary carbon credits) to all the stakeholders. People's participation in and benefits from protected areas will need to be strongly emphasized in conservation strategies. Buffer zones and their management will be key elements. The objective in this context will be to manage such zones so as to produce benefits for local people, thereby offsetting the effects of loss of access to resources in the related protected area. In this aspect, the incorporation of new policies as part of timber trade such as The Lacey Act provides a good entry point into forest governance issues and the reduction of illegal logging (Yeager 2008). There is a need to continue and leverage the benefits of exchange information and sharing of best current practices among the neighbours having similar situations.

Acknowledgement I am in debt and thankful to Prof. Dr. Normah Mohd Noor (INBIOSIS – UKM) for the opportunity in giving me the chance to write this chapter; Dr. Baskaran Krishnapillay for being a great teacher at FRIM; Ms. Noraliza Alias (FRIM) for all the assistance related to seed conservation; and the Seed Technology Unit (FRIM) for continuously investigating and collating seed information.

References

Ahmad D (2006) Vegetative propagation of dipterocarp species by stem cuttings using a very simple technique. In: Suzuki K, Ishii K, Sakurai S, Sasaki S (eds) Plantation technology in the tropical forest science. Springer, Tokyo, pp 60–77

Appanah S (1993) Mass flowering of dipterocarp forests in the aseasonal tropics. J Biosci 18(4):457–474

Appanah S, Turnbull JM (1998) A review of dipterocarps: taxonomy, ecology and silviculture. CIFOR, Bogor, 223 pp

Ashton PS (1982) Dipterocarpaceae. In: Flora Maleasiana series 1, vol 9. M. Nijhoff, Hague, pp 237–552

Ashton PS, Givnish TJ, Appanah S (1988) Staggered flowering in diptercarpaceae: new insights into floral induction and the evolution of mast fruiting in the aseasonal tropics. Am Nat 132:44–66

Baskaran K, Bariteau D, El-Kassaby M, Huoran YA, Kageyama W, Kigomo PN, Mesén F, Midgley S, Nikiema A, Patiño VF, Prado JA, Sharma MK, Ståhl PH (2002) Forestry genetic resources working papers. Published as Corporate Document. FAO

Bawa KS (1998) Conservation of genetic resources in the dipterocapaceae. Chapter 2 in: A review of dipterocarps: taxonomy, ecology and silviculture. Appanah & Turnbull (eds). CIFOR, Bogor, Indonesia, pp 45–56

Bonner FT (1992) Seed technology: a challenge for tropical forestry. Tree Plant Note 43(4): 142–145

Boontawee B (2001) The status of in situ and ex situ conservation of Dipterocarpus alatus Robx. in Thailand. In: In situ and ex situ conservation of commercial tropical trees. ITTO Project PD 16/96 Rev 4, pp 171–182

Butler RA (2006) Rainforests. http://rainforests.mongabay.com. Accessed on Feb 2011

Chai YM, Marzalina M, Krishnapillay B (1994) Effect of sucrose as a cryoprotective agent in the cryopreservation of some dipterocarp species. In: Proceedings of the 6th national biotechnology seminar. Published by Universiti Sains Malaysia

Chin HF, Roberts EH (1980) Recalcitrant crop seeds. Tropical Press, Kuala Lumpur, 152 pp

Chua LSL, Suhaida M, Hamidah M, Saw LG (2010) Malaysia plant red list: peninsular Malaysian Dipterocarpaceae. Research phamplet no. 129, 210 pp

Dennis RA, Meijaard E, Nasi R, Gustafssonm L (2008) Biodiversity conservation in Southeast Asian timber concessions: a critical evaluation of policy mechanisms and guidelines. Ecol Soc 13(1):25

Dereuddre J, Scottez C, Arnaud Y, Duron M (1990). Effets d'un endurcissement au froid des vitroplants de poirier (Pyrus communis L. cv. Beurré Hardy) sur la résistance des apex axillaires à une congélation dans l'azote liquide. CR Acad Sci Paris 310 Sér III, pp 317–323

Engelmann F (1992) Cryopreservation of embryos. In: Dattée Y, Dumas C, Gallais A (eds) Reproductive biology and plant breeding. Springer, Berlin, pp 281–290

Engelmann F, Dumet D, Chabrillangel N, Abdelnour-Esquivel A, Assy-Bah B, Dereuddre J, Duval Y (1995) Factors affecting the cryopreservation of coffee, coconut and oil palm embryos. Plant Genet Resour Newsl 103:27–31

FAO/IPGRI/DFSC (2001) Conservation of forest genetic resources. Part 3: conservation of genetic resources in planted forests. Draft. FAO and IPGRI, Rome/Danida Forest Seed Centre, Humlebaek

Fernando ES (2001) Genetic resources conservation strategies for timber trees in the Philippines. In: In situ and ex situ conservation of commercial tropical trees. ITTO Project PD 16/96 Rev 4, pp 69–82

Greathouse TE (1982) Tree seed and other plant materials: aspects of USAID-supported reforestation projects. Report to USAID S&T/FNRIF. Washington, DC, 27 pp

Guhardja E, Fatawi M, Sutisna M, Mori T, Ohta S (2000) Rainforest ecosystems of East Kalimantan: El Niño, drought, fire and human impact, vol 140, Ecological studies. Springer, Tokyo, 330 pp

Hamdan O, Khali Aziz H, Shamsuddin I, Raja Barizan RS (2012) Status of Mangroves in Peninsular Malaysia. FRIM, Kuala Lumpur, pp 8

Harikrishna K, Lee WW, Ong CH, Siti Suhaila AR (2002) Basic tools and methodologies supportive of biotechnology. In: Krishnapillay B (ed) Basic principles of biotechnology and their application in forestry. APAFRI, Kuala Lumpur, pp 33–48

Hoekstra FA (1995) Collecting pollen for genetic resources conservation. In: Guarino L, Rao VR, Reid R (eds) Collecting plant genetic diversity: technical guidelines. IPGRI/FAO/UNEP/IUCN/CAB International, Wallingford, pp 527–550

ITTO, RCFM (2000) Technical guidelines: establishment and management of ex situ conservation stands of tropical timber species. ITTO, Yokohama, 117 pp

Jalonen R, Choo KY, Hong LT, Sim HC (2009) Forest genetic resources conservation and management status in seven South and Southeast Asian countries. Published by APFORGEN, Kuala Lumpur

James PG (2007) Fundamentals of conservation biology, 3 th edn. Blackwell, Oxford

Jansen PCM, Lemmens RHMJ, Oyen LPA, Siemonsma JS, Stabast FM, van Valkenburg JLCH (1991) Basic list of species and commodity grouping (plant resources of Southeast Asia). Pudoc, Wageningen

Kjaer ED, Graudal L, Nathan I (2001) Ex situ conservation of commercial tropical trees: strategies, options and constraints. Paper presented at: ITTO international conference on ex situ and in situ conservation of commercial tropical trees, Yogyakarta

Krishnapillay B, Marzalina M (1995) Cryopreservation as a potential tool for the long term conservation of recalcitrant seeded forest species. In: Proceedings of the seventh national biotechnology seminar, Langkawi, 20–22 Nov 1995

Krishnapillay DB, Marzalina M, Pukittayacamee P, Kijkar S (1992) Cryopreservation of *Dipterocarpus alatus* and *Dipteracarpus intricatus* for long term storage. Poster presented at the 23rd International Seed Testing Association Congress (ISTA), Buenos Aires, 27 Oct–11 Nov 1992. Symposium abstract no. 54, 77 pp

Krishnapillay B, Marzalina M, Alang ZC (1994) Cryopreservation of whole seeds and excised embryos of *Pterocarpus indicus*. J Trop Forest Sci 7(20):313–322

Krishnapillay B, Schroeder C, Marzalina M (1996) Dipterocarp seed biology, handling and storage. Malaysian Forester 59(3):136–152

Krishnapillay B, Tsan FY, Marzalina M, Jayanthi N, Nashatul Zaimah NA (1998) Slow growth as a method to ensure continuous supply of planting material in recalcitrant species. Paper presented at the IUFRO seed symposium 1998, Nikko Hotel, Kuala Lumpur, 12–15 Oct 1998

Krishnapillay B, Marzalina M, Abdul Razak MA (2003) Research in the eco-physiology of tropical recalcitrant seeds-where do we go from here? In: Proceedings of 3 rd national seed symposium, 8–9 Apr 2003. UPM, Serdang, pp 71–77

Lapitan P (2002) Progress and prospects of in vitro conservation. In: Krishnapillay B (ed) Chapter 13: Basic principles of biotechnology and their application in forestry. APAFRI, Kuala Lumpur, pp 109–113

Lee DW, Krishnapillay B, Marzalina M, Haris M, Yap SK (1996) Irradiance and spectral quality affect asian tropical rain forest tree seedling development. Ecology 77(2):568–580

Marzalina M (1995) Penyimpanan Biji Benih Mahogany (Swietenia macrophylla King.). Thesis, Universiti Kebangsaan Malaysia

Marzalina M, Krishnapillay B (2002) Seed procurement and handling. In: Krishnapillay B (ed) A manual for forest plantation establishment in Malaysia. Malayan forest records 45. Forest Research Institute Malaysia, Kepong, pp 33–49

Marzalina M, Nashatul ZNA (2000) Cryopreservation of some Malaysian tropical urban forestry species. In: Engelmann F, Takagi H (eds) Cryopreservation of tropical plant germplasm – current research progress and application. JIRCAS international agriculture series no. 8. IPGRI Publication/Japan International Research Center for Agricultural Sciences, Rome/Tsukuba, pp 465–467

Marzalina M, Wan Tarmeze WA (2010) MYScrim Flagship Project: a FRIM's efforts towards sustainable wood-based industries. Paper presented at ISFFP. Legend Hotel, Kuala Lumpur, 5–6 Oct 2010

Marzalina M, Krishnapillay B, Yap SK (1993) Collecting seed of tropical rain forest tree: problems and solutions. In: Drysdale RM, John SET, Yapa AC (eds) Proceedings of the international conference on genetic conservation and production of tropical forest tree seeds, ASEAN-Canada Forest Tree Seed Centre, Saraburi, Thailand, pp 63–67

Marzalina M, Normah MN, Krishnapillay B (1994) Artificial seeds of *Swietenia macrophylla*. In: Proceedings of national seed symposium II, UPM Serdang, pp 132–134

Marzalina M, Krishnapillay B, Haris M, Siti Asha AB (1995) A possible new technique for seedling storage of recalcitrant species. In Ratnam W, Ahmad Zuhaidi Y, Amir Husni MS, Darus A (eds) Proceedings of the international workshop of Bio-Refor, Kangar, 28 November–1 December 1994. BIO-REFOR/FRIM Publications, KualaLumpur

Marzalina M, Normah MN, Krishnapillay B (1996) Collection and handling of mahogany (*Swietenia macrophylla* King) seeds for optimum viability. J Trop For Sci 9(3):398–410

Marzalina M, Abd Khalim AS, Siti Hasanah MS, Abd Rahman AJ, Kassim A (1998) Mobile seed-seedling chamber: mechanism to maintain germplasm viability over long journey. Paper presented at the IUFRO seed symposium 1998, Nikko Hotel, Kuala Lumpur, 12–15 Oct 1998

Mohd. Afendi H, Marzalina M, Tariq Mubarak H, Wan Tarmeze WA, Nashatul ZNA, Mohd. Nasir H (2011) Tumbuhan Pesisir Pantai – Fenologi & Penyediaan Bahan Tanaman. FRIM Special Publication, 135 pp

Ng FSP (1981) Vegetative and reproductive phenology of diptercarps. Malaysian Forester 44: 197–216

Normah MN, Marzalina M (1996) Achievements and prospects of in vitro conservation for tree germplasm. In: In vitro conservation of plant genetic resources. UKM, Bangi, pp 253–261

Norwati M, Lee SL, Kevin NKS, Lee CT, Siti Salwana H, Ng CH, Tnah LH (2010) DNA marker technologies: it's application to the conservation and sustainable utilisation of Malaysian forest tree species. In: International symposium on forestry and forest products (ISFFP),Legend Hotel, Kuala Lumpur, 5–6 Oct 2010

Palmberg-Lerche C (1994) FAO programmes and activities in support of the conservation and monitoring of genetic resources and biological diversity in forest ecosystems. Invited paper presented at the IUFRO/CIDA symposium on measuring biological diversity in tropical and temperate forests, Chiang Mai, Thailand, 28 Aug–2 Sept 1994

Primack R, Corlett R (2005) Tropical rainforests: an ecological and biogeographical comparison. Blackwell Science, Malden, 74 pp

Sakai C, Subiakto A, Nuroniah HS, Kamata N, Nakamura K (1994) Mass propagation method from the cutting of three dipterocarp species. J For Res 7:73–80

Schlegel FM, Tangan FT (1990) Ex situ conservation of Philippine bamboo species. J Forest Genet Res Info 18:28–31

Suzuki K, Ishii K, Sakirai S, Sasaki S (2006) Plantation technology in tropical forest science. Springer, Tokyo

Theilade I, Petri L (2003) Conservation of tropical trees ex situ through storage and use. Guidelines and technical notes no. 65. Danida Forest Seed Centre, Humlebaek

Theilade I, Luoma-aho T, Rimbawanto A, Nguyen HN, Greijmans M, Nashatul ZNA, SlothP A, Thea S, Burgess S (2006) An overview of the conservation status of potential plantation and restoration species in Southeast Asia. http://www.treeseedfa.org)

Thielges BA, Sastrapradja SD, Rimbawanto A (2001) In situ and ex situ conservation of commercial tropical trees. ITTO Project PD 16/96 Rev 4, 578 pp

Tompsett PB (1992) A review of the literature on storage of dipterocarp seeds. Seed Sci Technol 20:251–267

Towill LE (1985) Low temperature and freeze-/vacuum-drying preservation of pollen. In: Kartha KK (ed) Cryopreservation of plant cells and organs. CRC Press, Boca Raton, pp 171–198

Villalobos VM, Engelmann F (1995) Ex situ conservation of plant germplasm using biotechnology. World J Microb Biotechnol 11:375–382

Werner W (2003) Toasted forests – evergreen rain forests of tropical Asia under drought stress. ZEF – Discussion papers on development policy no. 76, Center for Development Research, Bonn, 46 pp

Withers LA, Englemann F (1997) In vitro conservation of plant genetic resources. In: Altman A (ed) Biotechnology in agriculture. Marcel Dekker, New York, pp 57–88

World Bank (2007) Annual report. http://web.worldbank.org/WBSITE/EXTERNAL/EXTABOUTUS/EXTANNREP/EXTANNREP2K7/0,,menuPK:4078239~pagePK:64168427~piPK:64168435~theSitePK:4077916,00.html

Yap SK, Marzalina M (1990) Gregarious flowering-myth or fact. In: Proceedings of the international conference on forest biology and conservation in Borneo. pp 47–49

Yeager C (2008) Conservation of tropical forests and biological diversity in Indonesia. USAID report no. 118, 83 pp

Chapter 15
The In Vitro Conservation of Plants Native to the Brazilian Amazon and Atlantic Forests

Eduardo da Costa Nunes, Fernanda Kokowicz Pilatti, Cristine Luciana Rescarolli, Thaise Gerber, Erica E. Benson, and Ana Maria Viana

15.1 Introduction

The production of this chapter coincides with the United Nations (UN) International Year of Forests 2011 (IYF-2011) proclaimed by the General Assembly to raise global awareness regarding strengthening the sustainable management, conservation and sustainable development of all types of forest for the benefit of current and future generations (UN-IYF 2011). The subjects of this review are two of the most significant and at risk forest ecosystems on the planet. The Amazon Forest (N Brazil) and the Atlantic Forest (S to NE Atlantic Coast) which comprise respectively ca. 49.29% and 13.04% of Brazilian territory (Instituto Brasileiro de Geografia e Estatística 2011). According to IBGE the Amazon region comprises the Amazon river basin (25,000 km of rivers) with a total area of ca. 6,900,000 km^2 of which 3,800,000 km^2 are in Brazil. The Amazon Forest has the world's highest level of biodiversity, it is one of Brazil's largest areas of forest biomass, comprising a third of the world's tropical humid forests and holding a fifth of the planet's drinkable water. Currently only 82% of the original Amazonian Forest remains, it has been at recurrent risk from deforestation, starting with extensive cattle farming in the 1970s and the expansion of soya plantations in the 1990s. The Amazon hardwood industry has expanded due to the low costs of production from the region, decreases in hard-

E. da Costa Nunes
EPAGRI, Estação Experimental de Urussanga, Caixa Postal 49, Urussanga, SC, 88840-000, Brazil

F.K. Pilatti • C.L. Rescarolli • T. Gerber • A.M. Viana (✉)
Departamento de Botânica, Centro de Ciências Biológicas, Universidade Federal de Santa Catarina, Florianópolis, SC, 88049-070, Brazil
e-mail: amarna@mbox1.ufsc.br

E.E. Benson
Damar Research Scientists, Drum Road, Cupar Muir, Fife, KY155RJ, Scotland, UK

M.N. Normah et al. (eds.), *Conservation of Tropical Plant Species*,
DOI 10.1007/978-1-4614-3776-5_15, © Springer Science+Business Media New York 2013

wood supplied by S Brazil and the demands of a growing national economy. The main causes of deforestation in the Amazon are population growth, due to immigration, expansion of hardwood industries and burning for grazing and agriculture (Instituto Brasileiro de Geografia e Estatística 2011). The impacts of climate change have special significance for the Amazon Forest as a provider of ecosystems services. This was raised by Lewis et al. (2011) who used data measured in 2005 for studying the relationship between dry seasons and forest biomass to predict the effects of the 2010 drought on carbon emissions. Their study predicted the Amazon forest would not absorb its usual billions of metric tons of CO_2 from the atmosphere during 2010–2011, conversely as a result of drought, dead and dying trees would release CO_2 in coming years.

The Atlantic Forest is one of the most endangered biomes on the planet as 70% of the Brazilian population lives in this territory this makes the Atlantic Forest the most endangered of Brazil's ecosystems. It has only 133,010 km^2 of remaining area representing <10% of original forest (Instituto Brasileiro de Geografia e Estatística 2011). The Brazilian conifers are native to the Mixed Ombrophyllous forests of the Atlantic Forest and the meridional forests of Brazil have been devastated since 1920 due to supply and demand for charcoal and wood from the Cerrado and Mid-North areas (Ministério do Meio Ambiente/Ibama 2003).

The goals of the IYF-2011 are to emphasize, at the international level the key role of forests and enhance public participation and awareness as to the importance of forest ecosystems, their biodiversity and the environmental services that they provide. Brazil's Minister of Environment launched the International Year of Forests by highlighting that Brazil is the country with the largest area of tropical forests in the world with the Amazon accounting for about 26% of the world's remaining tropical forests. This review focuses on the application of biotechnological tools for the in vitro conservation of plant species native to the Brazilian Amazon and south Atlantic Forests (Viana et al. 1999; Pilatti et al. 2011). It includes a summary of the use of molecular markers for the assessment of genetic diversity and conservation orientated research in order to enable the in situ, ex situ and in vitro conservation of forest plant genetic diversity. An overview of these technologies has been compiled for native forest plant species focusing on those of present or potential economic value to Brazil. Studies concerning the use of tissue culture and cryostorage for the ex situ conservation of species native to the Brazilian Amazon and Atlantic Forests are few in number. However, protocols are being developed and refined for economically important native taxa and this literature is highlighted. The justification being that techniques such as embryo rescue and micropropagation may be adapted for in vitro conservation, including medium term storage in slow growth and long-term storage in cryobanks. From another perspective in vitro technologies can be used to support the production of plants and forest products, circumventing the need to continually source from wild specimens in native forest habitats (Pilatti et al. 2011).

Over 1.6 billion people are dependent upon forests for their livelihoods; in 2004 the global trade in forest products was estimated to be $327 billion with 30% of forests being used for wood and non-wood forest products (UN-IYF 2011). One of the major applications of biotechnology is in the conservation of economically significant forest species to support the sustainable utilization of diverse forest products. In 2009 Brazilian wood production from native forests was higher than from planted forests, with total national wood production at 122,159,595 m^3, with 87.5% from cultivated forests and 12.5% harvested from native vegetation. Wood obtained from extractivist activities was 7.9% higher than in 2008 (15,248,187 m^3); the highest extractivist wood producers are located in the Amazon states: 39.2% in Para, 8.9% in Rondonia, 7.8% in Amazonas (Instituto Brasileiro de Geografia e Estatística 2011). The Amazon is the main tropical wood producing (28 million m^3) forest in the world with 86% being consumed in Brazil and 14% exported (Sobral et al. 2002).

Deforestation in the region has reached ca. 35 million hectares and one third of this area is degraded. This presents major challenges regarding how to exploit forest products sustainably and restore and establish productive processes in altered areas. Souza et al. (2008) present examples (*Virola surinamensis, Ceiba petandra, Bertholettia excelsa* and *Swietenia macrophylla)* of Amazonian tree species that have undergone irreversible genetic loss due to selective exploitation suggesting that the reforestation of degraded areas with highly valuable commercial wood species may be an option. Their studies show that the introduction of certain tree species (*Bertholettia excelsa, Carapa guianensis, Cedrela odorata, Copaifera multijuga, Hymenaea courbaril)* could help to recover degraded areas of the Amazon region as they adjust to full sunlight following deforestation. Restoration species were selected according to their rate of growth and utility in energy and wood production. Albeit, other biotic factors may need to be considered as *Swietenia macrophylla* had low survival rates due to the attack of the shoot borer *Hypsipyla grandella. Bertholettia excelsa* and *Swietenia macrophylla* are examples of other species on the verge of extinction in their native habitats.

According to INPA (The Amazon National Research Institute) there are ca. 300 species of edible fruits and ca. 10,000 plant species which have potential economic value in the pharmaceutical, cosmetic and biological control industries. Currently the forest products industry of the Amazon region is not economically significant but it has commercial prospects if the activities are adequately planned (Instituto Nacional de Pesquisas da Amazônia 2011). The Amazonian aromatic essences, fruits and oils that are used as raw materials by the chemical industries as cosmetics pharmaceuticals, insecticides and fungicides and many Amazonian species of the Lauraceae (*Aniba*, *Nectandra*, *Ocotea*, *Licaria* and *Dicypellium*) are renewable sources of biological actives. However their exploitation must be undertaken on a sustainable basis as *Aniba rosaeodora* and *Dicypellium caryophyllaceum* are on the red list of the species threatened by extinction. According to Marques (2001) the Lauraceae has 49 genera distributed mainly in the tropical and subtropical areas of the central and South American forests that are potentially important for wood production, essential oils and they are used in folk medicine. The genus *Aniba*, *Ocotea* and *Nectandra* have the highest number of economically important species; Rizzini and Mors (1995) report that the barks of *Ocotea cymbarum, Aniba citrifolia*, Ocotea *longifolia*, Clinostemon *mahuba* and Licaria *rigida* produce 0.1–0.5% yields of essential oils.

In the last century major commercial exploitation in the Amazon region focused on the essential oils of *Aniba rosaeodora* and *Copaifera* ssp. and on the seeds of *Dipteryx odorata* (Associação Brasileira da Indústria de Higiene Pessoal, Perfumaria e Cosméticos 2011). The author reports that, recently, the commercialization of forest products as aromatic sachets by local craft and artisan traders has been demonstrated for *Sagotia racemosa* (Euphorbiaceae), *Aniba fragrans* (Lauraceae) and *Vetiveria zizanioides* (Poaceae). Some species (*Aeollanthus suaveolens* [Lamiaceae], *Aniba canelilla* [Lauraceae], *Cyperus articulatus* [Cyperaceae] *Protium pallidum* [Burseraceae]) yield oils that are used by industries in cooperation with local communities to produce perfumes of commercial value. Extractivism is the main approach used for the commercial exploitation of the Amazonian aromatic species and only a small scale of domesticated production is achieved by local Brazilian producers, therefore, many species are now under pressure due to their intense exploitation, habitat deforestation and burning.

In the case of the south Atlantic Forest, there are several reports on the intensified extraction of medicinal plants leading to drastic reductions of natural populations (Pavan-Fruehauf 2000; Azevedo and Silva 2006); this is exacerbated by a lack of knowledge on reproductive physiology which impairs the development of sustainable propagation. Examples of medicinal species at the verge of extinction in the Atlantic Forest are *Maytenus muelleri*, *Ocotea odorifera* and *Ocotea catharinensis* (Figueiró-Leandro and Citadini-Zanette 2008). The High Montane Araucaria Moist Forest also has a major potential for medicinal plant exploitation with the most cited therapeutics including: antimicrobial, antioxidant, anti-inflammatory, antiviral, antifungal and anaesthetic properties. The most frequently extracted chemical component across all species are essential oils (60% of the species studied) especially from the families Asteraceae and Myrtaceae (Martins-Ramos et al. 2010). Examples of medicinal species native to the Araucaria Forest include: *Ilex paraguariensis*, *Ilex microdonta* (Aquifoliaceae), *Araucaria angustifolia (Araucariaceae), Dicksonia sellowiana (Dicksoniaceae), Mimosa invisa, Mimosa scabrella (Fabaceae), Ocotea pulchella (Lauraceae), Acca sellowiana, Myrceugenia euosma, Myrrhinium atropurpureum (Myrtaceae), Drimys angustifolia (Winteraceae)* and members of the families Asteraceae and Myrtaceae. *Araucaria angustifolia* and *Dicksonia sellowiana* are currently at risk of extinction in their native habitats.

The production of edible fruits and speciality food products is a significant component of the Brazilian forest products industry. The Brazilian Amazon Forest has a high diversity of native tree species that have been commercially exploited for the production of edible fruits for which extractivism was the main approach until the 1970s and until then only *Bertholettia excelsa* and *Euterpe oleracea* were exploited (Carvalho et al. 2003; Carvalho and Müller 2005). These authors report that in the last decade the expansion of cultivated areas and the adequate management of natural populations increased productivity by 100%; currently the expansion of cultivated areas of *Theobroma grandiflorum* was achieved. Other native species (*Platonia insignis*, *Spondias mombim* and *Byrsonima crassifolia)* have been incorporated in local agribusinesses and others show high potential (*Pouteria caimito*, *Rollinia mucosa, Myrciaria dubia, Garcinina mangostana, Bactris gasipaes, Nephelium lappaceum* and *Endopleura uchi)* for commercialisation.

In the South Atlantic Forest, *Euterpe edulis* (heart of palm) is the main species used intensively for commercial exploitation (Conte et al. 2008) it is on the verge of extinction in its natural habitat, as is *Araucaria angustifolia* (Schlögl et al. 2007), which is commercially important at the local level for its edible seeds and wood production.

15.2 Molecular Studies of Brazilian Forest Plant Species Genetic Diversity

Knowledge of the genetic structure and diversity between and within forest tree populations underpins in situ conservation, facilitates germplasm collection for ex situ preservation and informs sustainable management practices related to the exploitation of forest products (Hamrick and Godt 1996; Soares et al. 2008). Assessment of genetic diversity is central to conserving species at risk from non-sustainable exploitation and habitat loss, the consequences of which can rapidly and detrimentally change population structures, reduce biodiversity and ultimately lead to the extinction of local populations. The sustainable cultivation of many Brazilian forest tree species is difficult due to their long and complex life cycles, low germination rates in artificial conditions and low initial productivity. Therefore, it is essential that in situ wild populations are carefully managed and protected and to meet this challenge molecular technologies are employed to characterize tree diversity in natural forest populations and in ex situ germplasm collections. Table 15.1 summarizes the molecular genetic studies of plant taxa native to the Brazilian Amazon and South Atlantic forests.

Examples of how molecular technologies can be applied in forest conservation are provided by Soares et al. (2008), whilst not directed at species from the Amazon they can guide molecular genetic evaluations for other biomes. These researchers applied random amplified polymorphic DNA (RAPD) markers to assess differentiation and spatial patterns of genetic divergence among *Dipteryx alata* populations and found relatively high levels of genetic divergence, populations situated at less spatial distances tended to have slightly greater similarities than was expected by chance alone. The minimum distance between samples needed to assess and conserve genetic diversity with maximum efficiency was estimated. This type of evaluation is important to minimize redundancy within genetic resources collections and preservation areas and to make conservation programmes more cost-effective.

Kageyama et al. (2003) applied genetic markers to study the mating patterns, genetic diversity and structure of natural populations of tropical tree species at different successional stages: (a) pioneer species, (*Trema micrantha, Cecropia pachystachya*); (b) secondary species (*Cedrela fissilis, Cariniana legalis*) and (c) climax species (*Maytenus aquifolium, Esenbeckia leiocarpa*). Their results (population density/hectare) showed no clear pattern for the distribution of genetic diversity among and within populations of species from different successional stages. Thus, *Trema micrantha* and *Cecropia pachystachya* occurred in high densities and *Cedrela fissilis* an endangered species of the Brazilian Atlantic Forest

Table 15.1 Application of molecular tools to study the genetic diversity of plants native to the Amazon and Atlantic Forests

Family	Species/biome	Tools	Molecular assessment and evaluation rationale	References
Annonaceae	*Xylopia brasiliensis* Atlantic Forest	IS	Minimum population size for in situ conservation	Pinto and Carvalho (2004)
Araucariaceae	*Araucaria angustifolia* Atlantic Forest	PCR-RFLP	cpDNA non-coding regions in natural populations	Schlögl et al. (2007)
Arecaceae	*Bactris gasipaes* Amazon	RAPD	Molecular validation of morphologically defined landraces and phylogenetics	Rodrigues et al. (2004) Sousa et al. (2001)
	Elaeis oleifera Amazon	RAPD	Germplasm genetic diversity	Moretzsohn et al. (2002)
	Euterpe oleracea Amazon	RAPD, MS	Genetic differentiation in provenances in the Embrapa Germplasm Bank	Oliveira and Silva (2008) Oliveira et al. (2010)
	Euterpe edulis Atlantic Forest	MS, AZ	Genetic structure and mating systems	Conte et al. (2008)
	Euterpe edulis Atlantic Forest	MS	Forest fragmentation and population structure	Seoane et al. (2005)
	Euterpe edulis Atlantic Forest.	MS	Genetic structure, mating, long-distance gene flow	Gaiotto et al.(2003)
	Euterpe espiritosantensis Atlantic Forest	AZ	Seed differentiation	Martins et al. (2007)
Burseraceae	*Protium spruceanum* Amazon	IZ	Genetic diversity polymorphism studies	Fajardo et al. (2009)
Cecropiaceae	*Cecropia pachistachya* Atlantic Forest	IZ	Successional stages of natural populations	Kageyama et al. (2003)
Celesteraceae	*Maytenus aquifolium* Atlantic Forest	IZ, RAPD	Successional stages of natural populations	Kageyama et al. (2003)
	Maytenus ilicifolia Atlantic Forest	AFLP	Genetic diversity in ex situ germplasm bank accessions	Ribeiro et al. (2010)
	Maytenus aquifolium Atlantic Forest	RAPD	Genetic variability in natural populations	Sahyun et al. (2010)
	Maytenus ilicifolia Maytenus aquifolium Maytenus evonymoidis Atlantic Forest	RAPD	Genetic variability in 18 native populations of *M. ilicifolia*; inter-specific comparisons	Mossi et al. (2009)
Euphorbiaceae	*Hevea brasiliensis* Amazon	MS	Diversity, cross-amplification in wild species	Souza et al. (2009)
Leguminosae	*Hymenaea courbaril* var. *stilbocarpa* Atlantic Forest	MS	Genetic diversity analysis to determine the effective population size in a germplasm bank, recommendations for conservation	Feres et al. (2009)
	Hymenaea courbaril var. *stilbocarpa* Atlantic Forest	MS	Isolation and characterization of loci and transferability to *H. stigonocarpa*	Ciampi et al. (2008)

	Shizolobium parahyba (Vell.) Blake Atlantic Forest	RAPD	Structure, level, distribution of genetic variation within and among five populations	Freire et al. (2007)
	Copaifera langsdorffii Amazon, Atlantic Forest	MS	Population genetic structure in forest fragments	Martins et al.(2008)
	Caesalpinia echinata Atlantic Forest	AZ	Variability and genetic structure	Neto et al. (2005)
	Caesalpinia echinata Atlantic Forest	MS	Studies on mating system, gene flow, population structure and paternity in natural populations	Melo et al. (2007)
Lauraceae	*Ocotea catharinensis* Atlantic Forest	AZ	Genetic structure/diversity in four natural populations	Tarazi et al. (2010)
Lecythidaceae	*Cariniana estrellensis* Atlantic Forest	MS	Characterization of 15 MS loci for *C. estrellensis* and transferability to *C. legalis*	Guidugli et al. (2009)
	Cariniana legalis Atlantic Forest	IZ	Successional stages of natural populations	Kageyama et al. (2003)
Melastomataceae	*Tibouchina papyrus* Cerrado	RAPD	Genetic structure in two natural populations	Telles et al. (2010)
Meliaceae	*Swietenia macrophylla King* Amazon	MS	Genetic studies	Lemes et al. (2002)
	Swietenia macrophylla King Amazon	MS	Genetic structure of populations	Lemes et al. (2003)
	Swietenia macrophylla King Amazon	MS	Evaluation of mating systems in a logged population and implications for the management	Lemes et al. (2007a)
	Swietenia macrophylla King Amazon	cpDNA, MS	Phylogeographic structures from Amazon and Central America related to timber certification	Lemes et al.(2010)
	Carapa guianensis Amazon	MS	Fragmented population structure	Dayanandan et al. (1999)
	Carapa guianensis Amazon	MS	Genetic diversity and population structure	Raposo et al. (2007)
	Trichillia pallida Atlantic Forest	RAPD	Population structures	Zimback et al. (2004)
	Cedrela fissilis Atlantic Forest	MS	Genetic diversity and successional stages of natural populations	Kageyama et al. (2003)
Myrtaceae	*Eugenia uniflora L* Atlantic Forest	MS	Genetic analysis of three populations from forest fragments	Ferreira-Ramos et al. (2008)

(continued)

Table 15.1 (continued)

Family	Species/biome	Tools	Molecular assessment and evaluation rationale	References
	Eugenia involucrata Atlantic Forest			Golle (2010)
Piperaceae	*Piper hispidinervum C. DC.* Amazon	RAPD	Genetic structure and mating systems	Wadt and Kageyama (2004)
Rutaceae	*Esenbeckia leiocarpa* Atlantic Forest	IZ, RAPD	Successional stages of natural populations	Kageyama et al. (2003)
Sapotaceae	*Manilkara huberi* Amazon	MS	Spatial distribution, genetic structure and mating systems	Azevedo et al. (2007)
Sterculiaceae	*Theobroma grandiflorum* Amazon	MS	Genetic divergence, inbreeding in natural populations	Alves et al. (2007)
	Theobroma grandiflorum Amazon	MS	Mating systems in natural populations	Alves et al. (2003)
	Theobroma cacao Amazon	MS	Development, characterization, validation and mapping of SSRs and ESTs interactions	Lima et al. (2010)
	Theobroma grandiflorum, T. subincanum, T. sylvestre Amazon	MS	Cross-amplification, characterization of microsatellite loci	Lemes et al. (2007b)
Ulmaceae	*Trema micrantha* Atlantic Forest	IZ, RAPD	Genetic diversity and successional stages of natural populations	Kageyama et al. (2003)
Velloziaceae	*Vellozia gigantea* Atlantic Forest	MS	Genetic diversity at successional stages of natural populations	Lousada et al. (2011)

AFLP amplified fragment length polymorphism, *AZ* allozyme, *IZ* isozymes, *MS* microsatellite, *RAPD* random amplification of polymorphic DNA, *SSR* simple sequence repeats, *RFLP* restriction fragment length polymorphism, *PCR* polymerase chain reaction, *cpDNA* chloroplast DNA, *EST* expressed sequence tag and a valuable fast growing timber tree was found in low densities due to selective exploitation. *Cariniana legalis*, one of the largest trees (3 m in diameter, 40 m high) from the tropical humid forests of South America occurred at low densities as did *Maytenus aquifolium*, a native from the Brazilian Atlantic Forest and the semi-deciduous tree *Esenbeckia leiocarpa*. The genetic structure of species was associated with patterns of reproduction and demography within populations, those with higher population densities showed smaller genetic divergence among populations than species with higher selfing rates and lower population densities. Notably, *Cariniana ianeirensis* and *Cariniana parvifolia* are now extinct in the Atlantic Forest. The molecular marker technologies (Table 15.1) for conserving plant species native to the Amazon and Atlantic Forests are summarized, with examples as follows:

1. *Assessment of the genetic diversity of genebanks accessions.*
   Native palm tree *Euterpe oleracea, Elaeis oleifera* (Oliveira and Silva 2008; Oliveira et al. 2010; Moretzsohn et al. 2002); *Hymenaea courbaril* var. *stilbocarpa* (Feres et al. 2009) and *Maytenus ilicifoliun* (Ribeiro et al. 2010).
2. *Assessing impacts of fragmentation on natural population genetic structures.*
   Euterpe edulis (Seoane et al. 2005), *Eugenia uniflora* (Ferreira-Ramos et al. 2008), *Copaifera langsdorffii* (Martins et al. 2008), *Carapa guianensis* (Dayanandan et al. 1999); for an intensively logged forest, *Swietenia macrophylla* (Lemes et al. 2007a).
3. *Studying phylogenetic relationships amongst morphologically defined landraces.*
   Bactris gasipaes (Rodrigues et al. 2004; Sousa et al. 2001); phylogeographic structures of natural Amazonian and Central American populations of *Swietenia macrophylla* (Lemes et al. 2010).
4. *Studies of loci transferability.*
   Hymenaea courbaril var. *stilbocarpa* to *H. stignocarpa* (Ciampi et al. 2008); *Cariniana estrellensis* to *Cariniana legalis* (Guidugli et al. 2009)
5. *Studies of genetic structure, gene flow, mating, diversity of natural populations.*
   Comprehensive examples are shown in Table 15.1.

The studies on genetic structures of natural populations using molecular markers developed for a representative number of economically important species from the Amazon and Atlantic Forest is especially important for the development of strategies for the in situ conservation of species on the verge of extinction (e.g. *Araucaria angustifolia, Caesalpinia echinata, Euterpe edulis, Swietenia macrophylla* and *Ocotea catharinensis*). Molecular biomarkers applied to ex situ conservation programmes will enable the preservation and sustainable utilization of Brazil's native forest plants. The authors have reviewed (Pilatti et al. 2011) the status of in vitro conservation and cryopreservation methodologies as applied to plant biodiversity representing Brazil's six major biomes (the Amazon, Atlantic Forest, Cerrado, Caatinga and Pampa). Whilst focusing on the Amazon and Atlantic forests the same principles apply regarding their application for the protection of at risk species and including using insights gained from molecular marker studies.

15.3 Potential for Applying In Vitro Conservation to Native Forest Plant Species

Although there are numerous examples of in vitro techniques being applied to Brazilian Amazon and Atlantic forestry species their main focus concerns economically significant species (Tables 15.2 and 15.3). Examples include: Araucariaceae (Barbedo and Bilia 1998), Arecaceae (Carvalho and Muller 1998; Martins et al. 1999, 2000), Lauraceae (Carvalho et al. 2008), Myrtaceae (Carvalho et al. 2006; Delgado and Barbedo 2007) and Fabaceae (Faria et al. 2004) these taxa are of special relevance as they produce recalcitrant seeds and would benefit from in vitro conservation. Significantly, *Araucaria angustifolia, Euterpe edulis, Caesalpinia echinata, Ocotea catharinensisa* and *Ocotea odorifera* are on the verge of extinction in the Atlantic Forest, whilst *Aniba rosaeodora* and *Swietenia macrophylla* are at similar risk in the Amazon Forest.

The main motivation for adopting in vitro programmes in the Brazilian forestry sector is for forest products research and to enhance the sustainable production of wood and non-timber forest products, including the genetic characterization, manipulation and improvement of Brazil's native forest plant species. In contrast, there are few examples (e.g. Silva and Scherwinski-Pereira 2011; Higa 2006; Higa et al. 2011; Aguiar 2010; Nunes et al. 2002, 2003, 2007; Pilatti et al. 2011; Santa-Catarina et al. 2004; Catarina et al. 2005; Viana et al. 1999; Viana and Mantell 1999) where in vitro protocols and cryopreservation have been purposely created for forest genetic resources conservation. To facilitate their wider application it will be crucial to use regimes that complement in situ conservation practices *and* that maintain the totipotency and stability of tissue culture and seed-derived genetic resources held in in vitro genebanks/cryobanks.

15.3.1 Selecting Appropriate Regimes for Native Forest Plant In Vitro Conservation

Whilst most in vitro technologies applied to Brazil's native forest plants were developed in the framework of the forest products industries they provide a useful knowledge base as to how at risk species might respond to tissue culture (Tables 15.2 and 15.3). Nevertheless, for conservation purposes the decision to apply relevant technologies (e.g. embryo rescue, micropropagation, slow growth, and cryopreservation) will need to be based on criteria that justify using in vitro and cryopreservation practices in preference to more traditional approaches. These include: (a) a limitation in the use of existing (traditional) in situ and ex situ conservation methods and storage recalcitrance; (b) in vitro propagation and conservation helping to offset sampling from wild populations and (c) supporting sustainable measures associated with the exploitation of native plants.

From a biological perspective in vitro conservation and cryopreservation can also be justified on the basis that candidate species are highly vulnerable and their proliferation in the wild is compromised by one or more of the following factors: (a) dispersed, fragmented and small populations; (b) limited sexual reproduction, likely to be exacerbated by a lack of pollinators and seed dispersers; (c) low/infrequent seed production/viability and idiosyncratic storage behaviour and (d) limited or no capacity to proliferate vegetatively (asexually) in situ. To obtain a sufficient number of new plants for sustainable practices, conservation and restoration it is necessary to increase wild plant populations and viable seed production, in which case micropropagation, slow growth and cryopreservation of whole seeds, zygotic embryos/embryonic axes and vegetative propagules can help to safeguard genetic resources and produce new plants under controlled conditions. When choosing in vitro and cryostorage methods for conservation purposes, particularly for native species, it is precautionary to avoid manipulations that predispose cultures to genetic instability. In the case of micropropagation both organogenic and somatic embryogenic pathways provide mechanisms for the rapid proliferation of new plants, but they do have disadvantages as dedifferentiation and adventitious proliferation carries higher risks of genetic instability. Somaclonal variation (SCV) which is the heritable variation arising in high frequency in plants regenerated from tissue culture is associated with adventitious shooting and dedifferentiated (callus) proliferation (Scowcroft 1984). Therefore, when tissue culture is used for conservation purposes non-adventitious routes of proliferation (e.g. nodal shoot micropropagation) are usually recommended. Where this is not possible reducing the risks of instability and confirmation of trueness-to-type using morphological descriptors and genetic stability studies are advisable (Scowcroft 1984; Harding 2004; Harding et al. 2000, 2009). Somatic embryogenesis remains an option so long as precautions are taken that cultures are maintained in a dedifferentiated state for minimal periods and the morphogenetic pathway is without an intervening callus stage are preferentially manipulated to produce embryos via a direct route. It is also important to note that certain culture practices and genotype effects can influence the predilection to SCV (Scowcroft 1984; Harding 2004).

Where there is a paucity of technical know-how concerning culture manipulations it is necessary to use approaches that do not compromise a species by further narrowing its genetic base through clonal propagation. Applying biomolecular markers and knowledge of genetic diversity (Table 15.1) can help to inform in vitro conservation programmes (Tables 15.2 and 15.3) by monitoring the risks of: (1) clonal narrowing of the genetic base and (2) generating genetic instability via adventitious/callogenic manipulations (Scowcroft 1984; Harding 2004; Harding et al. 2009).

15.3.2 In Vitro Culture: Case Studies for Economically Significant Forest Plants

The following case studies demonstrate progress in tissue culture techniques as applied to economically important plant families native to the Brazilian Amazon and Atlantic Forests. They are collated with a view to demonstrating how a knowledge base developed for biotechnological purposes might be used to enable the in vitro conservation of native forest species.

Table 15.2 In vitro culture and conservation of native Brazilian plant species from the Amazon Forest

Family	Species	Explant	Technique	References
Annonaceae	*Rollinia mucosa*	Nursery grown seedlings	Micropropagation	Figueiredo et al. 2001
Apocynaceae	*Hancornia speciosa*	Seeds	In vitro germination	Pinheiro et al. (2001)
	Hancornia speciosa	Apical and nodal segments	Micropropagation	Grigoletto (1997)
	Hancornia speciosa	Nodal segments	Direct organogenesis	Soares et al. (2007)
		Micro-shoots	In vitro storage 6 months	Biondo et al. (2007)
Arecaceae	*Euterpe oleracea*	Zygotic embryos	In vitro germination and seedling establishment	Ledo et al.(2001)
Clusiaceae	*Hypericum brasiliense*	Axillary buds	In vitro germination	Velloso et al. (2009)
Fabaceae	*Schizolobium parahyba* var. *amazonicum*	Seeds	In vitro germination, callus	Reis et al. (2007)
Lauraceae	*Aniba rosaeodora*	Seeds, seedlings	Zygotic embryo culture, apical and axillary bud culture	Handa et al. (2005)
	Aniba rosaeodora	Apical and nodal segments from greenhouse grown seedlings	Micropropagation	Jardim et al. (2010)
Meliaceae	*Swietenia macrophylla*	Seeds, seedlings	In vitro germination, callus	Couto (2002)
Piperaceae	*Piper hispidinervum*	Leaf segments	Callus culture	Costa and Pereira (2005)
Rubiaceae	*Piper aduncum*	Seeds, seedlings	Callus culture	Pereira et al. (2008)
	Uncaria tomentosa	In vitro shoots	Slow growth	Silva and Scherwinski-Pereira (2011)
	Uncaria guianensis	In vitro shoots	Slow growth	Silva and Scherwinski-Pereira (2011)
Sterculiaceae	*Theobroma grandiflorum* × *T. obovatum*	Zygotic embryos, cotyledons	Callus culture	Venturieri and Venturieri (2004)
	Theobroma grandiflorum	Hypocotyl segments	Callus culture	Ferreira et al. (2004)

Table 15.3 In vitro culture and conservation of native Brazilian plant species from the Atlantic Forest

Family	Species	Explant	Technique	References
Apocynaceae	*Aspidosperma polyneuron*	Nodal segments	Micropropagation	Ribas et al. (2005)
	Hancornia speciosa	Seeds	In vitro germination	Pinheiro et al. (2001)
	Hancornia speciosa	Apical and nodal segments	Micropropagation	Grigoletto (1997)
	Hancornia speciosa	Nodal segments	Direct organogenesis	Soares et al. (2007)
Aquifoliaceae	*Ilex paraguariensis*	Microshoots	Ex vitro rooting	Quadros (2009)
		Seeds	Zygotic embryo culture	Hu (1975)
Araucariaceae	*Araucaria angustifolia*	Immature zygotic embryos	Somatic embryogenesis	Astarita and Guerra (2000)
Arecaceae	*Euterpe edulis*	Immature zygotic embryos	Somatic embryogenesis	Saldanha (2007)
		Immature zygotic embryos	Zygotic embryo culture	Saldanha (2007)
Bignoniaceae	*Tabebuia heptaphylla*	Immature zygotic embryos	Callus culture	Higa (2006)
		Seedling nodal segments	Shoot culture, in vitro conservation	Higa (2006)
	Tabebuia avellanedae	Seedling nodal segments	Shoot culture, callus culture	Aguiar (2010)
Boraginaceae	*Cordia trichotoma*	Seeds	In vitro germination	Fick et al. (2007)
Bromeliaceae	*Cryptanthus sinuosus*	Stem-root axis from greenhouse plants	Micropropagation	Arrabal et al. (2002)
	Vriesea reitzii	Young basal leaves	Micropropagation	Alves et al. (2006)
Fabaceae	*Caesalpinia echinata*	Leaf segments from greenhouse grown seedlings	Callus culture, embryogenesis	Werner et al. (2009, 2010)
Lauraceae	*Ocotea catharinensis*	Zygotic embryos	Somatic embryogenesis	Viana and Mantell (1999); Catarina et al. (2005)
		Air-dehydrated mature somatic embryos	Storage at 25°C in sealed petri dishes for 3 months	Santa-Catarina et al. (2004)
	Ocotea odorifera	Zygotic embryos	Cell culture	Maraschin et al. (2009)
	Ocotea odorifera	Zygotic embryos	In vitro germination, somatic embryogenesis	Santa-Catarina et al. (2001)
Meliaceae	*Cedrela fissilis*	Seeds, seedling nodal segments	In vitro germination, micropropagation	Nunes et al. (2002)

(continued)

Table 15.3 (continued)

Family	Species	Explant	Technique	References
	Cedrela fissilis	Alginate-encapsulated nodal segments	In vitro storage at 25°C for 6 months	Nunes et al. (2003)
Myrtaceae	*Eugenia pyriformis*	Seeds, seedling nodal segments	In vitro germination, micropropagation	Nascimento et al. (2008)
	Myrciaria aureana	Zygotic embryos	Somatic embryogenesis	Motoike et al. (2007)
Quillajaceae	*Quillaja brasiliensis*	Seedling nodal segments	Micropropagation	Fleck et al. (2009)

15.3.2.1 Araucariaceae

Araucaria angustifolia is a native Brazilian Coniferae in danger of extinction, for which Astarita and Guerra (2000) developed embryogenic cell suspension culture using immature zygotic embryos as the explant. Cell masses were cultured on liquid LP medium (Quorin et al. 1977) with 2,4-D (2,4-dichlorophenoxyacetic acid), kinetin and BAP (6-benzylaminopurine). Santos et al. (2002) similarly cultured *Araucaria angustifolia* using zygotic embryos excised from immature seeds and LP culture medium with or without 2,4-D, BAP and kinetin, they achieved genotype dependent induction rates of 14.2–15.5%. Globular and torpedo stage somatic embryos were obtained by adding PEG 3350, maltose (6% and 9%) and 1 μM BAP and Kinetin.

15.3.2.2 Arecaceae

Euterpe oleraceae zygotic embryos were isolated from mature seeds and complete seedlings were produced on MS (Murashige and Skoog 1962) medium with 0.6% (w/v) agar, 0.25% (w/v) activated charcoal, 3% (w/v) sucrose and 2.68 μM NAA (α-naphthaleneacetic acid) combined with 1.11–2.22 μM BAP (Ledo et al. 2001). Saldanha (2007) induced somatic embryogenesis in *Euterpe edulis* using immature zygotic embryos cultured on MS medium with 30–40% (w/v) sucrose; embryogenic cultures developed from the cotyledonary node of immature zygotic embryos and complete plantlets were produced.

15.3.2.3 Fabaceae

Schizolobium parahyba var. *amazoniucum*, a fast growing tree, is native of the Amazon forest and one of the most important forestation species in the region. Reis et al. (2007) germinated seeds of this species and produced normal seedlings on hormone free MS medium; callus was proliferated on medium with IBA (indole-3-butyric acid) and BAP. *Mimosa caesalpiniaefolia* a native species from NE Brazil is intensively exploited as a multipurpose tree and has a wide range of genetic variability. Its propagation by seed is not easy and micropropagation is an important tool for the multiplication of selected genotypes. Oliveira et al. (2007) developed a micropropagation protocol that can now be used for both germplasm conservation and multiplication. *Mimosa balduinii and Mimosa catharinensis* are native to the Atlantic Forest in extinction.

Werner et al. (2010) induced callus from *Caesalpinia echinata* leaf explants using 2 year-old green house plants as donors. Callus was successfully initiated on MS, B5 (Gamborg and Eveleigh 1968) and White's (White 1963) medium with optimal proliferation occurring on MS medium with 2.4 g/L ammonium nitrate and

1.35 g/L glutamine and 4.11 g/L potassium nitrate and 1 mg/L 2,4-D and 5 mg/L BAP. The combination of 0.5 mg/L 2,4-D and 5 mg/L BAP or 0.5 mg/IBA and 5 mg/L BAP with 2.4 g/L of ammonium nitrate enhanced and stimulated callus growth. The friable callus produced had high levels of phenolics but was non-embryogenic, although it comprised meristematic areas indicating that an optimized system may support somatic embryogenesis or organogenic development via adventitious buds and root primordia.

15.3.2.4 Lauraceae

Ocotea catharinensis Mez. (Lauraceae) is an endangered native tree of the Atlantic Forest its seeds demonstrate erratic and short viability characteristics therefore conservation measures are required because the species is important for the production of biologically active compounds (lignans, neolignans) and essential oils for the perfume and the cosmetic industries (Funasaki et al. 2009). Protocols for somatic embryogenesis (Viana and Mantell 1999; Catarina et al. 2005) and somatic embryo in vitro conservation (Santa-Catarina et al. 2004) were reported. Cotyledonary and mature somatic embryos air-dehydrated in sealed Petri dishes showed repetitive embryogenesis after 3 months storage at 25°C. This offers the possibility of using in vitro techniques for continuous, reliable somatic embryo production and short-term germplasm storage.

Santa-Catarina et al. (2001) developed protocols for in vitro germination of zygotic embryos of *Ocotea odorifera* on MS medium with 4.4 μM BAP, 3% (w/v) sucrose, 0.6% (w/v) agar and 0.15% (w/v) activated charcoal. Somatic embryogenic cultures were induced with immature zygotic embryos culture on MS medium with 72–144 μM 2,4-D and cultures maintained on MS medium without growth regulators.

Aniba rosaeodora is an economically important native tree from the Amazon forest for which its oil is used in the cosmetics industries; it can be seed and vegetatively propagated using root cuttings, however seed production is low and irregular. Handa et al. (2005) developed a protocol for zygotic embryo culture which produced a 53% germination rate, embryos were cultured on MS medium with 5% (w/v) sucrose and 1.8 g/L Phytagel. The survival of apical and axillary bud cultures was 48% on MS medium with 300 mg/L Agrymicin, 3% (w/v) sucrose and 0.7% (w/v) agar. Buds were excised from seedlings grown in the greenhouse and antibiotics were necessary to control endogenous contamination in explants. Jardim et al. (2010) developed a protocol for in vitro plant production using apical and nodal segments of greenhouse-grown seedlings on culture medium with 4 mg/L BAP and 6 mg/L IAA (indole-3-acetic acid), this culture regime supported shoot production and MS medium with 3 mg/L NAA stimulated rooting on shoot apices and nodal segments.

15.3.2.5 Meliaceae

Swietenia macrophylla a native tree from the Amazon is on the verge of extinction in the wild and has been intensively exploited for its valuable hardwood. Couto (2002) created a protocol using in vitro seed germination, callus culture and shoot formation from pre-existing buds. Shoot production was achieved when hypocotyl segments comprising original axillary buds were cultured on half-strength MS medium with BAP.

Nunes et al. (2002, 2003) reported the establishment of a consistent micropropagation protocol for *Cedrela fissilis* and a successful protocol for plant recovery and rooting from calcium-alginate encapsulated shoot tips, cotyledonary and epicotyledonary nodal segments stored on substrate containing 0.4% (w/v) agar, at 25°C. This protocol is used for the short term (6 months) in vitro storage of *C. fissilis*.

15.3.2.6 Myrtaceae

Eugenia pyriformis is a native species from the Atlantic Forest that produces valuable hardwood and edible fruits; although seedling production is restricted by limited seed storage longevity. Nascimento et al. (2008) developed protocols for in vitro germination, shoot proliferation, rooting and plantlet acclimatization. Shoot proliferation was achieved using nodal segments excised from axenic seedlings cultured on WPM (Woody Plant Medium, Lloyd and McCown 1981) medium with 1.0 mg/L BAP, 30 g/L sucrose and 0.7% (w/v) agar, to induce rooting WPM was with 1.0 mg/L IBA.

Myrciaria aureana (Brazilian grape tree) is a popular fruit tree native to the Atlantic Forest, of the nine species in this genus that are native to Brazil some are in danger of extinction. Motoike et al. (2007) formulated a somatic embryo production protocol for this species using zygotic embryo cotyledons cultured on medium with 2,4-D and indole-3 acetyl-L-aspartic acid. Several embryos germinated, but they did not develop apical shoots.

15.3.2.7 Piperaceae

Piper hispidinervum is important for safrole production and is commercially used by the cosmetics and insecticide industries (Wadt 2001); it has not been domesticated and there are many constraints regarding alternative methods of propagation. Costa and Pereira (2005) report callus induction and cell suspension establishment by initiating leaf explants on MS medium containing 5 mg/L NAA. A successful micropropagation protocol was developed using nodal explant from axenic seedlings germinated in vitro on MS medium cultured on WPM with 1 mg/L BAP.

Silva and Scherwinski-Pereira (2011) evaluated the slow-growth storage of the two Amazonian pepper species *Piper aduncum* and *Piper hispidinervum* with the objective of optimizing the best conditions for plant maintenance and recovery. Shoots were stored at reduced temperatures (10°C, 20°C and 25°C) on medium with osmotic agents (sucrose at 1% or 2% w/v and mannitol at 3% w/v) and abscisic acid (0, 0.5, 1.0, 2.0 and 3.0 mg/L), following 6 months of storage shoots were examined for survival and regrowth. Low temperature (20°C) was the most effective for the conserving the species; the presence of osmotica and abscisic acid was deleterious as shown by the low rate of recovered shoots.

15.3.2.8 Rubiaceae

Pereira et al. (2008) applied in vitro methods for the cultivation of *Uncaria tomentosa* and *Uncaria guianensis*. These are woody vines native to the Amazon and at risk of extinction due to the intensive harvesting and the increasing deforestation of their natural habitat. Callus was induced from leaf segments derived from glasshouse grown plants initiated on MS medium containing 1.5 mg/L picloram, 6 g/L agar and 30 g/L sucrose. In vitro plantlets were also induced from aseptic seeds cultured on MS medium with 30 g/L sucrose and 6 g/L agar, for *U. guianensis*, and on WPM medium with 1.0 mg/L of BAP, for *U. tomentosa*. Callus and plants were produced in a short time frame with alkaloid profiles similar to wild-grown plants.

15.3.2.9 Sterculiaceae

The culture of *Theobroma grandiflorum* is motivated by its economic importance for juice production and as a possible substitute for tree-grown cacao. Ferreira et al. (2004) tested protocols using media with several levels of auxin. Callus was initiated from embryonic axes and cotyledons placed on MS medium with 1 and 2 mg/L 2,4-D but the culture failed to regenerate plants. The cotyledonary and embryonic axis explants were further divided in the plumule, radicle and hypocotyl sections; the latter was found more responsive, forming friable callus, but did not result in somatic embryogenesis.

A *Theobroma grandiflorum* × *T. obovatum* hybrid demonstrated resistance to the witches-broom-disease, motivating the formulation of in vitro protocols. Venturieri and Venturieri (2004) produced callus from the seed tegument, cotyledons and embryos; cotyledon explants showed higher frequencies of callus development in the presence of TDZ.

15.3.3 Future Challenges: Developing and Harmonizing Cryobank Practices

Cryopreservation for the conservation of forestry genetic resources was pioneered by the temperate forestry community where it underpins commercial production and modern breeding programmes (Rajora and Mosseler 2001). Cryopreservation's application for the conservation of tropical forest species is considered by other contributors to this book. With respect to preserving Brazil's forest genetic resources, cryopreservation whilst considered important, remains limited (Viana et al. 1999; Pilatti et al. 2011) with most emphasis being placed on the traditional preservation of seed germplasm. However, cryostorage is also being explored as a complementary approach for preserving seed germplasm; an example is *Tabebuia heptaphylla* a hard wood tree of the south Atlantic Forest which produces orthodox seeds that are unable to withstand long-term storage in conventional seed banks. This highlights the unusual seed storage behaviour of certain Brazilian forest species and the need to develop alternative cryo-conservation strategies.

Higa et al. (2011) cryopreserved seeds of *T. heptaphylla* by direct immersion in liquid nitrogen achieving a germination range of 54–67%; no additional cryoprotective measures were required for seeds within the moisture content range of 0.08–0.09 $gH_2O\ gdw^{-1}$. This study also considered the logistical requirement of transferring thawed seeds to end users and the importance of harmonizing the technical aspects of cryobanking with down-stream processes. Seeds of *Cedrela fissilis*, a native of the Brazilian Atlantic Forest were cryopreserved, achieving 100% recovery after direct immersion in liquid nitrogen (Nunes et al. 2003). Pilatti et al. (2011) report the cryostorage of seeds representing members of the *Cecropiaceae*, *Bignoniaceae* and *Meliaceae (C. fissilis)* and achieving post-storage germination ranging from 20% to 84%.

Therefore, as well as optimizing protocols for vegetative germplasm propagated in vitro, future challenges will concern developing cryopreservation protocols, not only for recalcitrant seeds, but for seeds that display anomalous storage behaviours that do not conform to the usual classifications (Higa et al. 2011). A similar rationale applies to Brazilian forest species that have irregular or idiosyncratic reproductive cycles, or that are at high risk in their natural environments. In these cases in vitro preservation and cryostorage provide alternative and complementary options to traditional ex situ and in situ conservation strategies.

15.4 Conclusions

In vitro technologies offer the potential for conserving and sustainably utilizing plant species native to Brazil's Amazon and south Atlantic Forests. This chapter highlights the importance of integrating two biotechnological approaches, in which molecular markers are used for the study of genetics and diversity within the framework of in vitro and cryo-conservation technologies. Molecular techniques can inform evidence-based decisions regarding the selection of suitable genotypes for conservation, and help to ensure that the risks of SCV are minimised as a wide genetic base as possible is maintained when clonal propagation is applied. The conservation of plants native to Brazil's tropical and sub-tropical forests is a considerable task and it will benefit greatly from the timely integration of in vitro and cryostorage technologies with traditional conservation measures. Future challenges will be the development of routine micropropagation, slow growth and cryo-conservation protocols specifically designed for the purpose of conservation, applying these to a wider range of species and prioritizing those that are critically at risk in their native habitat.

Acknowledgments The authors acknowledge the support of the International Foundation for Science (Sweden), The British Council (UK), Capes (Ministry of Education, Brazil) and CNPq (National Research Council, Brazil).

References

Aguiar T (2010) Conservação de sementes zigóticas e cultura in vitro de espécies de *Tabebuia* Gomes Ex DC (Bignoniaceae). Dissertation, Universidade Federal de Santa Catarina

Alves RM, Artero AS, Sebbenn AM, Figueira A (2003) Mating system in a natural population of *Theobroma grandiflorum* (Willd. Ex. Spreng.) Schum., by microsatellite markers. Genet Mol Biol 26:373–379

Alves GM, Dal Vesco LL, Guerra MP (2006) Micropropagation of the Brazilian endemic bromeliad *Vriesea reitzii* through nodule clusters culture. Sci Hortic 110:204–207

Alves RM, Sebbenn AM, Artero AS, Clement C, Figueira A (2007) High levels of genetic divergence and inbreeding. Tree Genet Genome 3:289–298

Arrabal R, Amancio F, Carneiro LA, Neves LJ, Mansur E (2002) Micropropagation of endangered endemic Brazilian bromeliad *Cryptanthus sinuosus* (L. B. Smith) for in vitro preservation. Biodivers Conserv J 11:1081–1089

Associação Brasileira da Indústria de Higiene Pessoal, Perfumaria e Cosméticos (2011) http://www.abihpec.org.br/conteudo/UFPA/JOSEMAIA.pdf. Accessed 20 July 2011

Astarita LV, Guerra MP (2000) Conditioning of the culture medium by suspension cells and formation of somatic proembryos in *Araucaria angustifolia* (Coniferae). In Vitro Cell Dev Biol Plant 36:194–200

Azevedo SKS, Silva IM (2006) Plantas medicinais e de usos religioso comercializadas em mercados e feiras livres no Rio de Janeiro, RJ, Brazil. Acta Bot Bras 20(1):185–194. doi:10.1590/S0102-33062006000100016

Azevedo VCR, Kanashiro M, Ciampi AY, Grattapaglia D (2007) Genetic structure and mating system of *Manilkara huberi* (Ducke) A. Chev., a heavily logged Amazonian timber species. J Hered 98(7):646–654

Barbedo CJ, Bilia DAC (1998) Evolution of research on recalcitrant seeds. Sci Agric. doi:10.1590/s0103-90161998000500022

Biondo R, Souza AV, Bertoni BW, Soares AM, França SC, Pereira MAS (2007) Micropropagation, seed propagation and germplasm bank of *Mandevilla velutina* (Mart.) Woodson. Sci Agric 64(3):263–268

Carvalho JEU, Muller CH (1998) Níveis de tolerância e letal de umidade em sementes de pupunheira *Bactris gasipaes*. Rev Bras Frutic 20(3):238–283

Carvalho LR, Silva EAA, Davide AC (2006) Classificação de sementes florestais quanto ao comportamento no armazenamento. Rev Bras Sementes. doi:10.1590/s0101-31222006000200003328

Carvalho LR, Davide AC, Silva EAA, Carvalho MLM (2008) Classificação de sementes de espécies florestais dos gêneros Nectandra e Ocotea (Lauraceae) quanto ao comportamento no armazenamento. Rev Bras Sementes. doi:10.1590/s0101-31222008000100001

Carvalho JEU, Nazaré RFR, Nascimento WMA (2003) Características físicas e físico-químicas de um tipo de bacuri (*Platonia insignis* Mart.) com rendimento industrial superior. Rev Bras Frutic 25(2):326–328. doi:10.1590/S0100-29452003000200036

Carvalho JEU, Müller CA (2005) Biometria e rendimento Percentual de polpa de frutas da Amazônia. Comunicado Técnico 139, Embrapa. http://infoteca.cnptia.embrapa.br/bitstream/doc/404792/1/com.tec.139.pdf

Catarina CS, Moser JR, Bouzon ZL, Floh EIS, Maraschin M, Viana AM (2005) Protocol of somatic embryogenesis: *Ocotea chatarinensis* Mez (Lauraceae). In: Jain SM, Gupta PK (eds) Protocol for somatic embryogenesis in woody plants. Kluwer, Dordrecht, pp 427–443

Ciampi AY, Azevedo VCR, Gaiotto FA, Ramos ACS, Lovato MB (2008) Isolation and characterization of microsatellite loci for *Hymenaea courbaril* and transferability to *Hymenaea stigonocarpa*, two tropical timber species. Mol Ecol Res. doi:10.1111/j.1755-0998.2008.02159.x

Conte R, Reis MS, Mantovani A, Vencovsky R (2008) Genetic structure and mating system of *Euterpe edulis* Mart. populations: a comparative analysis using microsatellite and allozyme markers. J Hered 99(5):476–482. doi:10.1093/jhered/esn055

Costa FHS, Pereira JES (2005) Seleção de auxinas para a indução de calos friáveis em *Piper hispidinervum* visando o estabelecimento de cultivo de células em suspensão. Hortic Bras 23(2):656

Couto JMF (2002) Germination and morphogenesis in vitro of mahogany (*Swietenia macrophylla* King). Dissertation, Universidade Federal de Viçosa

Dayanandan S, Dole J, Bawa K, Kesseli R (1999) Population structure delineated with microsatellite markers in fragmented populations of a tropical tree, *Carapa guianenesis* (Meliaceae). Mol Ecol 8:1585–1592

Delgado LF, Barbedo CJ (2007) Desiccation tolerance of seeds of species of *Eugenia*. Pesq Agropec Bras. doi:10.1590/s0100-204x2007000200016

Fajardo CG, Vieira FA, Morais VN, Maracajá PB, Carvalho D (2009) Isozymes polymorphism in *Protium spruceanum* (Benth.) Engler (Burseraceae) as base for genetic diversity studies. Revista Verde 4(4):27–32

Faria JMR, Van Lammeren AAM, Hilhorst HWM (2004) Desiccation sensitivity and cell cycle aspects in seeds of *Inga vera* subsp. affinis. Seed Sci Res 14:165–178

Feres JM, Guidugli MC, Mestriner MA, Sebbenn AM, Ciampi AY, Alzate-Marin AL (2009) Microsatellite diversity and effective population size in germplasm bank of *Hymenaeae courbaril* var. *stilbocarpa* (Leguminosae), an endangered tropical tree: recommendations for conservation. Genet Resour Crop Evol 56:797–807

Ferreira MGR, Cárdenas FEN, Carvalho CHS, Carneiro AA, Filho CFD (2004) Indução de calos embriogênicos em explantes de cupuaçuzeiro. Rev Bras Frutic 26(2):372–374. doi:10.1590/S0100-29452004000200048

Ferreira-Ramos R, Laborda PR, Santos MO, Mayor MS, Mestriner MA, Souza AP, Alzate-Marin AL (2008) Genetic analysis of forest species *Eugenia uniflora* L. through newly developed SSR markers. Conserv Genet 9:1281–1285

Fick TA, Bisognin DA, Quadros KM, Horbach M, Reiniger LRS (2007) In vitro establishment and growth of louro-pardo (*Cordia trichotoma*) seedlings. Ciência Florestal 17(4):343–349

Figueiredo SFL, Albarello N, Viana VRC (2001) Micropropaation of *Rollinia mucosa* (Jacq.) Baill. In Vitro Cell Dev Biol Plant 37:471–475

Figueiró-Leandro ACB, Citadini-Zanette V (2008) Árvores medicinais de um fragmento florestal urbano do município de Criciúma, Santa Catarina. Brasil Rev Bras Pl Med 10(2):56–67

Fleck JD, Schwambach J, Almeida ME, Yendo ACA, Costa F, Gosmann G, Fett-Neto AG (2009) Immunoadjuvant saponin production in seedlings and micropropagated plants of *Quillaja brasiliensis*. In Vitro Cell Dev Biol Plant 45:715–720

Freire JM, Piña-Rodrigues FCM, Lima ER, Sodré SR, Corrêa RX (2007) Genetic structure of *Schizolobium parahyba* (Vell.) Blake (guapuruvu) populations by RAPD markers. Sci For 74:27–35

Funasaki M, Lordello AL, Santa-Catarina C, Viana AM, Floh EIS, Yoshida M, Kato MJ (2009) Neolignanas and sesquiterpenes from leaves and embryogenic cultures *of Ocotea catharinensis* (Lauraceae). J Braz Chem Soc 20:853–859

Gaiotto FA, Grattapaglia D, Vencovsky R (2003) Genetic structure, mating system, and long-distance gene flow in heart of palm (*Euterpe edulis* Mart.). J Hered 94(5):399–406. doi:10.1093/jhered/esg087

Gamborg OL, Eveleigh DE (1968) Culture methods and detection of glucanases in suspension cultures of wheat and barley. Can J Biochem 46:417–421

Golle DP (2010) Estabelecimento, multiplicação, calogênese, organogênese in vitro e análise da diversidade genética em acessos de *Eugenia involucrata* DC. Ph.D. thesis, Universidade Federal de Santa Maria

Grigoletto ER (1997) Micropropagação de *Hancornia speciosa* Gómez (Mangabeira). Dissertation, Universidade Estadual de Brasília

Guidugli MC, Campos T, Sousa ACBS, Feres JM, Sebbenn AM, Mestriner MA, Contel EPBC, Alzate-Marin AL (2009) Development and characterization of 15 microsatellite loci for *Cariniana estrellensis* and transferability to *Cariniana legalis*, two endangered tropical tree species. Conserv Genet 10:1001–1004

Hamrick JL, Godt MJW (1996) Conservation genetics of endemic plant species. In: Avise JC, Hamrick JL (eds) Conservation genetics. Chapman & Hall, New York, pp 281–304

Handa L, Sampaio PTB, Quisen RC (2005) Cultura in vitro de embriões e de gemas de mudas de pau-rosa (*Aniba rosaeodora* Ducke). Acta Amazônica 35(1):29–33

Harding K (2004) Genetic integrity of cryopreserved plant cells: a review. CryoLetters 25:3–22

Harding K, Johnston JW, Benson EE (2009) Exploring the physiological basis of cryopreservation success and failure in clonally propagated in vitro crop plant germplasm. Agric Food Sci 18:3–16

Harding K, Marzalina M, Krishnapillay B, Nashatul Z, Normah MN, Benson EE (2000) Molecular stability assessments of trees regenerated from cryopreserved mahogany seed germplasm using non-radioactive techniques to examine chromatin structure and DNA methylation status of the ribosomal genes. J Trop For Sci 12(1):149–163

Higa TC (2006) Estudos morfogenéticos e conservação in vitro de *Tabebuia heptaphylla* (Vellozo) Toledo (Bignoniaceae). Dissertation, Universidade Federal de Santa Catarina

Higa TC, Paulilo MTS, Benson EE, Pedrotti E, Viana AM (2011) Developing seed cryobanking strategies for *Tabebuia heptaphylla* (Bignoniaceae), a hardwood tree of the Brazilian South Atlantic Forest. CryoLetters 32:329–338

Hu CY (1975) In vitro culture of rudimentary embryos of eleven *Ilex species*. J Am Soc Hortic Sci 100(3):221–225

Instituto Brasileiro de Geografia e Estatística (2011) http://www.ibge.gov.br/home/presidencia/noticias/noticia_visualiza.php?id_noticia=169; http://www.ibge.gov.br/home/presidencia/noticias/noticia_visualiza.php?id_noticia=799; http://www.ibge.gov.br/home/noticias/noticia_visualiza.phd?id_noticia=1703&id_pagina=1; http://www.ibge.gov.br/home/presidencia/noticias/noticia_visualiza.php?id_noticia=1760&id_pagina=1. Accessed 31 Jul 2011

Instituto Nacional de Pesquisas da Amazônia (2011) http://projetos.inpa.gov.br. Accessed 20 Jul 2011

Jardim LS, Sampaio PTB, Costa SSC, Gonçalves CQB, Brandão HLM (2010) Effect of different growth regulators on in vitro propagation of *Aniba rosaeodora* Jucke. Acta Amazonica 40(2):275–279. doi:10.1590/s0044-59672010000200005

Kageyama PY, Sebbenn AM, Ribas LA, Gandara FB, Castellen M, Perecim MB, Vencovsky R (2003) Genetic diversity in tropical tree species from different successional stages determined with genetic markers. Sci For 64:93–107

Ledo AS, Lameira AO, Benbadis AK, Menezes IC, Ledo CAS, Oliveira MSP (2001) Cultura in vitro de embriões zigóticos de açaizeiro. Rev Bras Frutic 23(3):468–472

Lemes MR, Brondani RPV, Grattapaglia D (2002) Multiplexed systems of microsatellite markers for genetic analysis of mahogany, *Swietenia macrophylla* King (Meliaceae), a threatened neotropical timber species. J Hered 93(4):287–291

Lemes MR, Gibel R, Proctor J, Grattapaglia D (2003) Population genetic structure of mahogany (*Swietenia macrophylla* King, Meliaceae) across the Brazilian Amazon, based on variation at microsatellite loci: implications for conservation. Mol Ecol 12:2875–2883

Lemes MR, Grattapaglia D, Grogan J, Proctor J, Gribel R (2007a) Flexible mating system in a logged population of *Swietenia macrophylla* King (Meliaceae): implications for the management of a threatened neotropical tree species. Plant Ecol 192:169–179. doi:10.1007/s11258-007-9322-9

Lemes MR, Martiniano TM, Reis VM, Faria CP, Gribel R (2007b) Cross-amplification and characterization of microsatellite loci for three species of Theobroma (Sterculiaceae) from Brazilian Amazon. Genet Resour Crop Evol 54:1653–1657

Lemes MR, Dick CW, Navarro C, Lowe AJ, Cavers S, Gribel R (2010) Chloroplast DNA microsatellites reveal contrasting phylogeographic structure in mahogany (*Swietenia macrophylla* King, Meliaceae) from Amazonia and Central America. Trop Plant Biol 3:40–49. doi:10.1007/s12041-010-9042-5

Lewis SL, Brando PM, Phillips OL et al (2011) The 2010 Amazon drought. Science 331(6017):554. doi:10.1126/science.1200807

Lima LS, Gramacho KP, Pires JL, Clement D, Lopes UV, Carels N, Gesteria AS, Gaiotto FA, Cascardo JCM, Micheli F (2010) Development, characterization, validation, and mapping of SSRs derived from *Theobroma cacao* L. – *Moniliophthora perniciosa* interaction ESTs. Tree Genet Genome 6:663–676

Lloyd G, McCown B (1981) Commercially feasible micropropagation of mountain laurel, *Kalmia latifolia* by use of shoot tip culture. Int Plant Prop Soc Proc 30:421–427

Lousada JM, Borb EL, Ribeiro KT, Ribeiro LC, Lovato MB (2011) Genetic structure and variability of the endemic and vulnerable *Vellozia gigantea* (Velloziaceae) associated with the landscape in the Espinhaço Range, in southeastern Brazil: implications for conservation. Genetica. doi:10.1007/s10709-011-9561-5

Maraschin M, Pedrotti E, Viana AM, Wood K (2009) Metabolomic analysis of *Ocotea odorifera* cell cultures – a model protocol for acquiring metabolite data. In: Jain SM, Saxena PK (eds) Springer protocols, methods in molecular biology, protocols for in vitro cultures and secondary metabolite analysis of aromatic medicinal plants. Humana Press, New York, pp 347–358

Marques CA (2001) Economic importance of the family Lauraceae Lindl. Floresta e Ambiente 8(1):195–206

Martins CC, Nakagawa J, Bovi ML (1999) Tolerância à dessecação de sementes de palmito-vermelho (*Euterpe espiritosantensis* Fernandes). Rev Bras Bot. doi:10.1590/s0100-84041999000030000722

Martins CC, Nakagawa J, Bovi MLA (2000) Desiccation tolerance of four seed lots from *Euterpe edulis* Mart. Seed Sci Technol 28(1):101–113

Martins CC, Bovi MLA, Mori ES, Nakagawa J (2007) Isoenzimas na diferenciação de sementes de três espécies do gênero Euterpe. Rev Árvore 31(1):51–57

Martins K, Santos JD, Gaiotto FA, Moreno MA, Kageyama PY (2008) Estrutura genética populacional de *Copaifera langsdorffii* Desf. (Leguminosae – Caesalpinioideae) em fragmentos florestais do Pontal do Paranapanema, SP, Brasil. Rev Bras Bot 31(1):61–69

Martins-Ramos K, Bortoluzzi RLC, Mantovani A (2010) Plantas medicinais de um remanescente de Floresta Ombrófila Mista Altomontana, Urupema, Santa Catarina, Brasil. Rev Bras Pl Med 12(3):380–397

Melo SCO, Gaiotto FA, Cupertino FB, Corrêa RX, Reis AMM, Grattapaglia D, Brondani RPV (2007) Microsatellite markers for *Caesalpinia echinata* Lam. (Brazilwood), a tree that named a country. Conserv Genet 8:1269–1271

Ministério do Meio Ambiente/Ibama (2003) Desmatamento. Informativo Técnico 1:1–93

Moretzsohn MC, Ferreira MA, Amaral ZPS, Coelho PJA, Grattapaglia D, Ferreira ME (2002) Genetic diversity of Brazilian oil palm (*Elaeis oleifera* H.B.K.) germplasm collected in the Amazon Forest. Euphytica 124:35–45

Mossi AJ, Cansian RL, Leontiev-Orlov O, Chechet JL, Carvalho AZ, Toniazzo G, Echeverrigaray S (2009) Genetic diversity and conservation of native populations of *Maytenus ilicifolia* Mart. Ex Reiss Braz J Biol 69(2):447–453

Motoike SY, Saraiva ES, Ventrella MCO, Silva CV, Salomão LCC (2007) Somatic embryogenesis of *Myrciaria aureana* (Brazilian grape tree). Plant Cell Tiss Org Cult 89:75–81

Murashige T, Skoog F (1962) A revised medium for rapid growth and bioassays with tobacco tissue cultures. Physiol Plant 15:473–497

Nascimento AC, Paiva R, Nogueira RC, Porto JMP, Nogueira GF, Soares FP (2008) BAP e AIB no cultivo in vitro de *Eugenia pyriformis* Cambess. Rev Acad Ciênc Agr Ambient 6(2):223–228

Neto JDG, Sebbenn AM, Kageyama PY (2005) Sistema de reprodução em *Caesalpinia echinata* Lam. Implantada em arboreto experimental. Rev Bras Bot 28(2):409–418. doi:10.1590/S0100-84042005000200019

Nunes EC, Castilho CV, Moreno FN, Viana AM (2002) In vitro culture of *Cedrela fissilis* Vellozo (Meliaceae). Plant Cell Tiss Org Cult 70:259–268

Nunes EC, Benson EE, Oltramari AC, Araujo PS, Moser JR, Viana AM (2003) In vitro conservation of *Cedrela fissilis* Vellozo (Meliaceae), a native tree of the Brazilian Atlantic Forest. Biodivers Conserv J 12:837–848

Nunes EC, Laudano WLS, Moreno FN, Castilho CV, Mioto P, Sampaio FL, Bortoluzi JH, Benson EE, Pizolatti MG, Carasek E, Viana AM (2007) Micropropagation of *Cedrela fissilis* Vell. (Meliaceae). In: Jain SM, Häggmann H (eds) Protocols for micropropagation of woody trees and fruits. Springer, Dordrecht, pp 221–235

Oliveira FFM, Silva KMB, Oliveira GF, Dantas IM, Camacho RGV (2007) Micropropagação de *Mimosa caesalpiniaefolia* Benth. A partir de segmentos nodais e ápices caulinares. Rev Caatinga 20(3):152–159

Oliveira MP, Silva KJD (2008) Genetic differentiation in provenances of açaí tree by RAPD and SSR markers. Rev Bras Frutic 30(2):438–443

Oliveira MSP, Santos JB, Amorim EP, Ferreira DF (2010) Genetic variability among accessions of assai palm based on microsatellite markers. Ciênc Agrotec 34(5):1253–1260

Pavan-Fruehauf S (2000) Plantas medicinais da Mata Atlântica: manejo sustentado e amostragem. Annablume/FAPESP, São Paulo, 216 p

Pereira RCA, Valente LMM, Pinto JEBP, Bertolucci SKV, Bezerra GM, Alves FF, Santos PFP, Benevides PJC, Siani AC, Rosario SL, Mazzei JL, Avila LA, Gomes LNF, Aquino-Neto FR, Emmerick ICM, Carvalhaes SF (2008) In vitro cultivated *Uncaria tomentosa* and *Uncaria guianensis* with determination of the pentacyclic oxindole alkaloid contents and profiles. J Braz Chem Soc 19(6):1193–1200

Pilatti FK, Aguiar T, Simões T et al (2011) In vitro and cryogenic preservation of plant biodiversity in Brazil. In Vitro Cell Dev Biol Plant 47:82–98

Pinheiro CSR, Medeiros DN, Macedo CEC, Alloufa MAI (2001) Germinação in vitro de mangabeira (*Hancornia speciosa* Gomez) em diferentes meios de cultura. Rev Bras Frutic 23(2):413–416

Pinto SIC, Carvalho D (2004) Estrutura genética de populações de pindaíba (*Xylopia brasiliensis* Sprengel) por isoenzimas. Rev Bras Bot 27(3):597–605

Quadros KM (2009) Propagação vegetativa de erva-mate (*Ilex paraguariensis* Saint Hilaire). Dissertation, Universidade Federal de Santa Maria

Quorin MP, Lepoivre P, Boxus P (1977) Un premier bilan de 10 années de recherches sur les cultures de méristèmes et la multiplication in vitro de fruitiers ligneux. In: Compte Rendu des Recherches 1976–1977 et Rapports de Synthèse, Stat. Cult. Fruit. Et Maraich., Gembloux, pp 93–117

Rajora O, Mosseler A (2001) Challenges and opportunities for conservation of forest genetic resources. Euphytica 118:197–212

Raposo A, Martins K, Ciampi AY, Wadt LHO, Veasey EA (2007) Genetic diversity structure of crabwood in Baixo Acre, Brazil. Pesq Agropec Bras 42(9):1291–1298

Reis INRS, Lameira AO, Cordeiro IMCC (2007) Indução de calogênese em Paricá (*Schizolobium parahyba* var. amazonicum (Huber ex Ducke) Barneby) através da adição de AIB e BAP. Rev Bras Biociências 5(2):501–503

Ribas LLF, Zanette F, Kulchetscki L, Guerra MP (2005) Micropropagation of *Aspidosperma polyneuron* from single node culture of juvenile material. Rev Árvore 29(4). http://dx.doi.org/10.1590/S0100-67622005000400003

Ribeiro MV, Bianchi VJ, Rodrigues ICS, Mariot MP, Barbieri RL, Peters JA, Braga EJB (2010) Genetic diversity of espinheira santa (*Maytenus ilicifolia* Mart. ex Reis.) accessions collected in Rio Grande do Sul State, Brazil. Rev Bras Pl Med 12(4):443–451

Rizzini CT, Mors WB (1995) Botânica Econômica Brasileira. Âmbito Cultural Edições Ltda, Rio de Janeiro, 241 p

Rodrigues D, Filho AS, Clement CR (2004) Molecular marker-mediated validation of morphologically defined landraces of pejibaye (*Bactris gasipaes*) and their phylogenetic relationships. Genet Resour Crop Evol 51:871–882

Saldanha CW (2007) Conservação in vitro de *Euterpe edulis* Martius através da embriogênese somática. Dissertation, Universidade Federal de Santa Maria

Santa-Catarina C, Olmedo AS, Meyer GA, Macedo J, Amorim W, Viana AM (2004) Repetitive somatic embryogenesis of *Ocotea catharinensis* Mez. (Lauraceae): effect of somatic embryo developmental stage and dehydration. Plant Cell Tiss Org Cult 78:55–62

Santa-Catarina C, Maciel SC, Pedrotti EL (2001) Germinação in vitro e embriogênese somática a partir de embriões imaturos de canela sassafrás (*Ocotea odorífera* Mez). Rev Bras Bot 24(4):501–510, http://www.scielo.br/pdf/rbb/v24n4s0/9471.pdf

Santos ALW, Silveira V, Steiner N, Vidor M, Guerra MP (2002) Somatic embryogenesis in Paraná Pine (*Araucaria angustifolia* (Bert.) O. Kuntze). Braz Arch Biol Technol 45(1):97–106

Sahyun AS, Ruas EA, Ruas CF, Medri C, Souza JRP, Johansson LAPS, Miranda LV, Ruas PM (2010) Genetic variability of three natural populations of *Maytenus aquifolium* (Celesteraceae) from Telêmaco Borba, Paraná, Brazil. Braz Arch Biol Technol 53(5):1037–1042

Schlögl OS, Souza AP, Nodari RO (2007) PCR-RFLP of non-coding regions of cpDNA in *Araucaria angustifolia* (Bert.) O. Kuntze. Genet Mol Biol 30(2):423–427

Scowcroft WR (1984) Genetic variability in tissue culture: impact on germplasm conservation and utilization, International Board for Plant Genetic Resources. Report (AGPG: IBPGR/84/152), Rome

Seoane CES, Kageyama PY, Ribeiro A, Matias R, Reis MS, Bawa K, Sebbenn AM (2005) Efeitos da fragmentação florestal sobre a imigração de sementes e a estrutura genética temporal de populações de *Euterpe edulis* Mart. Rev Inst Flor São Paulo 17(1):25–43

Silva TL, Scherwinski-Pereira JE (2011) In vitro conservation of *Piper aduncum* and *Piper hispidinervum* under slow-growth conditions. Pesq Agropec Bras Brasília 46:384–389

Soares FP, Paiva R, Alavarenga AA, Nogueira RC, Emrich EB, Martinotto C (2007) Direct organogenesis in nodal explants of mangabeira (*Hancornia speciosa* Gomes). Ciênc Agrotec. doi:10.1590/s1413-70542007000400016

Soares TN, Chaves LJ, Telles MPC, Diniz-Filho JAF, Resende LV (2008) Landscape conservation genetics of *Dipteryx alata* ("baru" tree: Fabaceae) from Cerrado region of central Brazil. Genetica 132(1):9–19. doi:10.1007/s10709-007-9144-7

Sobral L, Veríssimo A, Lima E, Azevedo T, Smeraldi R (2002) Acertando o alvo 2: consumo de madeira amazônica e certificação florestal no Estado de São Paulo. Imazon, Belém, 72 p. http://www.imazon.or.gr/publicações/livretos/acertando-o-alvo-2-consumo-de-madeira-amazonica-e/at_download/file

Souza CR, Lima RMB, Azevedo CP, Rossi LMB (2008) Desempenho de espécies florestais para usos múltiplo da Amazônia. Sci For 36(77):7–14

Souza LM, Mantello CC, Santos MO, Gonçalves PS, Souza AP (2009) Microsatellites from rubber tree (*Hevea brasiliensis*) for genetic diversity analysis and cross-amplification in six Hevea wild species. Conserv Genet Resour 1:75–79. doi:10.1007/sl 2686-009-9018-7

Sousa NR, Rodrigues DP, Clement CR, Nagao EO, Astolfi-Filho S (2001) Discriminação de raças primitivas de pupunha (*Bactris gasipaes*) na Amazônia brasileira por meio de marcadores moleculares (RAPDS). Acta Amazônica 31(4):539–545

Tarazi R, Mantovani A, Reis MS (2010) Fine scale spatial genetic structure and allozymic diversity in natural populations of *Ocotea catharinensis* Mez. (Lauraceae). Conserv Genet 11:965–976

Telles MPC, Silva SP, Ramos JR, Soares TN, Melo DB, Resende LV, Batista EC, Vasconcellos BF (2010) Estrutura genética em populações naturais de *Tibouchina papyrus* (pau-papel) em áreas de campo rupestre no cerrado. Rev Bras Bot 33(2):291–300

United Nations International Year of Forests (2011) Fact sheets. What is the international year of forests; forests and people: a historical relationship. www.un.org/forests. Accessed 27 Jul 2011

Velloso MAL, Abreu IN, Mazzafera P (2009) Indução de metabólitos secundários em plântulas de *Hypericum brasiliense* Choisy crescendo in vitro. Acta Amazônica 39(2):267–272

Venturieri GA, Venturieri GC (2004) Calogênese do híbrido *Theobroma grandiflorum* × *T. obovatum* (Sterculiaceae). Acta Amazonica 34(4):507–511, http://www.scielo.br/pdf/aa/v34n4a04.pdf

Viana AM, Mazza C, Mantell SH (1999) Applications of biotechnology for the conservation and sustainable exploitation of plants from Brazilian rain forests. In: Benson E (ed) Plant conservation biotechnology. Taylor & Francis, London, pp 277–299

Viana AM, Mantell SH (1999) Somatic embryogenesis of *Ocotea catharinensis*: an endangered tree of Mata Atlantica (S. Brazil). In: Jain SM, Gupta P, Newton R (eds) Somatic embryogenesis in woody plants. Kluwer, Dordrecht, pp 3–30

Wadt LHO (2001) Estrutura genética de populações naturais de pimenta longa (*Piper hispidinervum* C. DC.) visando seu uso e conservação. Ph.D. thesis, Escola Superior de Agricultura Luiz de Queiroz, Universidade de São Paulo

Wadt LHO, Kageyama PY (2004) Genetic structure and mating system of *Piper hispidinervum*. Pesq Agropec Bras 39(2):151–157

Werner ET, Milanez CRD, Mengarda LHG, Vendrame WA, Cuzzuol GRF (2010) Meios de cultura, reguladores de crescimento e fontes de nitrogênio na regulação da calogênese do pau-brasil (*Caesalpinia echinata* Lam.). Acta Bot Bras 24(4):1046–1051. doi:10.1590/S0102-33062010000400019

Werner ET, Cuzzuol GRF, Pessoti KV, Lopes FP, Roger JA (2009) Controle da calogênese do pau-brasil in vitro. Rev Árvore 33(6):987–996. doi:10.1590/S0100-67622009000500020

White PR (1963) The cultivation of animal and plant cells, 2nd edn. Ronald Press, New York

Zimback L, Mori ES, Kageyama PY, Veiga RFA, Junior JRSM (2004) Genetic structure of *Trichilia pallida* Swartz (Meliaceae) populations by RAPD markers. Sci For 65:114–119

Chapter 16
Ex Situ Conservation of Plant Genetic Resources of Major Vegetables

Andreas Wilhelm Ebert

16.1 Introduction

Vegetables form a large and very diverse commodity group comprising a wide range of genera and species. They may be defined as the fresh edible portion of a herbaceous plant—roots, stems, leaves, flowers or fruit (Encyclopaedia Britannica 2011) or as usually succulent plants or plant parts that are consumed as a side dish with starchy staples (Siemonsma and Piluek 1994; Grubben and Denton 2004). Vegetables are an important component of a healthy diet, providing vitamins, antioxidants, minerals, fiber, amino acids and other health-promoting compounds for nutritional security (Tenkouano 2011).

The assignment of crops into the vegetable commodity group is often not easy. Green chilies and peppers are eaten fresh in relatively large quantities and are therefore considered vegetables, while milled chili products clearly fall into the category of spices and condiments. Similarly, garlic is treated as a vegetable in Asia, given its significant consumption in Asian dishes, while in other parts of the world it would be considered as a spice or medicinal plant. Some leguminous species grown mainly for their dry seeds constitute an important source of vegetables when harvested and consumed at the immature stage, such as immature seeds, green pods and/or leaves. This applies for vegetable soybean (*Glycine*), common bean (*Phaseolus*), winged bean (*Psophocarpus*), garden pea (*Pisum*), mungbean (*Vigna radiata*), Azuki beans (*Vigna angularis*), cowpea and yard-long bean (*Vigna unguiculata*), and black gram (*Vigna mungo*). Cassava leaves (*Manihot esculenta*) are an important leafy vegetable in many countries and should, therefore, be considered as such, although its edible starchy root constitutes the primary use of this crop. Among the Cucurbitaceae family, the group of melons and watermelons would fall under the category of fruit from the consumption point of view, but because of the annual growth pattern and

A.W. Ebert (✉)
AVRDC - The World Vegetable Center, P.O. Box 42, Shanhua, Tainan 74199, Taiwan
e-mail: andreas.ebert@worldveg.org

M.N. Normah et al. (eds.), *Conservation of Tropical Plant Species*,
DOI 10.1007/978-1-4614-3776-5_16, © Springer Science+Business Media New York 2013

type of cultivation of the plants, which are similar to the squashes and pumpkins, these are classed as vegetables in the Food and Agriculture Organization's (FAO) Production Statistics.

The FAO Statistics (FAOSTAT 2011) comprise a group of 27 commodities under the grouping of vegetables and melons with a total world production of over one billion tonnes in 2009. These are, in order of worldwide production in 2009: (1) fresh vegetables, not elsewhere specified (248.6 million tonnes); (2) tomatoes (153 million tonnes); (3) watermelons (98 million tonnes); (4) dry onions (73.2 million tonnes); (5) cabbages and other brassicas (64.3 million tonnes); (6) cucumbers and gherkins (60.6 million tonnes); (7) eggplants or aubergines (42.9 million tonnes); (8) carrots and turnips (33.6 million tonnes); (9) green chilies and peppers (28.1 million tonnes); (10) other melons, including cantaloupes (25.5 million tonnes); (11) lettuce and chicory (24.3 million tonnes); (12) garlic (22.3 million tonnes); (13) pumpkins, squash and gourds (22.1 million tonnes); (14) cauliflowers and broccoli (19.9 million tonnes); (15) spinach (19.6 million tonnes); (16) green beans (19 million tonnes); (17) green peas (16 million tonnes); (18) green maize (9.2 million tonnes); (19) asparagus (7.3 million tonnes); (20) mushrooms and truffles (6.5 million tonnes); (21) okra (6.5 million tonnes); (22) green onions, including shallots (3.7 million tonnes); (23) leeks and other alliaceous vegetables (2.1 million tonnes); (24) string beans (2 million tonnes); (25) artichokes (1.5 million tonnes); (26) leguminous vegetables, not elsewhere specified (1.4 million tonnes); and (27) cassava leaves (68,850 tonnes).

The production of vegetables including melons increased significantly in recent years (1999–2008) in area from 42.8 to 53.7 million ha, respectively—an increase of 26%—and in total production from 696 million tonnes in 1999 to 932 million tonnes in 2008, an even more pronounced increase of 34% (FAOSTAT 2011). Yield increase per unit area reached 6.7% during the same period, which is at least partially attributable to genetic gain due to advances in vegetable breeding.

With production of 522.7 million tonnes in 2009, China was by far the largest vegetable and melon producer accounting for close to 52% of world production, followed by India (9.2%), the United States of America (3.7%), Turkey, (2.6%), and Egypt (2%) (FAOSTAT 2011). Other major players in vegetable and melon production with a global share below 2% are: Iran (6th), Russian Federation (7th), Italy (8th), Spain (9th), and Republic of Korea (10th). The five largest producer countries worldwide of the 12 major vegetable crops are presented in Table 16.1. In all cases, China is the top producer with a world production share ranging from 29.7% in the case of tomatoes to 80.6% for garlic.

Given the vast array of vegetable crops grown worldwide, only a select group of vegetables is described in this chapter, namely the genera *Allium*, *Brassica*, *Raphanus*, *Capsicum*, *Solanum* section *Lycopersicon*, as well as African and Asian eggplant of the genus *Solanum*. Significant collections of all major vegetable crops are presented in Table 16.2 for reference. Among many of the described vegetable crops there are no clear boundaries in terms of climate zones. Many of

Table 16.1 Production of selected vegetable commodities in leading countries and the world in 2009

Rank	Commodity	Item code	Leading country/world	Rank	Production (tonnes)	%
1	Vegetables fresh (not elsewhere specified)	463	China	1	148,876,084	59.9
			India	2	28,006,300	11.3
			Vietnam	3	6,313,390	2.5
			Nigeria	4	4,536,380	1.8
			Philippines	5	4,382,810	1.8
			Others		56,476,917	22.7
			World		248,591,881	100
2	Tomatoes	388	China	1	45,365,543	29.7
			United States of America	2	14,141,900	9.2
			India	3	11,148,800	7.3
			Turkey	4	10,745,600	7.0
			Egypt	5	10,000,000	6.5
			Others		61,554,272	40.3
			World		152,956,115	100
3	Watermelons	567	China	1	65,002,319	66.3
			Turkey	2	3,810,210	3.9
			Iran (Islamic Republic of)	3	3,074,580	3.1
			Brazil	4	2,056,310	2.1
			United States of America	5	1,819,890	1.9
			Others		22,284,638	22.7
			World		98,047,947	100
4	Onions, dry	403	China	1	21,046,969	28.7
			India	2	13,900,000	19.0
			United States of America	3	3,400,560	4.6
			Turkey	4	1,849,580	2.5
			Egypt	5	1,800,000	2.5
			Others		31,234,721	42.7
			World		73,231,830	100
5	Cabbages and other brassicas	358	China	1	30,215,327	47.0
			India	2	6,869,600	10.7
			Russian Federation	3	3,312,090	5.2
			Republic of Korea	4	3,100,000	4.8
			Ukraine	5	1,509,300	2.3
			Others		19,320,440	30.0
			World		64,326,757	100
6	Cucumbers and gherkins	397	China	1	44,250,182	73.1

(continued)

Table 16.1 (continued)

Rank	Commodity	Item code	Leading country/world	Rank	Production (tonnes)	%
			Turkey	2	1,735,010	2.9
			Iran (Islamic Republic of)	3	1,603,740	2.6
			Russian Federation	4	1,132,730	1.8
			United States of America	5	888,180	1.5
			Others		10,945,730	18.1
			World		60,555,572	100
7	Eggplants (aubergines)	399	China	1	25,912,524	60.3
			India	2	10,377,600	24.2
			Egypt	3	1,250,000	2.9
			Iran (Islamic Republic of)	4	862,159	2.0
			Turkey	5	816,134	1.9
			Others		3,725,795	8.7
			World		42,944,212	100
8	Carrots and turnips	426	China	1	15,168,351	45.2
			Russian Federation	2	1,518,650	4.5
			United States of America	3	1,304,150	3.9
			Uzbekistan	4	995,000	3.0
			Poland	5	913,304	2.7
			Others		13,682,270	40.7
			World		33,581,725	100
9	Chillies and peppers, green	401	China	1	14,520,301	51.7
			Mexico	2	1,941,560	6.9
			Turkey	3	1,837,000	6.6
			Indonesia	4	1,100,000	3.9
			Spain	5	1,011,700	3.6
			Others		7,660,290	27.3
			World		28,070,851	100
10	Other melons (including cantaloupes)	568	China	1	12,224,801	47.9
			Turkey	2	1,679,190	6.6
			Iran (Islamic Republic of)	3	1,278,540	5.0
			United States of America	4	1,069,980	4.2
			Spain	5	1,007,000	4.0
			Others		8,243,184	32.3
			World		25,502,695	100
11	Lettuce and chicory	372	China	1	12,855,211	52.8

(continued)

Table 16.1 (continued)

Rank	Commodity	Item code	Leading country/world	Rank	Production (tonnes)	%
			United States of America	2	4,104,440	16.9
			Italy	3	945,800	3.9
			India	4	927,349	3.8
			Spain	5	875,000	3.6
			Others		4,619,517	19.0
			World		24,327,317	100
12	Garlic	406	China	1	17,967,857	80.6
			India	2	1,070,000	4.8
			Republic of Korea	3	380,000	1.7
			Russian Federation	4	227,270	1.0
			Myanmar	5	200,000	0.9
			Others		2,436,933	11.0
			World		22,282,060	100

Source: derived from data supplied by United Nations Food and Agriculture Organization, FAOStat (June 2011) http://faostat.fao.org/site/567/DesktopDefault.aspx?PageID=567#ancor

Table 16.2 Vegetable germplasm collections by major crop groups, genebank and country

		Genebank			Accessions	
Crop grouping	Genus	*Institute code*	*Acronym*	Country	*No.*	*(%)*
Allium	*Allium*	IND1457	NRCOG	India	2,050	7
Allium	*Allium*	RUS001	VIR	Russian Federation	1,888	6
Allium	*Allium*	JPN003	NIAS	Japan	1,352	5
Allium	*Allium*	USA003	NE9	USA	1,304	4
Allium	*Allium*	DEU146	IPK	Germany	1,264	4
Allium	*Allium*	GBR004	RBG	United Kingdom	1,100	4
Allium	*Allium*	TWN001	AVRDC	Taiwan	1,082	4
Allium	*Allium*		Others (168)	Others	19,858	66
Allium	***Allium***		**Total**	**World**	**29,898**	**100**
Oleracea	*Brassica*	GBR165	SASA	United Kingdom	2,367	12
Oleracea	*Brassica*	USA003	NE9	USA	1,625	8
Oleracea	*Brassica*	CHN004	BVRC	China	1,235	6
Oleracea	*Brassica*	DEU146	IPK	Germany	1,215	6
Oleracea	*Brassica*	FRA215	GEVES	France	1,200	6
Oleracea	*Brassica*	RUS001	VIR	Russian Federation	980	5
Oleracea	*Brassica*	JPN003	NIAS	Japan	672	3
Oleracea	*Brassica*	NLD037	CGN	Netherlands	631	3
Oleracea	*Brassica*		Others (98)	Others	10,257	51
Oleracea	***Brassica***		**Total**	**World**	**20,182**	**100**
Rape	*Brassica*	CHN001	ICGR-CAAS	China	4,090	16
Rape	*Brassica*	IND001	NBPGR	India	2,585	10
Rape	*Brassica*	BGD028	BINA	Bangladesh	2,100	8
Rape	*Brassica*	JPN003	NIAS	Japan	1,579	6
Rape	*Brassica*	AUS039	ATFCC	Australia	1,184	5
Rape	*Brassica*	TWN001	AVRDC	Taiwan	1,091	4

(continued)

Table 16.2 (continued)

		Genebank			Accessions	
Crop grouping	Genus	*Institute code*	*Acronym*	Country	*No.*	(%)
Rape	*Brassica*	PAK001	PGRI	Pakistan	682	3
Rape	*Brassica*	USA020	NC7	USA	645	3
Rape	*Brassica*		Others (80)	Others	11,610	45
Rape	***Brassica***		**Total**	**World**	**25,566**	**100**
Capsicum	*Capsicum*	TWN001	AVRDC	Taiwan	7,914[a]	11
Capsicum	*Capsicum*	USA016	S9	USA	4,698	7
Capsicum	*Capsicum*	MEX008	INIFAP	Mexico	4,661	6
Capsicum	*Capsicum*	IND001	NBPGR	India	3,835	5
Capsicum	*Capsicum*	BRA006	IAC	Brazil	2,321	3
Capsicum	*Capsicum*	JPN003	NIAS	Japan	2,271	3
Capsicum	*Capsicum*	PHL130	IPB-UPLB	Philippines	1,880	3
Capsicum	*Capsicum*	TWN005	TSS-PDAF	Taiwan	1,800	2
Capsicum	*Capsicum*	DEU146	IPK	Germany	1,526	2
Capsicum	*Capsicum*	CHN004	BVRC	China	1,394	2
Capsicum	*Capsicum*		Others (176)	Others	41,272	56
Capsicum	***Capsicum***		**Total**	**World**	**73,572**	**100**
Melon	*Citrullus*	RUS001	VIR	Russian Federation	2,412	16
Melon	*Citrullus*	USA016	S9	USA	1,841	12
Melon	*Citrullus*	CHN001	ICGR-CAAS	China	1,197	8
Melon	*Citrullus*	ISR002	IGB	Israel	840	6
Melon	*Citrullus*	UZB006	UzRIPI	Uzbekistan	805	5
Melon	*Citrullus*		Others (81)	Others	8,048	53
Melon	***Citrullus***		**Total**	**World**	**15,143**	**100**
Cantaloupe	*Cucumis*	USA020	NC7	USA	4,878	11
Cantaloupe	*Cucumis*	JPN003	NIAS	Japan	4,242	10
Cantaloupe	*Cucumis*	RUS001	VIR	Russian Federation	2,998	7
Cantaloupe	*Cucumis*	CHN001	ICGR-CAAS	China	2,892	7
Cantaloupe	*Cucumis*	BRA012	CNPH	Brazil	2,400	5
Cantaloupe	*Cucumis*		Others (127)	Others	26,888	60
Cantaloupe	***Cucumis***		**Total**	**World**	**44,298**	**100**
Cucurbita	*Cucurbita*	RUS001	VIR	Russian Federation	5,771	15
Cucurbita	*Cucurbita*	CRI001	CATIE	Costa Rica	2,612	7
Cucurbita	*Cucurbita*	BRA003	CENARGEN	Brazil	1,897	5
Cucurbita	*Cucurbita*	CHN001	ICGR-CAAS	China	1,767	4
Cucurbita	*Cucurbita*	MEX008	INIFAP	Mexico	1,580	4
Cucurbita	*Cucurbita*		Others (144)	Others	25,956	65
Cucurbita	***Cucurbita***		**Total**	**World**	**39,583**	**100**
Radish	*Raphanus*	JPN003	NIAS	Japan	877	11
Radish	*Raphanus*	DEU146	IPK	Germany	741	9
Radish	*Raphanus*	USA003	NE9	USA	696	9
Radish	*Raphanus*	RUS001	VIR	Russian Federation	626	8
Radish	*Raphanus*	IND001	NBPGR	India	458	6
Radish	*Raphanus*		Others (85)	Others	4,608	57
Radish	***Raphanus***		**Total**	**World**	**8,006**	**100**

(continued)

Table 16.2 (continued)

Crop grouping	Genus	Genebank Institute code	Acronym	Country	Accessions No.	(%)
Tomato	*Solanum*	TWN001	AVRDC	Taiwan	8,107[a]	10
Tomato	*Solanum*	USA003	NE9	USA	6,283	7
Tomato	*Solanum*	PHL130	IPB-UPLB	Philippines	4,751	6
Tomato	*Solanum*	DEU146	IPK	Germany	4,062	5
Tomato	*Solanum*	RUS001	VIR	Russian Federation	2,540	3
Tomato	*Solanum*	JPN003	NIAS	Japan	2,428	3
Tomato	*Solanum*	CAN004	PGRC	Canada	2,137	3
Tomato	*Solanum*	COL004	ICA/REGION 5	Colombia	2,018	2
Tomato	*Solanum*	ESP026	BGUPV	Spain	1,927	2
Tomato	*Solanum*	IND001	NBPGR	India	1,796	2
Tomato	*Solanum*		Others (154)	Others	48,230	57
Tomato	***Solanum***		**Total**	**World**	**84,279**	**100**
Eggplant	*Solanum*	TWN001	AVRDC	Taiwan	3,524[a]	16
Eggplant	*Solanum*	IND001	NBPGR	India	3,060	14
Eggplant	*Solanum*	JPN003	NIAS	Japan	1,223	6
Eggplant	*Solanum*	USA016	S9	USA	887	4
Eggplant	*Solanum*	BGD186	EWS R&D	Bangladesh	826	4
Eggplant	*Solanum*	PHL130	IPB-UPLB	Philippines	661	3
Eggplant	*Solanum*	NLD037	CGN	Netherlands	659	3
Eggplant	*Solanum*		Others (124)	Others	10,776	50
Eggplant	***Solanum***		**Total**	**World**	**21,616**	**100**
Soybean	*Glycine*	CHN001	ICGR-CAAS	China	32,021	14
Soybean	*Glycine*	USA033	SOY	USA	21,075	9
Soybean	*Glycine*	KOR011	RDAGB-GRD	Republic of Korea	17,644	8
Soybean	*Glycine*	TWN001	AVRDC	Taiwan	15,322[a]	7
Soybean	*Glycine*	BRA014	CNPSO	Brazil	11,800	5
Soybean	*Glycine*	JPN003	NIAS	Japan	11,473	5
Soybean	*Glycine*	RUS001	VIR	Russian Federation	6,439	3
Soybean	*Glycine*	IND016	AICRP-Soybean	India	4,022	2
Soybean	*Glycine*	CIV005	IDESSA	Ivory Coast	3,727	2
Soybean	*Glycine*	TWN006	TARI	Taiwan	2,745	100
Soybean	*Glycine*		Others (179)	Others	103,684	45
Soybean	***Glycine***		**Total**	**World**	**229,952**	**100**
Bean	*Phaseolus*	COL003	CIAT	Colombia	35,891	14
Bean	*Phaseolus*	USA022	W6	USA	14,674	6
Bean	*Phaseolus*	BRA008	CNPAF	Brazil	14,460	6
Bean	*Phaseolus*	MEX008	INIFAP	Mexico	12,752	5
Bean	*Phaseolus*	DEU146	IPK	Germany	8,680	3
Bean	*Phaseolus*	CHN001	ICGR-CAAS	China	7,365	3
Bean	*Phaseolus*	RUS001	VIR	Russian Federation	6,144	2
Bean	*Phaseolus*	MWI004	BCA	Malawi	6,000	2
Bean	*Phaseolus*	HUN003	RCA	Hungary	4,350	2

(continued)

Table 16.2 (continued)

Crop grouping	Genus	Genebank		Country	Accessions	
		Institute code	*Acronym*		*No.*	(%)
Bean	*Phaseolus*	IDN002	LBN	Indonesia	3,846	1
Bean	*Phaseolus*		Others (236)	Others	147,801	56
Bean	***Phaseolus***		**Total**	**World**	**261,963**	**100**
Bean	*Psophocarpus*	PNG005	DOA	Papua New Guinea	455	11
Bean	*Psophocarpus*	MYS009	DGCB-UM	Malaysia	435	10
Bean	*Psophocarpus*	CZE075	TROPIC	Czech Republic	413	10
Bean	*Psophocarpus*	LKA005	IDI	Sri Lanka	400	9
Bean	*Psophocarpus*	IDN002	LBN	Indonesia	380	9
Bean	*Psophocarpus*		Others (35)	Others	2,134	51
Bean	***Psophocarpus***		**Total**	**World**	**4,217**	**100**
Mungbean	*Vigna radiata*	PHL130	IPB-UPLB	Philippines	6,889	16
Mungbean	*Vigna radiata*	TWN001	AVRDC	Taiwan	6,035[a]	14
Mungbean	*Vigna radiata*	CHN001	ICGR-CAAS	China	5,181	12
Mungbean	*Vigna radiata*	IND885	AICRP-Mullarp	India	4,432	10
Mungbean	*Vigna radiata*	USA016	S9	USA	3,900	9
Mungbean	*Vigna radiata*	IND001	NBPGR	India	3,147	7
Mungbean	*Vigna radiata*	IND064	NBPGR	India	2,466	6
Mungbean	*Vigna radiata*	THA005	FCRI-DA/TH	Thailand	2,250	5
Mungbean	*Vigna radiata*	JPN003	NIAS	Japan	1,579	4
Mungbean	*Vigna radiata*		Others (54)	Others	7,286	17
Mungbean	***Vigna radiata***		**Total**	**World**	**43,165**	**100**
Azuki bean	*Vigna angul.*	CHN001	ICGR-CAAS	China	3,993	41
Azuki bean	*Vigna angul.*	TWN001	AVRDC	Taiwan	2,375	25
Azuki bean	*Vigna angul.*	JPN003	NIAS	Japan	2,177	22
Azuki bean	*Vigna angul.*		Others (17)	Others	1,180	12
Azuki bean	***Vigna ang.***		**Total**	**World**	**9,725**	**100**
Cowpea	*Vigna ung.*	NGA039	IITA	Nigeria	15,588	24
Cowpea	*Vigna ung.*	USA016	S9	USA	8,043	12
Cowpea	*Vigna ung.*	BRA003	CENARGEN	Brazil	5,501	8
Cowpea	*Vigna ung.*	IDN002	LBN	Indonesia	3,930	6
Cowpea	*Vigna ung.*	IND001	NBPGR	India	3,317	5
Cowpea	*Vigna ung.*	CHN001	ICGR-CAAS	China	2,818	4
Cowpea	*Vigna ung.*	JPN003	NIAS	Japan	2,431	4
Cowpea	*Vigna ung.*	PHL130	IPB-UPLB	Philippines	1,821	3
Cowpea	*Vigna ung.*	BWA002	DAR	Botswana	1,435	2
Cowpea	*Vigna ung.*	RUS001	VIR	Russian Federation	1,337	2
Cowpea	*Vigna ung.*		Others (114)	Others	19,185	30
Cowpea	***Vigna ung.***		**Total**	**World**	**65,406**	**100**
Black gram	*Vigna mung.*	IND885	AICRP-Mullarp	India	2,137	28
Black gram	*Vigna mung.*	IND001	NBPGR	India	1,535	20
Black gram	*Vigna mung.*	PAK001	PGRI	Pakistan	799	10

(continued)

Table 16.2 (continued)

Crop grouping	Genus	Genebank Institute code	Genebank Acronym	Genebank Country	Accessions No.	Accessions (%)
Black gram	*Vigna mung.*	TWN001	AVRDC	Taiwan	763[a]	10
Black gram	*Vigna mung.*	IND063	NBPGR	India	500	7
Black gram	*Vigna mung.*		Others (31)	Others	1,905	25
Black gram	***Vigna mun.***		**Total**	**World**	**7,639**	**100**
Okra	*Abelmoschus*	CIV005	IDESSA	Ivory Coast	4,185	19
Okra	*Abelmoschus*	USA016	S9	USA	2,969	13
Okra	*Abelmoschus*	IND001	NBPGR	India	2,651	12
Okra	*Abelmoschus*	PHL130	IPB-UPLB	Philippines	968	4
Okra	*Abelmoschus*	FRA202	ORSTOM-MONTP	France	965	4
Okra	*Abelmoschus*		Others (90)	Others	10,690	48
Okra	***Abelmoschus***		**Total**	**World**	**22,428**	**100**
Amaranth	*Amaranthus*	IND001	NBPGR	India	5,760	20
Amaranth	*Amaranthus*	USA020	NC7	USA	3,341	12
Amaranth	*Amaranthus*	BRA003	CENARGEN	Brazil	2,328	8
Amaranth	*Amaranthus*	PER027	UNSAAC/CICA	Peru	1,600	6
Amaranth	*Amaranthus*	CHN001	ICGR-CAAS	China	1,459	5
Amaranth	*Amaranthus*		Others (106)	Others	13,825	49
Amaranth	***Amaranthus***		**Total**	**World**	**28,313**	**100**
Jute	*Corchorus*	IND001	NBPGR	India	5,408	46
Jute	*Corchorus*	BGD001	BJRI	Bangladesh	4,110	35
Jute	*Corchorus*		Others (39)	Others	2,171	19
Jute	***Corchorus***		**Total**	**World**	**11,689**	**100**
			Grand total	**World**	**1,021,074**	

Source: derived from data supplied by the WIEWS database after genus-specific searches for the entire world, consulted in May 2011 (WIEWS 2011a) and FAO (2010)
[a]Actual data (May 2011) from Asian Vegetable Genetic Resources Information System

the crops that originated in temperate climates (leek, shallot, onion, headed cabbage, cauliflower, broccoli, Chinese cabbage, radish, bell pepper) are nowadays grown in tropical and subtropical countries, thanks to selection and breeding efforts as well as evolutionary adaptation processes. The same is true for vegetables of tropical origin (eggplant), which are now successfully grown in more temperate climates.

Other important vegetable crops not covered in this chapter are cucurbits and melons, legume crops with the genera *Glycine*, *Phaseolus*, *Psophocarpus* and *Vigna,* and a large group of leafy vegetables such as *Abelmoschus* (okra), *Amaranthus* (amaranth), *Cleome* (spider plant), *Corchorus* (jute), *Hibiscus sabdariffa* (roselle), *Ipomoea batatas* (sweet potato), *Momordica* (bitter gourd), *Moringa* (drumstick tree), *Solanum scabrum* (African nightshade) and other minor leafy vegetables.

16.2 Vegetable Germplasm Collections at the Global Level with Specific Reference to the Collections Maintained by AVRDC – The World Vegetable Center

As explained in the introduction section, there is a considerable overlap of vegetables and melons with other commodity groups such as cereals (*Amaranthus*), food legumes (*Phaseolus, Psophocarpus, Glycine, Pisum, Vigna radiata, V. unguiculata, V. angularis, V. mungo*), and fiber crops (*Corchorus*). Considering this overlap, there are approximately one million accessions of vegetables and melons conserved ex situ worldwide (Table 16.2). In a narrower sense as considered by FAO in the second report on the state of the world's plant genetic resources for food and agriculture, the genetic resources of vegetables conserved ex situ comprise about 500,000 accessions representing 7% of the globally held 7.4 million accessions of plant genetic resources (FAO 2010).

The figures presented in this chapter on vegetable germplasm collections held ex situ worldwide are mostly based on data provided by the World Information and Early Warning System (WIEWS) on plant genetic resources for food and agriculture (WIEWS 2011a) and the Second Report on the State of the World's Plant Genetic Resources for Food and Agriculture (FAO 2010). There are other, much more user-friendly databases available such as the System-wide Information Network for Genetic Resources (SINGER) or, more recently, GENESYS. SINGER is the germplasm information exchange network of the Consultative Group on International Agricultural Research (CGIAR) and its partners comprising 12 institutes (including AVRDC – The World Vegetable Center), 47 collections, and 746,611 accessions. Although of major importance concerning exchange of germplasm worldwide, the genebanks of the CGIAR and associated centers united under this portal are not representative of the worldwide holdings of vegetable genetic resources. GENESYS, a global portal developed by Bioversity International, provides accession level information pooled from a range of different portals such as the SINGER and EURISCO networks, GRIN of the United States Department of Agriculture-Agricultural Research Service (USDA-ARS), and national genebanks (Alercia and Mackay 2010). It is quite an ambitious undertaking as it aims to provide not only passport data, but also characterization, evaluation, and environmental data. It is designed for breeders and other scientists and includes GIS and mapping functions, allows queries across all data for accession-specific traits, and has a built-in automated ordering system. As of June 2011, GENESYS comprises information on 2,333,733 accessions, but the choice of crops is limited to the main staple crops. Selecting the bean crop with the genus *Phaseolus*, the SINGER portal indicates a total of 37,055 accessions (SINGER 2011), the GENESYS portal 96,433 accessions (GENESYS 2011), WIEWS a total of 259,971 accessions (WIEWS 2011b) and FAO (2010) a total of 261,963 accessions. Therefore, the latter two sources are referred to in this chapter.

Tomatoes, capsicums, melons and cantaloupe, brassicas, cucurbits, alliums, okra, and eggplant are well represented in ex situ collections at the global level, with a range between 22,000 and 84,000 accessions per vegetable group (Table 16.2). As genetic erosion continues in situ for various reasons such as population growth, expansion of human settlements, and climate change, complementary collecting efforts should be made with a major focus on crop wild relatives and poorly represented cultivated forms of some vegetable groups such as those described for the genera *Brassica* and *Capsicum* in this chapter.

AVRDC – The World Vegetable Center plays a major role in the conservation and distribution of vegetable germplasm held in the public domain. Headquartered in Taiwan, the institution initially was set up in 1971 as the Asian Vegetable Research and Development Center. With the geographical expansion of its activities in the 1990s and at the beginning of the new Millennium, the Center now operates with offices in Africa (established in 1992 in Tanzania), East and Southeast Asia (established in 1992 in Thailand), South Asia (established in 2004 in India), Central and West Asia and North Africa (established in 2010 in the United Arab Emirates) and Oceania (established in 2010; managed from headquarters). The Center holds about 59,500 accessions of vegetable germplasm comprising 170 genera and 435 species from 156 countries of origin, including some of the world largest vegetable collections held by a single institution, such as *Capsicum*, tomato, and eggplant (Table 16.2). The Center makes its germplasm accessions and materials derived from its breeding programs available to the world community.

16.2.1 *Allium*

The *Allium* group comprises a range of widely grown commercial vegetable crops. Among 27 vegetable and melon commodity groups grown worldwide in 2009, dry onions ranked 4th, garlic 12th, green onions, including shallots 22nd, and leeks and other alliaceous vegetables ranked 23rd in terms of global production (FAOSTAT 2011). Close to 30,000 *Allium* accessions are conserved worldwide, with individual institutions in India (7%), the Russian Federation (6%), Japan (5%), United States of America (4%), Germany (4%), United Kingdom (4%) and Taiwan (AVRDC, 4%) having the largest share (Table 16.2). Out of 300 *Allium* species reported in WIEWS (2011a), 55 are considered economically important (Wiersema and León 1999). Many of these species have potential as ornamentals: *Allium aflatunense*, *A. atropurpureum*, *A. caeruleum*, *A. caesium*, *A. carinatum* subsp. *pulchellum*, *A. cernuum* (also used as a flavoring additive), *A. cristophii*, *A. flavum* (has potential for weed control), *A. giganteum*, *A. karataviense*, *A. mairei*, *A. moly*, *A. neapolitanum*, *A. nigrum*, *A. oreophilum*, *A. rosenbachianum*, *A. schubertii*, *A. senescens*, *A. sphaerocephalon*, *A. stipitatum*, *A. triquetrum*, *A. tuberosum* (also used as vegetable, as flavoring additive, and in folklore medicine), and *A. unifolium*.

Some cultivated *Allium* species with major importance in vegetable production are briefly described. These are *A. ampeloprasum* var. *ampeloprasum*, known as great-head garlic or wild leek and *A. porrum*, leek, pearl onion; *A. cepa* var. *aggregatum*, shallot and *A. cepa* var. *cepa*, common onion; *A. chinense*, Chinese onion; *A. fistulosum*, Welsh onion; *A. sativum* var. *sativum*, garlic; and *A. tuberosum*, Chinese chives.

16.2.1.1 *Allium ampeloprasum* L. var. *ampeloprasum* or Great-Headed Garlic Group and *A. porrum*: Leek, Pearl Onion

A. ampeloprasum comprises a large variety of wild and cultivated plants originating from a wide geographical area, stretching from Iran to Portugal and northern Africa (Grubben and Denton 2004). The major cultivated vegetables derived from this group are leek, great-headed garlic, pearl onion and kurrat. While *A. porrum* is considered a synonym of *A. ampeloprasum* by Grubben and Denton (2004), *A. ampeloprasum* var. *porrum* is treated as synonym of *A. porrum* (leek, pearl onion) by Wiersema and León (1999) and the United States Department of Agriculture-Natural Resources Conservation Service (USDA-NRCS 2011). Great-headed garlic is still relatively close to the wild types of this group and is found in the highlands of northern India, Réunion, and Saudi Arabia. Kurrat is cultivated in the Middle East and its leaves are used in salads or served as a cooked vegetable. Pearl onions are grown in parts of Europe and are used for pickling.

Leek possibly has its Center of origin in the eastern Mediterranean region (Wiersema and Léon 1999) where several related types (Turkisk leek, kurrat, 'tarreé Irani') and related *Allium* species occur (Siemonsma and Piluek 1994). Leek has a long history of domestication and cultivation by the ancient Egyptians, Romans, Greeks, Welsh and Scots. Leek must have been very popular in the Roman cuisine, as can be judged from a first century cooking manual translated as *The Roman Cookery of Apicius* by John Edwards, which contains 17 leek recipes (Schneider 2001). Today, non-bulbous leek varieties are grown on all continents, except for areas with high temperatures and high levels of rainfall. In the tropics it is grown at higher elevations and propagated by seeds or tillers.

16.2.1.2 *Allium cepa* L. var. *aggregatum or A. cepa* Aggregatum Group (shallot) and *A. cepa* var. *cepa* or *A. cepa* Common Onion Group (Onion)

The division of *A. cepa* into two major cultivar groups i.e. the Common Onion Group with large, mostly solitary bulbs, reproduced from seeds or seed-grown bulblets or 'sets' and the Aggregatum Group, with smaller, numerous bulbs forming an aggregated cluster, reproduced vegetatively through lateral bulbs, is widely accepted (Hanelt 1990; Siemonsma and Piluek 1994; Wiersema and León 1999; Grubben and Denton 2004). Only cultivated forms of *A. cepa* are found widely distributed. The species might have originated from Central Asia (Turkmenistan and Afghanistan) where natural populations of some crop wild relatives such as *A. vavilovii* (southern Turkmenistan and northern Iran) and *A. asarense* (Iran) can still be found (Grubben and Denton 2004). *A. oschanii*, another ancestral species of *A. cepa* cannot be easily crossed with cultivated onion, but its domestication might have given rise to some European shallots.

Onion has been known since ancient times as a valuable vegetable crop and its domestication must have started before 2700–3000 BC as onion images were found on a mural from Egypt dating back to this time (Siemonsma and Piluek 1994; Shigyo and Kik 2008). From Egypt, onions reached the Mediterranean and the Roman Empire (Grubben and Denton 2004).

Vegetatively propagated shallots were derived from common onion by natural selection among onion variants and were first reported in France during the twelfth Century (Siemonsma and Piluek 1994). Probably from Europe, shallots spread around the world. Although less important than common onion, shallots are predominant in the tropical lowlands between 10°N and 10°S, where onions are difficult to grow due to the hot and humid climate, which causes problems with vernalization and seed production. Shallots are used as a vegetable and for seasoning. They are known to stimulate the appetite and were indeed called *appetíts* in France in the seventeenth and eighteenth centuries (Schneider 2001).

16.2.1.3 *Allium chinense* G. Don: Chinese Onion

Chinese onion or Japanese scallion is a little known vegetable native to central and eastern China (Siemonsma and Piluek 1994; ECOCROP 2011) and is cultivated elsewhere, especially in eastern Asia (Wiersema and León 1999). It is a biennial herb that reaches about 60 cm in height. Small-bulbed cultivars produce 20–25 small bulbs, while large-bulbed types yield 6–9 bulbs. The bulbs are usually pickled, either sweet or sour, but also eaten fresh or fried. Chinese onion is well adapted to intermediate latitudes of 30–40°N and S and requires 16 h light for bulb and flower formation. Phenolic compounds in this crop help prevent thrombosis and this finding might lead to a wider use and cultivation of the crop (Siemonsma and Piluek 1994).

Genetic resources of this vegetable crop are limited to farmers' selections. Although flowers are produced, these do not set seed due to female sterility (Siemonsma and Piluek 1994). The selection of seed-producing variants could enhance breeding of this crop.

16.2.1.4 *Allium fistulosum* L.: Welsh Onion

Only cultivated forms of Welsh onion are known and its origin is probably northwestern China (Grubben and Denton 2004) or eastern Asia (Wiersema and León 1999). Given this probable origin in eastern Asia, the common name 'Welsh onion' seems to hint at another origin of this onion species, namely Wales. However, it is likely that the term 'welsh' was derived from the German word 'welsche' meaning foreign (Siemonsma and Piluek 1994). DNA studies link this species to the wild species *A. altaicum*, wild populations of which are still found in Siberia and Mongolia and the plants are occasionally collected, either for home consumption or for export to China. As early as 200 BC, Welsh onion was cultivated in China from where it reached Japan before 500 AD and then spread to Southeast Asia and Europe (Grubben and Denton 2004). Welsh onion is still the most important *Allium* species in China, playing the same culinary role as common onion and leek in Europe. In

Japan, the Welsh onion ranks second in importance after the common onion. Although the crop is grown on a small scale throughout the world, the main production area ranges from Siberia to Indonesia.

Two cultivar groups of Welsh onion are known: (1) the Japanese Bunching Group, mainly grown in eastern Asia for its thick, blanched pseudostems which are used as pot-herbs; (2) the Welsh Onion Group, grown for its green leaves in Southeast Asia and parts of Africa (Grubben and Denton 2004). The leaves or whole plants of the latter group are used in salads or as herbs to flavor dishes. In Chinese traditional medicine, the crop has many therapeutic uses.

Concerning genetic importance, Bioversity International ranked the Welsh onion second in the genus *Allium* because of its disease resistance, ecological adaptability, and its compatibility with *A. cepa* (Grubben and Denton 2004). There are fertile commercial hybrids between *A. fistulosum* and *A. cepa* known as 'Beltsville Bunching,' 'Louisiana Evergreen' and 'Delta Giant' that are propagated by seed (Siemonsma and Piluek 1994).

16.2.1.5 *Allium sativum* L. var. *sativum*: Garlic

After onion, garlic is the second most widely cultivated *Allium* species and China is clearly dominating world production of this crop with a share of more than 80% in 2009 (Table 16.1). In ancient times garlic reached the Mediterranean and was known in Egypt by 1600 BC (Grubben and Denton 2004). Further transmigration introduced the crop to the New World. Today, this mostly sterile crop is grown all over the world, from the equator to latitudes of 50° in both the northern and southern hemisphere. The crop is very popular in China, the Mediterranean, and in Latin America, but is not suited for the hot and humid lowland tropics.

Central Asia is the likely center of origin of the crop (Wiersema and León 1999) and this was confirmed by phylogenetic studies that further revealed a secondary center of diversity in the Caucasus (Grubben and Denton 2004). *A. longicuspis* from Central Asia, which comprises sterile as well as seed-producing lines, is believed to be a wild ancestor of garlic (Wiersema and León 1999; Siemonsma and Piluek 1994; Klaas and Friesen 2002). Maas and Claas (1995) classified *A. sativum* into four groups based on phylogenetic analysis and morpho-physiological descriptors: (1) Sativum Group, (2) Ophioscorodon Group, (3) Longicuspis Group and (4) Subtropical Group. The authors considered the Longicuspis Group as the ancestral group from which all other groups evolved. Both, the Sativum and the Subtropical Groups comprise lines adapted to tropical climates. *Allium sativum* L. var. *ophioscorodon*, known as rocambole or serpent garlic, is mainly cultivated in Eastern Europe (Grubben and Denton 2004).

The mature bulbs of garlic are mainly used as a condiment to flavor meat, fish and salads, either in fresh or dehydrated form. In Asia, the green tops and immature bulbs are extensively used. Garlic is well-known for its medicinal values in lowering blood pressure, cholesterol levels, and inhibiting thrombosis, and is used in many different formulations.

16.2.1.6 *Allium tuberosum* Rottler ex Spreng.: Chinese Chives

The geographic distribution of Chinese chives includes China, Mongolia and the Indian Subcontinent (Wiersema and León 1999). This species grows wild in and is probably native to the central, northern and eastern parts of Asia (Siemonsma and Piluek 1994; Wiersema and León 1999) and is cultivated in East, Southeast and South Asia. Major production areas are found in China, Japan, and Taiwan. Ratoon cropping is common in Chinese chives and leaves are harvested several times from the same plants, sometimes up to 20 or 30 years as in China.

16.2.2 Brassica

The Brassicaceae (Cruciferae) family includes 338 genera and 3,709 species (Al-Shehbaz et al. 2006; Warwick et al. 2006) and is considered one of the ten economically most important plant families (Rich 1991). In terms of crop production, the genus *Brassica* is the most important among 50 other genera in the tribe Brassiceae, comprising 37 different species (Gomez-Campo 1980). Among the 37 *Brassica* species, there are many cultivated forms providing edible roots, leaves, stems, buds, flowers and seed. Apart from their use as vegetables, Brassicas are also used as mustard condiments, fodder, ornamentals, and as an oilseed crop. Crop wild relatives of *Brassica* have significant potential as sources for cytoplasmatic male sterility in hybrid seed production systems as well as sources of resistance genes to a range of insect pests and diseases. Among 27 vegetable and melon commodity groups grown worldwide in 2009, cabbages and other brassicas ranked 5th, and cauliflowers and broccoli 14th, a clear indication of the global importance of this genus in terms of world vegetable production (Table 16.1; FAOSTAT 2011).

The taxonomy of the genus *Brassica* is complex and is still subject to changes given the rapid advances of cytogenetic and molecular methods. A recent review of the taxonomy of the genus *Brassica* can be found in a book chapter written by Branca and Cartea (2011). There are six cultivated *Brassica* species of significant economic importance. Three of these species are monogenomic diploid species and occur wild and in cultivation: (1) *B. rapa* (AA, n = 10; formerly known as *B. campestris*); (2) *B. nigra* (BB, n = 8); and (3) *B. oleracea* (CC, n = 9) (Redden et al. 2009). The other three amphidiploid species are: (4) *B. napus* (AACC, n = 19; *B. oleracea* × *B. rapa*); (5) *B. carinata* (BBCC, n = 17; *B. nigra* × *B. oleracea*); (6) *B. juncea* (AABB, n = 18; *B. nigra* × *B. rapa*) (Rakow 2004).

The principal *Brassica* vegetable species is *B. oleracea*, which comprises a wide range of cole and cabbage types. More than 20,000 *B. oleracea* accessions are conserved worldwide, with individual institutions in United Kingdom (12%), United States of America (8%), China (6%), Germany (6%), France (6%), Russian Federation (5%), Japan (3%), and Netherlands (3%) holding the largest shares of these genetic resources (Table 16.2).

Another major vegetable species is *B. rapa,* which includes several subspecies and one botanical variety (Wiersema and León 1999). More than 25,000 *Brassica rapa* accessions are conserved worldwide, with individual institutions in China (16%), India (10%), Bangladesh (8%), Japan (6%), Australia (5%), Taiwan (AVRDC, 4%), Pakistan (3%), United States of America (3%), United Kingdom (2%), and Germany (2%) having the largest collections (Table 16.2).

16.2.2.1 *Brassica carinata* A. Braun: Abyssinian Cabbage

B. carinata (n=17) is an amphidiploid species derived from interspecific crosses between *B. nigra* (n=8) and *B. oleracea* (n=9). No wild forms of *B. carinata* have been reported, but it probably originated in Ethiopia, with incipient cultivation since about 4000 BC (Grubben and Denton 2004) from hybrids between kale, grown on the Ethiopian plateau since ancient times, and wild or cultivated forms of *B. nigra* (Rakow 2004). *B. carinata* grows slowly and its seed contains mustard oil, as in *B. nigra*. The species is used in Ethiopia as an oilseed crop as well as a leafy vegetable. It is also grown in other parts of Africa, especially in East and southern Africa and to a lesser extent in West and Central Africa, where the use as a leafy vegetable is predominant. In western and southern Asia, *B. carinata* is occasionally grown as an oilseed crop or for mustard (Grubben and Denton 2004). In Ethiopia, the seed is used as condiment to flavor raw meat or to treat stomach-ache. The crop can be used as fodder for livestock and the seed cake resulting from oil extraction can be fed to animals as it is very rich in protein.

It might be worthwhile to do further collecting of *B. carinata* as traditional landraces are gradually being replaced by new cultivars. Only 1,170 *B. carinata* accessions are conserved worldwide ex situ, with the Australian Temperate Field Crops Collection (AUS039; 128 accessions), the Institute of Biodiversity Conservation in Ethiopia (ETH085; 374 accessions) and the Centre for Genetic Research, the Netherlands Plant Research International (NLD037; 109 accessions) holding more than 100 accessions each (WIEWS 2011a).

16.2.2.2 *Brassica juncea* (L.) Czern.: Brown Mustard, Indian Mustard

Brassica juncea (n=18) is an amphidiploid species derived from interspecific crosses between *B. nigra* (n=8) and *B. rapa* (n=10). It is an important vegetable species comprising a group of leaf mustards widely grown and consumed in China and other Asian countries. This vegetable species includes a number of different botanical varieties such as *B. juncea* var. *crispifolia* (curled mustard), *B. juncea* var. *foliosa* (leaf mustard), *B. juncea* var. *japonica* (cut-leaf mustard), *B. juncea* var. *juncea* (brown mustard, Indian mustard), *B. juncea* var. *multiceps* (chicken mustard), *B. juncea* var. *napiformis* (root mustard), *B. juncea* var. *rugosa* (head mustard), *B. juncea* var. *strumata* (large petiole mustard), and *B. juncea* var. *tumida* (big-stem mustard) (Wiersema and León 1999).

The crop is grown as an oilseed in India (brown or Indian mustard) and as a leafy vegetable in China and southern Asia. Also, root-type mustards (var. *napiformis*) are grown in China. Although leaf mustards are very diverse in China, *B. juncea* cannot have originated here as wild forms of the parent species *B. nigra* and *B. rapa* do not exist in China. Wild forms of *B. juncea* have been reported from the Near East and southern Iran (Rakow 2004).

The Chinese *B. juncea* forms are yellow-seeded, while the Indian types have brown seed color and are larger in seed size (Rakow 2004). The two types also differ in their glucosinolate composition. The yellow-seeded types are grown as an oilseed crop in Ukraine. Brown mustard has been introduced to many African countries and it is an important leafy vegetable in West and southern Africa. In Europe, North America and Canada, *B. juncea* is grown for the production of condiment mustard, especially western Canada, which is known for its brown and oriental mustard.

A total of 22,646 accessions of *B. juncea* are conserved worldwide, with India having the largest share of 12,036 accessions (53%), followed by China (4,976 accessions; 22%), Australia (1,309 accessions; 6%), and the Russian Federation (1,113 accessions; 5%) (VIEWS 2012).

16.2.2.3 *Brassica napus* L.: Rutabaga, Rape Kale, Oilseed Rape, Canola

B. napus (n=19) is an amphidiploid species having originated from interspecific crosses between *B. oleracea* (n=9) and *B. rapa* (n=10). Wild forms of this species have been observed on the shores of Gothland, Sweden, Netherlands and Great Britain (Rakow 2004). Very distinct naturalized forms of *B. napus* were also found in coastal areas of New Zealand, where both parent species grow wild. The origin of *B. napus* is still being disputed and potential origins are the coastal zone of northern Europe, the Mediterranean, or western or northern Europe.

Rape kale (*B. napus* var. *pabularia*) is a minor leafy vegetable in western Europe and serves as an important fodder source during winter (Grubben and Denton 2004). It is also found in southern Africa where it was introduced during colonization, but due to its limited adaptation to tropical conditions, its potential is restricted in the tropics. Its use as a leafy vegetable is similar to leaf cabbage, but it has a more pungent taste. In Zimbabwe, rape kale leaves are dried to serve as vegetables during the dry season.

Rutabaga or swede (*B. napus* var. *napobrassica*) is grown as minor vegetable and fodder in western Europe and the United States of America (Grubben and Denton 2004).

Winter and summer forms of *B. napus* var. *napus* (oilseed rape, canola) are grown as major oilseed crops in several countries in Europe, in North America, Canada, China, Japan, and India. The highest seed yields of about 3 tonnes/ha are obtained in Europe (France, Germany, UK) with winter annual types, which are much more productive than the summer annual forms grown in Canada that yield at least 50% less (Rakow 2004). Canada and Australia are major rapeseed exporting countries; Japan and China are major importers. Oilseed rape is grown in the cool highlands of Kenya, Tanzania, and to a lesser extent in Ethiopia.

A total of 15,615 accessions of *B. napus* are conserved worldwide, with China having the largest share of 4,217 accessions (27%), followed by Japan (1,436 accessions; 9.2%), Germany (1,179 accessions; 7.6%), Australia (1,090 accessions; 7%), and the Russian Federation (1,012 accessions; 6.5%) (VIEWS 2012).

16.2.2.4 *Brassica nigra* (L.) W.D.J. Koch: Black Mustard

B. nigra (BB; n = 8) is native in Africa (northern Africa and northeast tropical Africa-Eritrea; Ethiopia), temperate Asia (western Asia; Caucasus; Middle Asia; China), tropical Asia (Indian subcontinent), and northern to southern Europe, but the exact native range is obscure (USDA-ARS 2011a). Black mustard is cultivated for its seeds, which are the source of table mustard (Duke 1983). The oil extracted from the seed is used as a condiment, lubricant, and illuminant, and serves as a constituent in soap making. Mustard flour is obtained from a mixture of black and white mustard (*Sinapis alba* L. subsp. *alba*) and is used in different condiments such as "English Mustard" when mixed with water, and "Continental Mustard" when mixed with vinegar (Duke 1983). Mustard flowers are of importance for honey production. Black mustard is also sown as a cover crop for soil improvement (USDA-ARS 2011a). Black mustard grows on poor soil and tolerates aluminum, low pH, smog, and weeds (Duke 1983), but is itself considered a noxious weed and may become a seed contaminant (USDA-ARS 2011a). Black mustard is also of medicinal value as it is a source of allyl isothiocyanate and it has multiple uses in folk medicine.

Black mustard is the progenitor of *B. juncea* (brown mustard; Indian mustard) and *B. carinata* (Abyssinian cabbage), is a secondary genetic relative of *B. oleracea* (kale), and a tertiary genetic relative of *B. napus* (rape). It serves as gene source for disease resistance breeding in rape, turnip, and Abyssinian cabbage (USDA-ARS 2011a).

16.2.2.5 *Brasssica oleracea* L.: Kales or Leaf Cabbage; Headed Cabbage; Cauliflower and Broccoli; Brussels Sprouts; Kohlrabi

This important vegetable species comprises many different botanical varieties such as Chinese kale (*B. oleracea* var. *alboglabra*), cauliflower (*B. oleracea* var. *botrytis*), cabbage (*B. oleracea* var. *capitata*), tronchuda cabbage (*B. oleracea* var. *costata*), Brussels sprouts (*B. oleracea* var. *gemmifera*), kohlrabi (*B. oleracea* var. *gongylodes*), broccoli (*B. oleracea* var. *italica*), marrow-stem kale (*B. oleracea* var. *medullosa*), Jersey kale (*B. oleracea* var. *palmifolia*), thousand-head kale or branching cabbage (*B. oleracea* var. *ramosa*), Savoy cabbage (*B. oleracea* var. *sabauda*), curly kale (*B. oleracea* var. *sabellica*), and collards (*B. oleracea* var. *viridis*) (Wiersema and León 1999).

Cole crops, the cultivated forms of *B. oleracea*, are grown all over the world, but are mostly restricted to higher elevations in the tropics. Domestication of *B. oleracea* started about 3000 BC. The cultivated forms are likely to have originated from wild cabbage (*B. oleracea* var. *oleracea*), which is found in the Mediterranean region, south-western Europe and southern England. Other wild relatives of *B. oleracea* are found in isolated areas, with very distinct phenotypes. Among these are *B. cretica,* a species found in the Aegean area on Kriti, Greece and in southwestern Turkey (Rakow 2004). It is a perennial plant with two subspecies: *nivea* and *cretica.* The *B. rupestris-incana* complex with distinct regional variants grows in Sicily, southern and central Italy and western Yugoslavia. *B. macrocarpa* is found on Isole Egadi, west of Italy. *B. insularis* is found in Corsica, Sardinia, and Tunisia. Other wild species are *B. montana,* found in northern Spain, southern France and northern Italy, and *B. hilaronis*, endemic in the Kyrenia Mountains, northern Cyprus.

Kales or Leaf Cabbage

Kales are ancient cole crops, derived from wild *B. oleracea* forms (Siemonsma and Piluek 1994). Tronchuda cabbage (*B. oleracea* var. *costata*), Jersey kale (*B. oleracea* var. *palmifolia*), thousand-head kale (*B. oleracea* var. *ramosa*), Savoy cabbage (*B. oleracea* var. *sabauda*), curly kale (*B. oleracea* var. *sabellica*) and collards (*B. oleracea* var. *viridis*) are mainly used as vegetable crops. Thousand-head kale and collards can be used as vegetables or as fodder crops, whereas marrow-stem kale (*B. oleracea* var. *medullosa*) is mostly used as a fodder crop. While tronchuda cabbage and marrow-stem kale are widely distributed, including tropical Africa (Grubben and Denton 2004), other kales are more restricted to Europe or Asia (Chinese kale). Kales are a main source of genes providing tolerance against abiotic stress for other *B. oleracea* types (Grubben and Denton 2004).

In contrast to the previously described kales, which all evolved in Europe, **Chinese kale** (*B. oleracea* var. *alboglabra*) probably originated in eastern Asia (Wiersema and León 1999) and is endemic in southern and central China. To date, it is widely cultivated in Southeast Asia. Stems, young leaves and young inflorescences of this crop are cooked or fried, sometimes also eaten raw. Chinese kale is widely grown in home gardens and ranks among the ten most important market garden vegetables in Thailand and other countries of Southeast Asia (Wiersema and León 1999). Chinese kale has great potential for cultivation in the tropics as it is very productive and provides a cheap source of leafy vegetables to meet the increasing demand of city dwellers. It is clearly more heat tolerant than the other botanical varieties of *B. oleracea* and is used to improve heat tolerance in broccoli.

Headed Cabbage: *B. oleracea* L. var. *capitata* L.: Red Cabbage, White/Green Cabbage and *B. oleracea* L. var. *sabauda* L.: Savoy Cabbage

Headed cabbage evolved from leafy, un-branched kales with thin stems introduced from the Mediterranean to northern and western Europe during the Roman Empire (Grubben and Denton 2004). During the sixteenth century, headed cabbage became one of the most important vegetables in Europe. Thereafter, headed cabbage was introduced worldwide, including the tropics, where the crop remained mostly confined to the cooler highlands. Headed cabbage is quite common in East Africa and Egypt. White cabbage with small, firm round to flat heads is also gaining ground in tropical Asia and is partially replacing the more perishable leafy vegetables (Siemonsma and Piluek 1994).

A total of 5,269 accessions of *B. oleracea* var. *capitata* are conserved worldwide, with China having the largest share of 1,129 accessions (21.4%), followed by the United States of America (1,007 accessions; 19.1%), United Kingdom (702 accessions; 13.3%), and France (680 accessions; 12.9%) (VIEWS 2012). For *B. oleracea* var. *sabauda*, only 224 accessions are conserved worldwide, of which Germany holds 122 accessions or 54.7%.

B. oleracea L. var. *botrytis* L.: Cauliflower and *B. oleracea* L. var. *italica* Plenck: Broccoli

A great diversity of cauliflower and broccoli-type vegetables evolved in Italy. White-headed cauliflower spread from there to central and northern Europe where secondary centers for diversity of annual and biannual types evolved (Grubben and Denton 2004). Biennial cauliflower from Europe adapted well in India and resulted in lines that perform well in hot and humid tropical climates. Italian immigrants took broccoli with them to the United States from where the crop spread worldwide. Both cauliflower and broccoli are grown in many countries of tropical Africa, but mostly in the highlands. In West Africa, the two crops can be found in the lowlands during the relatively cool harmattan season.

A total of 2,911 accessions of *B. oleracea* var. *botrytis* are held ex situ worldwide, with France conserving 637 accessions (21.9%), United Kingdom 530 accessions (18.2%), China 502 accessions (17.2%), and the United States of America 461 accessions (15.8%) (VIEWS 2012). Only 634 accessions of *B. oleracea* var. *italica* are conserved worldwide, of which the United Kingdom holds 410 accessions or 64.7%.

B. oleracea L. var. *gemmifera* Zenker: Brussels Sprouts and *B. oleracea* L. var. *gongylodes L.*: Kohlrabi

Brussels sprouts were first mentioned in the eighteenth century in Belgium and spread from there to other parts of Europe and occasionally to other parts of the world. In Japan, early-maturing Asian types have been developed.

Kohlrabi was developed from the fodder crop marrow-stem kale which has a thickened stem. It was first reported in the sixteenth century in northwestern Europe (Grubben and Denton 2004). Apart from the most common production areas under temperate climates in Europe, the United States of America and parts of Asia, kohlrabi is also grown under subtropical climates in India, northern Vietnam and China. It is occasionally grown at high elevations in East Africa (Grubben and Denton 2004).

Brussels sprouts and kohlrabi are quite poorly represented in genebanks. A total of 702 accessions of *B. oleracea* var. *gemmifera* are conserved worldwide, with United Kingdom conserving the majority, a total of 391 accessions (55.7%), followed by France with 152 accessions (21.7%) (VIEWS 2012). Only 240 accessions of *B. oleracea* var. *gongylodes* are conserved worldwide, of which the United Kingdom holds 55 accessions (22.9%).

16.2.2.6 *Brassica rapa* L.

Brassica rapa was formerly known as *B. campestris*. The species has been renamed to *B. rapa* according to the International Code of Botanical Nomenclature (Branca and Cartea 2011). It is believed that *B .rapa* (AA; n = 10) originated in the cold climate prevalent in the highlands close to the Mediterranean Sea (Tsunoda 1980). From here the species spread into Scandinavia and to Eastern Europe and Germany (Nishi 1980). It is likely that *B. rapa* was introduced into China through western Asia or Mongolia and from there or possibly from Siberia into Japan (Rakow 2004). *B. rapa* is grown as an oilseed crop in Sweden, Finland, Canada, and India.

There are seven *B. rapa* groups which are grown as vegetables (Rakow 2004). These are (Wiersema and León 1999): pak choi (*B. rapa* L. subsp. *chinensis* (L.) Hanelt; also called *B. rapa* Pak Choi Group); Chinese savoy, taatsai (*B. rapa* L. subsp. *narinosa* (L.H. Bailey) Hanelt; or *B. rapa* Taatsai Group); mizuna, kyona (*B. rapa* L. subsp. *nipposinica* (L.H. Bailey) Hanelt; or *B. rapa* Mizuna Group); Chinese cabbage, petsai (*B. rapa* L. subsp. *pekinensis* (Lour.) Hanelt; or *B. rapa* Chinese Cabbage Group); turnip (*B. rapa* L. subsp. *rapa*; or *B. rapa* Vegetable Turnip, Fodder Turnip, and Neep Greens (in part) Groups); caisin, false pak choi (*B. rapa* L. var. *parachinensis* (L.H. Bailey) Hanelt; or *B. rapa* Caisin Group); and spinach mustard, kabuna (*B. rapa* L. var. *perviridis* L.H. Bailey; or *B. rapa* Neep Greens Group, in part).

B. rapa L. subsp. *chinensis* (L.) Hanelt: Pak Choi

The crop evolved in China and has already been cultivated in the fifth century AD (Siemonsma and Piluek 1994). Now it is widely grown in China, Taiwan and two Southeast Asian countries, the Philippines and Malaysia, and to a limited extent in Indonesia and Thailand. In recent times it has been introduced to Japan, where it is still called ‘Chinese vegetable.’ Through emigrants from East Asia, pak choi has also found its way to North America, Europe and Australia.

Pak choi is non-headed and characterized by a loose rosette of leaves; the enlarged, fleshy petioles are highly appreciated. The vegetable constitutes the main ingredient for soup and stir-fried dishes. Specific cultivars are available for all four seasons: autumn, winter, spring, and summer.

Under the synonym *B. chinensis*, a total of 2,413 accessions are held ex situ worldwide with the vast majority of 1,789 accessions (74%) conserved in China

(WIEWS 2011a). The WIEWS database does not reflect the significant collection of pak choi held by AVRDC in Taiwan (757 accessions) as the AVGRIS database lists these entries under *B. rapa* subsp. *chinensis* (AVGRIS 2011). This fact highlights the difficulties faced in this genus due to often unclear taxonomic classification.

Pak choi breeding is concentrated in Japan and China where both open-pollinated and hybrid cultivars are being developed (Grubben and Denton 2004). Major breeding objectives are uniformity, high yield, early maturation, and adaptation to the humid, tropical lowlands.

B. rapa L. subsp. *pekinensis* (Lour.) Hanelt: Chinese Cabbage, Celery Cabbage, Pe-tsai, Nappa

The center of diversity of headed Chinese cabbage is found in northern China (Rakow 2004). It is believed that the crop evolved from a natural cross between non-headed pak choi and turnip (Siemonsma and Piluek 1994). Chinese cabbage was introduced into Korea in the thirteenth century, into Southeast Asia in the fifteenth century, and Japan in the nineteenth century. Today, Chinese cabbage is cultivated all over the world, but is of special importance in eastern Asia. In China it is among the most widely grown and consumed vegetables. It is a major constituent of Korean food and is mostly used in the preparation of kimchi, a fermented side dish eaten all year round (Opeña et al. 1988).

Only 141 accessions are reported worldwide in the WIEWS database (2011a) under the synonym *B. pekinensis*, the majority of which are held by Japan (83; 58.9%), followed by the Vavilov Research Institute in the Russian Federation (38; 27%). Similar to the case of pak choi, the 112 Chinese cabbage accessions held by AVRDC are unaccounted for in the WIEWS database due to confusing taxonomic classification.

Chinese cabbage has received major attention at AVRDC – The World Vegetable Center. The breeding program started in the early 1970s and focused on the adaptation of this temperate crop to hot and humid tropical lowlands. The extensive screening of accessions from the AVRDC genebank enabled the selection of open-pollinated heat-tolerant lines having the ability to form heads in tropical climates. Heat-tolerant and disease resistant lines were widely tested in collaboration with national programs and resulted in the successful cultivation of Chinese cabbage in the lowland tropics (Opeña et al. 1988; de la Peña et al. 2011).

Brassica rapa L. var. *parachinensis* (L.H. Bailey) Hanelt: Caisin, False Pak Choi

Caisin might have evolved in central China from the oil-yielding turnip rape (*Brassica rapa* L. subsp. *campestris* (L.) A. R. Clapham) introduced into China from the Mediterranean (Siemonsma and Piluek 1994) and pak choi (Rakow 2004). Caisin was selected for its inflorescences, which are superior in tenderness and smoothness compared to other leafy *Brassica* crops. Caisin cultivars have a maturity range from 40 to 80 days. The plants are cut for fresh consumption at early flowering stage. The produce is usually stir-fried, which preserves its tenderness and nutritional value. Caisin is common in tropical lowlands where headed Chinese cabbage is more difficult to grow. It is cultivated in southern and central China, Indonesia, Malaysia, Thailand, Vietnam, other parts of Indo-China, and in West India (Siemonsma and Piluek 1994).

Caisin collections seem hardly to exist. The WIEWS database reveals 8 accessions of *B. rapa* var. *parachinensis*, 38 accessions of *B. parachinensis* and 14 accessions of *B. rapa* Caisin group, held worldwide, respectively. The latter are held by AVRDC (WIEWS 2011a). Again, the heterogeneous taxonomic classification of this crop makes it difficult to retrieve the information on ex situ holdings in just one search session.

16.2.3 Raphanus L.: Radish

The genus *Raphanus* (n = 9) includes one cultivated species, *Raphanus sativus* L., and several wild species. *Brassica* is phylogenetically relatively close to *Raphanus* and nucleotide diversity studies among wild and cultivated radish species and several *Brassica* species revealed that *B. barrelieri* is the closest *Brassica* species to *Raphanus* and formed a sister clade with this genus, located between the Rapa/Oleracea and Nigra groups of *Brassica* (Lü et al. 2008). Cultivated radish has been classified into different botanical varieties (Wiersema and León 1999; Lü et al. 2008; Yasumoto et al. 2008) as follows: (1) *R. sativus* var. *hortensis* f. *raphanistroides* Makino – East Asian big long radish; (2) *R. sativus* L. var. *mougri* H.W.J. Helm or *R. sativus* Rat-Tailed Radish Group – mougri, rat-tail radish; (3) *R. sativus* L. var. *niger* J. Kern. or *R. sativus* Chinese Radish Group – Chinese radish, Japanese radish; (4) *R. sativus* L. var. *oleiformis* Pers. or *R. sativus* Leaf Radish Group – fodder radish, oil radish; and (5) *R. sativus* var. *sativus* L. or *R. sativus* Small Radish Group – radish, small radish.

The wild species *R. raphanistrum* L. (jointed charlock, Mediterranean radish, wild radish) and *R. raphanistrum* L. subsp. *landra* (Moretti ex DC.) Bonnier and Layens (Mediterranean radish, sea radish) are potential ancestors of the polymorphic cultivated radish (Lü et al. 2008). The greatest diversity of radish is found between the eastern Mediterranean and the Caspian Sea and, therefore, this region is considered as the primary gene center of radish (Grubben and Denton 2004). *R. raphanistrum* is used as a gene source for radish (primary gene pool), mustard (tertiary genetic relative of *B. juncea*), and rape (tertiary genetic relative of *B. napus*) and is reported to be native in Africa (*Macaronesia*: Madeira and Canary Islands; *Northern Africa*: Algeria, Egypt, Libya, Morocco, Tunisia); temperate Asia (*Western Asia*: Cyprus, Iran, Iraq, Israel, Jordan, Lebanon, Syria, Turkey; *Caucasus*: Armenia, Azerbaijan, Georgia, Russian Federation) and a large number of European countries (USDA-ARS 2011b).

The related wild species *R. raphanistrum* L. subsp. *landra* is used as gene source for radish and is slightly more restricted in its distributional range and is reported native in Africa (*Macaronesia* and *Northern Africa*) and Europe (USDA-ARS 2011c).

Surprisingly, cultivated radish displayed higher nucleotide diversity than wild species and this has been ascribed to frequent hybridization between cultivated radish and its wild relatives (Lü et al. 2008). The three cultivar groups *R. sativus* var. *sativus* (Small Radish Group), *R. sativus* var. *hortensis* f. *raphanistroides* (East Asian big long radish) and *R. sativus* var. *niger* (Chinese Radish Group) belonged to a different cluster of the phylogenetic tree constructed by Lü and co-workers (2008) and this could mean independent multiple origins of these cultivar groups.

Radish is an ancient crop and was cultivated in the Mediterranean before 2000 BC (Siemonsma and Piluek 1994). From this region it spread to China around 500 BC and to Japan about 700 AD. Radish has been cultivated since ancient times in the oases of the Sahara and in Mali and is now cultivated throughout the world (Grubben and Denton 2004). The Small Radish Group is mostly cultivated in temperate climates, while the larger-rooted cultivars from the Chinese Radish Group are important in East Asia (Japan, Korea, China) and Southeast Asia. The Leaf Radish Group is gaining importance in Europe and South Africa as a forage and green manure crop (Siemonsma and Piluek 1994; Grubben and Denton 2004). The Rat-Tailed Radish Group is of significant importance in India and eastern Asia, but is also grown in northern Thailand and Burma.

Radish is mainly grown for its thickened fleshy root. The Small Radish Group cultivars are pungent and used as appetizers and to add color to dishes (Grubben and Denton 2004). Cultivars of the Chinese Radish Group have a mild flavor and are crisp. After peeling, the roots are sliced and placed into soups and sauces or cooked with meat. They also can be preserved in salt. Chinese radish is also eaten fresh, usually mixed with other vegetables like tomatoes. The leaves of Chinese radish are served as a salad. Seedlings of Chinese radish, known as radish sprouts, are consumed fresh as appetizers just like garden cress (*Lepidium sativum* L.) or cooked as spinach. Cultivars of the Rat-Tailed Radish Group are grown for the immature crisp, fleshy fruits which are consumed raw, cooked or pickled (Grubben and Denton 2004). The roots of this cultivar group are not edible.

Only about 8,000 *Raphanus* accessions are maintained in ex situ collections around the globe (Table 16.2). Major collections are held by individual institutions in Japan (877 accessions; 11%), Germany (741 accessions; 9%), the Russian Federation (626 accessions; 8%), India (458 accessions; 6%), Great Britain (453 accessions; 6%), and the Netherlands (307 accessions; 4%). The breeding of heat tolerant cultivars will allow an expansion of this crop to the tropical lowlands.

16.2.4 Capsicum

Capsicum is an important vegetable crop and green chilies and peppers ranked 9th in terms of global production among 27 vegetable and melon commodity groups grown worldwide in 2009, with an output of 28 million tonnes (Table 16.1). China was by far the largest producer of this commodity (52% of world production), followed by Mexico (7%), Turkey (7%), Indonesia (4%), and Spain (4%).

Capsicum originated in the New World, in Central and South America where the five domesticated species and 26 wild or semi-domesticated *Capsicum* species are found (Eshbaugh 1993; Grubben and Denton 2004; Moscone et al. 2007). On his first voyage to the Americas, Christopher Columbus tasted *Capsicum* fruits and as their pungency reminded him of black pepper, *Piper nigrum*, he simply called it red pepper as the pods were of red color (Bosland 1996). Capsicums were widely distributed in pre-Columbian times, covering Mexico, Central America and parts of South America and were grown as early as 5200–3400 BC by Native Americans. Capsicums transmigrated with the Spanish and Portuguese explorers to Europe, and later to Africa and Asia and found their way into the cuisines of the respective continents. Some 500 years after the discovery of the New World, chili peppers started to dominate the trade of hot spices and are now widely grown in the tropics, subtropics as well as in temperate regions around the globe (Eshbaugh 1993).

The *Capsicum* terminology is quite confusing as pepper (bell pepper, capsicum pepper, Cayenne pepper, cherry pepper, chili pepper, cone pepper, green capsicum, green pepper, long pepper, red cone pepper, red pepper, sweet pepper), *chili*, *chilli*, *chile*, *aji* and paprika are used interchangeably for plants of the genus *Capsicum* (Bosland 1996; Wiersema and León 1999; Ebert et al. 2007). The Spanish term *chile* was derived from the Aztec word *chil* and simply means pepper, whether the fruits are pungent or sweet. The term 'chili' or 'chili pepper' usually refers to the pungent cultivars, while bell pepper is used for the non-pungent blocky pepper types.

The five domesticated *Capsicum* species are: *C. annuum* L. var. *annuum* – chili pepper, bell pepper, paprika; *C. baccatum* L. var. *pendulum* (Willd.) Eshbaugh – Peruvian pepper; *C. chinense* Jacq. – bonnet pepper, habanero pepper; *C. frutescens* L. – Tabasco pepper, bird pepper; and *C. pubescens* Ruiz and Pav. – rocoto, apple chile. In addition to these five domesticated species, another four species are considered of economic importance by Wiersema and León (1999): *C. annuum* L. var. *glabriusculum* (Dunal) Heiser and Pickersgill – American bird pepper; *C. baccatum* L. var. *baccatum* – locoto; *C. cardenasii* Heiser and P.G. Sm. – ulupica; and *C. eximium* Hunz. – ulapuca. These four species, as well as *C. baccatum* L. var. *praetermissum* (Heiser and P.G. Sm.) Hunz., *C. chacoense* Hunz., and *C. tovarii* Eshbaugh et al. are considered as semi-domesticated (Ribeiro et al. 2008). Another 20 species are referred to as wild species (Ribeiro et al. 2008). It remains unclear why as many as 133 *Capsicum* species are listed in the WIEWS database (WIEWS 2011a).

Chromosome numbers are known for 23 out of the 31 recognized species. The 23 species with known chromosome numbers form two groups. One group with a basic chromosome number of $n=12$ comprises 13 species, while the second group with $n=13$ includes 10 species (Moscone et al. 2007). The basic number $n=13$ is very rare in the Solanaceae family. Moscone and co-workers (2007) postulate that the *Capsicum* taxa with $n=13$ represent a derived form from the basic number $n=12$. *C. chacoense* ($n=12$), which is found in a semi-arid region in south-central

Bolivia, is considered the most primitive taxon, while the taxa with n = 13, which originated in Brazil, are the most advanced (*C. campylopodium* Sendtn., *C. cornutum* (Hiern.) Hunz., *C. friburgense* Bianchetti and Barboza, C. mirabile Mart., *C. pereirae* Barboza and Bianchetti, *C. recurvatum* Witas., *C. schottianum* Sendtn., and *C. villosum* Sendtn.).

With around 73,600 accessions, the genus *Capsicum* is well represented in ex situ collections around the globe (Table 16.2). The largest collection (7,914 accessions) is held by AVRDC in Taiwan (11% of accessions held globally), followed by individual institutions in the United States of America (4,698 accessions; 7%), Mexico (4,661 accessions; 6%), India (3,835 accessions; 5%), Brazil (2,321 accessions; 3%), and Japan (2,271 accessions; 3%). However, out of the five cultivated *Capsicum* species, *C. chinense* and *C. pubescens* are poorly represented in international ex situ collections and require additional collecting and conservation efforts. Many wild *Capsicum* species, such as *C. rhomboideum*, *C. flexuosum*, and *C. lanceolatum* are hardly found in ex situ collections. As some of the wild species are threatened by extinction, special attention is required for their effective in situ (Bosland and Gonzalez 2000) and ex situ conservation to preserve this potentially valuable material for future breeding efforts. New methods making use of geographic information systems (GIS) have been recommended for optimizing new collecting missions targeting rare wild pepper species (Jarvis et al. 2005).

16.2.4.1 *C. annuum* L. var. *annuum*: Bell Pepper, Chili Pepper, Paprika

C. annuum (n = 12) is the best known and economically most important *Capsicum* species in the world, most likely because it was the first pepper species discovered by Columbus and taken to Europe (Eshbaugh 1993). Mexico is the probable center of origin of *C. annuum,* producing both pungent and sweet pepper types (Grubben and Denton 2004). Bell or sweet pepper, the non-pungent form of *C. annuum* is widely grown and used as a green vegetable. Bell pepper grows better in temperate climates, while chili pepper does well in tropical climates. AVRDC's pepper breeding program is developing heat-tolerant sweet pepper lines for the humid lowland tropics and such lines are distributed to cooperators globally through AVRDC's International Sweet Pepper Nursery arrays (de la Peña et al. 2011).

A peculiarity of *C. annuum* is paprika. In a number of European countries the word paprika refers to vegetable bell peppers. The term originated from the Hungarian or Serbian word *paprika,* meaning pepper (Katzer 2011). The Hungarian word *paprika* is a diminutive form of the Serbo-Croatian *papar*, derived from the Latin *piper* or modern Greek *piperi* (Online Etymology Dictionary 2011). In general, the term paprika refers to a bright-red colored spice made from grinding together dried fruits of *C. annuum*, both bell and chili peppers. Paprika is used in many cuisines to add color and flavor to dishes. The subtle, sweet flavor of paprika goes well with hot and spicy dishes, but also with mild stews. Special care must be taken when frying paprika powder in hot oil as the sugar added to paprika quickly turns bitter (Katzer 2011).

A small group of *C. annuum* can be considered as ornamentals as they are mainly grown for their unusual pod shapes and colorful fruits, although they are also edible.

The hot pepper form of *C. annuum* is growing widely in Africa and is by some African people already considered a traditional African vegetable and spice, unlike the sweet pepper types, which are considered exotic (Grubben and Denton 2004). The production of green chilies and peppers in Africa reached 2.9 million tonnes in 2008, a significant increase compared with about 1 million tonnes in 2001 (FAOSTAT 2011). The five top producing countries in Africa are Nigeria, Egypt, Tunisia, Ghana, and Algeria.

16.2.4.2 *C. baccatum* L. var. *pendulum* (Willd.): Peruvian Pepper, *ají* (Spanish)

In the lowlands and mid-elevations of South America, *C. baccatum* (n=12) is the most widely grown species (Eshbaugh 1993; Ebert et al. 2007). It is found in cultivation from the northeastern part of South America, including Colombia, Ecuador, Peru and Bolivia to the Southeast of Brazil (Ribeiro et al. 2008). Within Brazil, it is mostly grown in the southern and southeastern region of the country. Known as *ají* or *escabeche* in Spanish (Wiersema and León 1999), the cultivated forms of this botanical variety are not only famous as a hot spice, but also renowned due to the subtle bouquet and distinct flavors they produce (Ribero et al. 2008).

A semi-domesticated form of this species is *C. baccatum* L. var. *baccatum*, known as *locoto* in Spanish (Wiersema and León 1999) or *cumari-verdadeira* or *pimento-de passarinho* in Brazil (Ribero et al. 2008). This form is found in central Peru, in Bolivia, northern Argentina and southern and southeastern Brazil.

Another semi-domesticated form is *C. baccatum* L. var. *praetermissum* (Heiser and P.G. Sm.) Hunz. This botanical variety is restricted to Brazil and is mainly cultivated in the central-eastern region, but is also found in the southeastern region of the country (Ribero et al. 2008). It is also called *cumari-verdadeira* or *pimento-de passarinho*. The fruits of *cumari-verdadeira* of both botanical varieties, *baccatum* and *praetermissum*, are characterized by a pronounced aroma and elevated pungency and are used in canned form (Ribero et al. 2008).

A fourth, but much lesser known botanical variety of this species is *C. baccatum* var. *umbilicatum* (Vell.) Hunz. and Barboza (USDA-ARS 2011d).

16.2.4.3 *C. chinense* Jacq.: Bonnet Pepper, Habanero Pepper

The Amazonian basin is the center of genetic diversity of *C. chinense* (n=12) (Eshbaugh 1993; Wiersema and León 1999), but it is also found in the central-eastern and northeastern regions of Brazil (Ribero et al. 2008). At a later stage, the species also spread into the Caribbean, but the diversity of taxa is much more limited there compared with the Amazonian basin (Eshbaugh 1993). The species name was wrongly assigned by the Austrian botanist Jacquin, dubbed the 'Austrian Linnaeus' (Erickson 2011) as he had obtained the seed of this species from China (Bosland 1996).

C. chinense is fairly common in West Africa and is preferred by farmers during the rainy season as it shows good resistance against anthracnose and viruses, unlike the *C. annuum* cultivars which are more susceptible (Grubben and Denton 2004).

The fruits of *C. chinense* are, in general, extremely pungent and aromatic. Bonnet or habanero peppers are well-known pungent chili varieties, measuring between 100,000 and 350,000 Scoville heat units (SHU) (Wikipedia 2011a). The 'Naga Viper pepper' cultivar, created by a chili farmer in Great Britain, is currently the official holder of the *Guinness World Records* of the "World's Hottest Chilli" as of 25 February 2011 with a measurement of 1.3 million SHU (Wikipedia 2011b).

16.2.4.4 *C. frutescens* L.: Tabasco Pepper, Bird Pepper

The geographic distribution of *C. frutescens* (n = 12) ranges from the lowlands of southeastern Brazil to Central America and the Antilles in the Caribbean (Ribero et al. 2008). In Brazil, it is mostly found in the northern region, but is also cultivated in the central-eastern and northeastern region.

Two cultivars of *C. frutescens* are widely known: the popular cultivar 'tabasco,' named after the Mexican state 'Tabasco' and being the main ingredient of Tabasco sauce and 'malagueta' (Ebert et al. 2007). Tabasco pepper is cultivated in the southeastern United States, while in Brazil, malagueta pepper (*pimento-malagueta*) is well known (Ribero et al. 2008).

16.2.4.5 *C. pubescens* Ruiz and Pav.: Rocoto, Apple Chile

The probable origin of *C. pubescens* (n = 12) is Brazil (Wiersema and León 1999; Ribero et al. 2008). It is well adapted to low temperatures, but does not tolerate frost. *C. pubescens* is grown in the highlands from Mexico to Argentina (León 2000) and can easily be distinguished from the other domesticated species through its leaf pubescence, purple flowers, and black seeds (Ebert et al. 2007). Common Spanish names for this species are: *chamburoto*, *chile japonés*, *escabeche*, *lacoto*, *siete caldos*, and *rocoto* (Bosland 1996; Wiersema and León 1999; León 2000). As the fruits often are apple- or pear-shaped, they are also called *chile manzana* in Central America or *perón* in South America.

16.2.5 Solanum lycopersicum L.: *Tomato*

Tomatoes are an economically very important vegetable crop and widely grown in 173 countries (FAOSTAT 2011). This crop ranked second in terms of global production among 27 vegetable and melon commodity groups grown worldwide in 2009, with an output of 153 million tonnes, notably up from 136.2 million tonnes in 2008 (FAOSTAT 2011). In 2009, China was by far the largest producer of this commodity (45.4 million tonnes; 30% of world production), followed by the United States of

America (14.1 million tonnes; 9%), India (11.1 million tonnes; 7%), Turkey (10.7 million tonnes; 7%), and Egypt (10 million tonnes; 7%) (Table 16.1).

Tomato belongs to the *Solanaceae* family, which comprises 95 accepted genera (USDA-ARS 2011e) and more than 3,000 species, having originated in both the Old (eggplant – southeastern Asia; Wiersema and León 1999) and New World (pepper, potato, and tomato – Central and South America; Knapp 2002). Although the *Solanaceae* family comprises a very diverse group of plants that are grown all over the globe, it is cytogenetically very conservative since most taxa have a basic chromosome number of n = 12 (Chiarini et al. 2010).

Botanically classified as a berry, tomato (n = 12) originated in the South American Andes, ranging from northern Chile in the south, through Bolivia, Peru to Ecuador and Colombia in the north (Grubben and Denton 2004; Bai and Lindhout 2007). One species, *S. galapagense*, is endemic in the Galapagos Islands. Initially, Peru had been proposed as the center of domestication of cultivated tomato (De Candolle 1886) which would coincide with its center of origin and genetic diversity, but Peru lacks depictions of this crop on textile or pottery artifacts during the pre-Columbian era (Rick 1995). Linguistic evidence pointed to Mexico and Central America as center of domestication of tomato as the word tomato has its origin in the Aztec word *xitomatl* (Cox 2000) and wild tribes in Central America called the crop *tomati* (Gould 1983). The ancient Peruvian tribes fail to mention a tomato-like fruit at all, while Aztec documents in Central America mention dishes comprised of peppers, salt and tomato (Cox 2000). Genetic evidence also pointed to Mexico as center of domestication as modern cultivars appeared to be more closely related to a cherry tomato-like cultivar grown widely in Mexico and throughout Central America at the time of the discovery by the Spanish than any wild species grown in Peru (Rick 1995). The cherry tomato (*S. lycopersicum* L. var. *cerasiforme* (Dunal) D.M. Spooner et al.) is thought to be the direct ancestor of cultivated tomato (Tanksley 2004) and this botanical variety is still found in a semi-wild state in Central America.

In the early sixteenth century, the Spanish explorers brought tomato to Spain under the name *pome dei Moro* (Moor's apple) and from here it spread to Italy and France where it was called *pomme d'amour* (love apple), possibly a corruption of the early Spanish name *pome dei Moro* (Cox 2000). Tomato was used to a limited extent as food in Mediterranean countries, whereas northern European countries took a more cautious approach as many people associated the tomato plant with poisonous members of the Solanaceae family such as the deadly nightshade, *Atropa belladonna* L., which slightly resembles the tomato plant (Cox 2000).

The Spanish distributed tomatoes throughout their colonies, including the Philippines, and from there it reached other parts of Asia. It was only in the nineteenth century that tomatoes became widely accepted in the United States and people started to use the fruit on a more regular basis. This culminated soon in a tomato mania as medicinal powers were attributed to the fruit and almost every pill and panacea consisted of tomato extract (Cox 2000). This 'tomato fever' lasted only a few years, but boosted the popularity of tomato enormously. Large-scale breeding of tomato started around 1870, both in Europe and the United States; about 10 years later, several hundred cultivars existed.

The generic status of tomato has been in flux since the sixteenth century. After the introduction of tomato into Europe during the early sixteenth century, botanists noted their close relationship with the genus *Solanum* and referred to this species as *Solanum pomiferum* (Luckwill 1943). Joseph Pitton de Tournefort (1694; cited by Peralta et al. 2006) was the first taxonomist to assign the generic name *Lycopersicon* to tomato. Carolus Linnaeus (1753) grouped tomato under the genus *Solanum*, but only one year later, Philip Miller (1754; cited by Peralta et al. 2006) followed the nomenclature of Tournefort and described tomatoes formally under the genus *Lycopersicon*. In a posthumous edition of the book *The Gardener's and Botanist's Dictionary* (Miller 1807; cited by Peralta et al. 2006), the editor decided to follow Linnaeus' nomenclature for tomato and merged the genus *Lycopersicon* with *Solanum*, describing the tomato species as *Solanum lycopersicum*. Today, there is a general acceptance by both taxonomists and breeders to describe tomato in the section *Lycopersicon* under the genus *Solanum*. This decision is based on evidence derived from phylogenetic studies using DNA sequences and more in-depth studies of plant morphology and distribution of the species (Peralta et al. 2006).

The section *Lycopersicon* in the genus *Solanum* comprises only one domesticated species, *S. lycopersicum* L. (Tanksley 2004) and 12 crop wild relatives (Peralta et al. 2006; Bai and Lindhout 2007). Another four species are closely related to this group of species in the section *Lycopersicon*, namely *S. juglandifolium* Dun., *S. lycopersicoides* Dun., *S. ochranthum* Dun., and *S. sitiens* I.M. Johnst. (synonym *S. rickii* Correll) (Rick 1988; Peralta et al. 2006). The dissimilarities of these four species with the section *Petota* (cultivated and wild potatoes) and their similarities with the section *Lycopersicon* are: the absence of tubers, presence of yellow corolla, articulated pedicels and pinnately segmented leaves. Charles M. Rick, founder of the Tomato Genetic Resource Center (TGRC) at the University of Davis, California divided the genus formerly known as *Lycopersicon* into two distinct complexes, the *Esculentum* and the *Peruvianum* Complex (Rick 1976) and the currently known 13 tomato species can be attributed to the two groups as follows (Peralta et al. 2006):

1. ***Solanum lycopersicum* complex** consisting of eight diploid species: (1) *S. lycopersicum* L., (2) *S. pimpinellifolium* L., (3) *S. cheesmaniae* (L. Riley) Fosberg, (4) *S. neorickii* D.M. Spooner et al. (synonym *L. parviflorum* C.M. Rick et al.), (5) *S. chmielewskii* (C.M. Rick et al.) D.M. Spooner et al., (6) *S. habrochaites* S. Knapp and D.M. Spooner (synonym *L. hirsutum* Dunal), (7) *S. pennelii* Correll, and (8) *S. galapagense* S.C. Darwin and Peralta. Most of these species can be easily crossed with the cultivated tomato and used for introgression of disease and insect pest resistance genes to create new cultivars with multiple disease resistance.
2. ***Solanum peruvianum* complex** consisting of five genetically very diverse diploid species: (1) *S. chilenese* (Dunal) Reiche, (2) *S. peruvianum* L., (3) *S. huaylasense* Peralta, (4) *S. corneliomulleri* J.F. Macbr., and (5) *S. arcanum* Peralta.

Most species within the *S. lycopersicum* complex can reciprocally hybridize with cultivated tomato, with the exception of *S. habrochaites* (Robertson and Labate 2006). *S. habrochaites* can act as pollen parent in crosses with cultivated tomato, but the reciprocal cross does not set fruit.

Within the *S. peruvianum* complex, compatibility with the cultivated tomato is rather limited. *S. chilense* can act as pollen parent for *S. lycopersicum*, but viable seeds are rare and embryo rescue is required (Robertson and Labate 2006). *S. chilense* does not accept pollen from the cultigen in a reciprocal cross. *S. peruvianum* is the most widespread and genetically and morphologically most diverse species within the section *Lycopersicon* (Peralta et al. 2005) and presents severe crossing barriers in hybridization attempts with cultivated tomato (Robertson and Labate 2006). *S. huaylasense* and *S. arcanum* have only recently been described as new wild tomato species (Peralta et al. 2005). *S. huaylasense* is endemic in northern Peru, Department of Ancash, on the rocky slopes of the Callejón de Huaylas along the Río Santa and in the adjacent Río Fortaleza drainage at an elevation of 1,700–3,000 m. Likewise, *S. arcanum* is also found in northern Peru, in coastal and inland Andean valleys as well as in lomas, dry valleys and dry rocky slopes at an elevation of 100–2,800 m (Peralta et al. 2005). There is considerable intraspecific variation in *S. arcanum* which led to the distinction of four morphotypes or groups.

It has been estimated that the genomes of the tomato cultigens contain less than 5% of the genetic diversity of their wild relatives (Miller and Tanksley 1990). Apparently, the domestication and transmigration process of tomato from the Andes to Central America and from there to Europe caused a major genetic drift in this inbreeding cultigen. Despite this narrow genetic base, cultivated tomato is extremely rich in shapes, colors and sizes, in contrast to the wild forms which bear only tiny fruit. It is likely that mutations associated with larger fruit were selected and accumulated during tomato domestication (Bai and Lindhout 2007). Only in the twentieth century did the genetic potential of wild tomato relatives become apparent in initial crosses made by Charles Rick with cultivated tomato. Interspecific crosses are now widely used to tap into the gene pool of wild tomato relatives when breeding for resistance to biotic and tolerance to abiotic stresses.

Genetic resistance is the cheapest and most effective control mechanism for many tomato diseases and pests and many tomato breeding programs make use of interspecific crosses to incorporate resistance-carrying genes from wild tomato relatives into cultigens. Tomato yellow leaf curl disease (TYLCD), which is linked to several begomovirus species and transmitted by the whitefly *Bemisia tabaci,* poses a major threat to tomato production in the tropics and subtropics. Distinct sources of resistance against this disease have been integrated into cultivated tomato from *S. chilense, S. habrochaites, S. peruvianum,* and *S. pimpinellifolium* (Hanson et al. 2006; Azizi et al. 2008; Vidavski et al. 2008). Resistance sources against TYLCD or Brazilian *Begomovirus* spp. have also been reported in *S. pimpinellifolium, S. cheesmaniae*, and *S. peruvianum.* Sources of combined resistance against nematodes (*Meloidogyne* spp.) and a range of begomovirus species were identified in five *S. peruvianum* accessions through tests carried out both in Spain and Brazil (Pereira-Carvalho et al. 2010). *S. peruvianum* is also a major source of resistance against *Tomato spotted wilt virus,* which is classified as a tospovirus (Gordillo et al. 2008). Broad-spectrum resistance against four tospovirus species were reported by Dianese and co-workers (2011) in seven *S. peruvianum* accessions. Further sources of resistance were found in *S. pimpinellifolium, S. chilense,*

S. arcanum, *S. habrochaites*, *S. corneliomulleri*, and *S. lycopersicum*. Resistance against the soil-borne pathogen bacterial wilt caused by *Ralstonia solanacearum* in cultivated tomato originated from *S. lycopersicum* var. *cerasiforme* or *S. pimpinellifolium* (Prior et al. 1994).

Similarly, sources of tolerance against abiotic stress can be found in many wild tomato relatives. *S. chilense* offers good prospects for raising levels of heat tolerance in tomato (de la Peña et al. 2011). Accessions of *S. chilense, S. pennellii, S. peruvianum,* and *S. sitiens* are tolerant against drought (Tal et al. 1979; Robertson and Labate 2006), accessions of *S. cheesmanii*, *S. chilense*, *S. esculentum* var. *cerasiforme*, *S. habrochaites*, *S. pennellii, S. pimpinellifolium,* and *S. peruvianum* against salinity-alkalinity (Tal et al. 1979; Foolad 2004; Robertson and Labate 2006), accessions of *S. esculentum* var. *cerasiforme*, *S. juglandifolium*, and *S. ochranthum* against flooding (Robertson and Labate 2006), accessions of *S. habrochaites*, *S. chilense*, and *S. lycopersicoides* against chilling (Robertson and Labate 2006).

With around 84,300 accessions, the section *Lycopersicon* in the genus *Solanum* is well represented in ex situ collections around the globe (Table 16.2). The largest collection (8,107 accessions; 10%) is held by AVRDC in Taiwan, followed by individual institutions in the United States of America (6,283 accessions; 7%), the Philippines (4,751 accessions; 6%), Germany (4,062 accessions; 5%), the Russian Federation (2,540 accessions; 3%), and Japan (2,428 accessions; 3%).

The tomato collection of AVRDC is composed of 6,069 *S. lycopersicum* accessions, 127 *S. lycopersicum* var. *cerasiforme* accessions, 722 accessions of wild tomato relatives, 595 accessions with still unidentified species name and a total of 595 hybrids, introgression lines (ILs), and recombinant inbred lines (RILs) (Table 16.3). Compared with interspecific hybrids, ILs and RILs are much more useful for quantitative trait loci (QTL) identification as they carry a single introgressed region and are otherwise identical in their genome. Phenotypic variation in the ILs can easily be linked with individual introgression segments. This allows the pyramiding of various quantitative traits from different ILs into new breeding lines to maximize yield, resistance to diseases, and tolerance to abiotic stress (Robertson and Labate 2006; Vidavski et al. 2008).

16.2.6 *S. aethiopicum* L.: *African Eggplant and Solanum melongena* L.: *Asian Eggplant and Other Related Species*

Eggplants or aubergines are an economically important vegetable crop which ranked seventh in terms of global production among 27 vegetable and melon commodity groups grown worldwide in 2009, with a total output of 42.9 million tonnes (Table 16.1). China was by far the largest producer of this commodity (25.9 million tonnes; 60% of world production), followed by India (10.4 million tonnes; 24%), Egypt (1.3 million tonnes; 3%), Iran (862,000 tonnes; 2%), and Turkey (816,000 tonnes; 2%).

Table 16.3 Genetic stocks of *Solanum* section *Lycopersicon* (tomatoes) maintained by AVRDC – The World Vegetable Center

Category	Description	No. of accessions
Wild species	*S. cheesmaniae*	17
	S. chilense	47
	S. chmielewskii	11
	S. galapagense	17
	S. habrochaites	60
	S. habrochaites f. glabratum	22
	S. neorickii	12
	S. pennellii	64
	S. pennellii var. *puberulum*	1
	S. peruvianum	135
	S. peruvianum f. glandulosum	11
	S. peruvianum var. *humifusum*	3
	S. pimpinellifolium	322
Sub-total		**722**
Unidentified	*Solanum* spp.	595
Cultivated forms	*S. lycopersicum*	6,069
Cultivated forms	*S. lycopersicum* var. *cerasiforme*	127
Sub-total		**7,513**
Introgression line (IL)	*S. lycopersicum* × *S. chilense*	100
Hybrids	*S. lycopersicum* × *S. habrochaites*	2
Hybrids	*S. lycopersicum* × *S. lycopersicum* var. *cerasiforme*	17
IL	*S. lycopersicum* × *S. pennellii*	79
Hybrids	*S. lycopersicum* × *S. peruvianum*	5
Hybrids	*S.lycopersicum* × *S. pimpinellifolium*	123
Recombinant inbred line (RIL)	*S. lycopersicum* × *S. pimpinellifolium*	75
RIL	*S. lycopersicum* (Hawaii 7996) × *S. pimpinellifolium* (WVa700) (F8)	188
Hybrids	*S. habrochaites* × *S. lycopersicum*	2
Hybrids	*S. pimpinellifolium* × *S. lycopersicum* var. *cerasiforme*	4
Sub-total		**595**
Total		**8,108**

A total of 21,616 accessions of *Solanum* (eggplant) are conserved ex situ worldwide, with AVRDC in Taiwan maintaining the largest eggplant collection of 3,524 accessions (16%), followed by individual institutions in India (3,060 accessions; 14%), Japan (1,223 accessions; 6%), United States of America (887 accessions; 4%), Bangladesh (826 accessions; 4%), Philippines (661 accessions; 3%), and the Netherlands (659 accessions; 3%) (Table 16.2).

16.2.6.1 *S. aethiopicum* L.: African Eggplant, Gilo, Kumba, Shum, Scarlett Eggplant

African eggplant is grown throughout tropical Africa and found its way through the slave trade to Brazil where cultivars of the Gilo Group are grown (Grubben and Denton 2004; Porcher 2010). In the humid tropics of West Africa it is mainly grown for its fruit which is harvested and consumed while still immature, while in the savanna both leaves and fruits are consumed. In East Africa, African eggplant is mainly grown as a leafy vegetable. The ancestors of *S. aethiopicum* are the wild species *S. anguivi* Lam. and the semi-domesticated species *S. distichum* Schumach. and Thonn., both of which are endemic in tropical Africa.

Four cultivar groups are known for *S. aethiopicum* (Grubben and Denton 2004; Porcher 2010):

- **Gilo Group** – The fruit is sub-globose to ellipsoid, 2.5–12 cm long. This group comprises cultivars with smooth fruits which are popular in West and East Africa, and cultivars with ribbed fruits. Fruit color can be pure white, creamy white, pale green, dark green, brown, or purple. There are also cultivars with striped fruits in two or more colors. Cultivars of this group are mostly grown in the humid tropics of Africa.
- **Kumba Group** – The fruit is depressed globose, furrowed, often multilocular, 5–10 cm in diameter. Some cultivars are used for both fruit and leaf consumption, while others are mainly grown as leafy vegetables. Cultivars of this group are mainly grown in the hot, semi-arid climate of the Sahel.
- **Shum Group –** The fruit is sub-globose, 1–3 cm in diameter. The cultivars of this group are mainly grown as leafy vegetables in the humid tropics or under irrigation. This cultivar group is widespread in Central Africa, but also very popular in Nigeria and Uganda.
- **Acelatum Group –** The fruit is sub-globose, furrowed, 3–8 cm in diameter and thus, twice the size of the fruit from cultivars of the Shum Group. The fruit is not eaten and cultivars of this group are not grown in Africa. Cultivars are mainly grown as ornamentals or used as rootstock for tomatoes and Asian eggplant. This group has been considered to be identical with *S. integrifolium* which carries resistance genes against *Fusarium* wilt caused by *Fusarium oxysporum* f. sp. *melongenae*, an important soil-borne disease of eggplant (Toppino et al. 2008). However, as many *Solanum* species were grown in botanical gardens in Europe and as early botanical descriptions were based on plants grown in these gardens, it could be that the ornamental *S. aethiopicum* Aculeatum group was produced in Europe by selecting progeny from crosses between *S. aethiopicum* Kumba group and *S. anguivi* (Lester 1986).

African eggplant from the Gilo, Kumba and Shum cultivar groups is an important vegetable in tropical Africa and is often sold to urban markets, as fruits can be easily transported and stored.

There are intermediate forms between the first three African cultivar groups and also between *S. aethiopicum* and its wild ancestor *S. anguivi* (Grubben and Denton

2004). Some of the intermediate forms are found in humid climates of southwestern Congo and northern Angola. The plants are shrub-like and produce sweet, but hollow fruits, the size of cultivars from the Gilo group; only the leaves are eaten.

Out of 21,616 eggplant accessions held ex situ worldwide (Table 16.2), only 960 accessions belong to *S. aethiopicum* (WIEWS 2011a). A total of 360 accessions (37.5%) are maintained by the Université Nationale de Côte d'Ivoire; 199 accessions by AVRDC, Taiwan, and 106 accessions by the Netherlands. In contrast to the slightly outdated records available under VIEWS (2012), the actual figure of *S. aethiopicum* holdings at AVRDC, Taiwan is currently 268 accessions (AVGRIS 2011). The large-scale commercialization of only a few cultivars of the Gilo and Kumba groups is causing genetic erosion among the local landraces (Grubben and Denton 2004), hence efforts should be made to collect and safeguard most of these landraces in ex situ collections for use by breeders.

AVRDC started selection and breeding of African eggplant at its Regional Center for Africa in 1993. The improved variety *Tengeru White* was released in 1999 in Tanzania. Since 2007, AVRDC expanded its work on African eggplant to Madagascar, Cameroon, and Mali. Early 2011, three African eggplant cultivars were officially released in Africa, namely *Soxna* and *L10* in Mali and *DB3* in Tanzania (AVRDC 2011).

16.2.6.2 *S. anguivi* Lam.

S. anguivi is native to Africa and widely distributed in Africa and the Arabian Peninsula (Wiersema and León 1999; Grubben and Denton 2004). It mostly grows wild, but also is found as semi-domesticated vegetable in Uganda ('Uganda Pea' eggplant) and Ivory Coast ('Gnangnan'). Wild and weedy forms have prickly leaves and stems and are usually not tolerated in cultivated areas. The species is dispersed by birds after feeding on the berries.

The bitter pea-size green fruits are eaten fresh or dried and are the equivalent to the pea eggplant *Solanum torvum* Sw., a species whose distribution range is Mexico and South America, but which is widely naturalized elsewhere (Wiersema and León 1999; Porcher 2010). The fruits are also ground and taken as medicine to control high blood pressure (Grubben and Denton 2004). The plants with the red berries are also appreciated as ornamentals.

S. anguivi can be easily crossed with all cultivar groups of *S. aethiopicum* and is used in breeding programs to increase the number of fruits in African eggplant and as source for disease resistance, e.g. bacterial wilt caused by *Ralstonia solanacearum* (Wiersema and León 1999; Grubben and Denton 2004).

16.2.6.3 *S. macrocarpon* L.

S. macrocarpon is known as African eggplant or gboma eggplant. The latter name is derived from *Gboma*, a village in Liberia, West Africa (Porcher 2010). It is a semi-domesticated species and spiny wild forms are found throughout non-arid, tropical Africa (Grubben and Denton 2004). The cultivated gboma eggplant is grown for its fruits, but also as leafy vegetable. The leafy types are common throughout West and Central Africa, while the fruit types are restricted to the humid coastal zones of West Africa. Several fruit types can be found in Suriname, South America and the Caribbean, most likely introduced from West Africa (Grubben and Denton 2004).

Four cultivar groups were recognized in *S. macrocarpon* (Bukenya and Carasco 1994; Porcher 2010): (1) Semi-wild Group (synonym *S. dasyphyllum* Thonn. Ex Schum.), (2) Mukuno Group, (3) Nabingo Group, and (4) Uganda Group.

Fertile interspecific crosses between *S. macrocarpon* and *S. aethiopicum* and between *S. macrocarpon* and *S. melongena* have been obtained. This allows the easy transfer of disease and insect pest resistance genes into African and Asian eggplant (Wiersema and León 1999; Grubben and Denton 2004).

16.2.6.4 *S. melongena* L.: Eggplant, Brinjal Eggplant, Aubergine

Southeastern Asia has been suggested as the probable origin of *S. melongena* (Siemonsma and Piluek 1994; Wiersema and León 1999). Wild populations of *S. melongena* are found in the Myanmar-Yunnan border region. These developed from the *Solanum incanum* complex having migrated from the Middle East and East Africa into Asia (Grubben and Denton 2004). Thus, the origin of *S. melongena* might indeed be the Middle East and East Africa, while the center of domestication could be defined as the region between India, Myanmar and China. Here primitive and weedy eggplant types are still abundant (Grubben and Denton 2004).

Eggplant was known in Iran as early as the sixth to seventh century AD (Grubben and Denton 2004). With the westward expansion of the Muslim Empire during the eighth and ninth century AD, eggplant reached the Maghreb and possibly tropical Africa as well as southern Europe. Today, Asian eggplant is cultivated worldwide, but major production areas are located in Asia.

In Southeast Asia, two cultivar groups are recognized (Siemonsma and Piluek 1994):

1. Cv. Group Common Eggplant – with robust growth habit, purple flowers, and persistent calices at the basis of the fruit. Fruits are highly variable in form (round, elongated oval), size, and color (purple, green, white). The fruit is usually eaten as a cooked vegetable, but in Indonesia, the light green, long fruits are also eaten raw.
2. Cv. Group Bogor Eggplant – with small, spreading growth habit, and greenish-white flowers. Fruits are round, 4–10 cm in diameter, green near the basis with the calices and marbled white toward the top. In Indonesia, the crispy, slightly bitter-tasting fruit of this cultivar group is often eaten raw. There are crossing barriers between both cultivar groups.

S. melongena is partially fertile in crosses with African eggplant, *S. aethiopicum* and its closely related species, *S. macrocarpon*. The latter has potential as a gene source for disease resistance in cultivated eggplant (Wiersema and León 1999).

The fruit of eggplant is usually harvested immature, when it is still glossy and attractively colored. The flesh of mature fruit is fibrous and bitter and the seeds are hard (Grubben and Denton 2004). The fruit is usually eaten grilled, fried or steamed, sometimes stewed with other vegetables, meat or fish, or roasted and seasoned. The fruits are also preserved as pickles, sweet jam, by air-drying, freeze-drying, canning or deep-freezing. Eggplant is also widely used for medicinal purposes.

16.3 Conclusion and Outlook

Vegetables form a large and economically important commodity group comprising a wide range of genera and species. World production of vegetables and melons composed of 27 distinct commodities reached over 1 billion tonnes in 2009. The ten major vegetable crop groups contributing to this impressive output were: fresh vegetables, not elsewhere specified; tomatoes; watermelons; dry onion; cabbages and other brassicas; cucumbers and gherkins; eggplants; carrots and turnips; green chilies and peppers, and other melons, including cantaloupes. Asia is the largest vegetable producer worldwide with China alone producing close to 52% of world output, followed by India with 9.2% of global production.

About one million accessions of crops used at least partially as vegetables are conserved ex situ worldwide. In a narrow sense of exclusive use of crops as vegetables, about 500,000 accessions of vegetables representing 7% of the globally-held 7.4 million accessions of plant genetic resources are maintained ex situ. Tomatoes, capsicums, melons and cantaloupe, brassicas, cucurbits, alliums, okra, and eggplant are well represented in ex situ collections at the global level, with a range between 84,000 and 22,000 accessions per vegetable group.

As genetic erosion continues in situ for various reasons such as population growth, expansion of human settlements, and climate change, complementary collecting efforts should be made with a major focus on crop wild relatives and poorly represented cultivated forms of some vegetable groups such as those described for the genera *Brassica* and *Capsicum* in this chapter.

Increased visibility and activity of nutrition-related NGOs such as 'Leaf for Life' (Mandell 2011) and many others enhance consumer awareness of the health benefits of fruits and vegetables and create a new demand for dietary diversity and nutrient-dense vegetables that has to be addressed by plant breeders. Future breeding goals might see a broadening of scope, aiming not only for high and stable yields through multiple disease resistance and tolerance against abiotic stress, but also for high nutrient content and tastiness. Nutritional aspects might have been overlooked in the recent past when developing new high-yielding varieties, as there is evidence of nutrient loss in modern fruit and vegetable varieties (Davis et al. 2004). Landraces with average or low productivity, but high nutrient content and unique flavors might be useful for breeders in this endeavor and such lines deserve special attention for additional collecting and ex situ conservation before they are replaced by modern cultivars and become extinct. Several seed companies are already expanding their offer of heirloom varieties and other niche or indigenous vegetables such as Chinese cabbage, cilantro, fennel, basil, etc. to address a new health-conscious consumer clientele.

The outlook for increased vegetable production and consumption is bright. Driving factors are global population growth, economic expansion of developing countries with concomitant demand for more and better food, and increased awareness of consumers of the importance of fruits and vegetables for a healthy diet. It has been estimated that worldwide per capita consumption of fruits and vegetables might grow 15%. Thanks to the highly dynamic Asian market, the vegetable seed sector is expected to almost double by 2020 to reach US$6.5 billion (The Context Network 2010). To sustain this growing demand for vegetables on limited farmland, vegetable producers will need to increasingly adopt high quality, primed seed with improved agronomic traits, often developed for precise growing conditions in locations with suitable microclimates for precision agriculture. This will also mean a shift away from open-pollinated to hybrid seed to enhance yield and meet quality standards of the evolving markets, not only in industrialized countries, but also in developing economies. A broad range of high quality vegetable produce needs to be available year-round for retailers and consumers.

There is an increasing demand for processed (vegetable juice, frozen and dehydrated vegetables), semi-prepared, and ready-to-eat vegetables. Apart from diversifying the range of produce in the market place, vegetable processing also helps to reduce postharvest losses. Among the lightly processed vegetables are those sold in fresh-cut form, such as bagged salads or carrot or celery sticks. Moreover, leafy greens, microgreens and sprouts are becoming quite popular. In this context, sanitary and phytosanitary concerns and produce traceability are becoming major issues as evidenced by the recent *E. coli* outbreak in bean sprouts in northern Germany which led to the death of 30 people, sickened 3,000, and caused dramatic economic losses for vegetable producers in many European countries (Channel News Asia 2011).

Locally grown vegetables as well as organic production are gaining in popularity and market share. This helps to address environmental issues and to reduce the carbon footprint, while at the same time saving on transportation costs, reducing postharvest losses and maintaining freshness of the produce. This can be of particular importance in countries like India, the second largest producer of vegetables in the world, where 30–35% of the total vegetable production is lost during postharvest (Ahsan 2006). These losses are due to the lack of suitable harvesting equipment and packing containers, lack of efficient collecting and distribution centers in major vegetable producing areas, lack of a cold chain and proper transportation systems, and lack of adequate commercial storage facilities. A significant reduction of these high postharvest losses and adequate handling of the produce from farmers' field to the consumer will substantially increase the availability of nutritious vegetables to satisfy the dietary needs of a growing population without the need to increase the land area grown to vegetables.

16.4 Appendix

16.4.1 Abbreviations

AICRP-Mullarp	All India Coordinated Research Project on Mullarp (India)
AICRP-Soybean	All India Coordinated Research Project on Soybean (India)
ATFCC	Australian Temperate Field Crops Collection (Australia)
AVRDC	AVRDC – The World Vegetable Center (former Asian Vegetable Research and Development Center [Taiwan])
BCA	Bunda College of Agriculture (Malawi)
BGUPV	Generalidad Valenciana, Universidad Politécnica de Valencia. Escuela Técnica Superior de Ingenieros Agrónomos, Banco de Germoplasma (Spain)
BINA	Bangladesh Institute of Nuclear Agriculture
BJRI	Bangladesh Jute Research Institute
BVRC	Beijing Vegetable Research Centre (China)
CATIE	Centro Agronómico Tropical de Investigación y Enseñanza (Costa Rica)
CENARGEN	Embrapa Recursos Genéticos e Biotecnologia (Brazil)
CGIAR	Consultative Group on International Agricultural Research
CGN	Centre for Genetic Resources (Netherlands)
CIAT	Centro Internacional de Agricultura Tropical (Columbia)
CNPAF	Embrapa Arroz e Feijão (Brazil)
CNPH	Embrapa Hortaliças (Brazil)
CNPSO	Embrapa Soja (Brazil)
DAR	Department of Agricultural Research, Ministry of Agriculture (Botswana)
DGCB-UM	Department of Genetics and Cellular Biology, University Malaya (Malaysia)
DOA	Department of Agriculture, Papua New Guinea University of Technology
EWS R&D	East West Seed Research and Development Division (Bangladesh)
FCRI-DA/TH	Field Crops Research Institute – Department of Agriculture (Thailand)
GEVES	Unité Expérimentale de Sophia-Antipolis, Groupe d'Étude et de Sophia-Antiopolis contróle des Variétés et des Semences (France)
IAC	Instituto Agronómico de Campinas (Brazil)
ICA/REGION 5	Centro de Investigación El Mira, Instituto Colombiano Agropecuario El Mira (Colombia)
ICGR-CAAS	Institute of Crop Germplasm Resources, Chinese Academy of Agricultural Sciences (China)
IDESSA	Institut des Savanes (Cóte d'Ivoire)
IDI	International Dambala (Winged Bean) Institute (Sri Lanka)
IGB	Israel Gene Bank for Agricultural Crops, Agricultural Research Organization, Volcani Centre
IITA	International Institute of Tropical Agriculture
INIFAP	Instituto Nacional de Investigaciones Forestales, Agricolas y Pecuarias (Mexico)
IPB-UPLB	Institute of Plant Breeding (IPB), University of the Philippines Los Baños (UPLB) (Philippines)

(continued)

Appendix (continued)

IPK (DEU146)	Genebank, Leibniz Institute of Plant Genetics and Crop Plant Research (Germany)
LBN	National Biological Institute (Indonesia)
NBPGR (IND001)	National Bureau of Plant Genetic Resources (India)
NBPGR (IND063)	Regional Station Hyderabad, National Bureau of Plant Genetic Resources (India)
NBPGR (IND064)	Regional Station Jodhpur, National Bureau of Plant Genetic Resources (India)
NC7	North Central Regional Plant introduction Station, United States Department of Agriculture, Agricultural Research Services
NE9	Northeast Regional Plant Introduction Station, Plant Genetic Resources Unit, United States Department of Agriculture, Agricultural Research Services, New York State Agricultural Experiment Station, Cornell University
NIAS	National Institute of Agrobiological Sciences (Japan)
NRCOG	National Research Centre for Onion and Garlic (India)
ORSTOM-MONTP	Laboratoire des Ressources Génétiques et Amélioration des Plantes Tropicales, ORSTOM (France)
PGRC	Plant Genetic Resources Centre (Sri Lanka)
PGRI	Plant Genetic Resources Institute (Pakistan)
RBG	Millennium Seed Bank Project, Seed Conservation Department, Royal Botanic Gardens, Kew, Wakehurst Place (United Kingdom)
RCA	Institute for Agrobotany (Hungary)
RDAGB-GRD	Genetic Resources Division, National Institute of Agricultural Biotechnology, Rural Development Administration (Republic of Korea)
RIPV	Research Institute of Potato and Vegetables (Kazakhstan)
S9	Plant Genetic Resources Conservation Unit, Southern Regional Plant Introduction Station, University of Georgia, United States Department of Agriculture, Agricultural Research Services
SASA	Science and Advice for Scottish Agriculture, Scottish Government (United Kingdom)
SOY	Soybean Germplasm Collection, United States Department of Agriculture, Agricultural Research Services
TARI	Taiwan Agricultural Research Institute
TROPIC	Institute of Tropical and Subtropical Agriculture, Czech University of Agriculture (Czech Republic)
TSS-PDAF	Taiwan Seed Service, Provincial Department of Agriculture and Forestry (Taiwan)
UNSAAC/CICA	Universidad Nacional San Antonio Abad del Cusco (Peru)
UzRIPI	Uzbek Research Institute of Plant Industry (Uzbekistan)
VIR	N.I. Vavilov All-Russian Scientific Research Institute of Plant Industry (Russian Federation)
W6	Western Regional Plant Introduction Station, United States Department of Agriculture, Agricultural Research Services, Washing State University

References

Ahsan H (2006) 5 India (1). In: Rolle RS (ed) Postharvest management of fruit and vegetables in the Asia-Pacific region. Asian Productivity Organization/Food and Agriculture Organization, Tokyo/Rome, pp 131–142

Alercia A, Mackay M (2010) Contribution of standards for developing networks, crop ontologies and a global portal to provide access to plant genetic resources. In: Scientific and technical information and rural development IAALD XIIIth world congress, Montpellier, 26–29 Apr 2010, 7 pp

Al-Shebaz A, Beilstein MA, Kellogg EA (2006) Systematics and phylogeny of the Brassicaceae (Cruciferae): an overview. Plant Syst Evol 259:89–120

AVGRIS (2011) AVRDC vegetable genetic resources information system. http://203.64.245.173/avgris/search_result.asp?VINO=&ACCNO=&TEMPNO=&SPECIE=EG%3A+SOLANUM AETHIOPICUM&PEDCUL=&SUBTAX=&COUNTR=&NOTES=; accessed in March 2011

AVRDC (2011) New vegetable variety releases expand market options for African farmers. http://www.avrdc.org/fileadmin/pdfs/media_releases/05_AVRDC_new_variety_releases_25March11.pdf; accessed in March 2011

Azizi A, Mozafari J, Shams-bakhsh M (2008) Phenotypic and molecular screening of tomato germplasm for resistance to *Tomato yellow leaf curl virus*. Iran J Biotechnol 6(4):199–206

Bai Y, Lindhout P (2007) Domestication and breeding of tomatoes: what have we gained and what can we gain in the future? Ann Bot 100:1085–1094

Bosland PW (1996) Capsicums: innovative uses of an ancient crop. In: Janick J (ed) Progress in new crops. ASHS Press, Arlington, pp 479–487

Bosland PW, Gonzalez MM (2000) The rediscovery of *Capsicum lanceolatum* (Solanaceae), and the importance of nature reserves in preserving cryptic biodiversity. Biodivers Conserv 9(10):1391–1397

Branca F, Cartea E (2011) Chapter 2 – Brassica. In: Kole C (ed) Wild crop relatives: genomic and breeding resources oilseeds. Springer, Berlin/Heidelberg, pp 17–36

Bukenya ZR, Carasco JF (1994) Biosystematic study of *Solanum macrocarpon–S. dasphyllum* complex in Uganda and relations with *Solanum linnaeanum*. East Afr Agric Forest J 59(3):187–204

Channel News Asia (2011) Bean sprouts source of killer E. coli outbreak. http://www.channelnewsasia.com/stories/afp_world/print/1134441/1/.html. Accessed 10 Jun 2011

Chiarini FE, Moreno NC, Barboza GE, Bernardello G (2010) Karyotype characterization of Andean Solanoideae (Solanaceae). Caryologia 63(3):278–291

Cox S (2000) I say tomayto, you say tomahto. http://www.landscapeimagery.com/tomato.html; accessed in May 2011

Davis DR, Epp MD, Riordan HD (2004) Changes in USDA food composition data for 43 garden crops, 1950 to 1999. J Am Coll Nutr 23(6):669–682

De Candolle A (1886) Origin of cultivated plants. Hafner, New York (1959 reprint)

de la Peña RC, Ebert AW, Gniffke PA, Hanson P, Symonds RC (2011) Genetic adjustment to changing climates: vegetables. In: Yadav SS, Redden B, Hatfiled JS, Lotze-Campen H, Hall A (eds) Chapter 18: crop adaptation to climate change 2011. Wiley-Blackwell, Oxford, UK, pp 396–410

Dianese EC, Fonseca MEN, Inoue-Nagata AK, Resende RO, Boiteux LS (2011) Search in *Solanum* (section *Lycopersicon*) germplasm for sources of broad-spectrum resistance to four *Tospovirus* species. Euphytica pp 1–13. doi:10.1007/s10681-011-0355-8. Online First 28 Jan 2011

Duke JA (1983) *Brassica nigra* (L.) Koch. In: Duke J (ed) Handbook of energy crops, unpublished. http://www.hort.purdue.edu/newcrop/duke_energy/brassica_nigra.html; accessed in May 2011

Ebert AW, Astorga C, Ebert ICM, Mora A, Umaña C (2007) Securing our future: CATIE's germplasm collections–Asegurando nuestro futuro: colecciones de germoplasma del CATIE. Technical series. Technical bulletin no. 26. Tropical Agricultural Research and Higher Education Center, CATIE. Litografía e Imprenta LIL SA, San José, 204 pp

ECOCROP (2011) Allium chinense. http://ecocrop.fao.org/ecocrop/srv/en/cropView?id=2994; accessed in May 2011

Encyclopaedia Britannica (2011) Vegetable. http://www.britannica.com/EBchecked/topic/624564/vegetable; accessed in April 2011

Erickson RF (2011) Nikolaus Joseph, Freiherr von Jacquin 1727–1817. Rare books from the Missouri botanical garden library. http://www.illustratedgarden.org/mobot/rarebooks/author.asp?creator=Jacquin,+Nikolaus+Joseph,+Freiherr+von&creatorID=80; accessed in April 2011

Eshbaugh WH (1993) History and exploitation of a serendipitous new crop discovery. In: Janick J, Simon JE (eds) New crops. Wiley, New York, pp 132–139

FAO (2010) The second report on the state of the world's plant genetic resources for food and agriculture. FAO, Rome

FAOSTAT (2011) http://faostat.fao.org/site/567/DesktopDefault.aspx?PageID=567#ancor; accessed in May 2011

Foolad MR (2004) Recent advances in genetics of salt tolerance in tomato. Plant Cell Tiss Org Cult 76:101–119

GENESYS (2011) http://www.genesys-pgr.org/; consulted on 18 May 2011

Gomez-Campo C (1980) Studies on Cruciferae: VI. Geographical distribution and conservation status of *Boleum* Desv., *Guiaroa* Coss. and *Euzomodendron* Coss. Anal Inst Bot Cavanilles 35:165–176

Gordillo LF, Stevens MR, Millard MA, Geary B (2008) Screening two *Lycopersicon peruvianum* collections for resistance to *Tomato spotted wilt virus*. Plant Dis 92:694–704

Gould WA (1983) Tomato production, processing and quality evaluation, 2nd edn. AVI, Westport, pp 3–50

Grubben GJH, Denton OA (2004) Plant resources of tropical Africa 2. Vegetables. PROTA Foundation/Backhuys Publishers/CTA, Wageningen/Leiden/Wageningen, 668 pp

Hanelt P (1990) Taxonomy, evolution, and history. In: Rabinowitch HD, Brewster JL (eds) Onions and allied crops, vol 1, Botany, physiology and genetics. CRC Press, Boca Raton, pp 1–26

Hanson P, Green SK, Kuo G (2006) Ty-2, a gene on chromosome 11 conditioning geminivirus resistance in tomato. Tomato Genet Coop Rep 56:17–18

Jarvis A, Williams K, Williams D, Guarino L, Caballero PJ, Mottram G (2005) Use of GIS for optimizing a collecting mission for a rare wild pepper (*Capsicum flexuosum* Sendtn.) in Paraguay. Genet Resour Crop Evol 52(6):671–682

Katzer G (2011) Paprika (*Capsicum annuum* L.) Gernot Katzer's spice pages. http://www.uni-graz.at/~katzer/engl/Caps_ann.html; accessed in April 2011

Klaas M, Friesen N (2002) Chapter 8 – molecular markers in Allium. In: Rabinowitch HD, Currah L (eds) Allium crop science: recent advances. CABI, Wallingford, pp 159–186

Knapp S (2002) Tobacco to tomatoes: a phylogenetic perspective on fruit diversity in the Solanaceae. J Exp Bot 53:2001–2022

León J (2000) Botánica de los cultivos tropicales, 3rd edn. IICA, San José, revisada y aum, 522 pp

Lester RN (1986) Taxonomy of scarlet eggplants, *Solanum aethiopicum* L. Acta Hortic 182:125–132

Lü N, Yamane K, Ohnishi O (2008) Genetic diversity of cultivated and wild radish and phylogenetic relationships among *Raphanus* and *Brassica* species revealed by the analysis of *trn*K/*mat*K sequence. Breed Sci 58:15–22

Luckwill LC (1943) The genus *Lycopersicon*: an historical, biological and taxonomical survey of the wild and cultivated tomatoes. Aberdeen Univ Stud 120:1–44

Maas HI, Klaas M (1995) Infraspecific differentiation of garlic (*Allium sativum* L) by isozyme and RAPD markers. Theor Appl Genet 91:89–97

Mandell T (2011) Program to focus on making better use of vegetables. http://richmondregister.com/localnews/x2134988776/Program-to-focus-on-making-better-use-of-vegetables; accessed in June 2011

Miller JC, Tanksley SD (1990) RFLP analysis of phylogenetic relationships and genetic variation in the genus Lycopersicon. Theor Appl Genet 80:437–448

Moscone EA, Scaldaferro MA, Grabiele M, Cecchini NM, García YS, Jarret R et al (2007) The evolution of chili peppers (*Capsicum* – Solanaceae): a cytogenetic perspective. Acta Hortic 745:137–170

Nishi S (1980) Differentiation of Brassica crops in Asia and the breeding of 'hakuran', a newly synthesized leafy vegetable. In: Tsunoda S, Hinata BK, Gómez-Ocampo C (eds) *Brassica* crops and wild allies. Japan Scientific Societies Press, Tokyo, pp 133–150

Online Etyomology Dictionary (2011) Paprika. http://www.etymonline.com/index.php?term=paprika; accessed in April 2011

Opeña RT, Kuo CG, Yoon JY (1988) Breeding and seed production of Chinese cabbage in the tropics and subtropics. Technical bulletin no 17. AVRDC, Shanhua, 92 pp

Peralta IE, Knapp S, Spooner DM (2005) New species of wild tomatoes (*Solanum* section *Lycopersicon*: Solanaceae) from northern Peru. Syst Bot 30:424–434

Peralta IE, Knapp S, Spooner DM (2006) Nomenclature for wild and cultivated tomatoes. Tomato Genet Coop Rep 56:6–12

Pereira-Carvalho RC, Boiteux LS, Fonseca MEN, Díaz-Pendón JA, Moriones E, Fernández-Muñoz R, Charchar JM, Resende RO (2010) Multiple resistance to *Meloidogyne* spp. and to bipartite and monopartite *Begomovirus* spp. in wild *Solanum (Lycopersicon)* acessions. Plant Dis 94:179–185

Porcher MH (2010) Know your eggplants – part 1. http://www.plantnames.unimelb.edu.au/new/Sorting/CATALOGUE/Pt1-African-eggplants.html; accessed in April 2011

Prior P, Grimault V, Schmit J (1994) Resistance to bacterial wilt (*Pseudomonas solanacearum*) in tomato: present status and prospects. In: Hayward AC, Hartman GL (eds) Bacterial wilt: the disease and its causative agent *Pseudomonas solanacearum*. CAB International, Wallingford, pp 209–223

Rakow G (2004) I.1 Species origin and economic importance of Brassica. In: Pua EC, Douglas CJ (eds) Biotechnology in agriculture and forestry, vol 54, Brassica. Springer, Berlin/Heidelberg, 344 pp

Redden R, Vardy M, Edwards D, Raman H, Batley J (2009) Genetic and morphological diversity in the Brassicas and wild relatives. In: Proceedings of the 16th Australian research assembly on Brassicas. Ballarat Victoria, 5 pp

Ribeiro CS da C, Carvalho SIC de, Henz GP, Reifschneider FJB (2008) Pimentas Capsicum. Embrapa Hortaliças, Brasília, Athalaia Gráfica e Editora Ltda, 200 pp

Rich TCG (1991) Crucifers of Great Britain and Ireland. Botanical Society of the British Isles, London, p 336 pp

Rick CM (1976) Tomato (family Solanaceae). In: Simmonds NW (ed) Evolution of crop plants. Longman, UK, pp 268–273

Rick CM (1988) Tomato-like nightshades: affinities, autoecology, and breeders' opportunities. Econ Bot 42:145–154

Rick CM (1995) Tomato. In: Smartt J, Simmonds NW (eds) Evolution of crop plants, 2nd edn. Wiley, New York, pp 452–457

Robertson LD, Labate JA (2006) Genetic resources of tomato (Lycopersicon esculentum Mill.) and wild relatives. In: Razdan MK, Mattoo AK (eds.) Genetic improvement of Solanaceous crops, Vol. 2: Tomato. Science Publishers Inc., Enfield, pp 25–75

Schneider E (2001) Vegetables from Amaranth to Zucchini: the essential reference. Harper Collins Publishers Inc. New York, 777 pp

Shigyo M, Kik C (2008) Onion. In: Prohens J, Nuez F (eds) Handbook of plant breeding-vegetables II: Fabaceae, Liliaceae, Umbelliferae and Solanaceae. Springer, New York, pp 121–159

Siemonsma JS, Piluek K (1994) Plant resources of South-East Asia. No 8. Vegetables. PROSEA Foundation, Bogor, 412 pp

SINGER (2011) http://singer.cgiar.org/index.jsp?page=showkeycount&search=gec=gephaseolus; consulted on 18 May 2011

Tal M, Katz A, Heikin H, Dehan K (1979) Salt tolerance in the wild relatives of the cultivated tomato: proline accumulation in *Lycopersicon esculentum* Mill., *L. peruvianum* Mill. and *Solanum pennellii* Cor. Treated with NaCl and polyethylene glycol. New Phytol 82:349–355

Tanksley SD (2004) The genetic, developmental, and molecular bases of fruit size and shape variation in tomato. Plant Cell 16:181–189

Tenkouano A (2011) The nutritional and economic potential of vegetables. In: State of the world's food and agriculture 2011 Worldwatch Institute/W. W. Norton & Company, New York, pp 27–38 (notes pp 190–193)

The Context Network (2010) Global seed sector 2020 outlook: major vegetable crops, 2nd edn. http://www.contextnet.com/Context%20Fact%20Sheet%202020%20MVC%20Outlook.pdf; accessed in August 2011

Toppino L, Valè G, Rotino GL (2008) Inheritence of *Fusarium* wilt resistance introgressed from *Solanum aethiopicum Gilo* and *Aculeatum* groups into cultivated eggplant (*S. melongena*) and development of associated PCR-based markers. Molecular Breeding 22(2):237–250

Tsunoda S (1980) Eco-physiology of wild and cultivated forms in *Brassica* and allied genera. In: Gómez-Ocampo C, Tsunoda S, Hinata bK (eds) Brassica crops and wild allies. Japan Scientific Societies Press, Tokyo, pp 109–120

USDA-ARS (2011a) Taxon: *Brassica nigra* (L.) W.D.J. Koch. Germplasm Resources Information Network–(GRIN) (online database). National Germplasm Resources Laboratory, Beltsville. http://www.ars-grin.gov/cgi-bin/npgs/html/tax_search.pl; accessed in March 2011

USDA-ARS (2011b) Taxon: *Raphanus raphanistrum* L. Germplasm Resources Information Network–(GRIN) (online database). National Germplasm Resources Laboratory, Beltsville. http://www.ars-grin.gov/cgi-bin/npgs/html/tax_search.pl; accessed in April 2011

USDA-ARS (2011c) Taxon: *Raphanus raphanistrum* L. subsp. *landra* (Moretti ex DC.) Bonnier & Layens. Germplasm Resources Information Network–(GRIN) (online database). National Germplasm Resources Laboratory, Beltsville. http://www.ars-grin.gov/cgi-bin/npgs/html/tax_search.pl; accessed in April 2011

USDA-ARS (2011d) Taxon: *Capsicum baccatum* L. var. *umbilicatum* (Vell.) Hunz. & Barboza. Germplasm Resources Information Network–(GRIN) (online database). National Germplasm Resources Laboratory, Beltsville. http://www.ars-grin.gov/cgi-bin/npgs/html/taxon.pl?448641; accessed in April 2011

USDA-ARS (2011e) Family: *Solananceae* Juss., nom. cons. National Genetic Resources Program. Germplasm Resources Information Network–(GRIN) (online database). National Germplasm Resources Laboratory, Beltsville. http://www.ars-grin.gov/cgi-bin/npgs/html/family.pl?1043; accessed in April 2011

USDA-NRCS (2011) *Allium porrum* L.–garden leek. National Resources Conservation Service (plants database). http://plants.usda.gov/java/profile?symbol=ALPO2; accessed in March 2011

Vidavski F, Czosnek H, Gazit S, Levy D, Lapidot M (2008) Pyramiding of genes conferring resistance to *Tomato yellow leaf curl virus* from different wild tomato species. Plant Breed 127(6):625–631

Warwick SI, Francis A, Al-Shehbaz IA (2006) Brassicaceae: species checklist and database on CD-Rom. Plant Syst Evol 259:249–258

Wiersema JH, León B (1999) World economic plants. A standard reference. CRC Press, Boca Raton

Wikipedia (2011a) Habanero chili. http://en.wikipedia.org/wiki/Habanero_chili; accessed in March 2011

Wikipedia (2011b) Naga Viper pepper. http://en.wikipedia.org/wiki/Naga_Viper_pepper; accessed in March 2011

World Information and Early Warning System on PGRFA (WIEWS) (2011a) WIEWS Germplasm Report (based on genus-specific searches worldwide; example Allium): http://apps3.fao.org/wiews/germplasm_report.jsp?i_STID=&i_RC=&i_VINST=&i_LT=N&i_d=false&i_j=&i_r=0&i_a=Navigate&i_t=&i_m=true&i_f=&i_op=&i_np1=&i_np2=&i_FC=&i_FG=&i_FP=&i_s=N&i_UP=N&i_TI=Y&i_TC=Y&i_TR=Y&i_SO=N&i_DA=&i_CHLE=&i_SELE=&i_CHGP=&i_SEGP=&i_CHPG=&i_SEPG=&i_CHGE=&i_SEGE=&i_CHOT=&i_SEOT=&i_All=&i_l=EN&query_CALLER=%2Fwiews%2Fgermplasm_query.htm&i_u=&i_p=&query_REGION=341&query_AREA=&query_INSTCODE=&query_SPECIES=Allium&query_SAMPLE=; accessed in April 2011

World Information and Early Warning System on PGRFA (WIEWS) (2011b) http://apps3.fao.org/wiews/germplasm_report.jsp?i_STID=&i_RC=&iWIEWSGermplasmReport._VINST=&i_LT=N&i_d=false&i_j=&i_r=0&i_a=Navigate&i_t=&i_m=true&i_f=&i_op=&i_np1=&i_np2=&i_FC=&i_FG=&i_FP=&i_s=N&i_UP=N&i_TI=N&i_TC=N&i_TR=Y&i_SO=N&i_DA=&i_CHLE=&i_SELE=&i_CHGP=&i_SEGP=&i_CHPG=&i_SEPG=&i_CHGE=&i_SEGE=&i_CHOT=&i_SEOT=&i_All=&i_l=EN&query_CALLER=%2Fwiews%2Fgermplasm_query.htm&i_u=&i_p=&query_REGION=362&query_AREA=&query_INSTCODE=&query_SPECIES=Phaseolus&query_SAMPLE=; consulted on 18 May 2011

World Information and Early Warning System (WIEWS) (2012) WIEWS Germplasm Report. http://apps3.fao.org/wiews/germplasm_report.jsp?i_STID=&i_RC=&i_VINST=&i_LT=N&i_d=false&i_j=&i_r=0&i_a=Navigate&i_t=&i_m=true&i_f=&i_op=&i_np1=&i_np2=&i_FC=&i_FG=&i_FP=&i_s=N&i_UP=N&i_TI=Y&i_TC=Y&i_TR=Y&i_SO=N&i_DA=&i_CHLE=&i_SELE=&i_CHGP=&i_SEGP=&i_CHPG=&i_SEPG=&i_CHGE=&i_SEGE=&i_CHOT=&i_SEOT=&i_All=&i_l=EN&query_CALLER=%2Fwiews%2Fgermplasm_query.htm&i_u=&i_p=&query_REGION=&query_AREA=&query_INSTCODE=&query_SPECIES=Brassica+juncea&query_SAMPLE=; accessed on 16 May 2012

Yasumoto K, Nagashima T, Umeda T, Yoshimi M, Yamagishi H, Terachi T (2008) Genetic and molecular analysis of the restoration of fertility (*Rf*) genes for Ogura male-sterility from a Japanese wild radish (*Raphanus sativus* var. *hortensis* f. *raphanistroides* Makino). Euphytica 164(2):395–404

Chapter 17
Conservation of Spices and Tree Borne Oil Seed Crops

Rekha Chaudhury and S.K. Malik

17.1 Introduction: Conservation of Spices

The aromatic vegetable parts and products chiefly of tropical origin are much in use the world over and are termed spices when used as a whole or in a pulverized state. Being characterized by pungency, strong odour, and sweet or bitter taste they are used for seasoning or garnishing foods and beverages. Hard or hardened parts of most important plants such as black pepper, small and large cardamom, cinnamon, clove, ginger, turmeric, nutmeg and mace, allspice and vanilla are the most known spices of tropical nature. Plant products from species of temperate regions are generally termed culinary herbs. Flavour and taste of spices is primarily due to the presence of essential oils. They are of little food and nutritive value but more for health, personal care and hygiene. They are used in the form of whole, paste or liquid and most spices increase shelf life of food especially dry ones. Some are added to improve texture and some to introduce colour and odour.

17.2 Historical Perspective

Since ancient times they are valued as basic components of incense, embalming, preservatives, ointments, perfumes, antidotes against poison, as cosmetics and medicines. The history of spices is detailed in several books and reviews (Purseglove et al. 1981; Weiss 2002). India is acclaimed as 'Home of Spices' and is the biggest producer and consumer of spices in the World since the ancient Egyptian, Greek and Roman times (Purseglove et al. 1981; Ravindran et al. 2005). Black pepper,

R. Chaudhury (✉) • S.K. Malik
National Bureau of Plant Genetic Resources, Pusa Campus, New Delhi, India
e-mail: rekha@nbpgr.ernet.in; skm@nbpgr.ernet.in

M.N. Normah et al. (eds.), *Conservation of Tropical Plant Species*,
DOI 10.1007/978-1-4614-3776-5_17,

cinnamon, turmeric and cardamom were known in India for thousands of years as early as 1000 B.C. as evident by their relics found from Indus Valley Civilization. In this context India has contributed to the World the 'King of Spices', black pepper (*Piper nigrum* L.) and 'Queen of Spices', cardamom (*Elettaria cardamomum* (L.) Maton) that originated in tropical evergreen forests of the Western Ghats of South India and their wild forms still exist there. There are well documented records of spice trade routes of the world which were dictated by high price attached to each spice. Garlic and onion, in fact, were recorded to be in use during the construction of the pyramids of Egypt i.e. from 2600 to 1000 B.C.

17.3 Genetic Resources

The International Standards Organization has listed about 112 plant species as spices. The richest variability is represented in peninsular India where more than 100 species of spices and herbs are grown in about 2 million hectares. India accounts for 48% of the global spice trade (Anonymous 2010a). India is the native home of spices like black pepper, cardamom, ginger, turmeric, cinnamon, tamarind, curry leaf, kokam and Malabar tamarind. Indian spices have obtained geographical indicators such as Malabar pepper, Alleppey Green Cardamom, Coorg Green Cardamom and Naga chilli. The origin of important spices along with present distribution in the world depicts a large movement of spices in the world (Shanmugavelu et al. 2002).

Spices can be broadly grouped as (1) major spices (2) tree spices (3) seed spices (4) herbal spices and (5) others. The major spices of the world are categorized according to these broad groups. It is interesting to note that Ginger (*Zingiber officinale*) is the earliest known oriental spice, Caraway (*Carum carvi*) is the oldest culinary spice and Saffron (*Crocus sativus*) is the most expensive spice in the world.

The Indian sub-continent has 7,500 species of higher plants and is a centre of crop diversity- the home land of 166 crop species and 320 wild relatives (Arora 1991), 27 of which are the wild relatives of spices. As observed recently, biodiversity of tropical regions is becoming rapidly depleted. The abundant reservoirs of adaptive variation such as crop wild relatives (CWR), landraces and wild species are recognized by global scientific community to be under-represented in the collections made (FAO 2010). In this context spices and their wild relatives that are in abundance in tropical zones especially in forest areas are now threatened. Apart from usual causes of loss of biodiversity, the advent of improved high yielding varieties is a serious threat to many old local cultivars of some spices. Although different spices are cultivated worldwide contributing immensely to species as well as germplasm diversity (Chadha and Rethinam 1994; Ravindran et al. 2005), there is immediate need for their protection and conservation.

Germplasm variability in seed spices mainly existing in the form of traditional varieties were subjected to natural selection for adaptation to local conditions. They harbor valuable genes for resistance against biotic and abiotic stresses. Newer improved varieties are slowly replacing the old cultivars necessitating the protection of traditional varieties. This is particularly observed for black pepper in India where more than 70 distinct cvs are under cultivation. Also wild pepper relatives *Piper hapnium, P. barberi* and *P. silentvalleyensis* are reported to be on the verge of extinction and *P. schmidtii* and *P. wightii* are also threatened (Ravindran and Nirmal Babu 1994). These species are now reportedly found in localized areas only. Need for exploration and collection of cardamom has also been emphasized since erosion of genetic resources was reported due to changes in the habitats of the Western Ghats. Additionally cardamom maintained in the field repositories is vulnerable to diseases such as 'Katte' and rhizome rot.

Narrow genetic base of some of the spices like clove and nutmeg necessitates exploration for variability or introduction of exotic germplasm for crop improvement. Similarly the primary genepool of *Vanilla planifolia* is narrow hence the secondary genepool possessing desirable traits like disease resistance is much required to combat the threat to diversity mainly caused by diseases and the destruction of its natural habitats (Divakaran et al. 2006). Several *Vanilla* species are considered rare and endangered hence requiring urgent attention for their protection and conservation.

17.4 Conservation

It is now well recognized that, for any given genepool of plant species including spices, a number of different approaches and methods are necessary for efficient and cost-effective conservation (Chaudhury and Malik 2004). In situ conservation efforts especially for crop wild relatives at global level are increasingly emphasized (Dulloo et al. 2010). Ex situ conservation aims at maintenance of seeds and other germplasm materials in viable condition up to maximum time periods using methodologies like maintenance in field genebanks, seed banks, in vitro repository and cryopreservation. Genetic resources of spices are primarily conserved in field genebanks where limitations are high maintenance cost and vulnerability to various biotic and abiotic stresses. It is, therefore, necessary to have alternative or complementary conservation strategies for their germplasm using traditional seed genebanking and/or in vitro conservation and cryopreservation methods depending upon explants to be stored and seed storage behaviour for short-, medium- and long-term conservation. Spices that are not seed propagated or produce recalcitrant seeds need to be conserved in vitro or maintained as live plants in fields. For safety purpose it is always advisable to conserve diverse type of the germplasm using more than one method.

Spices are propagated through seeds and also vegetatively by rhizomes, bulbs and stem cuttings. Hence conservation of germplasm can be achieved using most suitable explant or diverse explants depending on purpose and availability of suitable technique. Cultivar preservation is preferred by clonal explants; however, in absence of suitable clonal explants preservation techniques, seed conservation may be attempted leading to genepool conservation. However, several spices produce non-orthodox seeds and hence cryostorage needs to be attempted.

17.5 Seed Genebanking Using Conventional Storage

Several seed spices produce seeds of orthodox nature and hence their storage life can easily be extended by storage in conventional Seed Genebank. Spices belonging to families Brassicaceae, Apiaceae, Solanaceae, Lamiaceae and Fabaceae including wild species generally produce orthodox seeds. National Seed Genebank at the National Bureau of Plant Genetic Resources (NBPGR) conserving seeds at −20°C presently holds more than 2,894 accessions of seed spices namely fenugreek (*Trigonella foenum-graceum, T. corniculata*), chilli (*Capsicum annuum, C. baccatum, C. frutescens, C. nigra*), fennel (*Foeniculum vulgare*), coriander (*Coriandrum sativum*), celery (*Apium graveolens*), cumin (*Cuminum cyminum*), caraway (*Carum carvi, C. copticum, C. roxburghisnum*), *Alliums* etc. (Anonymous 2010b). The germplasm stored for various periods are regularly monitored for seed viability, seed quantity and seed health to ensure the status of the conserved germplasm as per the genebank standards.

National Research Centre for Seed Spices, Ajmer, Rajasthan, India maintains active collections of seed spices in Medium Term Storage modules (at 5°C) for short-term storage and for facilitating germplasm distribution to indenters. According to the current report of the All India Coordinated Research Project on Spices (AICRPS), the various centres spread over India maintain a total of 2,110 accessions of coriander, 590 accessions of cumin, 651 accessions of fennel and 1,118 accessions of fenugreek at their centres. Apart from these major Institutes, few Agricultural Universities and Traditional Universities in India are maintaining germplasm mainly of coriander, cumin, fennel and fenugreek a their Institutes.

17.5.1 Conservation of Non-Orthodox Seeds

For spices producing non-orthodox seeds with short viability, that may be intermediate or recalcitrant in seed storage behavior, cryopreservation was attempted as shown in the Table 17.1. Successful cryopreservation has so far been achieved with high success in seeds of *Piper nigrum, P. mullesua, P. attenuatum, P. argyrophyllum, P. trichostachyon, P. galeatum*, cardamom and with limited success in *Alpinia galanga*.

17.5.2 In Vitro Conservation of Vegetatively Propagated Spices

For clonally propagated species, short- to medium-term conservation can be achieved by in vitro conservation techniques using normal growth conditions, slow growth conditions, induction of storage organs etc. and their applications were reported for several spices. Emphasis is to utilize in vitro conservation for establish-

Table 17.1 Spices exhibiting non-orthodox seed storage behaviour and attempts for seed cryopreservation

Species	Seed storage behaviour	Method of propagation	Conservation attempted	References
Allium cepa	Orthodox albeit with very short viability	Seed	Seed genebank (–20°C) and Cryogenebank	Ellis et al. 1990; Chaudhury et al. 1989; Lakhanpaul et al. 1996
Alpinia galanga	Recalcitrant	Seed	Cryopreservation attempted with limited success	Personal communication
Amomum subulatum	Short lived	Vegetatively and also by seeds	–	
Cinnamomum verum	Non-orthodox, viability short upto 40 days	Vegetatively and also by seeds	–	Kannan and Balakrishnan 1967; Rai and Chandra 1987
Elettaria cardamomum	Intermediate	Vegetatively and also by seeds	Successful seed cryopreservation with 70–100% recovery	Chaudhury and Chandel 1995
Garcinia cambogia	Highly recalcitrant	Seeds	Cryo attempted but no success	Malik et al. 2005
G. indica	Highly recalcitrant	Seeds	Cryo attempted but no success	Malik et al. 2005
Murraya koenigii	Highly recalcitrant	Seeds	–	Nirmal Babu et al. 2000
Myristica fragrans	Highly recalcitrant	Seeds	–	Nirmal Babu et al. 1994
Piper nigrum	Intermediate	Mainly vegetatively propagated, some seed propagated	Successful seed cryopreservation using desiccation-freezing with 50–79% recovery	Chaudhury and Chandel 1994
P. attenuatum, P. mullesua, P. argyrophyllum, P. galeatum, P. trichostachyon	–	Mainly vegetatively propagated, some seed propagated	Successful seed cryopreservation with 63–85% recovery	Decruse and Seeni 2003
Syzygium aromaticum	Highly recalcitrant, viability falls within 2 days of harvest	Vegetatively and also by seeds	–	Nirmal Babu et al. 1994
Pimenta dioica	Recalcitrant, short viability	Seeds	–	Krishnamoorthy and Rema 1988
Illicium verum	Short viability	Seeds	–	Nybe et al. 2007

Table 17.2 Germplasm of spices maintained/conserved at Indian Institute of Spices Research (IISR), Calicut, India as clonal germplasm

Crop	Total no. of accessions
Black pepper	2,695 (1,286 wild, 1,400 cvs, 9 exotics)
Cardamom	550
Curcuma	1,026
Ginger	595
Turmeric	899
Vanilla	79 (and more than 500 seedlings/mutants/interspecific hybrids)
Clove	223
Cinnamon	408
Garcinia	61
Nutmeg	482
Paprika	130 (including 96 indigenous, 34 exotic)
Allspice	180

Source: IISR, research highlights, 2010–2011

ing active genebank for periods up to few decades and cryopreservation of various explants for establishing base genebank collection for long-term storage beyond hundreds of years.

In vitro genebank is important as a safety back up of field genebanks and clonal repositories. Tissue culture has helped in obtaining pathogen free materials of elite varieties. This has helped substantially in ginger as seed rhizomes of ginger harbor diseases which spread on field plantings and hence are completely avoided in tissue culture raised plants. Micropropagation of spices was initially done for disease eradication and for distribution, followed by use of these for conservation (Peter and Abraham 2007). Cardamom, curcuma and vanilla were the first spice crops where micropropagation was initiated and it was first commercialized in cardamom. The Indian Institute of Spices Research (IISR), Calicut, Kerala, a premiere Institute in India, possesses world's largest germplasm collection in various spices and the current status of germplasm holdings in field and in in vitro is shown in Table 17.2.

Biotechnological approaches to plant genetic resources conservation of the main spices like ginger, cardamom, turmeric, black pepper, other *Piper* species, vanilla, cinnamon, clove and nutmeg are underway (Table 17.3). National Bureau of Plant Genetic Resources (NBPGR) and IISR are undertaking in vitro conservation and cryopreservation of vegetatively propagated spices species. In vitro genebank at IISR holds more than 450 accessions comprising *Piper nigrum* and related species, *Elettaria cardamomum, Zingiber officinale, Curcuma longa, Kaempferia sp., Alpinia sp., Amomum sp., Hedychium* spp., *Vanilla* spp., *Murraya koenigii, Cinnamomum verum, C. camphora, Acorus calamus* and *Capsicum annuum.*

For cardamom, Indian Cardamom Research Institute at Myladumpara, Kerala reportedly is maintaining 600 accessions of cultivated and 12 of wild and related species at the centre. About 35 accessions are maintained in the in vitro genebank in the active state. All India Coordinated Project on Spices operational in India report maintenance of 750 accessions of black pepper, 322 of cardamom, 534 of

Table 17.3 In vitro conservation and cryopreservation of vegetatively propagated spices

Crop species	In vitro conservation	References	In vitro cryopreservation	References
Piper nigrum, P. barberi, P. colubrium, P. betle, P. longum	Slow growth on half strength WPM at low temp. (for 12 months) using in vitro plantlets and shoot cultures	Geetha et al. 1995; Nirmal Babu et al. 1996	Shoot tips	Ravindran et al. 2004
Piper (7 spp.)	Conservation for various periods (for 10–22 months)	Tyagi et al. 2000; Chaudhury et al. 2006	–	–
Piper longum	Shoot tips as explants	Nirmal Babu et al. 1999a; Peter et al. 2002	–	–
Allium sativum, Allium scorodoprasum	Conserved up to 16 months	El Gizawy and Ford Lloyd 1987; Chandel and Pandey 1992; Pandey et al. 2005	Vitrification, 40% survival using shoot tips from cloves	Anonymous 2010b; Keller 2002; Volk et al. 2004
Armoracia rusticana	–	–	Encapsulation-dehydration and encapsulation-vitrification using shoot tips and root tips	Hirata et al. 2002
Mentha spp.	Slow growth on MS medium with ½ nitrogen and no PGRs for 18–24 months	Reed 1999	Controlled rate cooling, vitrification, droplet vitrification, encapsulation-vitrification	Towill 1988; Ellis et al. 2006; Uchendu and Reed 2008
Zingiber officinale	Mannitol and low temperature (10°C, 15°C) incubation	Anonymous 1989	Shoot buds cryostored using encapsulation and vitrification	Ravindran et al. 2004
Zingiber spp.	Half strength MS + mannitol (conserved upto 8–24 months)	Dekkers et al. 1991; Geetha et al. 1995; Nirmal Babu et al. 1999a; Chaudhury et al. 2006	Vitrification, 80% recovery, encapsulation, 41% recovery and encapsulation and vitrification, 66% recovery using in vitro grown shoots	Yamuna et al. 2007
Kaemferia galanga	Medium-term conservation	Geetha et al. 1997; Geetha 2002; Shirin et al. 2000	–	–

(continued)

Table 17.3 (continued)

Crop species	In vitro conservation	References	In vitro cryopreservation	References
Elettaria cardamomum	Half strength MS + mannitol (upto 12 months)	Nirmal Babu et al. 1999b; Geetha 2002	Shoot tips cryostored using encapsulation and vitrification	Ravindran et al. 2004
Elettaria cardamomum	Conserved upto 15 months	Nirmal Babu et al. 1999b; Chaudhury et al. 2006		
	Half strength MS (upto 18 months)	Tyagi et al. 2009		
Curcuma longa		Balachandran et al. 1990; Tyagi et al. 1996; 1998; Geetha 2002; Ravindran et al. 2004	Shoot tips cryostored	Ravindran et al. 2004
Curcuma spp.	Conserved upto 6–10 months	Chaudhury et al. 2006		
Crocus sativus	–	–	Vitrification using PVS2 and PVS3 led to apparent survival but no regeneration	MalekZadeh et al. 2009
Vanilla planifolia	Slow growth method using mannitol (up to 12 months)	Divakaran et al. 2006	Apices from in vitro grown plants subjected to droplet vitrification	Gonzalez-Arnao et al. 2009; Ravindran et al. 2004
Vanilla spp.	Conserved up to 6 months	Chaudhury et al. 2006		

ginger, 1,372 of turmeric, 39 of clove, 122 of nutmeg, 54 of cinnamon, 52 of cassia and several other minor spices both in field genebank and in the in vitro repositories across India.

At the in vitro repository at NBPGR, slow growth techniques are routinely used for medium-term conservation (Mandal et al. 2000) and so far 360 accessions of ginger, turmeric, pepper, cardamom along with their wild species and more than 160 accessions of *Allium* species are conserved in vitro through periodic subculture (Anonymous 2010b). More than 820 accessions of spices were maintained in the Field Genebank of NBPGR Regional Stations spread in different parts of India. However, for the long-term conservation, cryopreservation of vegetative tissues such as shoot apices, tips and meristems are carried out (Table 17.3).

Research on in vitro conservation and cryopreservation of spices like black pepper, *Allium* species., Cardamom, *Zingiber* species, *Curcuma* species, Vanilla and *Capsicum* was reviewed (Chaudhury and Malik 2004; Chaudhury et al. 2006; Peter and Abraham 2007). In vitro conservation of seven species of *Piper* namely, *P. nigrum, P. colubrinum, P. arboreum, P. betle, P. barberi, P. hapnium* and *P. longum* was reported by Tyagi et al. (2000) and for *P. barberi, P. colubrinum, P. betle* and *P. longum* by Nirmal Babu et al. (1999a). The culture room temperatures of 22± 2°C are suitable for in vitro conservation of most of the spices germplasm (Dekkers et al. 1991; Geetha et al. 1995; Nirmal Babu et al. 1994, 1999b). El Gizawy and Ford Lloyd (1987) using high sucrose and low temperature achieved in vitro conservation of garlic (*Allium sativum*) germplasm up to 16 months. However, they recorded differences of garlic varieties to shoot tip culture on B5 medium in their studies and conservation period of 12 months was achieved. Manipulations of culture media like addition of osmotic inhibitors, reduced nutrient level, reduced temperatures, use of different enclosures and sealing materials were tested with good success in the above studies.

Induction of in vitro storage organs like bulblets and microrhizomes are yet other in vitro conservation strategies. In vitro microrhizomes are ideal systems for obtaining disease free materials and for use in conservation. Such organs in ginger are reported to be genetically more stable and were successfully induced by several researchers (Bhat et al. 1994; Sharma and Singh 1995; Ravindran et al. 2004) and in *Curcuma longa* (Parthasarthy and Sasikumar 2006). In *Allium* species storage of these organs in cryovials without medium and in ginger in minimal media have proven useful. Synthetic seeds using somatic embryos and shoot buds were successfully conserved in turmeric (Sajina et al. 1997; Nayak 2000; Gayathri et al. 2005) and using in vitro regenerated shoot buds in *Vanilla planifolia* (Divakaran et al. 2006).

In vitro storage organs increase the shelf life of conserved germplasm as *Allium cepa* and *A. ampeloprasum* could be conserved in vitro for medium-term storage using bulblets (Keller 1993). In vitro generated rhizomes were used for in vitro conservation of cultivated and wild species of ginger (Tyagi et al. 2006). Using synthetic seed technology 9 months storage could be achieved in encapsulated in vitro shoot buds with over 80% viability in black pepper and encapsulated embryogenic callus and in vitro developed shoot buds with 75% recovery in cardamom (Sajina et al. 1997; Ravindran et al. 2004). Still other methodologies like use of mineral oil overlay and low temperature storage especially for temperate species are available which can be tested for their applicability to spice crops. At NBPGR protocols were developed to reduce major costs of medium (about 40–70%) for in vitro conservation in ginger and turmeric, by use of inexpensive sources of carbon (ordinary sugar) and gelling agent (isabgol). Use of these substitutes did not affect the growth, conservation period and genetic stability of the plantlets (Tyagi et al. 2007).

17.5.3 Cryobanked Spices as Seeds and In Vitro Cultures

Cryopreservation technology holds much promise for conservation of wide range of plant germplasm, especially those which are difficult-to-store due to high moisture content and high desiccation and freezing sensitivity. It is now realized that cryopreservation methods can offer greater security for long-term, cost effective conservation of plant genetic resources, including orthodox seeds (Dulloo et al. 2010). Also, orthodox seeds with short life spans like onion are recommended for cryostorage (Panis and Lambardi 2006). Since seeds are the most preferred explants for storage, different wild species and varieties of seed spices germplasm totaling 334 accessions were cryostored at National Cryogenebank at NBPGR (Chandel et al. 1993; Anonymous 2010b; personal communication). Major genera cryostored are *Allium* (15 species), *Ocimum* (8 species), *Bunium* (2 species), *Carum* (2 species) and species of *Apium graveolens, Ammi majus, Coriandrum sativum* etc. Successful cryopreservation of seeds of *Apium graveolens* was reported (Gonzalez-Benito and Iriondo 2002). Twenty five germplasm banks of Europe reportedly store old and new cultivars of this species (Frison and Serwinski 1995).

However, many spices that are indigenous to the tropics show non-orthodox seed storage behaviour (Table 17.3). Research investigations at Cryolab at NBPGR led to categorization of several of these as intermediate and recalcitrant. Intermediate seeds of black pepper and cardamom withstood drying up to 10–14% and LN exposure (Chaudhury and Chandel 1994, 1995). Careful processing of such seeds resulted in extension of their storage life at ultra-low temperatures; as after 7 years of cryostorage *Piper nigrum* seeds retained their original viability (unpublished data). A total of 129 accessions of *Piper nigrum* are cryobanked at Cryo genebank at NBPGR. Limited attempts were made for other species; however no success was achieved in *Garcinia* species and *Alpinia.*

Cryobanking of meristematic tissues has been attempted in garlic (Kim et al. 2004; Volk et al. 2004) and in mint (Hirai and Sakai 1999; Towill and Bonnart 2003; Uchendu and Reed 2008). Studies on improving cryopreservation in garlic by change in the vitrification solution from PVS2 to PVS3 and by use of cloves and bulbils has led to very high success of 85–100% by various workers (Keller 2002; Kim et al. 2004). The genebank of Leibniz Institute of Plant Genetics and Crop Plant Research (IPK) Gatersleben, Germany works intensively on in vitro conservation and cryopreservation of spices like garlic and mint including their wild species. A total of 216 mint accessions are maintained in field and in vitro (Senula and Keller 2011). Droplet vitrification method led to cryobanking of 35 accessions from 12 *Mentha* species. In addition the garlic collection at this cryobank is the third largest in Europe after the collections of Spain and the Czech Republic (Zanke et al. 2011). At IPK, cultures of 36 accessions are established in vitro and using vitrification 26 accessions are cryostored. The cryoconserved accessions belong to *Allium sativum* and wild species *A. obliquum* and *A. hookeri* (Keller et al. 2011). Detailed investigations for cryobanking of garlic collections were undertaken at genebank of the Rural Development Administration (RDA), Korea. The developed protocols were applied to 1,651 *Allium* samples of different origins and a total of 1,158 accessions were stored in the cryobank of National Agrobiodiversity Centre (Kim, personal comm). At NBPGR in vitro cryobanking of two accessions of garlic was achieved as an initiative for larger cryobanking (Anonymous 2010b).

In vitro mint collections with over 100 accessions are maintained at NCGR, Corvallis, USA (Reed 1999). Vitrification, droplet vitrification and encapsulation dehydration were successfully used to store mint germplasm and 44 accessions of *Mentha* species were cryostored at NCGRP, Ft. Collins, USA (Staats et al. 2006). Additional accessions were stored by controlled rate cooling, vitrification and encapsulation dehydration (Uchendu and Reed 2008). Over 100 garlic accessions are now cryostored at NCGRP 5 years after the initial storage set of 27 was placed in liquid nitrogen (Ellis et al. 2006).

17.6 Introduction: Conservation of Tree Borne Oilseeds

Commercial utilization of tree borne oilseeds (TBOs) as biofuel resources and for edible and non-edible oils is becoming increasingly important for the world and especially for developing economies like India. Efforts are thus underway to assess the existing genetic diversity and to ensure its availability for the genetic improvement of these rather unconventional crops. Four of the most important TBOs are dealt herewith. Tree species bearing oil seeds are represented significantly in the Indian Gene Center. In this regard, oil yielding *Jatropha curcas* L. (Ratanjyot, physic nut, family Euphorbiaceae) is being promoted as a biofuel crop not only in India (Solanki 2005) but throughout the world. After its origin in Central America it was introduced into Asia and Africa by the Portuguese. Commercial cultivation of this species is being promoted in India, Nicaragua, Cape Verde Islands, Madagascar, parts of French West Africa and Zimbabwe (Paramathma et al. 2004). It is a wild growing plant well adapted to harsh conditions of soil and climate. Due to this, Jatropha plantations are mainly promoted under the wasteland development programs.

Pongamia pinnata (L.) Pierre (Karanj, pongam, family Leguminosae), another biofuel crop, is distributed along coasts and river banks in India and Myanmar. Native to the Asian subcontinent, it was introduced to humid tropical lowlands in the Philippines, Malaysia, Australia, the Seychelles, the United States and Indonesia. It is now increasing gaining popularity due to high oil yielding quality and being resistant to drought, moderately hardy to frost and also tolerant to salinity.

Prunus armeniaca (L.) wild apricot (Chuli, family Rosaceae) is an important bio-fuel crop of the northern and western Himalayan region of India. It is believed to have originated in the mountains of northern and north-eastern China, with the Dzhungar and Zachlag mountains of Central Asia as a secondary centre of origin (Zeven and De wet 1982). Wild apricots are found as isolated trees in the high altitudes of northern India, Afghanistan and Pakistan. In India it is found in states of Himachal Pradesh and Jammu and Kashmir at altitudes up to 3,500 m above sea level. Several local types/cultivars are identified as seedling selections and are being cultivated and maintained since several years by the farmers (Malik and Chaudhury 2010).

Neem (*Azadirachta indica* A. Juss) of family Meliaceae is believed to have emerged from the Indian sub-continent and Myanmar (Kundu and Tigersted 1997). From there it spread to dry regions of Asia, Africa, Americas, Australia and the South Pacific Islands. It now forms one of the important constituents of indigenous agro-biodiversity and stands prominently as one of most liked agroforestry species in India. Almost all plant parts of this tree are being used by common man for one or other purpose while seed kernel yields oil for use as bio-fuel. It is one of the most commonly planted species in arid and semi-arid regions of the world (Ahmed et al. 1989; Chamberlain et al. 2000).

17.6.1 Conservation Needs

TBOs plants are mostly present under natural conditions, in forests, barren-lands, wastelands, and roadsides etc. The population under the farm cultivation is very limited; however, some of these species were successful in the agroforestry. Besides this various biotic and abiotic stresses are continuous threat for the germplasm growing under natural conditions. Decline in neem due to unclear causes were reported (Chamberlain et al. 2000). There is also a realization that due to lack of resource assessment and high quality materials of Jatropha and pongam, there is an impediment in large scale plantation programmes with assured returns. So, there is an urgent need for collection, evaluation and screening of the germplasm grown under natural conditions. The elite germplasm can be used for direct cultivation and for crop improvement practices.

All the four species are seed propagated and clonal propagation methods of these species are still not developed. Therefore, the collection and conservation of seeds is primary step for further technology development in TBOs. The major collection constraint in TBOs seeds is the short collection period (4–6 weeks) that precedes or coincides with the rainy season. Hence immediate processing of seeds is the preferred method for long-term (infinite) conservation of germplasm. The seeds of several tropical tree species are highly sensitive to desiccation and freezing and hence are difficult to store in viable conditions for even few days. Based on detailed studies, seeds of Jatropha and pongam were found to be short-lived; viability was lost in less than 1 year (Chaudhury and Malik 2006).

Seeds of wild apricot and neem are reported to show intermediate seed storage behaviour (Chaudhury and Chandel 1991; Malik and Chaudhury 2010). In neem seeds success in conservation was found to be dependent on provenance, genetic variation, harvest method, dehydration procedure and seed maturity at time of harvest. Besides collection of germplasm at right maturity state and correct post harvest collecting method, conservation of germplasm by proper method is very important for sustainable use in future. The TBOs germplasm can be conserved by both in situ and ex situ conservation. There are no protection sites, as such, marked for the four TBO species detailed here. Ex situ conservation as plantations, in the genebank, using in vitro and cryopreservation techniques can be achieved for diverse germplasm for varying time durations.

Post harvest seed handling, seed storage studies, longevity and long-term storage using cryopreservation has been attempted in all the four species, namely in *Jatropha curcas* and *Pongamia pinnata* (Chaudhury and Malik 2006), in wild apricot (Malik and Chaudhury 2010) and in neem (Chaudhury and Chandel 1991; Dumet and Berjak 2002).

17.6.2 Desiccation and Freezing Sensitivity of Seed, Embryos and Embryonic Axes

Preliminary experiments have led to determination of critical moisture content and freezing tolerance in the four species investigated. Whole seeds and excised embryonic axes (EA) were desiccated to different moisture levels and critical moisture was determined at the level below which there was a sudden fall in germinability. Desiccated explants were exposed to liquid nitrogen (LN). Recovery growth of seeds was monitored by rolled paper towel and petri plate germination and of embryonic axes by in vitro culture on suitable culture media. High germinability values (100%) were observed for seeds and EA of jatropha and karanj after 24 h storage in LN. However, viability declined by 20% in case of wild apricot.

Whole seeds and embryonic axes of Jatropha and Pongamia, whole seeds, half seeds and excised embryonic axes of wild apricot (Malik and Chaudhury 2010) of more than 700 accessions of Jatropha, 366 accessions of Pongamia, 274 accessions of apricot and 1,500 accessions of neem were successfully cryostored under the category of agroforestry and industrial crops (Anonymous 2010b). Using excised embryonic axes, normal healthy plantlets of Jatropha and of wild apricot were recovered. Judged by the collections received from different districts of Indian states and cryostored, the large genetic variability representing whole of the country was represented in the cryogenebank.

17.7 Conclusions

Spices can be conveniently grouped as major spices, tree spices, seed spices, herbal spices and others. Spices traverse from families as far as Orchidaceae to Zingiberaceae. Being mostly of tropical origin, they were investigated for botany, propagation and utilization in the tropics. The immense variability available offers opportunity for exploitation of these resources for improvement although threats to diversity also exist in the wild and semi-wild. It has prompted investigations into short-, medium- and long-term conservation in in vitro genebanks, seed genebanks and cryogenebanks. Success achieved in few of the species is promising, awaiting application to all the other spices based on needs and priority. Cryobase collection using seeds as explants is practicable, and relatively easy, however, use of in vitro generated explants for creating cryobase collection is urgently needed. Tree borne oilseeds species, commercially important for produce of non-edible oils, are emerging as good alternative for fuel source. Although no defined varieties are available, the existing diversity is screened for traits of importance and long term cryoconservation is successful.

References

Ahmed S, Bamotech S, Munshi M (1989) Cultivation of neem in Saudi Arabia. Econ Bot 43:35–38

Anonymous (1989) National Bureau of Plant Genetic Resources, research highlights, Newsletter, Jan–Jun 1989, New Delhi

Anonymous (2010a) Indian horticulture database, National Horticulture Board, Min. Agril. Govt India

Anonymous (2010b) Annual report of the National Bureau of Plant Genetic Resources 2010–2011. NBPGR, New Delhi, pp 200

Arora RK (1991) Plant diversity in the Indian gene centre. In: Paroda RS, Arora RK (eds) Plant genetic resources–conservation and management. International Board for Plant Genetic Resources/Regional Office for South Asia and South East Asia, New Delhi, pp 25–54

Balachandran SM, Bhat SR, Chandel KPS (1990) In vitro clonal multiplication of turmeric (*Curcuma* spp.) and ginger (*Zingiber officinale Rosc.*). Plant Cell Rep 819:521–524

Bhat SR, Chandel KPS, Kackar A (1994) In vitro induction of rhizomes in ginger (*Zingiber officinale* Rosc.). Indian J Exp Biol 32(5):340–344

Chadha KL, Rethinam P (1994) Advances in horticulture, vol 9, Plantation and spice crops- part 1. Malhotra Publishing House, New Delhi, pp 653

Chamberlain JR, Childs FJ, Harris PJC (2000) An introduction to neem, its use and genetic improvement for forestry research programme of the renewable natural resources knowledge strategy, DFID

Chandel KPS, Pandey R (1992) Distribution diversity, uses and in vitro conservation of cultivated and wild Alliums–a brief review. Indian J Plant Genet Resour 5:7–36

Chandel KPS, Chaudhury R, Radhamani J (1993) Cryopreservation-potential and prospects for germplasm conservation. Seed Res 1:239–248

Chaudhury R, Chandel KPS (1991) Cryopreservation of desiccated seeds of neem (*Azadirachta indica* A. Juss.) for germplasm conservation. Indian J Plant Genet Resour 4:67–72

Chaudhury R, Chandel KPS (1994) Germination studies and cryopreservation of seeds of black pepper (Piper nigrum L.)-a recalcitrant species. CryoLetters 15:145–150

Chaudhury R, Chandel KPS (1995) Studies on germination and cryopreservation of cardamom (*Elettaria cardamomum* Maton.) seeds. Seed Sci Technol 23:235–240

Chaudhury R, Malik SK (2004) Genetic conservation of plantation crops and spices using biotechnological approaches. Indian J Biotechnol 3:348–358

Chaudhury R, Malik SK (2006) Post-harvest handling and germplasm storage of *Jatropha curcas* L. and Pongamia pinnata (L.) Pierre using seed cryopreservation. Indian J Agroforest 8(2):38–46

Chaudhury R, Lakhanpaul S, Chandel KPS (1989) Cryopreservation studies on plant germplasm. Indian J Plant Genet Resour 2:122–130

Chaudhury R, Sharma N, Pandey R, Mandal BB, Malik SK, Gupta S, Hussain Z (2006) Biotechnological approaches to conserve plant genetic resources. In: Singh AK, Srinivasan K, Saxena S, Dhillon BS (eds) Hundred years of plant genetic resources management in India. National Bureau of Plant Genetic Resources, New Delhi, Ankur graphics, B-62/8, Naraina Indl. Area Phase-II, New Delhi pp 211–226

Decruse SW, Seeni S (2003) Seed cryopreservation is a suitable storage procedure for a range of *Piper* species. Seed Sci Technol 31:213–217

Dekkers AJ, Rao AN, Ghosh CJ (1991) In vitro storage of multiple shoot culture of gingers at ambient temperature of 24–29°C. Sci Hortic 47:157–167

Divakaran M, Nirmal Babu K, Peter KV (2006) Conservation of vanilla species, in vitro. Sci Hortic 110(2):175–180

Dulloo ME, Hunter D, Borelli T (2010) Ex situ and in situ conservation of agricultural biodiversity: major advances and research needs. Not Bot Hortic Agrobot Cluj 38(2):123–135

Dumet D, Berjak P (2002) Biotechnology in agriculture and forestry. In: Towill LE, Bajaj YPS (eds) Cryopreservation of neem (*Azadirachta indica* A. Juss.) seeds, vol 50, Cryopreservation of plant germplasm- II. Springer, Berlin/Heidelberg/New York, pp 213–219

El Gizawy AM, Ford Lloyd BV (1987) An in vitro method for the conservation and storage of garlic (*Allium sativum*) germplasm. Plant Cell Tiss Org Cult 9:147–150

Ellis RH, Hong TD, Roberts EH, Tao KL (1990) Low moisture content limits to relations between seed longevity and moisture. Ann Bot 65:493–504

Ellis D, Skogerboe D, Andre C, Hellier B, Volk G (2006) Implementation of garlic cryopreservation techniques in the national plant germplasm system. CryoLetters 27:99–106

FAO (2010) The second report on the state of the world's plant genetic resources for food and agriculture. Food and Agriculture organization of the United Nations, Rome

Frison EA, Serwinski J (1995) Directory of European Institutions holding crop genetic resources collections, vol 1, 2, 4th edn. IPGRI, Rome

Gayatri MC, Roopa Darshini V, Kavyasree R, Kumar CS (2005) Encapsulation and regeneration of aseptic shoot buds of turmeric (*Curcuma longa* L.). Plant Cell Biotechnol Mol Biol 6:89–94

Geetha SP (2002) In vitro technology for genetic conservation of some genera of Zingiberaceae. Ph.D. Thesis, University of Calicut, Calicut

Geetha SP, Manjula C, Sajina A (1995) In vitro conservation of genetic resources of spices. In: Proceedings of the seventh Kerala science congress, State Committee on Science Technology and Environment, Kerala, pp 12–16

Geetha SP, Manjula C, John CZ, Minoo D, Nirmal Babu K, Ravindran PN (1997) Micropropagation of *Kaempferia galangal* L. and *K. rotunda* L. J Spices Aromat Crops 6(2):129–135

Gonzalez-Arnao MT, Lazaro-Vallijo CE, Engelmann F, Gamez-Pastrana R, Martinez-ocampo YM, Pastelin-Solano MC, Diaz-Ramos C (2009) Multiplication and cryopreservation of vanilla (*Vanilla planifolia* 'Andrews'). In vitro Cell Develop Biol – Plant 45:574–582

Gonzalez-Benito ME, Iriondo JM (2002) Cryopreservation of *Apium graveolens* L. (celery) seeds. In: Towill LE, Bajaj YPS (eds) Biotechnology in agriculture and forestry, vol 50, Cryopreservation of plant germplasm- II. Springer, Berlin/Heidelberg/New York, pp 48–56

Hirai D, Sakai A (1999) Cryopreservation of in vitro grown axillary shoot-tip meristems of mint (*Mentha spicata*) as encapsulation- vitrification. Plant Cell Rep 19:150–155

Hirata K, Monthana P, Sakai A, Miyamoto K (2002) Cryopreservation of *Armoracia rusticana* P. Gaert., B. Mey. Et Scherb. (Horseradish) hairy root cultures. In: Towill LE, Bajaj YPS (eds) Biotechnology in agriculture and forestry, vol 50, Cryopreservation of plant germplasm- II. Springer, Berlin/Heidelberg/New York, pp 57–65

Kannan K, Balakrishnan S (1967) A note on viability of cinnamon seeds. Madras Agric J 54:78–79

Keller ERJ (1993) Sucrose, cytokinin, and ethylene influence formation of in vitro bulblets in onion and leek. Genet Resour Crop Evol 40(2):113–120

Keller ERJ (2002) Cryopreservation of garlic (*Allium sativum* L.). In: Towill LE, Bajaj YPS (eds) Biotechnology in agriculture and forestry, vol 50, Cryopreservation of plant germplasm- II. Springer, Berlin/Heidelberg/New York, pp 37–47

Keller ERJ, Senula A, Zanke C (2011) Alliaceae in cryopreservation, achievements and constraints. Acta Hortic ISHS 908:495–508

Kim HH, Cho EG, Baek HJ, Kim CY, Keller ERJ, Engelmann F (2004) Cryopreservation of garlic shoot tips by vitrification: effects of dehydration, rewarming, unloading and regrowth conditions. CryoLetters 25(1):59–70

Krishnamoorthy B, Rema J (1988) Nursery technique in tree spices. Indian Cocoa Arecanut Spices J 11(3):83–84

Kundu S, Tigersted PMA (1997) Geographical variation in seed and seedlings traits of neem (*Azadirachta indica* A. Juss) among ten populations studied in growth chamber. Silvae Genet 46:129–137

Lakhanpaul S, Babrekar PP, Chandel KPS (1996) Monitoring studies in onion (*Allium cepa L.*) seeds retrieved from storage at −20°C and − 180°C. CryoLetters 17:219–232

MalekZadeh S, Khosrowshahli M, Taeb M (2009) Cryopreservation of the axial meristem of *Crocus sativus L.* Cryobiology 59:412

Malik SK, Chaudhury R (2010) Cryopreservation of seeds and embryonic axes of wild apricot (*Prunus armeniaca* L.). Seed Sci Technol 38:231–235

Malik SK, Chaudhury R, Abraham Z (2005) Desiccation – freezing sensitivity and longevity in seeds of *Garcinia indica,* G. cambogia and G. xanthochymus. Seed Sci Technol 33:723–732

Mandal BB, Tyagi RK, Pandey R, Sharma N, Agrawal A (2000) In vitro conservation of germplasm of agri-horticultural crops at NBPGR: an overview. In: Razdan MK, Cocking EC (eds) Conservation of plant genetic resources in vitro, vol 2, Application and limitations. Oxford/IBH, New Delhi, pp 279–308

Nayak S (2000) In vitro microrhizome production in four cultivars of turmeric (*Curcuma longa* L.) as regulated by different factors. In: Proceedings of the centennial conference on spices and aromatic plants, challenges and opportunities in the new century, Calicut 20–23 Sept 2000, pp. 3–9

Nirmal Babu K, Rema J, Ravindran PN (1994) Biotechnology research in spice crops. In: Chadha KL, Rethinam P (eds) Advances in horticulture, vol 9, Plantation crops and spices. Malhotra Publishing House, New Delhi, pp 633–653

Nirmal Babu K, Geetha SP, Manjula C, Sajina A, Minoo D, Samsudeen K, Ravindran PN, Peter KV (1996) Biotechnology- its role in conservation of genetic resources of spices. In: Das MR, Satish M (eds) Biotechnology for development. State committee on Science, Technology and Environment, Thiruvananthapuram, pp 198–212

Nirmal Babu K, Geetha SP, Minoo D, Ravindran PN, Peter KV (1999a) In vitro conservation of germplasm. In: Ghosh SP (ed) Biotechnology and its application in horticulture. Narosa Publishing House, New Delhi, pp 106–129

Nirmal Babu K, Geetha SP, Minoo D, Ravindran PN, Peter KV (1999b) In vitro conservation of cardamom (*Elettara cardamomum Maton*) germplasm. Plant Genet Resour Newsl 119:41–45

Nirmal Babu K, Anu A, Ramashree AB, Praveen K (2000) Micropropagation of curry leaf tree. Plant Cell Tiss Org Cult 61:199–203

Nybe EV, Raj NM, Peter KV (2007) In: Peter KV (ed) Spices: horticulture science series-5. New India Publishing Agency, New Delhi, p 316

Pandey R, Das A, Apte SR, Negi KS (2005) In vitro conservation of exotic *Allium scorodoprasum* germplasm. Indian J Plant Genet Resour 18:99–100

Panis B, Lambardi M (2006) Status of cryopreservation technologies in plants (crops and forest trees). In: Ruane J, Sonnino A (eds) The role of biotechnology in exploring and protecting agricultural genetic resources. Food and Agriculture Organization of the United Nations, Rome, pp 61–78

Paramathma M, Parthiban KT, Neelakantan KS (2004) *Jatropha curcas*. Forest College and Research Institute, Tamil Nadu Agricultural University, Mettupalayam, pp 48

Parthasarthy VA, Sasikumar B (2006) Biotechnology of curcuma, CAB reviews: perspectives in agriculture, veterinary science, Nutrition and National Resources 1 no.020:1–9
Peter KV, Abraham Z (2007) Biodiversity in horticultural crops, vol I. Daya Publishing House, New Delhi
Peter KV, Ravindran PN, Nirmal Babu K, Sasikumar B, Minoo D, Geetha SP, Rajalakshmi K (2002) Establishing In vitro conservatory of spices germplasm. ICAR project report. Indian Institute of Spices Research, Calicut, p 131
Purseglove JW, Brown EG, Green CL, Robbins SRJ (1981) Spices, vol 1. Longman Group Limited, Harlow, p 439
Rai VRS, Chandra KSJ (1987) Clonal propagation of *Cinnamomum zeylanicum* Breyn. by tissue culture. Plant Cell Tiss Org Cult 9(1):81–88
Ravindran PN, Nirmal Babu K (1994) Chemotaxonomy of south Indian piper. J Spices Aromat Crops 3:6–13
Ravindran PN, Nirmal Babu K, Saji KV, Geetha SP, Praveen K, Yamuna G (2004) Conservation of spices genetic resources in in vitro gene banks. ICAR project report. Indian Institute of Spices Research, Calicut, p 81
Ravindran PN, Nirmal Babu K, Peter KV, Abraham Z, Tyagi RK (2005) Spices. In: Dhillon BS, Tyagi RK, Saxena S, Randhawa GJ (eds) Plant genetic resources: horticultural crops. Indian Society of Plant Genetic Resources/Narosa Publishing House, New Delhi, pp 190–227
Reed BM (1999) In vitro storage conditions for mint germplasm. HortScience 34:350–352
Sajina A, Minoo D, Geetha P, Samsudeen K, Rema J, Nirmal Babu K, Ravindran PN, Peter KV (1997) Production of synthetic seeds in few spice crops. In: Ramana KV, Sasikumar B, Nirmal Babu K, Eapen SJ (eds) Biotechnology of spices, medicinal and aromatic plants. Indian Society for Spices, Calicut, pp 65–69
Senula A, Keller ERJ (2011) Cryopreservation of mint- routine application in a genebank, experience and problems. Acta Hortic ISHS 908:467–475
Shanmugavelu KG, Kumar N, Peter KV (2002) Production technology of spices and plantation crops. Agrobios, India, p 546
Sharma TR, Singh BM (1995) In vitro microrhizome production in *Zingiber officinale* Rosc. Plant Cell Rep 15:274–277
Shirin F, Kumar K, Mishra Y (2000) In vitro plantlet production system for *Kaempferia galanga*, a rare Indian medicinal herb. Plant Cell Tiss Org Cult 63(3):193–197
Solanki KR (2005) *Jatropha curcas* (Ratanjyot): an important alternative and renewable source of biodiesel. Indian Farm 54:19–22
Staats E, Towill LE, Laufmann J, Reed B, Ellis D (2006) Genebanking of vegetatively propagated crops – cryopreservation of forty-four *Mentha* accessions. In Vitro Cell Dev Biol Plant 42:45A
Towill LE (1988) Survival of shoot tips from mint species after short term exposure to cryogenic conditions. HorticSci 23:339–341
Towill LE, Bonart R (2003) Cracking in a vitrification solution during cooling or warming does not effect growth of cryopreserved mint shoot tips. CryoLetters 24:341–346
Tyagi RK, Bhat SR, Chandel KPS (1996) In vitro conservation strategies for spices crop germplasm-*Zingiber*, *Curcuma* and *Piper* species. In: Proceedings of the 12th symposium plantation crops, (PLACROSYM-XII), Kottayam, 27–29 Nov 1996, pp 77–82
Tyagi RK, Bhat SR, Chandel KPS (1998) In vitro conservation strategies for spices crop germplasm: *Zingiber, Curcuma* and *Piper* species. In: Mathew NM, Jacob CK (eds) Developments in plantation crops research. Allied Publishers Ltd., New Delhi, pp 77–82
Tyagi RK, Yusuf A, Dua P (2000) In vitro conservation of *Zingiber, Curcuma* and *Piper* species at NBPGR. In: Proceedings of the National symposium on prospects and potentials of plant biotechnology in India in the 21st century, Department of Botany, Jai Narain Vyas University, Jodhpur, p 172
Tyagi RK, Agrawal A, Yusuf A (2006) Conservation of cultivated and wild species of *Zingiber* through in vitro rhizome formation. Sci Hortic 108:210–219
Tyagi RK, Agrawal A, Mahalakshmi C, Zakir H, Tyagi H (2007) Low-cost media for in vitro conservation of turmeric (*Curcuma longa* L.) and genetic stability assessment using RAPD markers. In vitro Cell Develop Biol Plant 43:51–58

Tyagi RK, Goswami R, Sanayaima R, Singh R, Tandon R, Agrawal A (2009) Micropropagation and slow growth in vitro conservation of cardamom (*Elettaria cardamomum* Maton). In vitro Cell Develop Biol Plant 45:721–729

Uchendu EE, Reed BM (2008) A comparative study of three cryopreservation protocols for effective storage of in vitro-grown mint (*Mentha* spp.). CryoLetters 29:181–188

Volk GM, Maness N, Rotindo K (2004) Cryopreservation of garlic (*Allium sativum*) using plant vitrification solution 2. CryoLetters 25(3):219–226

Weiss EA (2002) Spice crops. CAB International, Wallingford, p 411

Yamuna G, Sumathi V, Geetha SP, Praveen K, Swapna N, Nirmal Babu K (2007) Cryopreservation of in vitro grown shoots of ginger (*Zingiber officinale* Rosc.). CryoLetters 28(4):241–252

Zanke C, Zamecnik J, Kotlinska T, Olas M, Keller ERJ (2011) Cryopreservation of garlic for the establishment of a European core collection. Acta Hortic ISHS 908:431–438

Zeven AC, de Wet JMJ (1982) Dictionary of cultivated plants and their regions of diversity: excluding most ornamentals, forest trees and lower plants. Centre for Agricultural Publishing and Documentation (Pudoc), Wageningen, pp 1–11

Chapter 18
Conservation of Medicinal Plants in the Tropics

Neelam Sharma and Ruchira Pandey

18.1 Introduction

Medicinal plants play pivotal role in the treatment of a number of diseases for a large population in the developing world. Regarding the total number of medicinal plants on earth, there is no authentic information. However, according to a recent report, the numbers of species used medicinally include: 53,000 worldwide: 10,000–11,250 in China, 7,500 in India; 2,237 in Mexico etc. (see Sharma et al. 2010). Of about 250,000 plant species known to occur in the world, about 150,000 are distributed in the tropics of which approximately, 6,000 have medicinal value (Ratnam and Teik 1999). The usage of plants by human beings for medicines dates back to prehistoric times. According to World Health organization (WHO), 80% of the world population relies upon plants for primary health care (Vieira 1999).

Approximately two-thirds of the biological diversity of the world is found in tropical zones, mainly in developing countries. According to an earlier estimate, over 50% of all plant species occur in 25 biodiversity hotspots and 16 of these are located in the tropics (Sarasan et al. 2006). Known for the biodiversity richness and endemism of several species, many of these hotspots harbour extremely valuable species of medicinal value. With increasing demand coupled with expanding international trade, many wild populations of medicinal plants throughout tropics are under pressure and have succumbed to over-exploitation. Additionally, an increasing population, rapid urbanization and habitat destruction coupled with the uncontrolled collection from the wild, have further aggravated the loss. In case of medicinal plants, the use of root, bark, wood, stem and/or whole plant (in some cases), leads to destructive harvesting which poses a major threat to the genetic stocks and the diversity of medicinal plants.

N. Sharma (✉) • R. Pandey
Tissue Culture and Cryopreservation Unit, National Bureau of Plant Genetic Resources, Pusa Campus, New Delhi 110012, India
e-mail: neelam@nbpgr.ernet.in; ruchira@nbpgr.ernet.in

M.N. Normah et al. (eds.), *Conservation of Tropical Plant Species*,
DOI 10.1007/978-1-4614-3776-5_18, © Springer Science+Business Media New York 2013

In an all out effort towards saving the world's biodiversity from extinction, it is of utmost concern to realize the importance of the medicinal plants in providing humankind with a variety of life-saving medicines. As these comprise an important renewable natural resource, their scientific conservation involves an integration of various activities such as protection, preservation, maintenance, conservation and sustainable utilization (Chandel et al. 1996; Joy et al. 1998; Natesh 1999; Sarasan et al. 2006). There has to be a major emphasis on providing high-quality planting material for use on a sustainable basis for which saving the medicinal plant diversity in the wild, is crucial.

18.2 The Need for Conservation

Arguments were put forward a few years ago to sensitize the public about destruction of tropical rain forest and seasonally dry monsoon forests. No heed was given to these arguments earlier but today, an all out effort is being made to conserve biodiversity at the global level including serious initiatives in many tropical countries. As cited by Okigbo et al. (2008), the current cry of ecological genocide, genetic erosion, environmental degeneration as well as extinction of our biological heritage, has to be attributed to the inaction in the past. It has also been opined that with continued destruction of tropical forests, medicinal plants and their natural habitats, will remain under the threat of overexploitation more than ever before. The ultimate goal of conservation is to preserve the natural habitats of vulnerable medicinal plant species and to achieve their sustainable exploitation in less vulnerable areas (Cunningham 1993). Despite the importance of medicinal plants, they are mercilessly exploited with little concern for their future maintenance. With increasing consumer demand, there is indiscriminate harvesting of wild plants. Consequently, both, the ecosystems and their precious biodiversity are critically affected. The damage is especially significant with the removal of bark, roots, seeds and flowers – all essential for the species survival. The preservation of this medicinal plant diversity is, therefore, of utmost importance and needs to be addressed seriously on high priority. Not all medicinal plants are threatened by extinction and not all methods of harvesting are endangering their survival. To ensure the availability of raw material for industries and to explore the possibility of future development, sustainability of medicinal plants and preservation of the variability of germplasm are absolutely necessary (Chandel et al. 1996; Sharma et al. 2005, 2010). Conservation of medicinal plants is concerned mainly with activities to protect them against human disturbance, hence public participation is mandatory.

18.3 Distribution of Medicinal Plants in the Tropics

The tropical regions of the world have an important role to play in maintaining human health as they meet with the current global demand for natural medicine. Around 70% of world's medicinal plants are located in the tropics where threat to their survival is greatest and conservation resources are minimal (Myers et al. 2000). The mass extinction of species that is occurring currently can be significantly reduced by providing conservation support within these hotspots (Myers et al. 2000). Although plant-based drugs contribute significantly to the pharmacopeia of modern medicine, yet systematic investigation for their pharmacological activity has been carried out in only 5–10% of all plants in the world (Ratnam and Teik 1999). Awareness about important uses of medicinal plants in many countries has led to the accelerated destruction of the natural resources. It is, therefore, desirable that the exploitation of medicinal plants be stopped with suitable measures adopted to conserve the precious diversity. Many of the plants, known to occur in the threatened tropical forests, are still uninvestigated. Besides, complete inventory of medicinal plants is also lacking. In most of the tropical countries which are in possession of such useful resources, there is no systematic record of the information, till date. In view of the above, the information presented below has been compiled from available sources and restricted to countries wherein there is substantial diversity and suitable measures being adopted to conserve it.

18.3.1 Africa

Medicinal plants constitute an important part of natural biodiversity endowment of many countries in Africa. The last decade has witnessed an increasingly important application of medicinal plants in healthcare, local economies, cultural integrity and the well-being of people, particularly the rural population (Okigbo et al. 2008). Results of the ethnobotanical studies carried out throughout Africa indicate that native plants contribute majorly towards traditional African medicines (Okigbo et al. 2008). In Africa traditional medical practioners (TMPs) are consulted for healthcare by a large population. The collection for traditional medicinal plants is generally done from the wild (Cunningham 1993). The demand by the local population of Africa has led to indiscriminate harvesting of medicinal plants, including those in forests which resulted in many medicinal plants and other genetic materials become extinct before they could even be properly documented (Okigbo et al. 2008). According to a report about 70% of the wild plants in North Africa are known to be of potential value in fields such as medicine, biotechnology and crop improvement.

The North Africa Biodiversity Programme (NABP) initiated in 1994, aimed at promoting biodiversity conservation and sustainable use of natural resources. Emphasis was also laid on cultivation of medicinal and aromatic plants in nurseries, and the knowledge acquired was then transmitted to the local communities. Initiatives undertaken in Algeria had pilot projects wherein cultivation of medicinal plants, rural development and gender empowerment were interlinked (http://www.iucn.org/about/union/secretariat/offices/iucnmed/iucn_med_programme/species/medicinal_plants_in_north_africa/). The herbal medicine trade is gaining popularity worldwide including Africa. In case of medicinal plants, the species specific nature of demand has led to generating long distance trade across international boundaries. In Africa, the gathering of medicinal plants was traditionally being done by TMPs or by people trained by them. The number of traditional practitioners in Tanzania has been estimated to be 30,000–40,000 in comparison to 600 medical doctors. In Malawi, there were an estimated 17,000 TMPs and only 35 medical doctors in practice in the country (see Cunningham 1993).

18.3.2 Brazil

Brazil is a country which harbors the greatest biodiversity on the planet, with around 55,000 native species distributed over six major biomes: Amazon (30,000); Cerrado (10,000); Caatinga (4,000); Atlantic rainforest (10,000), Pantanal (10,000) and the subtropical forest (3,000) (Vieira 1999). The Brazilian Amazon Forest has about 800 plant species which include hundreds of medicinal plants. The 'Cerrado' is the second largest ecological dominion of Brazil, where a continuous herbaceous stratum is joined to an arboreal stratum, with variable density of woody species. The cerrados covers approximately 25% of Brazilian territory and has around 220 species used in the traditional medicine. In the 'Caatinga' region, several medicinal plants have their centers of genetic diversity and find their use as folk medicines. The Atlantic Forest extending over the Brazilian coastline has many important medicinal species such as *Mikania glomerata, Bauhinia forficata, Psychotria ipecacuanha*, and *Ocotea odorifera.* The territory of the Meridional Forests and Grasslands is reported to have several medicinal plants, such as chamomile (*Matricaria recutita*), calendula (*Calendula officinalis*), lemon balm (*Melissa officinalis*), rosemary (*Rosmarinus officinalis*), basil (*Ocimum basilicum*) and oregano (*Origanum vulgare*). There have been efforts during last two decades towards collection and preservation of medicinal plants in Brazil. The National Center for Genetic Resources and Biotechnology – Cenargen, in association with other research centers of Embrapa (Brazilian Agricultural Research Corporation), and several universities, have been actively involved in establishing germplasm banks for medicinal and aromatic species (Vieira 1999).

18.3.3 India

India has a glorious tradition of health care system based on plants, which dates back to 5000 B.C. Representing one of the world's mega biodiversity centres, India is 10th amongst gene-rich countries of the world and 4th amongst the countries of Asia. India possesses about 45,000 species of plants of which 20,000 represent higher plants, one-third of it being endemic and about 750 species are categorized to have proven clinical activity (Chandel et al. 1996). Interestingly, a very small proportion of medicinal plants represent lichens, ferns, algae etc. From amongst 34 biodiversity hotspots in the tropics, three are located in India- namely in Indo-Burma (earlier Eastern Himalayas), Western Ghats and Himalayas (http://www.kerenvis.

nic.in/isbeid/biodiversity.htm.). Around 70% of India's medicinal plants are found in the tropical areas, mostly in the various forest types spread across the western and eastern Ghats, Vindhyas, Chotta Napier plateau. Aravalis, the Terai region in the foothills of the Himalayas and the North East.

India is a treasure house of medicinal plants which are commonly used in allopathy, in Indian system of medicine (ISM) and in tribal and folk practices. The rich heritage of Indian system of medicine includes Ayurveda (about 5,000 years old); Unani, about 3,000 years old and Sidha, also equally old (Natesh 1999). Indian subcontinent possesses a rich and varied cultural and ethnic diversity, which includes over 550 tribal communities belonging to 227 ethnic groups spread over 5,000 forest villages. Since long, these tribes and communities have been using these naturally occurring medicinal resources for their primary health care. The associated indigenous knowledge has been conserved and used by these folklores for treating their day-to-day illnesses.

In India, the Western Ghats, represent one of the major repositories of tropical medicinal plants. It has around 4,000 species of higher plants of which 500 species have medicinal value (Krishnan et al. 2011). Of these, about 50 species find use in traditional and folk medicinal practices. Many valuable species such as *Coscinum fenestratum, Nothapodytes foetida, Coleus forskohlii, Mucuna pruriens* and *Rauvolfia serpentina* are the sources of important phytochemicals such as berberine, camptothecin, forskolin, L-Dopa and reserpine, respectively. These valuable species are known to occur in Western Ghats. Majority of medicinal plants from Western Ghats, used extensively in traditional medicine, are seldom cultivated. Figure 18.1 illustrates some tropical medicinal plants.

In the absence of cultivation practices, there is destructive harvesting for various plant parts such as stem, roots or tubers in more than 70% of the plant collections and this leads to their depletion from the wild. In addition, with expanding global market, many important medicinal plants are becoming scarce and some are on the brink of extinction. In recent years, the growing demand for herbal products has also led to a quantum jump in volume of plant material traded within and outside the country. Consequently, the danger is looming large with respect to threat to genetic stocks and to the diversity of medicinal plants. Therefore, it is important to conserve the extensively traded medicinal plants in their natural environment or cultivate them in favorable environments. According to the IUCN designed Conservation Assessment and Management Plan (CAMP) methodology, about 112 species from southern India, 74 species from Northern and Central India and 42 species from the high altitude of Himalayas are threatened in the wild (Krishnan et al. 2011, see Sharma et al. 2010).

18.3.4 Malaysia

Malaysia is one of the mega diversity-rich countries of the world and possesses an estimated 1,85,000 species of fauna and 12,500 species of flowering plants, many of

Fig. 18.1 (**a–g**) Some tropical medicinal plants. (**a**) *Acorus calamus*. (**b**) *Kaempferia galanga*. (**c**) *Rauvolfia serpentina*. (**d**) *Alpinia purpurata*. (**e**) *Costus speciosus*. (**f**) *Aristolochia indica*. (**g**) *Psoralea corylifolia*

which are endemic to tropical forests in this region. Out of 12,500 species of seed plants and 5,000 species of cryptogams, about 2,000 species have medicinal properties (Latiff 1999). Many of these medicinally important species grow in the wild with some cultivated near houses. Many of these need to be investigated for chemical analyses with some having been tested for their chemical content. Traditionally, medicinal plants have been collected for domestic use but the growing awareness about traditional medicines and the international demand for local crude material has led to their over harvesting from forests. Though genetic diversity estimation of these medicinally important species is still being worked upon, genetic erosion has already taken place. The need of the hour is to safeguard this biodiversity which is an important and invaluable national asset for both present and future generations. Thus, both in situ and ex situ conservation have to be practiced for their sustainable utilization (Normah et al. 2011).

18.3.5 Sri Lanka

In Sri Lanka, the traditional system of health care and *ayurveda* (a holistic system of medicine and health care which originated in India) have been practiced for 3,000 years and about 30–35% of the people use these to treat illnesses. As per available information, total of 1,414 plant species have been used for this purpose and some of these are endemic species under threat of extinction (Mahindapala 2004). Of the 200 species of medicinal plants in common use, 50 are routinely used in *ayurvedic* and traditional health care systems (Russell-Smith et al. 2006). It is interesting to note that medicinal plants are also used in rituals, cultural activities and in religious functions in Sri Lanka. The increased demand both for local use as well as for export has resulted in the decline of the natural populations of medicinal plants with local people collecting these precious medicinal resources in an unsustainable manner. Nearly 68% of the national demand for medicinal plants is met with collections made from the wild which has compounded the problem and led to near threatened status of around 80 medicinal plant species (Mahindapala 2004). Regarding the knowledge about traditional medicinal systems, only a small part is documented while the majority remains recorded in ancient, obscure *ola* (palm leaf) manuscripts scattered around the country or in the memory of elderly practitioners (Mahindapala 2004).

18.4 Conservation Strategies

Biodiversity conservation of medicinal plants is an important activity not only for preservation, procurement and sustainability but also for popularization, education and environmental conservation. A scientific approach is required to conserve these threatened species and it mainly involves the in situ and ex situ methods of conservation. Subsequently top priority should be accorded to in situ conservation of such naturally occurring forest and non-forest areas by clearly demarcating them as 'Medicinal Plant Conservation Reserves'. Ex situ conservation should be done preferably within the natural areas of distribution and the selected germplasm should be widely spaced and broad based including all the varieties and strains existing within the entire range spread over the geographical distribution of the particular species. In order to make conservation sustainable, involvement of local communities is an important strategy. Effective conservation strategy for medicinal plants should thus take place within four main areas:- in-situ conservation, ex-situ conservation, education and research. Although species conservation is achieved most effectively through the management of wild populations and natural habitats (in situ conservation), ex situ techniques can be used to complement in situ methods and, in some instances, may be the only option for some species (Maunder 2001).

18.4.1 In Situ Conservation

The in situ or on site conservation is the best way of conserving wild species or stock of biological community as these are protected and preserved in their natural habitat. Many species survive and perpetuate optimally only in their own niche or the micro-climate available in the wild habitats. For this, saving the ecosystem and natural habitat is very important (Pareek et al. 2005). Thus, in situ conservation involves preservation of entire habitat rich in biodiversity in their pristine purity (Natesh 1999). The in situ methodology for conserving the medicinal plant flora is adopted to conserve the flora on a long term sustainable basis for use by future generations.

A holistic approach is required for systematic management of in situ conservation zones. In situ conservation activities had been initiated by Government of India and accomplished through Man and Biosphere Programme, under the aegis of National Committee on Environmental Planning and Co-ordination, and United Nations Educational Scientific and Cultural Organization (UNESCO) wherein out of the 14 Biosphere reserves, 12 were declared. In addition, an integrated network of protected areas including National Parks (89), wild life sanctuaries (504) and wetland and mangroves (39) providing in situ conservation of many medicinal plants, had also been demarcated. According to Dubey and Dubey (2010), in situ conservation of medicinal plants, has been further promoted by the Government of India through The Ministry of Environment and Forests. Establishment of Protected Area Network of over 1,50,000 sq. km. (including 34,819 sq. km. under National Parks and 1,15,903 sq. km. under sanctuaries) has been accomplished under the Wildlife (Protection) Act, 1972. In many other tropical countries, initiative has been taken to implement conservation efforts by designating various types of natural ecosystems as natural reserves, wild life sanctuaries, national parks, biosphere reserves, protection of forests, recreational forests and other types of protected areas (Khan et al. 2005).

To strengthen the in situ conservation of the medicinal plant resources, three factors are critical and these need to be incorporated in all conservation-oriented programmes: (1) Prevention of the destruction of populations and their habitats, (2) Maintainance and/or enhancement of the population level and variability, and (3) Prevention of collection and excessive commercial exploitation (Chandel et al. 1996; Natesh 2000).

In this connection, the Foundation for Revitalization of Local Health Traditional (FRLHT) at Bangalore, a non- governmental foundation, is coordinating a major medicinal plant conservation activity. This foundation has set up 200 ha area to conserve cross sections of diverse ecosystems containing 30 medicinal plant species and their genetic diversity. Pioneering effort include the establishment of the world's first in situ gene bank network of 55 conservation sites. The focus of these in situ sites is the conservation of wild medicinal plant germplasm resources. It has also made a chain of medicinal plants conservation parks (MPCPs) which are being established in the western and eastern Ghats across the states of Karnataka, Kerala and Tamil Nadu. This has been a major step towards in situ conservation of medicinal plant genetic resources in India. In fact, these 55 MPCA sites represent models for other communities worldwide to implement them for maintaining their own indigenous health traditions along with biological and cultural diversity (Khan et al. 2005). In Sri Lanka, there are five medicinal Plant Conservation Areas (MCPAs) in different ecological zones. Each MCP houses a medicinal plant garden, a medicinal plant processing center for the use of communities, an ayurvedic dispensary, and an information center (Mahindapala 2004).

Several indigenous communities/ethnic societies found in India live in complete harmony with natural surroundings. They possess their own habitats and natural resources, beliefs and diverse ways of conservation and use of plants. These communities have been responsible for identifying the plant species and evolved innovations practices. Some indigenous communities have developed a traditional system of conservation called 'Sacred Groves' (the forests of God and Goddess) which is linked with certain faiths and beliefs. These are untouched, virgin forests and conserve the endangered flora and fauna. Such types of forests have been left untouched by the tribal and local people and any interference into it is "Taboo". In these forests, the dead wood or the fallen foliage may be picked up, but never the live plant parts or tree branches can be extracted. There are more than 500 sacred groves in the tribal belt of the North-eastern region, Maharashtra, Western Ghats, Orissa, and Nilgiris in Tamil Nadu and Kerala in India and many in Sri Lanka. The Government of India has also protected these forests by declaring them as Reserve areas. The Namdapha Biosphere Reserve in Arunachal Pradesh for *Coptis teeta* and Dombeyang Valley in North Sikkim for *Panax pseudo ginseng* are two examples of the conservation of natural habitat. For this, saving habitat itself is very important and tribal communities are helping in this approach by naming the forest as sacred groves (Chandel et al. 1996; Natesh 2000).

In situ conservation has been made almost impossible due to the disappearance of large wild areas. Ongoing efforts at the global level include both in situ and ex situ approaches to conservation. While in situ conservation should be the preferred choice, it is unlikely that pressures on land would permit more than 4% of the geographical area to be set aside as protected area. Hence, it is essential to complement in situ approaches through ex situ measures. In case of medicinal plants in situ conservation should be further strengthened by ex situ methods through cultivation and maintenance in botanical gardens, arboreta, parks, and in any other possible sites (Natesh 2000).

18.4.2 Ex Situ Conservation

Ex situ approaches involve removal of species for conservation to an alien environment through: (1) Whole plants for cultivation in botanical gardens, herbal gardens, and arboreta or (2) reproductive parts for storage in different types of 'banks'.

Thus, available diversity is maintained in seed gene banks, in vitro gene banks, field gene banks, botanical and herbal gardens. Ex situ conservation under gene banks has a limited role as it lacks natural evolutionary processes that are always operative in nature. This problem can be partly addressed by linking ex situ interventions with in situ programmes, wherever possible. Among the ex situ methods high priority needs to be given to maintenance of collections in fields, herbal gardens and botanic gardens.

18.4.2.1 Field Genebanks

Field gene bank (FGB) represents the most common ex situ method of conserving genetic resources with recalcitrant seeds and vegetatively propagated plants.

With this method of conservation, genetic variation is maintained away from its original location and samples of a species, subspecies or variety are transferred and conserved as living collections. The establishment of a FGB assures existence of collections for long term and ensures supply of genuine material or propagule for use as well as for research purposes. The practical advantage of this method is accessibility of the germplasm along with constant evaluation of the plants being grown. Cultivated material along with wild related species are conserved and these are available for breeding, reintroduction, research and other purposes as and when required. These are particularly appropriate for long-lived perennial trees and shrubs which can not be adequately conserved in the wild and which may take decades to produce seeds.

Though economically useful plants are good material for conservation yet medicinal plants with known biological activities and chemical constituent need to be conserved in ideal situations to avoid loss of essential compounds responsible for biological actions. However, prioritization of species needs to be kept in mind to completely utilize any particular strategy. In this context it is important to mention about the establishment of Bale Goba medicinal plants Field Gene Bank in Ethiopia, wherein a total of 248 germplasm were collected and priority was set for collection of rare, endangered and endemic medicinal plants. This was based on the data obtained from National Herbarium and graduate students of Biology Department, Addis Ababa University (www.ibc.gov.et accessed on May 13, 2011). Some of the proposed species, which were given priority for collection, included: *Dodonaea angustifolia*, *Lupinus albus*, *Glinus lotides*, *Plumbago zeylanica*, *Rumex abyssinicus* and *Rumex nervosus*. According to one report, Ethiopia has 9,000 accessions in the field gene bank (Anonymous 2010a). In India, National Bureau of Plant Genetic Resources (NBPGR), New Delhi which is under Indian Council of Agricultural Research (ICAR), Ministry of Agriculture, Government of India, is actively involved in collection and conservation of medicinally important species which are being maintained at its regional stations at Akola, Cuttack, Hyderabad, Ranchi and Thrissur (NBPGR 2010). As is evident from Table 18.1, in India medicinal plants are being maintained by other organizations also (DMAPR 2009–2010; TBGRI 2009–2010). High cost of plant establishment and maintenance coupled with risk of exposure to pests, natural calamities and vandalism are only some of the constraints.

18.4.2.2 Botanical Gardens

A botanical garden refers to a place where plants, especially ferns, conifers and flowering plants, are grown and maintained for the purpose of research and education. This makes them different from parks and pleasure gardens where plants, usually with showy flowers, are grown for general public. Arboretum also represents a garden which houses only trees and these are occasionally associated with zoos. There are approx 2,500 botanical gardens world wide which house approximately 80,000 plant species; with about 150 in India; 30 in Mexico; and 12 in Indonesia (Anonymous 2010a, b). The number in most of the other countries is less than 10.

Table 18.1 Status of tropical medicinal plants in field genebanks

	Field genebank		
Institution	Genera	Species	Accessions
NBPGR Farm, New Delhi, India	10	10	454
NBPGR Regional Station Cuttack, India	72	75	175
NBPGR Regional Station Akola, India	19	20	33
NBPGR Regional Station Thrissur, India	20	22	329
NBPGR Regional Station Hyderabad, India		88	175
NBPGR Regional Station Ranchi, India	99	111	331
TBGRI, India		842	
DMAPR, India	11		641
Wendo Genet Field Gene Bank, Ethiopia	51	118	320
Bale Goba Medicinal Plants Field Genebank, Ethiopia			248
Brasilia Botanical Garden, Brazil			165
Embrapa-Genetic Resources and Biotechnology, Brazil			335
Embrapa-Oriental Amazon, Brazil			109
Univ. of Campinas Cpqba, Brazil			330
Univ. of Ceara, Brazil			224

Source: Vieira 1999, annual reports of respective Institutes

Of the 150 botanical gardens in India, Tropical Botanical Garden and Research Institute (TBGRI), at Thiruvananthapuram, Kerala, India, is a unique garden especially devoted to tropical medicinal plants. Founded in 1979, it is mainly involved in conservation of tropical plant genetic resources and their sustainable utilization. It has 300 acre conservatory garden for the wild tropical plant genetic resources of the country as well as representatives from other tropical regions with a well integrated multidisciplinary research and development system. It is one of the biggest conservatory gardens in Asia with over 50,000 accessions belonging to about 3,500 plant species. The collections of the garden include 842 species of medicinal plants, 700 species of trees, 65 species of bamboos and an orchidarium with 600 wild species (TBGRI 2009–2010). Interestingly, with the network of 150 botanical gardens in the country, 33 are attached to the Universities and hardly 30 are carrying out the conservation related programmes. In Indonesia, medicinal plants are being maintained in at least four botanical gardens (Anonymous 2010b).

Botanical gardens play a special role in conservation of threatened medicinal and aromatic plants but always suffer from the inadequate population and gene pool concept and criteria as they have limited individuals to represent genetic diversity.

18.4.2.3 Herbal Gardens

Herbal gardens specialize mainly in cultivation and maintenance of medicinal plants. These herbal gardens can also act as experimental farms for breeding and selection of new cultivars in any specific species as also in the development of field production technology. In India, a number of herbal gardens are being maintained at various universities, states and private sectors and also at personal level. The Union Ministry of Agriculture (Horticulture Division) has sanctioned funds to establish 16 Medicinal Plant Gardens in State Agricultural Universities and Research Institutes all over the country (Pareek et al. 2005).

Tropical Botanical Garden and Research Institute is one of the first such gardens established in 1979, with a gene pool of more than 850 species in an area spread over 20 ha. A high priority has also been assigned for their conservation in national programmes and some of the species are also being maintained by the Botanical Survey of India.

Apart from producing true seeds for utilization, these gardens are also utilized to create awareness in the public through short term training programmes. Though a large number of herbal gardens exist in India, the information is very scattered and it is not available in the public domain. A project on Networking of herbal gardens in India has been initiated by Directorate of Medicinal and Aromatic Plants Research (DMAPR), Anand, Gujarat, which is under ICAR, Ministry of Agriculture, Government of India, with a view to provide information about the available herbal gardens in the country, important species of medicinal plants being used in primary health care and also about species under various degrees of threat in India (Rao et al. 2010). According to an estimate, 237 species of tropical medicinal plants are being maintained in 74 herbal gardens spread across India (www.herbalgardenindia.org). Some of the Botanical Gardens imparting training to public are TBGRI, Lal Bagh Botanical Garden, Bangalore and Royal Botanical Garden, Kolkata.

Requirements for seed viability, growth, and reproduction are extremely diverse for medicinal and aromatic plants. For many wild-harvested and vegetatively propagated medicinal species, seed conservation may not be possible. Therefore, maintenance of living collections in field gene banks and in vitro techniques could be more appropriate. This is one of the principal advantages to be realized from the involvement of botanic gardens in conservation of medicinal and aromatic plant through ex situ conservation efforts.

18.5 Seed Conservation

Conservation of seeds is the most popular, safe and economic strategy for plant species producing orthodox seeds, i.e., the seeds which do not lose viability if dried to a moisture content of 3–7%. Such seeds are expected to remain viable for 5–25 years during medium-term storage (0–5°C and 35% RH) and for 50–100 years during long-term storage (−10°C to −20°C). However, in several medicinal species, bulb, corm, rhizome, tubers or some other vegetative part may be the site of storage of active ingredients and often, such species do not set seeds. If seeds are set, they are either sterile or intermediate/recalcitrant. The latter category, are desiccation sensitive and are generally killed if dried below a critical moisture content, usually between 12% and 35%. Recalcitrant seeds do not undergo maturation drying and cannot withstand water loss to the magnitude of that experienced by orthodox seeds.

Hence, these seeds are shed at relatively high moisture content. Some important tropical medicinally important species exhibiting recalcitrant or intermediate seed storage behaviour are: *Murraya koenigii, Murraya paniculata, Madhuca indica, Eugenia spp., Garcinia* spp., *Cola nitida, Theobroma cacao, Camellia sinensis*, and *Flacourtia indica* (Parihar and Dadlani 2010).

Storage of seeds in a genebank is accomplished following seed genebank standards for germination. For most of the tropical medicinal plants standards for laboratory germination are not yet defined. The seeds should be available in sufficient quantity with >85% seed germination. Amongst tropical medicinal plants, ca. 40% plant species are seed-producing; whereas10% reproduce via seeds as well as by vegetative means. *Aristolochia* spp., *Asparagus racemosus, Commiphora mukul, Gloriosa superba, Oroxylum indicum* and *Rauvolfia serpentina* are some of the species which belong to the latter category. Since most of the collections are made from the wild, first and foremost limitation is quantity of seeds for conservation. Seed dormancy, non-uniform maturity of seeds and lack of literature on germination/propagation/storage behavior are some of the other constraints faced while working with seed conservation. The limitation is further intensified keeping in view the diverse forms and ecological niche of these plants.

Seed germination is a pre-requisite for seed conservation. As is evident from the foregoing paragraph, germination of tropical medicinal plant seed needs to be standardized. The germination could be inhibited due to physical factors such as hard seed coat, seed dormancy or physiological factors such as immaturity of embryos. Therefore, such species require an additional, dormancy breaking treatments to achieve higher germination percentage. Experimental work done at NBPGR revealed varying degree of success with respect to enhanced germination percentages (Table 18.2) following sand paper scarification, acid scarification, cutting/piercing the seed coat, pre-soaking in water, treatment with GA_3 or 0.2% KNO_3.

Limited attention has been paid regarding storage behavior of seeds of tropical medicinal plants and their sensitivity to low temperature and desiccation. Fundamental research on desiccation and low temperature tolerance is required for the development of successful conservation protocols. In the case of a species with orthodox seeds, sufficient quantity of seeds is required for long-term conservation at −20°C. Successful examples of conserved germplasm at National seed genebank at NBPGR include *Abelmoschus moschatus*, *Euryale ferox*, *Abrus precatorius* and *Andrographis paniculata*.

Seeds of *Aristolochia tagala* (under ambient conditions) and *Oroxylum indicum* (at 20°C) could be stored for 18 and 16 months, with 85% and 75% recovery, respectively (Anandalakshmi et al. 2007). For species with orthodox seeds, but insufficient quantity of seeds, there is a need to regenerate/multiply these for conservation. Propagation of seed material by simulating conditions for each plant species is a herculean/challenging task keeping in view a large number of species coupled with low or no available know-how of their propagation/cultivation.

Plant species with recalcitrant seeds are conserved in field genebanks. It is necessary to highlight that all wild plant species cannot be cultivated. Botanical gardens and herbal gardens will be a very crucial link with seed genebanks to accomplish

Table 18.2 Effect of some physico-chemical dormancy breaking treatments in tropical medicinal plants

Plant species	Treatments for dormancy breaking	Germination (%)	
		Control	After treatment
Abelmoschus moschatus	Sand paper scarification	40	82
Abrus precatorius	25% H_2SO_4 for 20 min and sand paper scarification	42	86
Abutilon indicum	Hot water treatment at 70°C for 5 min	8	82
Argemone mexicana	Presoaking in water at 40°C overnight	44	100
Caesalpinia bunduc	Crack in seed coat	Nil	86
Caesalpinia sappans	Crack in seed coat	Nil	86
Cardiospermum halicacabum	Sand paper scarification	Nil	72
Cassia occidentalis	Sand paper scarification	40	82
Costus speciosus	Presoaking in 0.2% KNO_3 and GA_3 for 24 h	Nil	78
Datura metel	Sand paper scarification	56	92
Embelia ribes	GA_3 100 ppm or 0.2% KNO_3	48	92
Helicteres isora	Conc. H_2SO_4 for 5 min	40	86
Indigofera tinctoria	100% acid scarification for 5 min	10	76
Mucuna pruriens	Cut seed	24	96
Rauvolfia serpentina	Presoaking in water overnight	40	84
Rubia cordifolia	Removing seed coat	56	92

Source: Gupta 2002

this task. Owing to problems faced in the field gene banks, in vitro slow growth and/or cryopreservation using embryonic axes are the methods of choice (for details see Sects. 18.6.3 and 18.6.4). There are no standard protocols for species with highly recalcitrant seeds. Modification of existing protocols or development of new one is required but this can be accomplished only when a detailed understanding of the recalcitrance nature of explants is achieved (Normah et al. 2011). According to published reports, it is clear that work on storage of seeds has mainly been carried out at NBPGR and TBGRI, in India.

18.6 In Vitro Conservation

There are two ways in which tissue culture technique can be used for conservation of tropical medicinal plants. The first one is the use of in vitro multiplication technique in species with reproductive problems and/or with extremely low population to increase number of individuals. The material thus generated can be of great value for conservation, utilization and also for plant reintroduction programmes. The second strategy is the development of an in vitro storage technique which is particularly useful when conservation of seeds is not possible. In vitro conservation involves either normal growth or slow growth for short- to medium-term conservation, and cryopreservation for long-term conservation. The main objective of in vitro conservation is to reduce frequent demand of subculturing which can be accomplished in two ways: by culturing them under normal growth conditions or by subjecting them to growth limiting conditions. Literature survey revealed that currently there is little documented information on in vitro conservation of tropical medicinal plants.

18.6.1 In Vitro Propagation

Development of efficient plant regeneration protocols is a prerequisite for successful conservation of any plant species. In vitro propagation can be accomplished by three types of regeneration pathways: -axillary bud proliferation, adventitious regeneration or somatic embryogenesis. Usually, shoot tips or nodal segments are cultured on a nutrient medium containing specific combination of cytokinin and/or auxin to stimulate bud break. Other propagules such as leaf segments, roots, etc. can also be used for propagation.

The last two decades have witnessed an increasing application of tissue culture techniques for multiplication of tropical medicinal plants. As evident from Table 18.3, success to varying degree has been achieved in more than 50 tropical medicinal plant species with respect to in vitro propagation. The method for propagation include enhanced axillary branching (e.g. *Bacopa monnierii, Holostemma annulare, Plumbago zeylanica* and *Rauvolfia serpentina*) (Fig. 18.2), induction of somatic embryogenesis using explants from juvenile or mature plants (e.g. *Rotula aquatica*) depending on the availability of material, and adventitious regeneration directly from explants (e.g. *Aristolochia indica*) or through intervening callus (e.g. *Urginia indica*). Callus-mediated shoot regeneration has also been reported in *Aristolochia* (Siddique and Bari 2006), *Bacopa* (Tejavathi and Shailaja 1999) and *Plumbago* (Rout et al. 1999a). In our laboratory, direct adventitious shoot regeneration (in *Curculigo orchioides*) and somatic embryogenesis (in *Pogostemon patchouli*) were obtained using leaf segment explants (Chandel et al. 1996; Sharma et al. 2009b).

With respect to conservation, axillary bud proliferation is the preferred mode of propagation as it ensures greater genetic stability of the species being worked upon. The rate of multiplication varies from 2 or 3 shoots in 4 weeks in *Kaempferia galanga* (shoot base explant; Rahman et al. 2005) and *Hemidesmus indicus* (Sreekumar et al. 2000), to as high as 150 shoots every 4 months in *Coleus* node explant; Sen and Sharma 1991) and 129 shoots in *Bacopa* (leaf explant; Tiwari et al. 2001). High multiplication rates have major advantages for raising plants for nurseries and commercial plantings. However for conservation programme, a very high rate is not desirable.

Amongst tropical countries, tissue culture studies in this group of plants have been successfully reported from Bangladesh (*Aloe barbadensis, Aristolochia indica, Boerhaavia diffusa, Gloriosa superba, Hemidesmus indicus, Kaempferia galanga, Plumbago indica*), Brazil (*Alpinia purpurata*), India (*Aloe barbadensis, Aristolochia indica, Boerhaavia diffusa, Gloriosa superba, Gymnema sylvestre, Hemidesmus*

Fig. 18.2 (**a–f**) In vitro conservation and cryopreservation of tropical medicinal plants. (**a–c**) *Holostemma* sp. (**a**) In vitro multiple shoots. (**b**) Post-thaw recovery of cryopreserved shoot tips. (**c**) Regrowth of cryopreserved shoot tip. (**d–f**) *Bacopa monnieri*. (**d**) In vitro conserved germplasm. (**e**) Recovery growth of cryopreserved shoot tips. (**f**) Cryopreserved plantlets established in soil

indicus, Kaempferia spp., *Plumbago* spp., *Tinospora cordifolia* etc.) Sri Lanka (*Coscinium fenestratum*), Mexico (*Aloe barbadensis*) and South Africa (*Aloe polyphylla, Dioscorea bulbifera*) (see Table 18.3).

From a conservation and pharmaceutical point of view, propagation from pre-existing meristem is the method of choice as plants with desired traits are obtained. However, certain species are recalcitrant to tissue culture and this is a major obstacle in using tissue culture for germplasm conservation. In most of the species rooting and ex vitro transfer of plantlets has been applied successfully without much difficulty and with high survival (see Table 18.3). Most of the in vitro-propagated medicinal plants (both at NBPGR and TBGRI, and in literature) transferred directly to pots and maintained at high humidity/mist chambers (without any pretreatment procedure) exhibited 70–100% survival and establishment.

The main advantage of these techniques is generation of large number of material in a short time from a minimum starting material. In order to meet ever-increasing demand of quality material, in vitro technique offers a great potential by virtue of mass cloning of elite types. Since it accelerates clonal propagation, the method is particularly beneficial in plants like *Aloe vera, Curculigo orchioides, Dioscorea floribunda* and *Rauvolfia serpentina*, wherein clonal propagation by vegetative

Table 18.3 Status of in vitro multiplication in tropical medicinal plants

Plant species	Explant	Rate of shoot multiplication	Multiplication medium	Mode of multiplication	Rooting (%)	Rooting medium	Establishment in soil (%)	Reference
Acorus calamus	Rhizome bud	Multiple shoots in 98% explant	MS +2.0 mg/l Kn +0.05 mg/l NAA	Multiple shoots	40	MS +1.2 mg/l IBA	–	Ahmed et al. 2010
Acorus calamus	Apical meristem	26 shoots/explant	MS + 1.0 mg/l BAP	Multiple shoots	–	MS basal	75	Hettiarachchi et al. 1997
Acorus calamus	Rhizome bud	8–10 shoots in 8 weeks	MS + 8.87 µM BA + 5.37 µM NAA	Multiple shoots	~100	MS basal	90–95	Anu et al. 2001
Adhatoda beddomei	Nodal segments	15–27 shoots in 6 weeks	SH + 3.0 mg/l BAP + 0.2 mg/l 2iP + 1.0 mg/l IAA	Axillary shoots	–	SH + 0.2 mg/l IBA or 0.2 mg/l IAA	95	Sudha and Seeni 1994
Aloe barbadensis	Apices	–	MS + 1.1 µM 2,4-D + 2.3 µm Kn for 15–30 days then transferred to MS + 0.11 µM 2,4-D + 2.2 µM BAP	Micropropagation	–	–	–	Natali et al. 1990
Aloe barbadensis	Decapitated shoot	Both adventitious and axillary bud formation	MS + 5 µM IBA	Adventitious buds	–	MS + 5 µM IBA + IAA	–	Meyer and Van Staden 1991
			MS + IAA	Axillary buds	–	–	–	
Aloe barbadensis	Shoot tip	10 shoots/explant	MS + 2.0 mg/l BAP + 0.5 mg/l NAA	Multiple shoots	95	1/2MS + 0.5 mg/l NAA	70	Baksha et al. 2005
Aloe barbadensis	Shoot tip	80% shoot regeneration and 24.4 shoots/shoot tip	MS + 4 mg/l BA + 1 mg/l NAA	Shoots	–	MS + 1 mg/l IBA + 0.5% AC	Successful	Supe 2007

(continued)

Table 18.3 (continued)

Plant species	Explant	Rate of shoot multiplication	Multiplication medium	Mode of multiplication	Rooting (%)	Rooting medium	Establishment in soil (%)	Reference
Aloe barbadensis	Cut pieces of apical meristem	8.8 shoots in 4 weeks	MS+2.5 mg/l BA+2.0 mg/l Kn+0.1 mg/l IBA+0.01 mg/l IAA+15 mg/l AdS	–	93.2	MS+0.5 mg/l NAA	90–95	Bhatt et al. 2007
Aloe polyphylla	Seeds	Maximium shoot formation	MS+1.0 mg/l BA	Axillary and adventitious buds	–	MS+0.5 mg/l IBA	Acclimatized in greenhouse	Abrie and van Staden 2001
Aloe vera	Rhizomatous stem	–	MS+0.25 mg/l NAA +1.5 mg/l BAP	Multiple bud break	Yes	MS+0.25 mg/l NAA +1.5 mg/l BAP	–	Gantait et al. 2010
Aloe vera	Shoot explant	–	MS+1.5 mg/l+BAP+5 0 mg/l AdS	Enhanced axillary branching	90% micro-shoots produced roots	MS+1.0 mg/l NAA	Successful	Kalimuthu et al. 2010
Aloe vera	Axillary shoot segment	–	MS+1.0 mg/l 2,4-D+0.2 mg/l Kn	Organogenesis	–	–	–	Roy and Sarkar 1991
Aloe vera	Shoot tip	9.67 shoot/explant in 8 weeks	MS+0.5 mg/l BA+0.5 mg/l NAA	–	–	MS+0.5 mg/l BA +0.5 mg/l NAA	95	Hashemabadi and Kaviani 2008
Alpinia purpurata	Inflorescence buds	15–20 new shoots each 4 weeks	MS+10 µM BA+5 µm NAA	–	NM	NM	–	Illg and Faria 1995
Alpinia galanga	Emerging buds of rhizome	8 shoots within 8 weeks	MS+3.0 mg/l Kn	Shoot bud	100	MS+3.0 mg/l Kn	80	Borthakur et al. 1998
Alpinia officinarum	Rhizome bud	11 shoots/explant	MS+3.0 mg/l Kn+1.0 mg/l NAA	Multiple shoots	–	1/2 MS+0.5 mg/l BA	93	Selvakkumar et al. 2007

Alpinia purpurata	Rhizome bud	6.4 ± 0.32 shoots/explants	MS + 3.0 mg/l BA + 2.0 mg/l Kn	Multiple shoots	100	MS + 3.0 mg/l BA + 2.0 mg/l Kn	100	Kochuthressia et al. 2010
Alpinia zerumbet	Rhizome bud	(7.9 shoots/explant) 6–7 weeks	MS + 1.5 mg/l BA + 0.5 mg/l Kn	Multiple shoots	–	1/2 MS + 0.5 mg/l IBA	Successful	Rakkimuthu et al. 2011
Aristolochia indica	Shoot tip and nodal segments	Multiple shoots	MS + BAP	Axillary shoot multiplication	–	MS + 1.0 mg/l IBA	Transferred to soil	Theriappan et al. 2010
Aristolochia indica	Axillary shoots	Organogenesis	MS + 2.0 mg/l Kn + 1.0 mg/l BAP for callusing; MS + 2.05 mg/l Kn + 1.0 mg/l BA for shoot regeneration	Callus mediated shoot regeneration	–	MS + 1.0 mg/l Kn	Successful	Siddique and Bari 2006
Aristolochia indica	Shoot tip; nodal segment	45–50 shoots	MS + 0.54 μM NAA + 13.31 μM BA	Regeneration via axillary and adventitious shoots	More than 85	White's medium + 2.46 μM IBA	85	(Manjula et al. 1997)
Aristolochia indica	Internodal segment	Callus mediated shoots	MS + 2.69 μm NAA + 1.0 mg/l PG	Organogenesis	–	–	–	
Aristolochia indica	Leaf segment	Direct de novo shoots	MS + 13.31 μM BA + 50 mg/l AC	Adventitious shoots	–	–	–	

(continued)

Table 18.3 (continued)

Plant species	Explant	Rate of shoot multiplication	Multiplication medium	Mode of multiplication	Rooting (%)	Rooting medium	Establishment in soil (%)	Reference
Aristolochia indica	Shoot tip and nodal segments	12–14 shoots	MS + 5.0 mg/l 2iP	Adventitious directly from leaf bases and internodes + organogenesis	–	MS + 1.0 mg/l IBA	Successful	Soniya and Sujitha 2006
Asparagus racemosus	Nodal segments	40–45 shoot buds/explant	MS + 3.0 mg/l BA + 0.5 mg/l NAA	Adventitious shoot bud regeneration and callus	In vitro rooting	1/2 MS + NAA	75	Kumar 2009
Asparagus racemosus	Nodal segment	8 shoots/node	MS + 0.5 mg/l BA	–	–	MS + 0.1 mg/l NAA	60% survival upto 3 months	Pant and Joshi 2009
Bacopa monnieri	Nodal segments, internodes and leaf from 4-week-old cultures	129 shoots in leaf after 3 sub-cultures of 4 weeks	MS + 6.8 μM TDZ (initiation), MS + 2.2 μM BA	Adventitious shoots	100	MS + 4.9 μM IBA	100	Tiwari et al. 2001
Bacopa monnieri	Leaves and stem segments bet 2 and 6 node from apex from field	Shoot bud initiation in 90–100% explants after 3–4 weeks	MS + 2.0 μM BA + 0.2% gelrite	Adventitious shoots	–	–	–	Shrivastava and Rajani 1999

Bacopa monnieri	Leaf segments, node and internode from field	81.7% shoot bud initiation from nodes	MS basal	Adventitious shoots	70	MS basal	97	Mathur and Kumar 1998
Bacopa monnieri	1. internodal segments, 2. leaf and 3. flower buds	1. 7.5 shoots, 2. 6.2 shoots 3. 10.4 shoots	1. and 2. MS + 1.0 mg/l Kn + 0.1 mg/l IAA 3. MS + 0.1 mg/l 2iP + 0.1 mg/l IAA	Callus mediated multiple shoot formation	–	Rooting on same medium	–	Tejavathi and Shailaja 1999
Bacopa monnieri	Nodal segments	~22 shoots in 8 weeks	MS + 0.2 mg/l BA	Axillary shoots	100	Same medium	100	Sharma et al. 2007a
Boerhaavia diffusa	Nodal explant	12 shoots/culture in 93% cultures	MS + 2.0 mg/l BAP + 0.2 mg/l NAA	Multiple shoots	100	MS + 1.0 mg/l IBA	90	Biswas et al. 2009
Boerhaavia diffusa	Shoot tip; nodal segments	14.4 shoots/node, 7.83 shoots/ shoot tip	MS + 3.0 mg/l BAP	Shootlets regeneration	–	1/2 MS medium + 1.0 mg/l IBA	–	Wesely et al. 2010
Boerhaavia diffusa	Shoot tips and nodal segments	Maximum multiple shoots	MS + 1.5 mg/l BAP + 0.5 mg/l NAA	Shoots	90	1/2 MS + 1.0 mg/l IBA + 1.0 mg/l IAA	80	Roy 2008
Centella asiatica	Nodal segment	4–5 shoots/ explants in 91% cultures	MS + 22.2 μm BA + 2.68 μm NAA	Axillary branching	90	MS + 2.46 μm IBA	95	Tiwari et al. 2000
Centella asiatica	Shoot tip	16.8 shoots/shoot tip in 88% cultures	MS + 17.76 μm BA + 1.44 μm GA_3	25,000 plantlets in 160 days	27.66 roots/ culture	1/2 MS + 10.74 μm NAA	95	Sivakumar et al. 2006

(continued)

Table 18.3 (continued)

Plant species	Explant	Rate of shoot multiplication	Multiplication medium	Mode of multiplication	Rooting (%)	Rooting medium	Establishment in soil (%)	Reference
Commiphora wightii	Hypocotyl region	–	Modified MS +0.5 g/l AC + 10% sucrose	Somatic embryogenesis	–	–	–	Kumar et al. 2006
Commiphora wightii	Nodal segment	–	MS + 17.8 µm BA + 18.6 µM Kn +100 mg/l Glutamine + 10 mg/l thiamine HCL + 0.3% AC	Forced axillary branching	–	MS + IAA + IBA in dark, transfer to low salt medium + AC	Successful	Barve and Mehta 1993
Coleus forskohlii	Nodal segments and shoot tips	12-fold multiplication	MS + 2.0 mg/l Kn + 1.0 mg/l IAA	Axillary shoots	80–100	MS + 1.0 mg/l IAA	80–100	Sharma et al. 1991b
Coleus forskohlii	Shoot tip and bud initials of nodal segments	150 shoots in 4 weeks	MS + 2.0 mg/l BA	Axillary shoots	90	MS basal	60	Sen and Sharma 1991
Coleus forskohlii	Mature leaves	–	MS + 4.6 µM Kn + 0.54 µM NAA for regeneration	Callus mediated organogenesis	–	1/2 MS basal	–	Reddy et al. 2001
Coscinium fenestratum	Epicotyl	5 shoots/explant in 75 day	MS + 1.0 µm Kn + 0.25 µm 2,4 -D	Shoot formation	100	1/2 MS + 2.5 µM IBA	66.7% after acclimatization 100% in field	Senarath 2010

Costus speciosus	Zygotic embryos	22.7 ± 1.92 shoots	SH +250 mg/ 1Casamino acids (CA)	Multiple shoots leading to rhizome formation	–	–	Successful	Roy and Pal 1991
Costus speciosus	Shoot tips	10 Adventitious shoots after 60 days; 90 plants/shoot tip/year	MS + 0.5 mg/l BAP or 1.0 mg/l Kn + 15 mg/ lAdS + 1.0 mg/l IAA	Rhizome	100	SH + 1 mg/l IAA	Complete plants grew normally in soil	Chaturvedi et al. 1984
Costus speciosus	Single axillary buds	40 propagules	MS + 10 µM AdS + 1 µm NAA + 50 g/l sucrose + 7 µM BAP	Shoot multiplication	100	MS + 10 µMAdS + 1 µM NAA + 50 g/l sucrose + 7 µM BAP	95	Punyarani and Sharma 2010
Costus speciosus	Pseudostem	Shoot proliferation	MS + 0.05 mg/l BAP	Shoot proliferation	–	MS + 0.1 mg/l IBA	75	Robinson et al. 2009
Curculigo orchioides	Rhizomes and leaves	Poor results with rhizome explants	MS + 2.0 mg/l 2,4 D	Direct organogenesis in leaf explants	–	MS + 2.0 mg/l 2,4 D	82.5	Prajapati et al. 2003
Curculigo orchioides	Rhizome buds, shoot base	7 shoots per explant	MS + 0.2 mg/l BA + 2.5 mg/l spermidine	Multiple shoots	100	MS basal	86	Sharma et al. 2007b
Cymbogon flexuosus	Somatic embryogenesis through callus	30–35 plantlets/ 100 mg callus in 2 months	MS + 5.0 mg/l 2,4-D + 0.1 mg/l NAA + 0.5 mg/l Kn	–	–	MS + 3.0 mg/l BA + 1.0 mg/l GA_3 + 0.1 mg/l NAA	Transplanted to soil	Nayak et al. 1996

(continued)

Table 18.3 (continued)

Plant species	Explant	Rate of shoot multiplication	Multiplication medium	Mode of multiplication	Rooting (%)	Rooting medium	Establishment in soil (%)	Reference
Dioscorea bulbifera	Nodel stem	9 shoots/node	MS+1.0 mg/l BA		Yes	–	Successful	Forsyth and van Staden 1982
Dioscorea bulbifera	Nodal segments	–	MS+0.25 mg/l Kn+0.25 mg/l NAA	Axillary shoots	100	MS+0.15 mg/l NAA	–	Mandal et al. 2000
Dioscorea floribunda	Node and internode segments	–	MS+modified White's medium+2,4-D or NAA+BAP or Kn	Indirect organogenesis	Aerial tuber formation	1/2 MS basal +0.5 mg/l NAA	70	Sengupta et al. 1984
Dioscorea floribunda	Nodal segments	–	MS+0.25 mg/l Kn+0.25 mg/l NAA	Axillary shoots	100	MS+0.15 mg/l NAA	–	Mandal et al. 2000
Gloriosa superba	Nodal explants	Large scale induction of root tubers	MS+3.0 mg/l NAA+1.0 mg/l TDZ	Root tuber induction	Rooting	1/2 MS +NAA+IBA	90	Madhavan and Joseph 2010
Gloriosa superba	In vitro root tuber	Mass propagation	MS+2.0 mg/l Kn+1.0 mg/l NAA	Shoot initiation and multiplication	Rooting	1/2 MS +NAA+IBA	–	Madhavan and Joseph 2010
Gloriosa superba	Non dormant tubers	In vitro tubers in 12 weeks	MS basal	In vitro tuberization	Sprouting and rooting	Modified MS basal	–	Ghosh et al. 2007
Gloriosa superba	Apical and axillary buds of young sprouts	15 shoots per culture	MS+1.5 mg/l BA+0.5 mg/l NAA+15% coconut water+2 g/l AC	Shoot multiplication	–	1/2 MS+1.0 mg/l IBA+0.5 mg/l IAA	85–90	Hassan and Roy 2005
Glycyrrhiza glabra	Apical and axillary buds	83.3% sprouting and 1:10 multiplication rate	3/4 MS+4.44 μM BA	Shoot multiplication	–	1/2 MS+2.85 μM IAA +4.90 μm IBA	90% in greenhouse	Thengane et al. 1998

Glycyrrihiza glabra	Stem segments with axillary buds	Multiple shoots 40 days culture	MS + 1.0 mg/l BAP	Multiple shoots	–	MS + 0.01–0.50 mg/l NAA	Shoots developed in whole plants	Kohjyouma et al. 1995
Gymnema sylvestre	30 day old seedling axillary node explants	57.2 Shoots	MS + 1.0 mg/l BA + 0.5 mg/l Kn + 0.1 mg/l NAA + 100 mg/l ME + 100 mg/l citric acid	Multiple shoots	50	1/2 MS + 3.0 mg/l IBA	Successful	Komalavalli and Rao 2000
Gymnema sylvestre	Node	7 shoots/explant	MS + 5.0 mg/l BA + 0.2 mg/l NAA	Enhanced axillary sprout	–	1/2 MS	75	Reddy et al. 1998
Gymnema sylvestre	Apical buds from mature plant	Highest shoot frequency	MS + 4.44 µm BA + 4.64 µM Kn + 3% sucrose	Shoot induction	–	1/2 MS + IAA	85	Sharma and Bansal 2010
Gymnema sylvestre	Seedling	Shoot bud differentiation	MS + 1.0 mg/l BA + 0.5 mg/l IAA + 100 mg/l Vitamin B2 + 100 mg/l citric acid	Organogenesis	53	1/2 MS	–	Subathra Devi and Srinivasan 2008

(continued)

Table 18.3 (continued)

Plant species	Explant	Rate of shoot multiplication	Multiplication medium	Mode of multiplication	Rooting (%)	Rooting medium	Establishment in soil (%)	Reference
Hemidesmus indicus	Leaf/stem segments	–	MS + 1.0 mg/2,4-D + Kn for callus; MS + 2.0 mg/l NAA + 0.5 mg/l Kn for organogenesis	Organogenesis	–	1/2 MS basal	Successful	Sarasan et al. 1994
Hemidesmus indicus	Axillary shoots	85% organogenic callus induction; 90% differentiation	MS+ + 1.0 mg/l NAA + 2.0 mg/l Kn for callus induction; MS + 1.0 mg/l NAA + 2.5 mg/l Kn for differentiation	Indirect organogenesis	Optimum rooting	MS + IBA + Kn	Plantlets transferred to pots	Siddique and Bari 2006
Hemidesmus indicus	Nodes (0.5 cm)	9.37 shoots in 4 weeks	1/2 MS + 2.22 μM BA + 1.07 μm NAA	Multiplication upto 25 passages	–	1/4 MS + 9.8 μM IBA	96	Sreekumar et al. 2000
Hemidesmus indicus	Roots segment (0.5 cm)	2.6 shoots in 4 weeks	MS + 4.44 μM BA + 2.69 μM NAA	–	–	–	–	Sreekumar et al. 2000
Holarrhena antidysenterica	Seedling cotyledonary nodes	Multiple shoots	MS + 0.5 mg/l BA	–	Yes	In vitro – MS basal; ex vitro MS + 2 mg/l IBA	90	Mallikarjuna and Rajendrudu 2009
Holostemma annulare	Chlorophyllus root segments (3–4 cm)	Shoots with 13–14 cm with 8–9 nodes in a period of 3–4 weeks	MS + 0.2 mg/l BA	Adventitious shoots	–	1/2 MS basal	80	Sudha et al. 2000

Holostemma ada-kodien	Nodal segment	8 shoots/node	MS + 2.0 mg/l BAP + 0.5 mg/l IBA	Axillary sprouting	–	1/2 MS + 0.05 mg/l IBA	90	(Martin 2002)
Holostemma ada-kodien	Basal callus	15 shoots	MS + 1.5 BA	Callus mediated organogenesis	–			
Holostemma ada-kodien	Leaf; internode; root	40 embryos/10 mg callus	MS + 1.0 mg/l 2,4-D	Induction of somatic embryos	–	50% embryo maturation and conversion upon transfer to1/10 MS basal	90	Martin 2003a
Kaempferia galanga	Rhizome	13-fold in 4 weeks	0.75 MS + 12 μM BA + 3 μM NAA	Plantlet	–	0.75 MS + 12 μM BA + 3 μM NAA	1,000 plantlets hardened	Shirin et al. 2000
Kaempferia galanga	Rhizome tip and lateral bud	85%	MS + 2.0 mg/l BA + 0.2 mg/l NAA	Organogenesis and multiple shoot regeneration	96	MS + 1.0 mg/l IBA	81	Kalpana and Anbazhagan 2009
Kaempferia galanga	Rhizome tip and lateral bud	2 or 3 shoot/ explant	MS + 1.0 mg/l BA + 0.1 mg/l NAA	Multiple shoots	100	Modified MS + 0.2 mg/l IBA	85	Rahman et al. 2005
Kaempferia galanga	Rhizome with vegetative buds	NM	MS + 0.1 mg/l BA + 1.0 mg/l NAA	Callus induced embryogenesis	NM	MS basal	+	Vincent et al. 1992a
Kaempferia galanga	Axillary buds	NM	MS + 13.9 μM Kn + 2.2 μM BA	Axillary shoots	95	MS + 13.9 μM Kn + 2.2 μM BA	90	Vincent et al. 1992b

(continued)

Table 18.3 (continued)

Plant species	Explant	Rate of shoot multiplication	Multiplication medium	Mode of multiplication	Rooting (%)	Rooting medium	Establishment in soil (%)	Reference
Kaempferia galanga	Rhizome buds	13 shoots	MS+0.57 μM IAA+4.65 μM Kn	Multiple shoots	–	MS+6–9% sucrose +22.2 μM BAP or 23.25 μM Kn	80–90	(Chirangini et al. 2005)
Kaempferia galanga	Microshoots	Microrhizome induction	MS+22.2 μM BA or 23.25 μM Kn +6/9% sucrose	Microshoots mediated microrhizome	–			
Kaempferia rotunda	Rhizome buds	9 shoots	MS+2.69 μM NAA+2.85 μM BAP	Multiple shoots	–			
Kaempferia rotunda	Rhizome with vegetative buds	7 shoots	MS+2.0 mg/l BA+0.5 mg/l IAA	Multiple shoots	6 roots/shoots	MS+2.0 mg/l BA +0.5 mg/l IAA	70–80	Anand et al. 1997
Oroxylum indicum	Cotyledonary node explants	11 shoots/explant	MS+8.87 μM BA+2.85 μM IAA	Mass multiplication	91.6	MS+2.69 μM NAA +5.71 μM IAA	70–72	Dalal and Rai 2003
Oroxylum indicum	Apical and axillary buds	Large-scale propagation	MS+BAP	Multiple shoot regeneration	–	1/2 MS+4.92 μM IBA	Successful	Gokhale and Bansal 2009
Pogostemon patchouli	Nodal segments	>40	MS+0.2 mg/l Kn+0.1 mg/l IAA	Axillary branching	–	MS basal	Successful	Sharma et al. 1991a
Plumbago indica	Nodal segments; leaf segments	Multiple shoot buds; callus	For nodal segments-MS+2.0 mg/l BA+1.0 mg/l IAA; for callus from leaf MS+3.0 mg/l BA+1.5 mg/l Kn+1.0 mg/l NAA	Multiple shoot buds or callus	–	1/4 MS+0.5 mg/l IAA+0.75% sucrose	90	Bhadra et al. 2009

Plumbago indica	Encapsulated clump of shoots	–	MS + 2.0 mg/l BAP +3% sucrose	4–6 plantlets/bead	–	MS + 1.0 mg/l putrescine	Successful transplant-ing	Bhattacharyya et al. 2007
Plumbago indica	Nodal segments	–	MS + 3.0 mg/l BA + 0.1 mg/l IAA	Adventitious shoot initiation and multiplication	–	–	–	Chetia and Handique 2000
Plumbago rosea	Stem segments	–	MS + 2.5 mg/l 2,4-D + 1.5 mg/l Kn for callusing; 2.0 mg/l BAP + 1.0 mg/l NAA for shoot formation	Indirect organogenesis	–	MS + 1.5 mg/l IBA	60	Satheesh and Seeni 2003
Plumbago zeylanica	Leaf and stem	82.3% response in shoot cultures	MS + 4.44 μM BAP + 4.2 μM IAA	Callus mediated shoot formation	–	MS + 0.51 μM IAA + 2% sucrose	90	Rout et al. 1999a
Plumbago zeylanica	Leaves from 4 weeks old cultures	85% leaf formed shoots	MS + 0.7 μM BA + 4 μM IAA + 370 μM AdS	Direct shoots (adventitious)	92.7	MS + 1.2 μM BA + 2% sucrose	90	Das and Rout 2002
Plumbago zeylanica	Nodal explant	–	MS + 0.5–1.0 mg/l BA + 0.01 mg/l IAA	Multiple shoot formation	–	1/2 MS + 0.25 mg/l IBA + 2% sucrose	90	Rout et al. 1999b
Plumbago zeylanica	Nodal explants from field	8 plantlets in 5 months	MS + 2.46 μM IBA + 27.2 μM AdS	Direct shoot formation (adventitious)	92.7	MS + 4.92 μM IBA + 2% sucrose	90	Selvakumar et al. 2001
Plumbago zeylanica	Nodal explants from field	4–8 shoots/explant	MS + 1.0 mg/l BA + 0.1 mg/l NAA	Axillary shoots	–	MS basal	–	Sharma et al. 2002
Rauvolfia serpentina	Nodal segments	40 shoots per 8 weeks	MS + 1.0 mg/l BA + 0.1 mg/l NAA	Multiple shoot formation	–	MS + 1.5 mg/l NAA	100	Mathur et al. 1987

(continued)

Table 18.3 (continued)

Plant species	Explant	Rate of shoot multiplication	Multiplication medium	Mode of multiplication	Rooting (%)	Rooting medium	Establishment in soil (%)	Reference
Rauvolfia serpentina	Shoot tips	15–20 shoots per shoot tip	MS + 2.0 mg/l BA + 0.5 mg/l NAA	Multiple shoot formation	–	MS + 0.5 mg/l NAA + 2.0 mg/l Kn	60	Mukhopadhyay et al. 1991
Rauvolfia serpentina	Nodal segments and Shoot apices	–	0.5–0.1 mg/l BA + 0.1 mg/l NAA	Callus mediated	–	–	–	Sarkar et al. 1996
Rauvolfia serpentina	Nodal segments	15–20 shoots	MS + 1.0 mg/l BA + 0.1 mg/l NAA	Multiple shoot formation	–	MS + 1.5 mg/l NAA	–	Sharma and Chandel 1992a
Rauvolfia tetraphylla	Shoot tips	–	MS + 1.0 mg/l NAA + 5.0 mg/l BA or MS + 1.0 mg/l IAA + 5.0 mg/l BA	Multiple shoots	–	MS + 2.0 mg/l NAA	70	Ghosh and Banerjee 2003
Rotula aquatica	Leaf and internode	25.6 somatic embryos per 100 mg callus	MS + 0.45 μM 2,4-D for callus induction; 1/2 MS + 0.23 μm 2,4-D (liquid) for somatic embryo induction	Indirect somatic embryogenesis	–	1/2 MS semi solid	95	Chithra et al. 2005
Rotula aquatica	Mature nodal segments	11.2 shoots/ explant	MS + 0.5 mg/l Kn + 600 mg/l charcoal	Multiple shoots	–	1/2 MS basal	70	Sebastian et al. 2002
Rotula aquatica	Axillary bud	15 shoots/node	MS +1.0 mg/l BAP+0.5 mg/l IBA	Enhanced axillary branching	In vitro 80% Exvitro 75%	1/2 MS + 0.5 mg/l NAA	80	(Martin 2003b)
Rotula aquatica	Basal callus	Shoot regeneration	1/2 MS + IAA + 2,4-D + BAP and Kn	Indirect organogenesis	–	–	–	

Rubia cordifolia	Nodes; split vertical halves of nodes from seedling	5.9–5.2 shoots/explant in 2 weeks	MS + 1.0 mg/l BA + 0.02 mg/l IAA	Shoot multiplication	98	MS + 1.0 mg/l IBA + 1.0 mg/l IAA	89	Radha et al. 2011
Tinospora cordifolia	Nodal segments	4-fold multiplication	MS + 5 μM BA +150 μM glutamine	Axillary shoots	100	1/2 MS + 0.5 μM IBA	100	Mishra et al. 2010
Tinospora cordifolia	Mature nodes	6.3 shoots/explant	WPM + 8.87 μM BA for induction; MS + 2.22 μM BA + 4.65 μM Kn for elongation	Axillary shoots	–	1/2 MS + 2.85 μM IAA	80	Raghu et al. 2006
Tylophora indica	Axillary buds	5.3 shoots every 6 weeks	MS + 5.0 mg/l BAP + 0.5 mg/l NAA + 100 mg/l ascorbic acid	Axillary shoots	100	MS + 1.0 mg/l IAA	90–100	Sharma and Chandel 1992b
Tylophora indica	Leaf segment	85% from surface of callus	MS + 10 μM 2,4,5-T for callusing, 1/2 MS + 5 μM Kn for regeneration	Adventitious shoots	–	MS + 0.5 μM IBA	Successful	Faisal and Anis 2003

(continued)

Table 18.3 (continued)

Plant species	Explant	Rate of shoot multiplication	Multiplication medium	Mode of multiplication	Rooting (%)	Rooting medium	Establishment in soil (%)	Reference
Urginea indica	Bulb explants	400 bulblets in 18 weeks	MS+2 mg/l 2,4-D+15% CM or 4 mg/l 2,4-D+ 2 mg/l NAA +2 mg/l Kn+1 g/l YE	Callus mediated shoots	400 bulblets	MS+0.5% sucrose	90	(Jha et al. 1984)
	Outer scale	400 bulblet/scale leaf in 18 weeks	Modified MS+1 mg/l 2,4-D	Adventitious shoots	400 bulblets	–		
Vetiveria zizanioides	Shoot tips	75% and 3.76 shoots/explant	MS+3.0 mg/l paclobutrazol (PBZ)	Shoot multiplication	NM	NM	NM	Moosikapala and Te-chato 2010
Vitex negundo	Nodal segments	–	MS+4.44 μM BAP	Shoot proliferation	–	1/2 MS+4.92 μM IBA	Successful	Johnson et al. 2008
Vitex negundo	Shoot tip	6.3 shoots in 4 weeks	MS+8.87 μM BA+2.69 μM NAA	Multiple shoots	–	1/2 MS+4.90 μM IBA+2.85 μm IAA+2% sucrose+AC	85% after 1 month	Usha et al. 2007
Woodfordia fruiticosa	Shoot tips	26–35 shoots/ explants in 4–5 weeks	SH+5.0 mg/l BA+0.5 mg/l NAA	Axillary shoots	NM	MS+0.2 mg/l IAA	89	Krishnan and Seeni 1994

AC activated Charcoal, *AdS* adenine sulphate, *BA/BAP* 6-benzyladanine, *CA* casamino acid, *2,4-D* 2,4-dichlorophenoxy acetic acid, *GA3* gibberillic acid, *IAA* indole- 3- acetic acid, *IBA* indole-3-butyric acid, *2iP* N-isopentyl adenine, *Kn* Kinetin (6-furfurylamino Purine), *μM* micromolar, *ME* malt extract, *MS* Murashige and Skoog's medium, *NAA* 1-naphthaleneacetic acid, *NM* not mentioned, *PBZ* paclobutrazol, *PG* phloroglucinol, *SH* Schenk and Hildebrandt medium, *TDZ* thidiazuron, *WPM* Woody Plant Medium, *YE* yeast extract means is inadequate/slow. The attraction of in vitro propagation lies in its ability to rapid multiplication and rooting of shoots, and plantlet establishment. Such plants produced in large numbers can also be reintroduced in nature in case of rare or threatened plant species as well as in plants with nonviable or difficult to germinate seeds. It also helps in genetic manipulation of cultivars of medicinal plants for enhanced active ingredients. Another advantage is that in vitro genebank can be a viable alternative to seed banks for species wherein seed conservation is constrained due to reasons discussed in section on seed conservation.

18.6.2 Normal Growth

With renewed interest in conservation of tropical medicinal plants, the number of species under in vitro conservation are increasing rapidly and Table 18.4 enumerates the progress of in vitro conservation in this group of plants. Work on in vitro conservation of tropical medicinal plants has been primarily undertaken in India at TBGRI and NBPGR. Although reports exist regarding in vitro slow growth of tropical medicinal plants, the purpose has merely been an academic exercise. Shoot cultures of *Rauvolfia* can be maintained without recourse to any growth inhibitory treatment, on a simple tissue culture medium (MS + 1.0 mg/l BA) for 12–24 months at 25°C (Gautam et al. 2000; Sharma et al. 2000). At NBPGR, in authors laboratory a single medium has been standardized for maintenance of shoot cultures of a number of tropical medicinal plants – *Centella asiatica, Costus speciosus, Curculigo orchioides, Plumbago* spp., etc.

18.6.3 Slow Growth

Slow growth strategy allows cultures to be held for 1–2 years without subculture. In the case of tropical medicinal plants, various strategies like minimal media, culture tube enclosure, use of osmotic a and low temperature incubation have been tested, and success achieved is detailed in Table 18.4.

Using axillary buds, 35–40% survival in *Rauvolfia serpentina* was reported after 6 months of storage at 25°C on half-strength and full-strength MS basal medium (Sharma and Chandel 1992a).

Type of enclosure of the culture vessel has been shown to influence the rate of evaporation of the medium. According to Sharma and co-workers, shoot cultures of *Coleus forskohlii, Rauvolfia serpentina* and *Tylophora indica* could be conserved for 12–20 months at 25°C using polypropylene caps (Sharma and Chandel 1992a, b; Sharma et al. 1995). Similarly, shoot tip cultures of *Baliospermum montanum* and *Geophila reniformis* could be maintained for 12 months using polypropylene caps as enclosures (see Krishnan et al. 2011).

Table 18.4 Status of in vitro conservation of tropical medicinal plants

Plant species	Culture system	Strategy	Response	Institute/country of conservation	Reference
Acorus calamus	Shoot culture	MS + 1.0 mg/l BAP, 20°C	12 months	SriLanka	Hettiarachchi et al. 1997
Allium sativum	Bulblets	10°C and 10% sucrose	9–14 months	NBPGR, India	Chandel and Pandey 1992
Aloe vera	Shoot culture	MS + 0.25 mg/l NAA + 1.5 mg/l BAP	5 months	India	Gantait et al. 2010
Alpinia purpurata	Shoot culture	Deionized water with 0.5% sucrose	6 months	State University Campinas, Brazil	Kochuthressia et al. 2010
Bacopa monnieri	Shoot cultures	Polypropylene caps at 25°C	12 months	NBPGR, India	Sharma et al. 2007a, c
Bacopa monnieri	Multiple shoot clumps	1/2 MS + 2% sucrose; polypropylene caps	20 months	Center for Medicinal Plant Research, India	Satheesh et al. 2003
Baliospermum montanum	Shoot tip cuttings	1/2 MS + 1.33 µM BA + 1% agar, polypropylene caps at 25°C	12 months	TBGRI, India	See Krishnan et al. 2011
Coleus forskohlii	Axillary shoots	Polypropylene caps at 25°C	12–18 months	NBPGR, India	Sharma et al. 1995; Chandel and Sharma 1996, 1997
Curculigo orchioides	Shoot cultures	Polypropylene caps at 25°C	8–12 months	NBPGR, India	Sharma et al. 2009b
Geophila reniformis	Shoot tip cuttings	1/2 MS, polypropylene caps, 25°C	12 months	TBGRI, India	See Krishnan et al. 2011
Gloriosa superba	In vitro root tuber	MS + 3.0 mg/l NAA + 1 mg/l TDZ	24 months	India	Madhavan and Joseph 2010
Hemidesmus indicus	Shoot cultures	1/2 MS + 2% sucrose, polypropylene caps	18–22 months	Center for Medicinal Plant Research, India	George et al. 2010
Holostemma annulare	In vitro nodes	1/2 MS + 2% mannitol, polypropylene caps, 25°C	12 months	TBGRI, India	See Krishnan et al. 2011
Kaempferia galanga	Shoot cultures	Polypropylene caps at 25°C	12 months	NBPGR, India	Sharma et al. 2000
Pogostemon patchouli	Shoot cultures	Polypropylene caps, minimal media	18 months	NBPGR, India	Sharma 1999
Plumbago indica	Shoot cultures	MS + BAP 2.0 mg/l + 3% mannitol	12 months	India	Bhattacharyya et al. 2007

Plumbago indica	Encapsulated clump of shoots	20–24°C	3 months	India	Bhattacharyya et al. 2007
Plumbago zeylanica	Shoot culture	Polypropylene caps at 25°C	8–12 months	NBPGR, India	Sharma et al. 2005
Rauvolfia serpentina	Axillary shoots	MS + 4.44µM BA +0.54 µM NAA 15°C	15 months	NBPGR, India	Sharma and Chandel 1992a; Chandel et al. 1996
Rauvolfia serpentina	Shoot cultures	Minimal media	18–24 months	NBPGR, India	Chandel et al. 1996
Rauvolfia serpentina	Shoot cultures	Polypropylene caps, osmoticum 25°C	15–20 months	NBPGR, India	Sharma et al. 1995; Chandel et al. 1996
Rauvolfia serpentina	Root cultures	25°C	16 years with periodic subculture	National Botanical Research Institute, India	Chaturvedi et al. 1991
Tylophora indica	Axillary shoots	LT 15°C and 25°C	12 months	NBPGR, India	Chandel and Sharma 1996; Sharma et al. 1995
Utlaria salicifolia	Shoot cultures	1/2 MS + 4% sucrose	24 months	Center for Medicinal Plant Research, India	George et al. 2010
Vetiveria zizanioides	Shoot tips	MS + paclobutrazol(PBZ) 14 h photoperiod, 27 ± 1°C	12 months	Faculty of Natural Resources, Thailand	Moosikapala and Te-Chato 2010

BA/BAP 6-benzyladanine, *MS* Murashige and Skoog's medium, *NAA* 1-naphthaleneacetic acid, *PBZ* paclobutrazol, *YE* yeast extract

Low temperature, in general, is not beneficial for conservation of tropical species. Shoot cultures of *R. serpentina* have been successfully conserved for over 15 months at 15°C, while 10°C and 5°C were deleterious to growth of cultures (Sharma and Chandel 1992a). There are limited reports regarding use of osmotica and there is none on the use of growth retardants. Mineral oil overlay has proved beneficial in extending storage duration in *Bacopa monnierii* and *Curculigo orchioides* (Neelam Sharma, Unpublished data). In vitro induction of storage organs has been advocated as a promising strategy to increase the storage period in plant species. In tropical medicinal plants in vitro induction of storage organs in *Allium sativum* (bulblets, Chandel and Pandey 1992) *Kaempferia galanga* (rhizome, Chirangini et al. 2005) and *Gloriosa superba* (tuber, Ghosh et al. 2007; Madhavan and Joseph 2010) (see Table 18.4) has been reported. Though successful in extending storage duration in *A. sativum* for 12 months (Chandel and Pandey 1992; Mandal et al. 2000), its applicability in conservation of other tropical medicinal plants is yet to be demonstrated.

An important observation made in authors' laboratory is that rooted cultures survive longer probably due to roots being capable of absorbing water and nutrients from medium more efficiently than shoot cultures especially during later stages of growth (Chandel et al. 1996). It is worthwhile to mention here that in the in vitro genebank of NBPGR, germplasm of various tropical medicinal plants has been maintained for the last 20–25 years with periodic subculture of 8–18 months.

Short-and medium-term conserved cultures in vitro, act as an active gene bank, and facilitate maintenance, distribution, evaluation, exchange and utilization of the germplasm. Base collections should be maintained in duplicates for long-term conservation. This acts as a source of material in case of loss from the active gene banks (Chandel et al. 1996).

18.6.4 Cryopreservation

Cryopreservation is potentially the most appropriate strategy for long-term conservation of vegetatively-propagated plants and those with recalcitrant seeds. As mentioned earlier, it is also considered a safe method for intermediate and orthodox seeds in special cases such as those involving threatened germplasm or for those species for which very few seeds can be collected from habitats and which are on the verge of extinction. Table 18.5 provides information regarding cryopreservation of tropical medicinal plants. It is evident that most of the work has been carried out in India. This is mainly due to establishment of four National Genebanks by Department of Biotechnology, Government of India with special emphasis on conservation of medicinal and aromatic wealth of the country. Though limited yet encouraging success has been achieved in tropical medicinal plants and there is an increasing interest in cryopreservation as this technique is now considered suitable for general application.

It is important to note that of the nine tropical medicinal species tested for cryopreservation, seeds of *Emelia ribes* could not survive cryopreservation and failed to germinate after storage in LN (Decruse et al. 1999a). Success has been achieved in cryopreservation of *Rauvolfia micrantha*, an intermediate type of seed producing species, and seven orthodox seed-bearing species (Decruse et al. 1999a). In case of recalcitrant species, such as *Celastrus paniculatus*, excised zygotic embryos proved useful for cryopreservation. Simple desiccation under laminar airflow resulted in 60% regeneration of whole plants after cryopreservation (Radha et al. 2010). In contrast, in *Coscinium fenestratum*, cryopreserved embryos exhibited only 34% germination. At TBGRI, embryos of 8 accessions belonging to four species such as *Coscinium fenestratum, Myristica beddomei, M. malabarica* etc. have been maintained in the cryobank (TBGRI 2009–2010).

There is very little information regarding successful application of classical cryopreservation techniques on tropical medicinal plants. It clearly indicates lack of interest and thrust on conservation of this group of plants across countries. During the mid-1990s 'G-15 Gene Banks for Medicinal and Aromatic Plants' (G-15 GEBMAP) project was commissioned in India and in many other countries with Department of Biotechnology, Government of India acting as the coordinator. With the establishment of four National Genebanks for Medicinal and Aromatic Plants in India, with funding support of DBT, there has been stimulation of interest regarding thrust on cryopreservation of medicinal plants. With the development of widely applicable new protocols such as vitrification and encapsulation-dehydration, success has been recorded in some species of this group. In vitro cryopreservation is a promising tool for long-term germplasm conservation but it is still at an experimental stage. Successful cryopreservation of shoot tips, nodal segments and embryogenic cultures has been reported for various species by different laboratories (Table 18.5, Fig. 18.2). Success has been achieved to varying degrees with respect to cryopreservation and plantlets regeneration from in vitro shoot tips of *Bacopa* and *Dioscorea* spp., and *Holostemma* at NBPGR and TBGRI, respectively (Decruse et al. 1999b; Mandal et al. 2000). Plantlets thus generated could be successfully transferred to soil in *Bacopa* and *Dioscorea* spp. (Sharma et al. 2009a, 2011; Mandal et al. 2000). Further studies using morphological, molecular and biochemical parameters indicated maintenance of genetic stability in cryopreserved plants (Ahuja et al. 2002; Dixit et al. 2003).

Application of in vitro cryopreservation for long-term conservation (Cryobanking) of germplasm has not yet been reported for any tropical medicinal plant. However, the use of cryopreservation is limited to small laboratory collections and its use on a large scale is currently exceptional.

Although moisture content of the excised embryos and embryonic axes is the most critical factor for a successful cryopreservation protocol, manipulations of desiccation conditions, particularly the rate of desiccation, physiological status of the plant material, preculture and cryoprotective treatments, cooling/rewarming rates and recovery medium require optimization for the recovery of vigorous plant-

Table 18.5 Status of cryopreservation of tropical medicinal plants

Plant species	Propagule cryopreserved	Strategy	Regrowth of cryopreserved propagule	Reference
Allium sativum	Apical meristems (2 mm) from post-dormant bulbs	Vitrification	100% regrowth in 12 garlic cultivars	Niwata 1995
Allium sativum	Apices from bulbs and bulbils	Vitrification	Success depended on the size of bulbils	Makowska et al. 1999
Abrus precatorius	Seed (orthodox)	Desiccation	88% germination	Decruse et al. 1999a
Andrographis paniculata	Seed (orthodox)	Desiccation	86.4% germination	Decruse et al. 1999a
Bacopa monnieri	In vitro shoot tips	Vitrification	20% regrowth	Sharma et al. 2009a, 2011
Celastrus paniculatus	Zygotic embryo (recalcitrant seed)	Desiccation	65% regeneration	Radha et al. 2010
Coleus forskohlii	Seed (orthodox)	Desiccation	92.7% germination	Decruse et al. 1999a
Coscinium fenestratum	Zygotic embryo (intermediate seed)	Desiccation	34% regeneration	See Krishnan et al. 2011
Dioscorea bulbifera	In vitro shoot tips	Encapsulation-dehydration	Plantlet regeneration	Malaurie et al. 1998
Dioscorea bulbifera	Somatic embryos/embryogenic tissue	Encapsulation-dehydration	Regeneration of plants	Mandal et al. 1999; Mukherjee et al. 2009
Dioscorea floribunda	In vitro shoot tips	Encapsulation-dehydration	75% survival; 25% regeneration	Mandal et al. 2000; Mandal and Ahuja-Ghosh 2007
Dioscorea floribunda	In vitro shoot tips	Vitrification	87% survival; 30% regeneration	Mandal et al. 2000; Mandal and Ahuja-Ghosh 2007
Dipteracanthus patulus	Seed (orthodox)	Desiccation	87.7% germination	Decruse et al. 1999a
Hemidesmus indicus	Seed (orthodox)	Desiccation	79% germination	Decruse et al. 1999a
Holostemma annulare	In vitro shoot tips	Pregrowth; encapsulation-dehydration	54.2% regeneration	Decruse et al. 1999b, 2004; Decruse and Seeni 2002
Kaempferia galanga	In vitro shoot tips	Pregrowth; vitrification	50% regeneration	See Krishnan et al. 2011
Kaempferia galanga	Somatic embryos	Pregrowth; desiccation	42.8% regeneration	See Krishnan et al. 2011
Ocimum gratissimum	Seed (orthodox)	Desiccation	86.8% germination	Decruse et al. 1999a
Rauvolfia serpentina	Nodal segments	Pregrowth; vitrification	66% regeneration	Ray and Bhattacharya 2008
Tylophora indica	Seed (orthodox)	Desiccation	80.5% germination	Decruse et al. 1999a

lets from the cryogen (see Normah et al. 2011). As a result of detailed study carried out by Decruse and coworkers on cryopreservation of shoot tips of *Holostemma annulare* it was concluded that several factors such as culture medium, preparative procedures and ammonium ions play crucial role influencing the success of post-thaw regeneration (Decruse and Seeni 2002; Decruse et al. 2004).

Researchers face many problems due to lack of basic studies and this is the reason why the results obtained are often erratic and the rate of repeatability is low even with the established techniques (Normah et al. 2011). With better understanding of mechanism of cryopreservation and cryobanking by various research teams worldwide on plant species including tropical plant species, it is expected that application of cryopreservation of genetic resources of tropical medicinal plants will receive due attention and success in near future.

18.7 Genetic Stability

Monitoring of genetic stability is an important aspect of in vitro conservation and cryopreservation. The technique of monitoring stability will depend on the need, type and nature of species and its economic product (see Chandel et al. 1996). As is evident from the foregoing account, medicinal plant conservation per se has invited attention, only since the mid 1990s. The number of species under in vitro conservation is increasing every day. Very little information exists on the topic and the available information deals mainly with comparisons between in vitro regenerants and their putative parents (Table 18.6). Limited information exists regarding maintenance of genetic stability using morphological, biochemical and molecular analyses for example in *Dioscorea* spp. (Ahuja et al. 2002; Dixit et al. 2003). It is difficult to devise general strategy for screening active principle (secondary metabolites), which are as diverse and as less worked out as the tropical medicinal plants themselves. This, however, is one of the major areas which need attention.

18.8 Status of Genebanks

In most cases, in situ conservation is the ideal approach. However, due to various types of pressures in land use, it is unlikely that more than 4% of world's land area can be set aside for this purpose. Hence, in situ conservation efforts need to be complemented with ex situ conservation methods. Seed banking is probably the easiest and cheapest way of preserving a wide range of genotypes of species. Limited data are available on medicinal plants held in Seed Genebanks.

There are almost 1,750 genebanks world over conserving 7.4 million collections (Anonymous 2010b). Most of the Seed Genebanks are used mainly for conserving genetic diversity of agricultural crops, and in general medicinal plants are very poorly represented in Seedbanks.

Table 18.6 Genetic stability studies in tropical medicinal plants

Plant species	Culture system	Strategy	Response	Reference
Bacopa monnieri	Regenerated plants	Molecular (RAPD)	No change observed in regenerants	Ceasar et al. 2010
Coleus forskohlii	Regenerated plants	Cytological analysis	Seven plants showed diploids while three did not	Sen and Sharma 1991
Coleus forskohlii	6-month old regenerated plants	Chemical analysis	No change in forskolin content	Sharma et al. 1991b
Dioscorea. bulbifera	Explant-cryopreserved derived in vitro plantlets	Molecular (RAPD) biochemical (HPLC) and morphological analysis	Cryopreserved derived plants maintained genetic stability	Dixit et al. 2003
Dioscorea. floribunda	Explant-cryopreserved derived in vitro plantlets	Molecular (RAPD) biochemical (HPLC) and morphological analysis	Cryopreserved derived plants maintained genetic stability	Ahuja et al. 2002
Pogostemon patchouli	In vitro regenerants	Chemical analysis	Stability of essential oil	Sharma et al. 1991a
Plumbago zeylanica	In vitro regenerants	Molecular (RAPD) analysis	No change obsesrved in regenerants	Rout and Das 2002
Rauvolfia serpentina	In vitro regenerants	Electrophoretic analysis	No change obsesrved in regenerants	Chandel et al. 1996
Vetiveria zizanioides	Conserved shoot cultures	Molecular (RAPD) analysis	No variation between non-conserved and conserved plants	Moosikapala and Te-chato 2010

In India, four genebanks have been established specifically for medicinal and aromatic plants at TBGRI, Central Institute of Medicinal and Aromatic Plants (CIMAP), NBPGR and Regional Research Laboratory, Jammu (RRL) under the G-15 GEBMAP programme. These institutions are actively engaged in the collection and conservation of genetic resources in the Indian sub-continent. Conservation of tropical medicinal plants has primarily been undertaken by NBPGR and TBGRI. Seeds of 1,546 accessions of 110 species belonging to 95 genera of tropical medicinal plants have been stored at –20°C at the National Seed Genebank at NBPGR. Seedbank at TBGRI maintains a reference collection of 2,387 accessions and an active collection of 275 accessions.

At TBGRI, 38 accessions belonging to 26 medicinal plant species are maintained as shoot cultures in the In Vitro Bank and include *Acorus calamus, Adhatoda beddomei, Alpinia calcarata, Baliospermum solanifolium, Celastrus paniculatus, Geophila reniformis, Holostemma annulare, Plumbago rosea, Rauvolfia micrantha R. serpentina, Rubia cordifolia* and *Utleria salicifolia* (TBGRI 2009–2010).

At NBPGR, 127 accessions comprising 14 genera and 19 species of tropical medicinal plants are maintained as shoot cultures in the In Vitro Genebank. Some of these include *Acorus calamus, Centella asiatica, Holostemma ada-kodien, Coleus* spp., *Costus speciosus, Curculigo orchioides, Plumbago* spp., *Rauvolfia* spp. and *Tylophora indica.*

The Cryo Genebank at NBPGR maintains ~ 850 accessions of medicinal and aromatic plants but information regarding tropical medicinal plants is not available. At TBGRI, embryos of eight accessions belonging to four species namely *Coscinium fennestratum, Myristica beddomei, M. malabarica* etc. have been maintained in the Cryobank (TBGRI 2009–2010). In Ethiopia, the Medicinal Plants Genetic Resources Department in collaboration with the two stakeholders collected and conserved 320 accessions of medicinal plants including 31 accessions as seeds in the cold room. A total of 72,000 seed samples are also being maintained at Embrapa Genetic Resources and Biotechnology (www.cenargen.embrapa.br/). In Indonesia some endangered medicinal plants like *Alyxia reinwardtii, Rauvolfia serpentina, Ruta graveolens* and *Pimpinella purpruatian* have been conserved using in vitro culture technique and cryopreservation (Anonymous 2010a).

18.9 Pollen Preservation

Pollen preservation could complement seed storage or clonal preservation. Through not a conventional strategy for germplasm conservation, it may be used for hybridizing materials that flower at different times. Additionally, the explorer at times may be able to collect pollen which may be useful in future crop improvement programmes through upcoming biotechnological approaches. The preservation of pollen ensures availability of at least part of genome for future utilisation. Techniques involved in pollen preservation are simple and require low input. Successful cryopreservation, as reported for many crop plants, led to establishment of 'Pollen

Cryobanks'. There is very limited information regarding applicability of stored pollen in tropical medicinal plants (see Chandel et al. 1996).

18.10 DNA Conservation

DNA, the basic unit of heredity, can also be used for conservation of plant species. This is particularly useful in case of species available in limited number or which may have been lost before their potential was realised. Additionally, DNA isolated from dead tissues e.g. herbarium specimens, can be a source of information that would otherwise be lost if living tissues are not available.

The strategy of conserving the genomic DNA, as back-up to the storage of living tissues, can be of high relevance in case of the endangered, endemic and threatened medicinal plants. Additionally, this technique may be useful for many undescribed or unstudied plant species whose seeds cannot be stored and genes of which would be otherwise lost if living tissues are not available. There is very little information in the literature regarding its applicability in case of tropical medicinal plants, probably due to technical and/or economic reasons (see Chandel et al. 1996).

18.11 Other Conservation Strategies

18.11.1 Conservation Through Cultivation

The world over, few tropical medicinal species have been brought into cultivation as most of these are harvested from the wild. Primary component of any conservation programme basically addresses human needs and problems hence it is imperative that systematic cultivation of medicinal plants be undertaken to conserve biodiversity and protect threatened species. One of the major constraints to medicinal plant cultivation is lack of land and/or quality planting material. It is recommended that there should be rapid development of alternative supply sources through cultivation in large enough quantities and at low enough price to compete with prices obtained by gatherers of wild stocks. This will satisfy market demands, resulting in more secure jobs and provide fewer incentives to gatherers from the wild. If this does not occur, key species will disappear from the wild, thereby undermining the local medicinal resource base. One of the possibilities to curb overexploitation of these valuable resources, which threatens forests and the people dependent upon them, is to establish community medicinal plant farms as a form of participatory approach (Cunningham 1997). Growing of medicinal plants in agricultural fields should be encouraged for fast growing as well as slow growing species as it may increase supply of material and reduce collection pressure on wild resources (Oladele et al. 2011). Cultivation of these medicinally important species reduces pressure on natural population and facilitates the demand for plant raw drugs which makes it one of the most effective ways of conservation.

18.11.2 Conservation Through Reintroduction

Reintroduction refers to the establishment of individuals of an extinct/endangered species into an area where it has become extinct. The idea is to establish a self-sustaining population for conservation purposes. When conventional propagation is not successful or constrained, micropropagation and reintroduction are recommended. Experimental reintroduction of 8 medicinal plants including the endemic species of Western Ghats such as *Decalpes arayalpathra, Mahonia leschenaultia, Heracleum candolleanum, Calophyllum apetalum* and *Blepharistemma membranifolia* were carried out (Krishnan et al. 2011). Consequently, eco-restoration of threatened medicinal plants of western Ghats could be attempted using micropropagation techniques. A national programme on recovery of endangered taxa has been initiated by the DBT, Govt of India, using biotechnological tools. This has given considerable boost to the necessity of saving endangered species through the use of in vitro propagation technology. About 10 medicinal plants, which had restricted distribution, were multiplied in vitro and reintroduced in the natural habitat as well. In species where natural propagation is slowed down owing to destructive harvesting or reproductive barriers, micropropagation and ecorestoration play a crucial role in supporting the in situ conservation and facilitating population enhancement.

Initiative of individuals or a community can also help in restoration of medicinal plants as illustrated by the following example. *Warburgia salutaris* (Pepper-bark tree), a highly prized medicinal plant of southern Africa, was on the point of extinction due to excessive collection. The initiative by the scientists at Southern Alliance for Indigenous Resources resulted in reintroduction of this species from nurseries in South Africa to Zimbabwe (tropical climate). Though reintroduction was not back to natural forest but it was in the home gardens of local farmers, who knew and valued the species and were willing to guard the plants (Hamilton 2004).

18.12 Future Perspective

The alarming rate at which loss of medicinal plant species is occurring, particularly in the tropics, is a matter of great concern. A holistic approach involving both in situ as well as ex situ methods in combination with emerging technologies is required to be applied for saving this valuable medicinal plant wealth. No single technology holds all the answers and no single technique is suitable for all the situations. It is amply clear that large scale use of medicinal plants in primary health care in developing countries will continue to increase. It is, therefore, of utmost importance to include conservation of medicinal plants as an integral approach to develop traditional medicine and to ensure that required plant material continues to be available. Much of the work on medicinal plants appears haphazard. The need of the hour is to document the information in public domain and link various organizations – government institutes, NGO's and Pharmaceutical sector for a meaningful conservation strategy for sustainable utilization.

Acknowledgements Authors thank Director, NBPGR for encouragement. Special thanks are due to Dr. S William Decruse, TBGRI and Drs. R K Tyagi, Z Abraham, V Kamala, M A Nizar, D R Pani, J B Tomar, NBPGR for readily sharing information and photographs. Thanks are also due to all those authors, whose published work has been extensively used.

References

Abrie AL, van Staden J (2001) Micropropagation of the endangered *Aloe polyphylla*. Plant Growth Regul 33:19–23

Ahmed A, Shashidhara S, Rajasekharan PE, Kumar RV, Honnesh NH (2010) In vitro regeneration of *Acorus calamus* – an important medicinal plant. J Curr Pharma Res 2:36–39

Ahuja S, Mandal BB, Dixit S, Srivastava PS (2002) Molecular, phenotypic and biosynthetic stability in *Dioscorea floribunda* plants derived from cryopreserved shoot tips. Plant Sci 163:971–977

Anand MPH, Hariharan KN, Martin KP, Hariharan M (1997) In vitro propagation of *Kaempferia rotunda* Linn. "Indian Crocus"- a medicinal plant. Phytomorphology 47:281–286

Anandalakshmi R, Warrier RR, Sivakumar V, Singh BG (2007) Investigation on seeds of threatened wild medicinal plants for ex-situ conservation. In: Shukla PK, Chaubey OP (eds) Threatened wild medicinal plants: assessment, conservation and management. Anmol Publications, New Delhi, pp 212–223

Anonymous (2010a) The second report on state of the world's plant genetic resources for food and agriculture. Commission on Genetic Resources for food and agriculture/FAO, Rome

Anonymous (2010b) Country report on the state of plant genetic resources for food and agriculture: Indonesia. FAO, Rome

Anu A, Babu K, Nirmal John CZ, Peter KV (2001) In vitro clonal multiplication of *Acorus calamus* L. J Plant Biochem Biotechnol 10:53–55

Baksha R, Jahan M, Khatun R, Munshi JL (2005) Micropropagation of *Aloe barbadensis* Mill. through in vitro culture of shoot tip explants. Plant Tiss Cult Biotechnol 15:121–126

Barve MD, Mehta AR (1993) Clonal propagation of mature elite trees of *Commiphora wightii.* Plant Cell Tiss Org Cult 35:237–244

Bhadra SK, Akhter T, Hossain MM (2009) In vitro micropropagation of *Plumbago indica* L. through induction of direct and indirect organogenesis. Plant Tiss Cult Biotechnol 19:169–175

Bhatt BS, Vadodaria HK, Vaidya RP (2007) Clonal propagation of *Aloe barbadensis* Mill.: an important medicinal plant via apical meristem culture. J Phytol Res 20:123–127

Bhattacharyya R, Ray A, Gangopadhyay M, Bhattacharyya S (2007) In vitro conservation of *Plumbago indica* – a rare medicinal plant. Plant Cell Biotechnol Mol Biol 8:39–46

Biswas A, Bari MA, Roy M, Bhadra SK (2009) Clonal propagation through nodal explants culture of *Boerhaavia diffusa* L.- a rare medicinal plant. Plant Tiss Cult Biotechnol 19:53–59

Borthakur M, Hazarika J, Singh RS (1998) A protocol for micropropagation of *Alpinia galanga.* Plant Cell Tiss Org Cult 55:231–233

Ceasar AS, Maxwell SL, Prasad KB, Karthigan M, Ignacimuthu S (2010) Highly efficient shoot regeneration of *Bacopa monnieri* (L.) using a two stage culture procedure and assessment of genetic integrity of micropropagated plants by RAPD. Acta Physiol Plant 32:442–452

Chandel KPS, Pandey R (1992) Distribution, diversity, uses and in vitro conservation of cultivated and wild Alliums- a brief review. Indian J Plant Genet Resour 5:7–36

Chandel KPS, Sharma N (1996) In vitro conservation of diversity of medicinal plants. In: Handa SS, Kaul MK (eds) Supplement to cultivation and utilization of medicinal plants, RRL Jammu. CSIR, India, pp 741–752

Chandel KPS, Sharma N (1997) Micropropagation of *Coleus forskohlii* (Willd.) Briq. In: Bajaj YPS (ed) Biotechnology in agriculture and forestry 40: high-tech and micropropagation VI. Springer, Berlin

Chandel KPS, Shukla G, Sharma N (1996) Biodiversity in medicinal and aromatic plants in India. National Bureau of Plant Genetic Resources, New Delhi, 239 pp

Chatuevedi HC, Misra P, Jain M (1984) Proliferation of shoot tips and clonal multiplication of *Costus speciosus* in long- term culture. Plant Sci Lett 35:67–71

Chaturvedi HC, Sharma M, Sharma AK, Sane PV (1991) Conservation of plant genetic resources through excised root culture. In: Zakri AH, Normah MN, Abdul Karim AG, Senawi MT (eds) Conservation of plant genetic resources through in vitro methods. Forest Research Institute/ Malaysian National Committee on Plant Genetic Resources, Malaysia

Chetia S, Handique PJ (2000) High frequency in vitro shoot multiplication of *Plumbago indica,* a rare medicinal plant. Curr Sci 78:1187–1188

Chirangini P, Sinha SK, Sharma GJ (2005) In vitro propagation and microrhizome induction in *Kaempferia galanga* Linn. and *K. rotunda* Linn. Indian J Biotechnol 4:404–408

Chithra M, Martin KP, Sunandakumari C, Madhusoodanan PV (2005) Somatic embryogenesis, encapsulation, and plant regeneration of *Rotula aquatica* Lour., a rare rhoeophytic woody medicinal plant. In Vitro Cell Dev Biol Plant 41:28–31

Cunningham AB (1993) African medicinal plants: setting priorities at the interface between conservation and primary healthcare. People and plants working paper1, pp 34, United Nations Educational, Scientific and Cultural Organization, UNESCO Press

Cunningham AB (1997) An Africa-wide overview of medicinal plant harvesting, conservation and healthcare, non-wood forest products. In: Medicinal plants for forest conservation and healthcare. FAO, Italy

Dalal NV, Rai RV (2003) In vitro propagation of *Oroxylum indicum* Vent. A medicinally important forest tree. J Forest Res 9:61–65

Das G, Rout GR (2002) Direct plant regeneration from leaf explants of *Plumbago* species. Plant Cell Tiss Org Cult 68:311–314

Decruse SW, Seeni S (2002) Ammonium nitrate in the culture medium influence regeneration potential of cryopreserved shoot tips of *Holostemme annulare*. Cryoletters 23:55–60

Decruse SW, Seeni S, Pushpangadan P (1999a) Effect of cryopreservation on seed germination of selected rare medicinal plants of India. Seed Sci Technol 27:501–505

Decruse SW, Seeni S, Pushpangadan P (1999b) Cryopreservation's of alginate coated shoot tips of in vitro grown *Holostemma annulare* (Roxb.) K.Schum, an endangered medicinal plant: influence of preculture and DMSO treatment on survival and regeneration. Cryoletters 20:243–250

Decruse SW, Seeni S, Nair GM (2004) Preparative procedures and culture media effect on the success of cryostorage of *Holostemma annulare* shoot tips. Plant Cell Tiss Org Cult 76:179–182

Dixit S, Mandal BB, Ahuja S, Srivastava PS (2003) Genetic stability assessment of plants regenerated from cryopreserved embryogenic tissues of *Dioscorea bulbifera* L. using RAPD, biochemical and morphological analysis. Cryoletters 24:77–84

DMAPR (2009–2010) Annual report, Directorate of Medicinal and Aromatic Plants Research, Anand, Gujrat. www.dmapr.org.in/research/E-book. Accessed 16 Mar 2011

Dubey KP, Dubey K (2010) Conservation of medicinal plants and poverty alleviation. In: National conference on biodiversity, development and poverty alleviation. Souvenir pp 89–92, Uttar Pradesh State Biodiversity Board, India

Faisal M, Anis M (2003) Rapid mass propagation of *Tylophora indica* Merrill. via leaf callus culture. Plant Cell Tiss Org Cult 75:125–129

Forsyth C, van Staden J (1982) An improved method of in vitro propagation of *Dioscorea bulbifera*. Plant Cell Tiss Org Cult 1:275–281

Gantait S, Mandal N, Bhattacharyya S, Das PK (2010) A novel strategy for in vitro conservation of *Aloe vera* L. through long term shoot culture. Biotechnology 9:326–331

Gautam PL, Ray Choudhuri SP, Sharma N (2000) Conservation, protection and sustainable use of medicinal plants. In: Vienna Jandl R, Devall M, Khorchidi M, Schimpf E, Wolfrum G, Krishnapillay B (eds) Forests and society: the role of research (abstracts of group discussions), vol 11, XXI IUFRO World Congress 2000, Malaysia, pp 197–198

George S, Geetha SP, Anu A, Indra B (2010) In vitro conservation studies in *Hemidesmus indicus*, *Decalpis hamiltoni* and *Ulteria salicifolialin*. In: Proceedings of the 22nd Kerala science congress, Kerala State Council for Science Technology and Environment, Thiruvanthapuram, pp 258–259

Ghosh KC, Banerjee N (2003) Influence of plant growth regulators on in vitro micropropagation of *Rauvolfia tetraphylla* L. Phytomorphology 53:11–19

Ghosh S, Ghosh B, Jha S (2007) In vitro tuberisation of *Gloriosa superba* L. on basal medium. Sci Hortic 114:220–223

Gokhale M, Bansal YK (2009) Direct in vitro regeneration of a medicinal tree *Oroxylum indicum* (L.) Vent. through tissue culture. Afr J Biotechnol 8:3777–3781

Gupta V (2002) Seed germination and dormancy breaking techniques for medicinal and aromatic plants germplasm. J Med Arom Plant Sci 25:402–407

Hamilton AC (2004) Medicinal plants, conservation and livelihoods. Biodivers Conserv 13:1477–1517

Hashemabadi D, Kaviani B (2008) Rapid micropropagation of *Aloe vera* L. via shoot multiplication. Afr J Biotechnol 7:1899–1902

Hassan AKM, Roy SK (2005) Micropropagation of *Gloriosa superba* L. through high frequency shoot proliferation. Plant Tiss Cult 15:67–74

Hettiarachchi A, Fernando KKS, Jayasuriya AHM (1997) In vitro propagation of wadakaha (*Acorus calamus* L.). J Natl Sci Coun Srilanka 25:151–157

Illg RD, Faria RT (1995) Micropropagation of *Alpinia purpurata* from inflorescence buds. Plant Cell Tiss Org Cult 40:183–185

Jha S, Mitra GC, Sen S (1984) In vitro regeneration from bulb explants of Indian squill, *Urginea indica* Kunth. Plant Cell Tiss Org Cult 3:91–100

Johnson M, Das S, Yasmin N, Pandian RM (2008) Micropropagation studies on *Vitex negundo* L. – a medicinally important plant. Ethnobot Leaflet 12:1–5

Joy PP, Thomas J, Mathew S, Skaria BP (1998) Medicinal plants. In: Bose TK, Kabir J, Das P, Joy PP (eds) Tropical Horticulture. Naya Prokash, Calcutta, pp 449–632

Kalimuthu K, Vijayakumar S, Senthilkumar R, Sureshkumar M (2010) Micropropagation of *Aloe vera* Linn. – a medicinal plant. Int J Biotech Biochem 6:405–410

Kalpana M, Anbazhagan M (2009) In vitro production of *Kaempferia galanga* (L.)- an endangered medicinal plant. J Phytol 1:56–61

Khan SK, Karnat M, Shankar D (2005) India's foundation for the revitalization of local health traditions pioneering in situ conservation strategies for medicinal plants and local cultures. HerbalGram 68:34–48

Kochuthressia KP, Britto SJ, Raj MLJ, Jaseentha MO, Senthilkumar SR (2010) Efficient regeneration of *Alpinia purpurata* (Vielli.) K.Schum. plantlets from rhizome bud explants. Int Res J Plant Sci 1:043–047

Kohjyouma M, Kohda H, Tani N, Ashida K, Sugino M, Yamamoto A, Horikoshi T (1995) In vitro propagation from axillary buds of *Glycyrrhiza glabra* L. Plant Tiss Cult Lett 12:145–149

Komalavalli N, Rao MV (2000) In vitro micropropagation of *Gymnema sylvestre* – a multipurpose medicinal plant. Plant Cell Tiss Org Cult 61:97–105

Krishnan PN, Seeni S (1994) Rapid micropropagation of *Woodfordia fruiticosa* (L) Kurz (Lytheraceae), a rare medicinal plant. Plant Cell Rep 14:55–58

Krishnan PN, Decruse SW, Radha RK (2011) Conservation of medicinal plants of Western Ghats, India and its sustainable utilization through in vitro technology. In vitro Cell Dev Biol-Plant 47:110–122

Kumar A (2009) In vitro plantlet regeneration in *Asparagus racemosus* through shoot bud differentiation on nodal segments. http://www.science20.com. Accessed 21 Mar 2011

Kumar S, Mathur M, Jain AK, Ramawat KG (2006) Somatic embryo proliferation in *Commiphora wightii* and evidence for guggulsterone production in culture. Indian J Biotechnol 5:217–222

Latiff A (1999) Medicinal and aromatic plants of Malaysia: approaches to exploitation and conservation. In: Salleh K, Natesh S, Osman AF, Kaldir AA (eds) Conservation of medicinal and aromatic plants: strategies and technologies. Forest Research Institute, Malaysia, pp 20–31

Madhavan M, Joseph JP (2010) In vitro root tuber induction from leaf and nodal explants of *Gloriosa superba* L: an endangered medicinal plant of India. Plant Arch 10:611–615

Mahindapala R (2004) Medicinal plants: conservation and sustainable use in Sri Lanka. IK Notes 66

Makowska Z, Keller J, Engelmann F (1999) Cryopreservation of apices isolated from garlic (*Allium sativum*) bulbils and cloves. Cryoletters 20:175–182

Malaurie B, Trouslot MF, Engelman F, Chabrillange N (1998) Effect of pretreatment on the cryopreservation of in vitro cultured yam (*Dioscorea alata* "Brazo Fuerte" and *D. bulbifera* "Noumea Imboro") shoot apices by encapsulation-dehydration. Cryoletters 19:15–26

Mallikarjuna K, Rajendrudu G (2009) Rapid in vitro propagation of *Holarrhena antidysenterica* using seedling cotyledonary nodes. Biol Plant 53:569–572

Mandal BB, Ahuja-Ghosh S (2007) Regeneration of *Dioscorea floribunda* plants from cryopreserved encapsulated shoot tips: effects of plant growth regulators. Cryoletters 28:329–336

Mandal BB, Ahuja S, Dixit S (1999) Conservation of yam (*Dioscorea bulbifera*) somatic embryos by encapsulation dehydration. Cryobiology 39:377–378

Mandal BB, Tyagi RK, Pandey R, Sharma N, Agarwal A (2000) In vitro conservation of germplasm of agri-horticultural crops at NBPGR: an overview. In: Razdan MK, Cocking EC (eds) Conservation of plant genetic resources in vitro. Science, USA, pp 297–307

Manjula S, Thomas A, Daniel B, Nair GM (1997) In vitro plant regeneration of *Aristolochia indica* through axillary shoot multiplication and organogenesis. Plant Cell Tiss Org Cult 51:145–148

Martin KP (2002) Rapid propagation of *Holostemma ada-kodien* Schult., a rare medicinal plant, through axillary bud multiplication and indirect organogenesis. Plant Cell Rep 21:112–117

Martin KP (2003a) Plant regeneration through somatic embryogenesis of *Holostemma add-kodien*, a rare medicinal plant. Plant Cell Tiss Org Cult 72:79–82

Martin KP (2003b) Rapid in vitro multiplication and ex vitro rooting of *Rotula aquatica* Lour., a rare rhoeophytic woody medicinal plant. Plant Cell Rep 21:415–420

Mathur S, Kumar S (1998) Phytohormone self sufficiency for regeneration in the leaf and stem explants of *Bacopa monnieri*. J Med Arom Plant Sci 20:1056–1059

Mathur A, Mathur AK, Kukreja AK, Ahuja PS, Tyagi BR (1987) Establishment and multiplication of colchi-autotetraploids of *Rauvolfia serpentina* L. Benth.ex Kurz. through tissue culture. Plant Cell Tiss Org Cult 10:129–134

Maunder M (2001) Plant conservation encyclopedia of biodiversity, vol 4. Academic Press, San Diego, pp 645–657

Meyer HJ, van Staden J (1991) Rapid in vitro propagation of *Aloe barbadensis* Mill. Plant Cell Tiss Org Cult 26:167–171

Mishra Y, Usmani GM, Mandal AK (2010) Micropropagation and field evaluation of *Tinospora cordifolia*: an important medicinal climber. Indian J Plant Physiol 15:359–363

Moosikapala L, Te-chato S (2010) Application of in vitro conservation in *Vetiveria zizanioides* Nash. J Agri Tech 6:401–407

Mukherjee P, Mandal BB, Bhat KV, Biswas AK (2009) Cryopreservation of Asian *Dioscorea bulbifera* L. and *D. alata* L. by vitrification: importance of plant growth regulators. Cryoletters 30:100–111

Mukhopadhyay S, Mukhopadhyay MJ, Sharma AK (1991) In vitro multiplication and regeneration of cytologically stable plants of *Rauvolfia serpentina* Benth. through shoot tip culture. Nucleus 34:170–173

Myers N, Mittermeier CG, da Fonseca GAB, Kent J (2000) Biodiversity hotspots for conservation priorities. Nature 403:853–858

Natali L, Sanchez C, Cavallini A (1990) In vitro culture of *Aloe barbadensis* Mill.: micropropagation from vegetative meristems. Plant Cell Tiss Org Cult 20:71–74

Natesh S (1999) Conservation of medicinal and aromatic plants in India- an overview. In: Salleh K, Natesh S, Osman AF, Kaldir AA (eds) Conservation of medicinal and aromatic plants: strategies and technologies. Forest Research Institute, Malaysia, pp 1–12

Natesh S (2000) Biotechnology in the conservation of medicinal and aromatic plants. In: Ravindran PN, Sahijram L, Chadha KL (eds) Biotechnology in horticultural and plantation crops. Malhotra Publishing House, New Delhi, pp 548–561

Nayak S, Dabata BK, Sahoo S (1996) Rapid propagation of lemongrass (*Cymbopogon flexuosus* (Nees) Wats.) through somatic embryogenesis in vitro. Plant Cell Rep 15:367–370

NBPGR (2010) Annual report. National Bureau of Plant Genetic Resources, New Delhi

Niwata E (1995) Cryopreservation of apical meristems of garlic (*Allium sativum* L.) and subsequent high plant generation. Cryoletters 16:102–107

Normah MN, Kean CW, Vun YL, Mohamed-Hussein ZA (2011) In vitro conservation of Malaysian biodiversity-achievements, challenges and future directions. In Vitro Cell Dev Biol-Plant 47:26–36

Okigbo RN, Eme UE, Ogbogu S (2008) Biodiversity and conservation of medicinal and aromatic plants in Africa. Biotechnol Mol Biol Rev 3:127–134

Oladele AT, Alade GO, Omobuwago OR (2011) Medicinal plants conservation and cultivation by traditional medicine practitioners (TMPs) in Aiyedaade Local Government Area of Osun state. Nigeria Agric Biol J N Am 2:476–487

Pant KK, Joshi DS (2009) In vitro multiplication of wild Nepalese *Asparagus racemosus* through shoots and shoots induced callus cultures. Bot Res Intl 2:88–93

Pareek SK, Gupta V, Bhat KC, Negi KS, Sharma N (2005) Medicinal and aromatic plants. In: Dhillon BS, Tyagi RK, Saxena S, Randhawa GJ (eds) Plant genetic resources: horticultural crops. Narosa Publishing House, New Delhi, pp 279–308

Parihar SS, Dadlani M (2010) Significance of seed storage behavior studies in conservation of biodiversity of medicinal and aromatic plants. In: Abstracts of the national conference on biodiversity of medicinal and aromatic plants: collection, characterization and utilization, Medicinal and Aromatic plants Association of India, Anand, Gujrat, pp 38–39

Prajapati HA, Patel DH, Mehta SR, Subramanian RB (2003) Direct in vitro regeneration of *Curculigo orchioides* Gaertn., an endangered anticarcinogenic herb. Curr Sci 84:747–749

Punyarani K, Sharma JG (2010) Micropropagation of *Costus speciosus* (Koen.) Sm. using nodal segment culture. Notulae Scientia Biologicae 2:58–62

Radha RK, Decruse SW, Amy MV, Krishnan PN (2010) Zygotic embryo cryopreservation of *Celastrus paniculatus*. In: Golden jubilee national symposium on plant diversity utilization and management, Department of Botany, University of Kerala, Keriavatom

Radha RK, Shereena SR, Divya K, Krishanan PN, Seeni S (2011) In vitro propagation of *Rubia cordifolia* Linn., a medicinal plant of the Western Ghats. Int J Plant Bot 7:90–96

Raghu AV, Geetha SP, Martin G, Balachandran I, Ravindran PN (2006) In vitro clonal propagation through mature nodes of *Tinospora cordifolia* (Willd.) Hook. f. & Thoms.: an important ayurvedic medicinal plant. In Vitro Cell Dev Biol Plant 42:584–588

Rahman MM, Amin MN, Ahamed T, Ahmed S (2005) In vitro rapid propagation of black thorn (*Kaempferia galanga* L.): a rare medicinal and aromatic plant of Bangladesh. J Biol Sci 5:300–304

Rakkimuthu R, Jacob J, Aravinthan KM (2011) In vitro micropropagation of *Alpinia zerumbet* Variegate, an important medicinal plant, through rhizome bud explants. Res Biotech 2:7–10

Rao NS, Geetha KA, Maiti S, Patel V (2010) Networking of herbal gardens in India: a promising web application for medicinal plants information. In: Abstracts of the national conference on biodiversity of medicinal and aromatic plants: collection, characterization and utilization. Medicinal and Aromatic plants Association of India, Anand, Gujrat, p 87

Ratnam W, Teik NL (1999) Practical application of biotechnology in conservation of medicinal and aromatic plants. In: Salleh K, Natesh S, Osman A, Kaldir AA (eds) Conservation of medicinal and aromatic plants: strategies and technologies. Forest Research Institute, Malaysia, pp 53–62

Ray A, Bhattacharya S (2008) Cryopreservation of in vitro grown nodal segments of *Rauvolfia serpentina* by PVS2 vitrification. Cryoletters 29:321–328

Reddy S, Gopal GR, Lakshmi SG (1998) In vitro multiplication of *Gymnema Ssylvestre* R.Br.- an important medicinal plant. Curr Sci 75:843–845

Reddy S, Rodrigues R, Rajasekharan R (2001) Shoot organogenesis and mass propagation of *Coleus forskholii* from leaf derived callus. Plant Cell Tiss Org Cult 66:183–186

Robinson JP, Britto SJ, Balkrishan V (2009) Micropropagation of *Costus speciosus* (Koem.ex.retz) Sm., antidiabetic plant by using explants of pseudostems. Bot Res Int 2:182–185

Rout GR, Das G (2002) An assessment of genetic integrity of micropropagated plants of *Plumbago zeylanica*. Biol Plant 45:27–32

Rout GR, Saxena C, Samantaray S, Das P (1999a) Rapid plant regeneration from callus cultures of *Plumbago zeylanica*. Plant Cell Tiss Org Cult 56:47–51

Rout GR, Saxena C, Samantaray S, Das P (1999b) Rapid clonal propagation of *Plumbago zeylanica*. Plant Growth Regul 28:1–4

Roy PK (2008) Rapid multiplication of *Boerhaavia diffusa* L. through in vitro culture of shoot tip and nodal explants. Plant Tiss Cult Biotechnol 18:49–56

Roy A, Pal A (1991) Propagation of *Costus speciosus* (Koen.) Sm. through in vitro rhizome production. Plant Cell Rep 10:525–528

Roy SC, Sarkar A (1991) In vitro regeneration and micropropagation of *Aloe vera* L. Sci Hortic 47:107–113

Russell-Smith J, Karunaratne NS, Mahindapala R (2006) Rapid inventory of wild medicinal plant populations in Sri Lanka. Biol Conserv 132:22–32

Sarasan V, Soniya EV, Nair GM (1994) Regeneration of Indian sarasaparilla, *Hemidesmus indicus* R.Br., through organogenesis and somatic embryogenesis. Indian J Exp Biol 32:284–287

Sarasan V, Cripps R, Ramsay MM, Atherton C, McMichen M, Prendergast G, Rowntree JK (2006) Conservation in vitro of threatened plants – progress in the past decade. In vitro Cell Dev Biol Plant 42:206–214

Sarkar KP, Islam A, Islam R, Hoque A, Joarder OI (1996) In vitro propagation of *Rauvolfia serpentina* through tissue culture. Planta Med 62:358–359

Sateesh KK, Seeni S (2003) In vitro mass multiplication and production of roots in *Plumbago rosea*. Planta Med 69:83–86

Satheesh G, Geetha SP, Sudhakar Raja S, Balachandran I, Ravindran PN (2003) In vitro medium-term conservation of *Bacopa monnieri* (L.) Pennell – the memory plus plant – under slow-growth conditions. PGR Newsl FAO Biovers 151:49–55

Sebastian DP, Benjamin S, Hariharan M (2002) Micropropagation of *Rotula aquatica* Lour. – an important woody medicinal plant. Phytomorphology 52:137–144

Selvakkumar C, Balakrishan A, Lakshmi SB (2007) Rapid in vitro micropropagation of *Alpinia officinarum* Hance, an important medicinal plant, through rhizome bud explants. Asian J Plant Sci 6:1251–1255

Selvakumar V, Anbudurai PR, Balakumar T (2001) In vitro propagation of the medicinal plant *Plumbago zeylanica* L. through nodal explants. In vitro Cell Dev Biol Plant 37:280–284

Sen J, Sharma AK (1991) In vitro propagation of *Coleus forskholii* Briq. for forskolin synthesis. Plant Cell Rep 9:696–698

Senarath WTPSK (2010) In vitro propagation of *Coscinium fenestratum* (Gaertn.) Colebr. (Menispermaceae)- an endangered medicinal plant. J Natl Sci Foundation Sri Lanka 38:219–223

Sengupta J, Mitra GC, Sharma AK (1984) Organogenesis and tuberization in cultures of *Dioscorea floribunda*. Plant Cell Tiss Org Cult 3:325–331

Sharma N (1999) Conservation of Patchouli through in vitro method. Indian Perfumer 43:19–22

Sharma B, Bansal YK (2010) In vitro propagation of *Gymnema sylvestre* Retz. R.Br through apical bud culture. J Med Plant Res 4:1473–1476

Sharma N, Chandel KPS (1992a) Low temperature storage of *Rauvolfia serpentina* Benth. Ex Kurz – an endangered, endemic medicinal plant. Plant Cell Rep 11:200–203

Sharma N, Chandel KPS (1992b) Effects of ascorbic acid on axillary shoot induction on *Tylophora indica* (Burm. f.) Merill. Plant Cell Tiss Org Cult 29:109–113

Sharma N, Chandel KPS, Maheshwari ML (1991a) Feasibililty of mass production of *Pogostemon patchouli* Hook. F. through tissue culture for commercial cultivation. Indian Perfumer 36:70–73

Sharma N, Chandel KPS, Srivastava VK (1991b) In vitro propagation of *Coleus forskohlii* Briq., a threatened medicinal plant. Plant Cell Rep 10:67–70

Sharma N, Chandel KPS, Paul A (1995) In vitro conservation of threatened plants of medicinal importance. Indian J Plant Genet Res 8:107–112

Sharma N, Sharma B, Gautam PL (2000) Tissue culture – an effective method to conserve diversity of medicinal plants. In: International conference on managing natural resources for sustainable agricultural production in the 21st century, vol 2. Indian Society of Soil Sciences, New Delhi, pp 393–394

Sharma N, Pandey A, Vimla Devi S (2002) In vitro establishment of *Plumbago zeylanica* Linn. for conservation In: International symposium on plant biodiversity: conservation and evaluation, Kolkata

Sharma N, Vimala Devi S, Satsangi R, Pandey R (2005) In vitro clonal propagation of *Plumbago zeylanica* Linn. for conservation. In: Abstracts of the national symposium on plant sciences research in India: challenges and prospects, Botanical Survey of India, Dehradun, p 111

Sharma N, Satsangi R, Pandey R, Vimala Devi S (2007a) In vitro clonal propagation and medium term conservation of Brahmi [*Bacopa monnieri* (L.) Wettst.]. J Plant Biochem Biotechnol 16:139–143

Sharma N, Vimala Devi S, Pandey R (2007b) In vitro propagation of a threatened, anticarcinogenic, herb, *Curculigo orchioides* Gaertn. J Plant Biochem Biotechnol 16:63–65

Sharma N, Vimala Devi S, Satsangi R (2007c) Biotechnological approaches in multiplication and conservation of plants of medicinal value. In: Shukla PK, Chaubey OP (eds) Threatened wild medicinal plants: assessment, conservation and management. Anmol Publications, New Delhi, pp 248–257

Sharma N, Satsangi R, Pandey R (2009a) Cryopreservation of shoot tips of *Bacopa monnieri* (L.) Wettst. – a commercially important medicinal plant by vitrification technique. In: Abstracts of the 1st international symposium on cryopreservation in horticultural species, Leuven, p 125

Sharma N, Vimala Devi S, Meena R, Bhat KC, Pandey R (2009b) In vitro regeneration of *Curculigo orchioides* using various explants. In: Abstracts of the national symposium on recent global developments in the management of plant genetic resources, National Bureau of Plant Genetic Resources, New Delhi, pp 303–304

Sharma S, Rathi N, Kamal B, Pundir D, Kaur B, Arya S (2010) Conservation of biodiversity of highly important medicinal plants of India through tissue culture technology- a review. Agric Biol J N Am 1:827–833

Sharma N, Satsangi R, Pandey R (2011) Cryopreservation of shoot tips of *Bacopa monnieri* (L.) Wettst by vitrification technique. Acta Hortic 908:283–288

Shirin F, Kumar S, Mishra Y (2000) In vitro plantlet production system for *Kaempferia galanga*, a rare Indian medicinal herb. Plant Cell Tiss Org Cult 63:193–197

Shrivastava N, Rajani M (1999) Multiple shoot regeneration and tissue culture studies on *Bacopa monnieri* (L.) Pennell. Plant Cell Rep 18:919–923

Siddique NA, Bari MA (2006) Plant regeneration from axillary shoot segments derived callus in *Hemidesmus indicus* (L.) R. Br. (Anantamul) an endangered medicinal plant in Bangladesh. J Plant Sci 1:42–48

Sivakumar G, Alagumanian S, Rao MV (2006) High frequency in vitro multiplication of *Centella asiatica*; an important industrial medicinal herb. Eng Life Sci 6:597–601

Soniya EV, Sujitha M (2006) An efficient in vitro propagation of *Aristolochia indica*. Biol Plant 50:272–274

Sreekumar S, Seeni S, Pushpangadan P (2000) Micropropagation of *Hemidesmus indicus* for cultivation and production of 2-hydroxy 4-methoxy benzaldehyde. Plant Cell Tiss Org Cult 62:211–218

Subathra Devi C, Srinivasan VM (2008) In vitro propagation of *Gymnema sylvestre*. Asian J Plant Sci 7:660–665

Sudha CG, Seeni S (1994) In vitro multiplication and field establishment of *Adhatoda beeddomei* C.B Clarke, a rare medicinal plant. Plant Cell Rep 13:203–207

Sudha CG, Krishnan PN, Seeni S, Pushpgandan P (2000) Regeneration of plants from in vitro root segments of *Holostemma annulare* (Roxb.) K.Schum., a rare medicinal tree. Curr Sci 78:503–506

Supe UJ (2007) In vitro regeneration of *Aloe barbadensis*. Biotechnology 6:601–603

TBGRI (2009–2010) Annual report. Tropical Botanic Garden and Research Institute, Thiruvananthapuram

Tejavathi DH, Shailaja KS (1999) Regeneration of plants from the cultures of *Bacopa monnieri* (L.) Pennel. Phytomorphology 49:447–452

Thengane SR, Kulkarni DK, Krishnamurthy KV (1998) Micropropagation of Licorice (*Glycyrrhiza glabra* L.) through shoot tip and nodal cultures. In Vitro Cell Dev Biol Plant 34:331–334

Theriappan P, Devi KS, Dhasarathan P (2010) Micropropagation studies of a medicinal plant *Aristolochia indica*. Int J Curr Res 11:200–204

Tiwari KN, Sharma NC, Singh BD (2000) Micropropagation of *Centella asiatica* (L.), a valuable medicinal herb. Plant Cell Tiss Org Cult 63:179–183

Tiwari V, Tiwari KN, Singh BD (2001) Comparative studies of cytokinins on in vitro propagation of *Bacopa monniera*. Plant Cell Tiss Org Cult 66:9–16

Usha PK, Benjamin S, Mohanan KV, Raghu AV (2007) An efficient micropropagation system for *Vitex negundo* L., an important woody aromatic medicinal plant through shoot tip culture. Res J Bot 2:102–107

Vieira RF (1999) Conservation of medicinal and aromatic plants in Brazil. In: Janick J (ed) Perspectives on new crops and new uses. ASHS Press, Alexandria, pp 152–159

Vincent KA, Hariharan M, Mathew KM (1992a) Embryogenesis and plantlet formation in tissue culture of *Kaempferia galanga* L.- a medicinal plant. Phytomorphology 42:253–256

Vincent KA, Mathew KM, Hariharan M (1992b) Micropropagation of *Kaempferia galanga* L. – a medicinal plant. Plant Cell Tiss Org Cult 28:229–230

Wesely EG, Johnson M, Regha P, Kavitha MS (2010) In vitro propagation of *Boerhaavia diffusa* L. through direct and indirect organogenesis. J Chem Pharm Res 2:339–347

Section III
Future Perspectives

Chapter 19
Global Challenges for Agricultural Plant Biodiversity and International Collaboration

Mohammad Ehsan Dulloo

19.1 Introduction

Biological diversity (also known as biodiversity) is "the diversity of life on Earth" and is essential for the functioning of ecosystems that underpin the provisioning of ecosystem services that ultimately affect human well being (Millennium Ecosystem Assessment (MEA) 2005). By definition, Biodiversity has three main components: it includes diversity among ecosystems, among species and within species (known as genetic diversity or genetic resources). An important component of biodiversity is Agricultural Biodiversity which has a more direct link to the wellbeing and livelihoods of people than other forms of biodiversity. According to the Convention on Biological Diversity (CBD), Agricultural Biodiversity is a broad term that includes all components of biological diversity of relevance to food and agriculture, and all components of biological diversity that constitute the agricultural ecosystems: the variety and variability of animals, plants and micro-organisms, at the genetic, species and ecosystem levels, which are necessary to sustain key functions of the agro-ecosystem, its structure and processes (COP decision V/5, appendix). All crops, livestock and fish, their wild relatives and all interacting species of pollinators, symbionts, pests, parasites, predators and competitors form part of agricultural biodiversity.

In this chapter we focus on the tropical plant diversity, more precisely on crop genetic diversity, that is important for humankind as food crops, medicinal, horticultural and forestry. The tropical world harbours the greater part of the world's

M.E. Dulloo (✉)
Bioversity International, via dei tre Denari 472/a,
Maccarese, Rome 00057, Italy

Food and Agriculture Organization of the United Nations,
Viale delle terme di Caracalla, Rome 00153, Italy
e-mail: ehsan.dulloo@fao.org

M.N. Normah et al. (eds.), *Conservation of Tropical Plant Species*,
DOI 10.1007/978-1-4614-3776-5_19, © Springer Science+Business Media New York 2013

biodiversity. It is known that tropical forests takes up less than 2% of the world area and yet contains more than 50% of its biodiversity (Collen et al. 2008). The distribution of plants species is also not uniform and tends to be concentrated in specific diversity rich areas. Myers et al. (2000) noted that as many as 44% of all species of vascular plants and 35% of all species in four vertebrate groups are confined to 25 hotspots comprising only 1.4% of the land surface of the Earth mostly located in tropical areas. These biodiversity hotspot areas are targeted for priority conservation actions (Mittermeier et al. 1998). Also most of the centres of origin and diversity of crop plants, as defined by Nicolai Vavilov (1931), are located within the tropical and sub tropical regions of the world and represent significant diversity rich areas for crop plants.

19.2 Plant Diversity and Its Importance

There is still much debate about the total number of plant species in the world. Estimates range from 250,000 to more than 400,000 (Mabberley 1997; Prance et al. 2000; Govaerts 2001; Bramwell 2002). It is not known exactly how many of these plants occur in the tropical and subtropical world, but it is fair to say that a majority of them do. Of the total number of identified plants, humans have used around 7,000 plant species for food over the course of history and another 70,000 plants are known to have edible parts (Wilson 1989). Furthermore, Prescott-Allen and Prescott-Allen (1990) calculated that the world's food comes just from 103 plant species based on calories, protein and fat supply. However, only four crop species (maize, wheat, rice and sugar) supply almost 60% of the calories and proteins in the human diet (Palacios 1998). Plant diversity is one of our most fundamental and essential resources for humankind, one that has enabled farming systems to evolve since the birth of agriculture about 10,000–12,000 or more years ago (Harlan 1992). Plant and animal species for food have been collected, used, domesticated and improved through traditional systems of selection over many generations. This has resulted in even more intraspecific diversity developed by early farmers in terms of crop varieties and local landraces and breeds. These form the basis on which modern high yielding and disease resistant varieties have been produced in the past (Plucknett et al. 1987) and will continue to do so to feed the growing human population, expected to reach 9.1 billion by 2050 (FAO 2010).

Given its enormous dependence on relatively few food species, as mentioned above, humanity depends heavily on the diversity within these species for its survival. Breeding and agronomic improvements have exploited the diversity in local varieties of these few crops to improve yield. On average, it achieved a linear increase in food production globally, at an average rate of 32 million metric tons per year (Tester and Langridge 2010). The Food and Agriculture

Organization of the United Nations (FAO) (2010) estimates that there is a need to increase food production by 70% in order to meet the demand for food by 9.1 billion people by 2050. As the land is going to be scarce, this increase will have to be met by the further use of crop and animal diversity and by reviewing current agricultural and husbandry practices to make them more sustainable. Along with diversity still extant in farmer's cultivars/landraces, crop wild relatives (CWR) represent an important reservoir of genetic resources for breeders (Maxted and Kell 2009). Many useful traits from CWR, such as pest and disease resistance, abiotic stress tolerance or quality improvements, have been introgressed in today's crops (Hajjar and Hodgkin 2007). For example, genes from *Oryza nivara* S.D. Sharma and Shastry, a wild relative of rice, are providing strong and extensive resistance to grassy stunt virus on millions of hectares of rice fields in South and Southeast Asia (Barclay 2004).

Crop diversity also plays an important role in food security by helping to reduce genetic vulnerability, whereby diversity within a field or within a production system helps to ensure stability in overall food production and thus reduces the risks to agricultural production. A more diverse cropping system helps to buffer against the spread of pests and diseases and the vagaries of weather, likely to occur in uniform monoculture cultivation. In the global context, the phenomenon of genetic vulnerability represents a major risk with regard to the capacity of our agricultural systems to ensure sustainable food security, as well as the livelihoods of farmers.

Maintaining crop diversity is also a key strategy for farmers around the world to guarantee their sustenance. Many studies have shown that crop genetic diversity in the form of traditional varieties continues to be maintained on farm by poor, small-scale farmers who rely on traditional crop varieties to meet their livelihood needs (Bezançon et al. 2009; Kontoleon et al. 2009; Sadiki et al. 2007). Such diversity is vital to cope with climate uncertainty for farmers who depend on rainfed agriculture. It is common to find poor farmers growing many varieties of the same crop to increase the likelihood of producing a crop to feed their families, regardless of the specific weather conditions in a given year. For example, a farmer in Papua New Guinea mentioned that he grew 50 different varieties of sweet potatoes on his farm (Prem Mathur, personal communication). In addition, many plant species growing in wild ecosystems are valuable for food and agriculture and often play an important cultural role in local societies. They can provide a safety net when food is scarce and are increasingly marketed locally and internationally, providing an important contribution to household incomes (FAO 2010). Box 19.1 gives some of the major benefits of crop genetic diversity.

Therefore, an effective conservation and use of agricultural biodiversity is very important in ensuring sustainable increases in the productivity and production of healthy food by and for mankind, as well as contributing to increased resilience of agricultural ecosystems that can help in achieving the Millennium Development Goals.

Box 19.1 Benefits of Crop Genetic Diversity

- Provides broader range of food, medicines, forages and others uses for humankind
- Provides a reservoir of genetic resources for breeders –many useful traits, such as pest and disease resistance, abiotic stress tolerance, adapted genes or quality improvements for crop improvement
- Supports evolutionary potential for crops to generate new variability
- Helps to reduce genetic vulnerability by broadening genetic base
- Helps farmers around the world to avert risk of crop failures to guarantee their livelihoods and sustenance

19.3 Threats to Biodiversity

It is widely recognized that biodiversity continues to be lost at an unprecedented rate (MEA-Millennium Ecosystem Assessment 2005). The recent Global Biodiversity Outlook (GBO-3) which provides an assessment of the state of the world's biodiversity in 2010, reports that the current target of reducing the rate of biodiversity loss by 2010 has not been met, either internationally or nationally in any part of the world. Moreover, it warns that the principal drivers of biodiversity loss are in many cases intensifying as a result of human actions. The Millennium Ecosystem Assessment (2005) identified five major drivers of biodiversity loss (climate change, habitat change, invasive alien species, overexploitation, and pollution) responsible for the continuing decline in biodiversity in all three of its components – ecosystems, species and genes. It concluded that 60% of ecosystem services worldwide have become degraded in the past 50 years, primarily due to unsustainable use of land, freshwater and ocean resources. Most major habitats have declined in this time and at the species level, The IUCN Red List of Threatened Species™ tells us that 22% of the world's mammals are threatened and at risk of extinction worldwide, as well as nearly one third of amphibians, one in eight birds, 27% of reef building corals and 28% of conifers. Species extinction rates are up to 1,000 times greater than the average rates in pre-human times and are increasing (IUCN 2010).

Crop and livestock diversity also continues to decline in most agricultural systems. The most important threats include changes in land use, replacement of traditional varieties by modern cultivars, agricultural intensification, increased population, poverty, land degradation and environmental change (including climate change) (FAO 2010; van de Wouw et al. 2009). It is predicted that climate change will have a significant impact on agriculture with temperatures rising on average by 2–4 °C over the next 50 years, causing significant changes in regional and seasonal patterns of precipitation (IPCC 2007; Burke et al. 2009). Model projections carried out by

Lane and Jarvis (2007), based on global distribution of suitable cultivated areas of 43 crops, highlight that more than 50% of these crops may decrease in extent. Climate change will also impact agricultural biodiversity in a major way. Evidence based on bioclimatic modelling suggests that climate change could cause a marked contraction in the distribution ranges of crop wild relatives. In the case of wild populations of peanut (*Arachis* spp.), potato (*Solanum* spp.) and cowpea (*Vigna* spp.), studies suggest that 16–22% of these species may become extinct by 2055, with most species possibly losing 50% of their range size (Jarvis et al. 2008).

Information regarding the threat and rate of genetic erosion among various components of agricultural biodiversity is important, yet very little work has been carried out to quantify the magnitude of any trends (Laikre et al. 2010). Laikre et al. (2010) provide some examples of empirical work that demonstrates how populations and even species can collapse due to loss of genetic diversity. It also provides evidence supporting the importance of maintaining genetic variation to sustain species and ecosystems. Policy makers and scientists require a better understanding of how the genetic diversity varies over time and space to make informed decisions on which populations to prioritize for conservation. However there is no routine global scale monitoring of genetic diversity over time (Frankham 2010; Laikre et al. 2010), except for a few target species at national level (Laikre et al. 2008). A major challenge remains to develop simple inexpensive means to monitor genetic diversity on a global scale (Frankham 2010). Several efforts have been undertaken in past couple of decades to address the monitoring of genetic diversity in crops plants. FAO has been involved in the identification and development of indicators for monitoring the status and trends of agricultural biodiversity and other components of biodiversity in agricultural ecosystems within the context of the Global Plan of Action for the Conservation and Sustainable Utilization of Plant Genetic Resources for Food and Agriculture (GPA) and the Global Strategy for the Management of Farm Animal Genetic Resources (Global Strategy) in order to identify, and assess, the extent of genetic diversity, genetic erosion, and genetic vulnerability in agricultural ecosystems (Collette 2001; Diulgheroff 2004). Brown and Brubaker (2002) published a list of desirable properties of indicators for managing genetic resources. FAO in collaboration with International Plant Genetic Resources Institute (now Bioversity International) organised an expert consultation in 2002 to review the rationale for the use of indicators, the criteria for good indicators for genetic diversity, genetic erosion and genetic vulnerability, as well as prepared technical advice concerning indicators for these three concepts (FAO/IPGRI 2002). More recently, in preparation of the second State of the World Report on Plant Resources for Food and Agriculture (FAO 2010), a thematic background paper on indicators of Genetic Diversity, Genetic Erosion and Genetic Vulnerability for Plant Genetic Resources for Food and Agriculture has been prepared, in which summary measures of genetic diversity in cultivated plants and wild relatives are provided (Brown 2008). In the context of the Convention on Biological Diversity (CBD), world leaders committed to achieve a significant reduction in the rate of biodiversity loss by 2010 and established framework of 22 cross disciplinary headline indicators, including one for genetic diversity, to track progress against the 2010 target (Walpole et al. 2009).

A partnership called "the 2010 Biodiversity Indicators Partnership" (http://www.twentyten.net) was established to identify further useful indicators under each of the headline indicators to detect changes in species, ecosystem and genetic diversity. There are only two initiatives that are explicitly working on developing indicators that deal with genetic variation for agricultural biodiversity (Laikre et al. 2010; Walpole et al. 2009). These include an indicator on ex situ crop collections and the number of food production breeds of domestic animals. These initiatives are still under development (Walpole et al. 2009) under the 2010 Biodiversity Indicators Partnership and the Pan European initiative "Streamlining European 2010 Biodiversity Indicators," which deals exclusively with the number of domestic livestock breeds within countries and not plants (Bubb et al. 2005; EEA 2007 cited in Laikre 2010). There is a strong need to further develop indicators for crop genetic diversity and this is currently being undertaken under both the implementation of the FAO GPA and CBD Strategic Plan for the post 2010 period, as adopted at the CBD Conference of Parties (COP) in Nagoya, Japan in 2010.

19.4 Conservation of Plant Diversity

It is imperative that the plant diversity be properly conserved to provide the benefits it brings to its users. For this, there are two main complementary approaches for conserving plant diversity: *in situ* and in situ conservation. The latter conservation approach also include on farm conservation and the conservation of wild habitats.

19.4.1 Ex Situ Conservation

Ex situ conservation is usually carried out in genebanks and botanic gardens. Different types of genebanks have been established for the storage of plant diversity, depending on the type of plant material conserved. These include seed banks (for seeds), field genebanks (for live plants), in vitro and cryo genebanks (for plant tissue and cells), pollen and DNA banks (Engelmann and Engels 2002; Thormann et al. 2006; Dulloo et al. 2006). According to the Second Report of the State of the World's Plant Genetic Resources for Food and Agriculture there are about 7.4 million accessions that are conserved in over 1,750 genebanks around the world in the different types of genebanks described above (FAO 2010), representing in excess of 52,000 species of PGRFA (Stefano Diulgheroff, personal communication). The ex situ collection is heavily biased towards major crops, and minor crops and CWR are under-represented, despite the fact there has been a large increase in the number of minor crop species observed/recorded in the last 10 years (FAO 2010). In addition to the crop genebanks, there are also over 2,500 botanic gardens maintaining samples of some 80,000 plant species which represents mostly wild species (FAO 2010). These global accessions represent the world's attempt at conserving the genetic diversity of major food crops. According to the Subsidiary Body on Scientific Technical and Technology Advice for the CBD (SBSTTA 2010), this is estimated to be around 70% of the total genetic diversity of some 200–300 crops. However the actual extent of genetic diversity of these crops is still unknown. This diversity occurs in farmers' fields, semi-wild and wild habitats and needs to be documented and adequately conserved in situ.

19.4.2 In Situ Conservation (Wild Habitats)

The aim of in situ conservation is to ensure that populations of targeted species are maintained in the natural habitats where they evolved and that their continued survival is not threatened. The most commonly used in situ conservation method is the establishment of protected areas and taking appropriate measures to ensure their conservation. There are different types of protected areas, depending on the purpose of conservation, as defined by the International Union for Conservation of Nature (IUCN); and different levels of management interventions (Heywood and Dulloo 2005; Maxted et al. 2008). In the broader context of conserving plant genetic diversity, in situ conservation may involve the creation of "genetic reserves" where the ultimate goal is to ensure that the maximum possible genetic diversity is maintained and available for potential utilization (Maxted et al. 2008). From an agricultural biodiversity perspective, the focus of in situ conservation is on the CWR. It should be emphasized here that most protected areas around the world have been established to preserve particular ecosystems, exceptional scenery or habitats for particular charismatic species, but very seldom for crop wild relatives (Maxted et al. 2008). Notable examples includes Erebuni State Reserve in Armenia, which was established in 1981 specifically to protect wild cereals (Avagyan 2008); and the Sierra de Manantlán Reserve in Mexico, established specifically for the conservation of the endemic perennial wild relative of maize, *Zea mays* (FAO 2010). Whatever the type of protected area, in general, in situ conservation involves a range of activities (Heywood and Dulloo 2005; Iriondo et al. 2008) which includes:

- Setting priorities for target species and their populations and extent of genetic diversity;
- Planning, design and setting up of conservation areas;
- Management and monitoring of in situ populations; and
- Policy and legal support.

Although there has been a heightened awareness of the importance of CWR, as shown by the second Report on the State of the World's Plant Genetic Resources for Food and Agriculture (FAO 2010), there has been very little concrete action taken to establish genetic reserves for their conservation. Some progress has been made with regard to prioritization of species and areas, assessments of distributions, diversity and threat status, in situ management in protected areas, development of CWR national plans and strategies and raising awareness and understanding of their importance. For example, a UNEP/GEF supported a global project on in situ conservation of crop wild relatives in five countries (Armenia, Bolivia, Sri Lanka, Madagascar and Uzbekistan). As a result of this project, a manual on in situ conservation has been published by Earthscan (Hunter and Heywood 2011), which provides useful guides and lessons learnt on in situ conservation of crop wild relatives. The FAO is also commissioning a plant genetic resources toolkit which will provide guidance on how to develop national conservation strategies for CWR and landraces.

19.4.3 On-Farm Conservation

On-farm conservation concerns the maintenance of useful species (such as cultivated crops, forages and agroforestry species), as well as their wild and weedy relatives in the agro-ecosystem along with the cultivation and selection processes on local varieties and landraces. Substantial evidence exist that farmers maintain significant crop genetic diversity on farm for a wide diversity of uses (Jarvis et al. 2011). Practices that support the maintenance of diversity within agricultural production systems include agronomic practices, farmers' seed production and informal seed exchange systems, as well as the management of the interface between the wild and cultivated ecosystems. Farmers also often use home gardens as a site for experimentation, for introducing new cultivars, or for the domestication of wild semi-domesticated species. Home gardens are important locations for agricultural biodiversity conservation, providing microenvironments that serve as refuges for crops and crop varieties that were once more widespread in the larger agroecosystem. Home gardens can serve as buffer zones around protected areas, as is the case with the Sierra del Rosario Biosphere Reserve in Cuba (Herrera and Garcia 1995). Studies of the genetic diversity of key home garden species in Cuba, Guatemala, Ghana, Indonesia, Sri Lanka, Venezuela, and Viet Nam have demonstrated that significant crop genetic diversity does exist in home gardens, and that home gardens can be a sustainable in-situ conservation system (Watson and Eyzaguirre 2002). Informal seed systems are a key element in the maintenance of crop diversity on farm, which in some countries can account for up to 90% of seed movement (FAO 2010). According to the Second Report on the State of the World's Plant Genetic Resources for Food and Agriculture (FAO 2010) significant progress has been achieved in developing tools to support the assessment, conservation and management of on-farm diversity in domesticated species, with countries reporting increasing numbers of national surveys and inventories documenting the status of conservation efforts targeting these genetic resources and priorities for further action. The role of farmers and of traditional knowledge in understanding and managing crop diversity has increasingly been recognized as essential for the maintenance of PGRFA, with a significant amount of crop genetic diversity in the form of traditional varieties continuing to be maintained on farm (Jarvis et al. 2011; Bezançon et al. 2009; Sadiki et al. 2007). More often than not these are poor, small-scale farmers (Kontoleon et al. 2009) who rely on traditional crop varieties to meet their livelihood needs.

19.4.4 Complementarity Between Ex Situ and In Situ Conservation

It is important to emphasize that the concept of ex situ conservation is fundamentally different to that of in situ conservation, although both are important approaches to conservation of plant genetic resources. The principal difference (and hence the reason for the complementarity) between the two approaches lies in the fact that ex situ conservation implies the maintenance of genetic material outside of the "normal" environment where the species has evolved and aims to maintain the genetic integrity of the material at the time of collecting, whereas in situ conservation (maintenance of viable populations in their natural surroundings) is a dynamic system which allows the biological resources to evolve and change over time through natural or human-driven selection processes. It should be noted that in situ conservation can be either on farm, requiring the maintenance of the agro-ecosystem along with the cultivation and selection processes on local varieties and landraces; or in the wild, which involves the maintenance of the ecological functions that allow species to evolve under natural conditions (Dulloo et al. 2010; Heywood and Dulloo 2005; Maxted et al. 1997).

A conservation strategy that uses a combination of ex situ and in situ techniques, taking into account their respective advantages and disadvantages is therefore most likely to secure the most diversity for future use. The concept of a complementary conservation strategy involves the combination of different conservation actions, which together lead to an optimum sustainable conservation of genetic diversity in a target genepool, now and in the future. Dulloo et al. (2005) provides a step by step process for developing a complementary conservation strategy.

19.5 International Conventions, Treaties and Agreements

There are a number of conventions, treaties, agreements and other international and regional initiatives that pertain to biodiversity conservation and which provide the global context for the conservation of biodiversity. When it comes to agricultural biodiversity, it is important to emphasize that with the global challenges ahead (especially need to increase food production due to increasing world population and impacts of climate change), no country in the world will be self sufficient in genetic resources and countries will rely on each other for access to genetic resources for agricultural development (Halewood and Nnadozie 2008). Palacios (1998) carried out a study to assess the degree of dependence of a country's main food crops on genetic diversity in areas of origin and primary diversity located elsewhere. Although his study focused on primary centers of diversity, he showed that for each of the six regional groups namely Africa, Asia and the Pacific, Near East, Europe, Latin America, and the Caribbean and North America, there were, in general, high dependencies of the countries in places where their main staple crops came from primary centres of diversity located elsewhere. For example in Southern Africa the level of dependencies of countries ranged from 51% to 100%, given that large calorie intake from these countries comes from maize, cassava, wheat, rice, beans, plantain, banana and potatoes, all of which have centres of diversity outside southern Africa. On the other hand countries which depend more on their indigenous crops for food security (e.g. wheat and barley in Ethiopia, Mali and Somalia) or are primary centres of diversity for their main crops (e.g. rice in Bangladesh and India), have low dependencies (Palacios 1998). This interdependency calls for increased collaboration among countries to fully support, participate and implement the conventions, treaties and agreements which aim at the effective conservation and sustainable use of the biological diversity essential for food and agriculture.

19.5.1 Convention on Biological Diversity (CBD)

The CBD is one of the major international conventions on biodiversity conservation with the principal objectives of conservation and sustainable use of biological diversity, and the fair and equitable sharing of benefits arising from its utilization of genetic resources. It came into force in 1993 and currently has 194 parties (http://www.cbd.int/convention/parties/list/). Its substantive provision (articles 6–20) provides for measures for the in situ and ex situ conservation of biodiversity, incentives for conservation and use, research and training, public awareness and education, regulating access to genetic resources among others. The CBD represented a major paradigm shift in the access of genetic resources. Before the convention, there was a universally accepted principle that genetic resources were considered as a common heritage of mankind and countries had free access to them. The CBD however changed all this and established the national sovereignty right over genetic resources. This shift has had the effect of restricting the movement of germplasm around the world and countered the objectives of the International Undertaking on Plant Genetic Resources, whose objective was 'to ensure that plant genetic resources … particularly for agriculture, will be explored, preserved, evaluated and made available for plant breeding and scientific purposes'.

The Conference of Parties (COP) of the CBD has over the years developed a series of thematic programmes, cross-cutting issues, protocols and other important initiatives. In particular in the year 2000, the CBD-COP established the *Cartagena Protocol on Biosafety*, an international treaty governing the movement from one country to another of living modified organisms (LMOs) resulting from modern biotechnology (Decision EM-I/3).

The tenth Conference of Parties (COP-10) of the CBD was held in Nagoya, Japan and coincided with International Year of Biodiversity (2010). A number of key decisions were taken, including:

The *CBD Strategic Plan for the post 2010 period*, including a revised biodiversity target was adopted. As we may recall, the 2010 biodiversity target – to achieve a significant reduction of the current rate of biodiversity loss at the global, regional and national level as a contribution to poverty alleviation and to the benefit of all life on Earth – has not been met in full and the third Global Biodiversity Outlook (GBO-3) provides an assessment of the progress made and scenarios for the future of biodiversity. The vision of the new strategic plan is a world of "Living in harmony with nature": where "By 2050 biodiversity is valued, conserved, restored and wisely used, maintaining ecosystem services, sustaining a healthy planet and delivering benefits essential for all people." The Plan includes 20 headlines target for 2020 which are organized under five strategic goals (see Box 19.2).

COP-10 also adopted a revised *Global Strategy on Plant Conservation* first established in 2002, aimed at halting the current and continuing loss of plant diversity, and considered issues of sustainable use and benefit-sharing, and aims to contribute to poverty alleviation and sustainable development. It included 16 outcome-oriented global targets set for 2010. These targets were updated for the period 2011–2020 and the updated GSPC was adopted at COP-10 in Nagoya, Japan. COP-10 invited parties to develop and update national targets and incorporate them in relevant plans, programmes and initiatives including national biodiversity strategies and action plans.

Another important decision taken at COP-10 was the historic adoption of a *Protocol on Access and Benefit-sharing* on the basis of a proposal developed by the *Ad Hoc* Open-ended Working Group on Access and Benefit-sharing. Article 3 of the protocol defines the scope as applying to genetic resources within the scope of Article 15 of the Convention and to the benefits arising from the utilization of such resources. The Protocol also applies to traditional knowledge associated with genetic resources within the scope of the Convention and to the benefits arising from the utilization of such knowledge. In this respect, the Protocol goes further than the CBD by including (qualified) commitments by parties to take measures to ensure that genetic resources and associated traditional knowledge held by indigenous and local communities are accessed subject to the prior informed consent (PIC) and on mutually agreed terms (MAT) agreed with those communities.

Box 19.2 Five Strategic Goals of the Convention on Biological Diversity Strategic Plan 2011–2020. Source: http://www.cbd.int/sp/elements/

Strategic Goal A: Address the underlying causes of biodiversity loss by mainstreaming biodiversity across government and society;
Strategic Goal B: Reduce the direct pressures on biodiversity and promote sustainable use;
Strategic Goal C: To improve the status of biodiversity by safeguarding ecosystems, species and genetic diversity;
Strategic Goal D: Enhance the benefits to all from biodiversity and ecosystem services;
Strategic Goal E: Enhance implementation through participatory planning, knowledge management and capacity building.

19.5.2 International Treaty on Plant Genetic Resources for Food and Agriculture (ITPGRFA)

The International Treaty on Plant Genetic Resources for Food and Agriculture (henceforth The Treaty) was adopted by the FAO conference in 2001 and came into force in 2004. Currently, 127 State parties (including the European Union) have ratified or acceded to it. The objectives of the Treaty are the conservation and sustainable use of plant genetic resources for food and agriculture and the fair and equitable sharing of benefits arising out of their use, in harmony with the CBD for sustainable agriculture and food security. Like the CBD, The Treaty sets out general provision for the ex situ and in situ conservation and use of PGRFA (Articles 5 and 6). The Treaty also recognizes Farmers' rights in view of the enormous contribution that farmers and their communities have made and continue to make to the conservation and development of plant genetic resources. This also includes the protection of traditional and indigenous knowledge and the right to participate equitably in benefit-sharing and in national decision making about plant genetic resources. Governments are responsible for realising such rights.

Most importantly, The Treaty creates the *'Multilateral system of access and benefit-sharing'* (MLS), which is a virtual, international pool of PGRFA from 35 crops and 29 forage species (commonly referred to as Annex 1 crops to the Treaty) as well as related information. It makes provision for all PGRFA listed in Annex 1 that "are under the management and control of the Contracting Parties and in the public domain" to be included automatically in the multilateral system. Materials can also voluntarily be placed in the MLS by public and private players and by international organization signing article 15 of the Treaty. The Treaty is thus an important framework for international cooperation for use and conservation of PGRFA. It establishes rules and administrative supports for the common pooling and low-transaction cost use of listed crops and forages through the MLS. While the Treaty recognizes the national sovereignty rights over PGRFA, contracting parties have agreed to provide facilitated access to PGRFA of most of the important crops, by placing the Annex 1 crops under their management and control in the MLS, thus helping to balance the need of free the flow of germplasm with national sovereignty and Intellectual Property Rights (IPR).

Germplasm held in collections will be made available to users under the conditions of a *Standard Material Transfer Agreement (SMTA)* of The Treaty. Under the SMTA, the recipient undertakes that the material made available shall be used or conserved only for the purposes of research, breeding and training for food and agriculture. Such purposes shall not include chemical, pharmaceutical and/or other non-food/feed industrial uses. The recipient shall also not claim any intellectual property or other rights that limit the facilitated access to the material provided under this agreement, or its genetic parts or components, in the form in which the genetic material is received under the Multilateral System. The SMTA also provides for a mandatory financial benefit-sharing in the event that the recipient commercializes a product that is derived from PGRFA and that incorporates material as referred to in Article

3 of this Agreement, and where such a product is not available without restriction to others for further research and breeding, the recipient shall then pay a fixed percentage of the sales of the commercialized product into the mechanism established by the Governing Body of The Treaty for this purpose, in accordance with Annex 2 to the SMTA Agreement. The Treaty also provides for a Funding Strategy, which aims to enhance the availability, transparency, efficiency and effectiveness of the provision of financial resources for the implementation of the Treaty. It includes a Benefit-sharing Fund which holds the financial resources under the direct control of the Governing Body of the Treaty. In 2010 the Governing Body of The Treaty opened a call for proposals focusing on the three agreed priorities on:

- Information exchange, technology transfer and capacity-building;
- Managing and conserving plant genetic resources on farm; and
- The sustainable use of plant genetic resources.

The Benefit sharing Fund is helping to ensure sustainable food security by assisting farmers to adapt to climate change through a targeted set of high impact activities on the conservation and sustainable use of plant genetic resources for food and agriculture.

There are many steps that a Contracting Party can take to implement the Treaty at the national level, such as improving the national systems for the conservation and sustainable use of plant genetic resources for food and agriculture (PGRFA) in line with the general direction set out in Articles 5 and 6 of the Treaty, and providing for the realization of farmers' rights. In addition, a Contracting Party should review its national legislation and/or administrative procedures to ensure that the Treaty can be implemented and ensures compatibility with other laws and regulations such as those adopted for the implementation of the CBD. It should take measures to encourage holders of collections to bring their collection in to the multilateral system. So far, apart from the CGIAR Centres' collections, and a few national exceptions, most contracting parties have not functionally or meaningfully clarified what materials fall within the MLS. Countries are encouraged to notify the Executive Secretary of the Treaty of materials that they have placed in the MLS. It is also important that a Contracting Party clearly designates a competent authority or authorities to sign SMTAs to facilitate the exchange of genetic resources. A learning module has been developed by the CGIAR System-wide Genetic Resources Programme (SGRP) under the leadership of Bioversity International to help relevant players develop knowledge and skills to implement the Treaty and to use The Treaty's Standard Material Transfer Agreement (Moore and Goldberg 2010).

19.5.3 FAO Global Plan of Action (GPA)

A rolling Global Plan of Action (GPA) for the conservation and sustainable utilization of plant genetic resources for food and agriculture, first adopted in 1996 at the Fourth International Technical Conference on Plant Genetic Resources Leipzig,

Germany, is one of the main pillars of the FAO Global system for the conservation and sustainable use of PGRFA (FAO 1996). It represents the most comprehensive strategy for in situ and ex situ conservation and sustainable use of PGRFA and allows the FAO Commission on Genetic Resources to prioritize and promote rationalization and coordination of efforts on their programme of work on plant genetic resources (http://www.fao.org/agriculture/crops/core-themes/theme/seeds-pgr/gpa/en/). In this way it facilitates international collaboration among the member states of the Commission and helps countries to formulate their own national PGRFA policies and strategies. The GPA is also one of the supporting components of the International Treaty on PGRFA, as recognized under article 14 of the Treaty. The GPA originally contained 20 priority areas of action including four main areas, covering in situ conservation and development, ex situ conservation, plant genetic resources utilization, and institutions and capacity-building, reflecting the recommendations of the First State of the World report on PGRFA (FAO 1998). The GPA is a rolling plan meaning that it should periodically be reviewed in light of new development and new challenges facing PGRFA. The recent Second State of the World Report on PGRFA (FAO 2010) identified a number of gaps and needs, which are forming the basis for the current update of the GPA.

19.5.4 Global Crop Diversity Trust

The Global Crop Diversity Trust was established with support from the CGIAR Centres and FAO, with the objective of developing a multimillion-dollar endowment fund to provide support, in perpetuity, for the ex situ conservation of important PGRFA. The Trust has signed a 'relationship agreement' with the governing body of The Treaty, which recognizes the Trust as an 'essential element of the Funding Strategy of the International Treaty in relation to ex situ conservation and availability of PGRFA.' To date, the Trust has raised over US $110 million in its endowment fund and has provided grants to support the long-term conservation of 17 ex situ crop collections. It has also entered into contracts with over 100 organizations, in 74 countries, to support the regeneration of 85,000 accessions of unique, 'at risk' materials. Duplicates of these materials will be saved in long-term storage facilities and will be made available using the SMTA.

19.5.5 Intergovernmental Platform on Biodiversity and Ecosystem Services (IPBES)

Another important development for biodiversity is the recent adoption on 21 December 2010, by the 65th United Nations General Assembly (UNGA), of an Intergovernmental Platform on Biodiversity and Ecosystem Services (IPBES). This new international body is aimed at catalyzing a global response to the loss of biodiversity and world's economically important forests, coral reefs and other ecosystems, in a similar manner as the Intergovernmental Panel on Climate change (IPCC) does for the United Nations Framework Convention on Climate Change (UNFCCC).

19.5.6 Other Legal Instruments

In addition to the above there are a number of legal instruments which interplay with the Treaty and CBD and must be taken in to account. These include:

- Agreement between Governing Body of The Treaty and CGIAR Centres and other International Organizations;
- CGIAR policies and instruments;
- FAO code of conduct to plant germplasm collecting and transfer
- International Plant Protection Convention;
- WTO-TRIPS Agreement;
- UPOV;
- Regional agreements;
- Networks agreements;
- National Laws.

With so many different instruments (agreements, conventions, treaties and so on) which govern the conservation and use of plant biodiversity at the international level, some legally binding and others not, it has become very challenging for countries to reconcile together the different instruments that they have signed up to. As discussed above, the two main instruments, namely CBD and the Treaty complement each other and both have similar objectives. The main difference between them is that while, on the one hand, the CBD recognizes the sovereign rights of countries on their biological diversity, it empowers countries to exercise control and provide access to its genetic resources. On the other hand, the overall aim of the Treaty aims is to ensure food security and to establish a multilateral system to facilitate access to genetic resources for food and agriculture. The Global Plan of Action in turn provide a framework for the implementation for the provisions of the Treaty and help countries to develop their own strategies for an effective and efficient conservation and use of their plant genetic resources. The other instruments are designed to help protect and ensure the resources as well as the interest and rights of the owners and users of the resources in a fair and equitable way.

19.6 Conclusions

In conclusion, agricultural biodiversity is of vital importance in contributing to food and nutrition security and agricultural sustainability, upon which human survival depends, and as such must be effectively safeguarded, conserved, promoted and used. Their effective conservation and use requires increased focus and investment, through greater involvement and collaboration of all stakeholders. This is all more important considering that, as discussed in this chapter, there is no country that can consider itself as self sufficient for all genetic diversity in all its crops and breeds for all time and all countries are completely interdependent in terms of genetic resources. This interdependency calls for increased collaboration among countries and between national and international research institutions, such as CGIAR, UN Organizations, International NGOs like the International Union for the Conservation of Nature (IUCN) and other advanced research organizations as well as with civil societies and farmers' organizations to fully support, participate and implement the conventions, treaties and agreements all of which aim at the effective conservation and sustainable use of the biological diversity essential for food and agriculture. There is also a need for greater public awareness and policy advocacy for greater support, to achieve a holistic understanding of the importance of agro-biodiversity and their effective and efficient use to address food security and poverty issues.

19.7 Appendix

19.7.1 Abbreviations

CBD	Convention on Biological Diversity
CWR	Crop Wild Relatives
FAO	Food and Agriculture Organization of the United Nations
GBO-3	Global Biodiversity Outlook – 3
GCDT	Global Crop Diversity Trust
GPA	Global Plan of Action
IPCC	Intergovernmental Panel on Climate Change
IPBES	Intergovernmental Platform on Biodiversity and Ecosystem Services
IUCN	International Union for Conservation of Nature
LMOs	Living Modified Organisms
MEA	Millennium Ecosystem Assessment
PGRFA	Plant Genetic Resources for Food and Agriculture
SBSTTA	Subsidiary Body on Scientific Technical and Technology Advice for the CBD
SGRP	CGIAR System-Wide Genetic Resources Programme
UNFCCC	United Nations Framework Convention on Climate Change
UNGA	United Nations General Assembly

References

Avagyan A (2008) Crop wild relatives in Armenia: diversity, legislation and conservation issues. In: Maxted N, Ford-Lloyd BV, Kell SP, Iriondo J, Dulloo E, Turok J (eds) Crop wild relative conservation and use. CAB International, Wallingford, pp 58–66

Barclay A (2004) Feral play: crop scientists use wide crosses to breed into cultivated rice varieties the hardiness of their wild kin. Rice Today 3(1):14–19

Bezançon G, Pham JL, Deu M, Vigouroux Y, Sagnard F, Mariac C, Kapran I, Manadou A, Gérard B, Ndjeunga J (2009) Changes in the diversity and geographic distribution of cultivated millet (*Pennisetum glaucum* [L.] R. Br.) and sorghum (*Sorghum bicolor* (L.) Moench) varieties in Niger between 1976 and 2003. Genet Resour Crop Evol 56:223–236

Bramwell D (2002) How many plant species are there? Plant Talk 28:32–34

Brown AHD (2008) Indicators of genetic diversity, genetic erosion and genetic vulnerability for plant genetic resources for food and agriculture. Thematic background study. Food and Agriculture Organization of the United Nations, Rome, 26 pp

Brown AHD, Brubaker CL (2002) Indicators for sustainable management of plant genetic resources: how well are we doing? In: Engels JMM, Rao Ramanatha V, Brown AHD, Jackson MT (eds) Managing plant genetic diversity. CABI, UK, pp 249–262

Bubb, P., J. Jenkins, and V. Kapos. 2005. Biodiversity indicators for national use: experience and guidance. United Nations Environment Programmes World Conservation Monitoring Centre (UNEPWCMC), Cambridge, United Kingdom

Burke MB, Lobell DB, Guarino L (2009) Shifts in African crop climates by 2050 and the implications for crop improvement and genetic resources conservation. Global Environ Change 19(3):317–325. doi:10.1016//j.gloenvcha.2009.04.003

Collen B, Ram M, Zamin T, McRae I (2008) The tropical biodiversity data gap: addressing disparity in global monitoring. Trop Conserv Sci 1(2):75–88, Available on line tropicalconservationscience.org

Collette L (2001) Indicators of agricultural genetic resources: FAO's contribution in monitoring agricultural biodiversity. Paper presented at OECD expert meeting on agri-biodiversity indicators, Zürich, 5–8 Nov 2001. Available online at http://www.oecd.org/dataoecd/9/55/40351057.pdf

Diulgheroff S (2004) A global overview of assessing and monitoring genetic erosion of crop wild relatives and localvarieties using WIEWS and other elements of the FAO global system of PGR. In: Ford-Lloyd BV, Dias SR, Bettencourt E (eds) Genetic erosion and pollution assessment methodologies. Proceedings of PGR forum workshop 5, Terceira Island, Autonomous region of the Azores, Portugal, 8–11 Sept 2004, pp 6–14. (Published on behalf of the European crop wild relative diversity assessment and conservation forum, Bioversity International, Rome, 100 pp. Available at http://www.bioversityinternational.org/fileadmin/bioversity/publications/pdfs/1171.pdf)

Dulloo ME, Ramanatha Rao V, Engelmann F, Engels J (2005) Complementary conservation of coconuts. In: Batugal P, Rao VR, Oliver J (eds) Coconut genetic resources. IPGRI-APO, Serdang, pp 75–90

Dulloo ME, Nagamura Y, Ryder O (2006) DNA Storage as a complementary conservation strategy. In: de Vicente MC (ed) DNA banks- providing novel options for genebanks? Topical reviews in agricultural biodiversity. International Plant Genetic Resources Institute, Rome, pp 11–24

Dulloo ME, Hunter D, Borelli T (2010) Ex situ and in situ conservation of agricultural biodiversity: major advances and research needs. Not Bot Hortic Agrobot Cluj 38(2):123–153 (special issue)

Engelmann F, Engels JMM (2002) Technologies and strategies for ex situ conservation. In: Engels JMM, Rao Ramanatha V, Brown AHD, Jackson MT (eds) Managing plant genetic diversity. CABI, UK, pp 89–103

FAO (1996) Global plan of action for the conservation and sustainable utilization of plant genetic resources for food and agriculture. Food and Agriculture Organization of the United Nations, Rome

FAO (1998) The state of the world's plant genetic resources for food and agriculture. Food and Agriculture Organization of the United Nations, Rome

FAO (2010) The second report on the state of the world's plant genetic resources for food and agriculture. Food and Agriculture Organization of the United Nations, Rome, 370 pp

FAO/IPGRI (2002) Review and development of indicators for genetic diversity, genetic erosion and genetic vulnerability (GDEV): summary report of a joint FAO/IPGRI workshop, Rome, 11–14 Sept 2002, 14 pp

Frankham R (2010) Challenges and opportunities of genetic approaches to biological conservation. Biol Conserv 143:1919–1927

Govaerts R (2001) How many species of plants of plants are there? Taxon 50:1085–1090

Hajjar R, Hodgkin T (2007) The use of wild relatives in crop improvement: a survey of developments over the last 20 years. Euphytica 156:1–13

Halewood M, Nnadozie K (2008) Giving priority to the commons: the international treaty on plant genetic resources for food and agriculture. In: Tanse G, Rajotte T (eds) The future control of food: a guide to international negotiations and rules on intellectual property biodiversity and food security. Earthscan, London, pp 115–140

Harlan J (1992) Crops and man, 2nd edn. American Society of Agronomy, Madison

Herrera M, Garcia M (1995) La Reserva de la Biosfera Sierra del Rosario, Cuba. Doc. de Trabajo. Programa Sur-sur. Unesco, Paris

Heywood VH, Dulloo ME (2005) In situ conservation of wild plant species – a critical global review of good practices. Technical bulletin no 11. International Plant Genetic Resources Institute

Hunter D, Heywood VH (2011) Crop wild relatives. A manual of in situ conservation. Earthscan, London/Washington, DC

IPCC (2007) Summary for policymakers. In: Solomon S, Qin D, Manning M, Chen Z, Marquis M, Averyt KB, Tignor M, Miller HL (eds) Climate change (2007) The physical science basis. Contribution of working group I to the fourth assessment report of the intergovernmental panel on climate change, Cambridge University Press, Cambridge, UK/New York

Iriondo J, Maxted N, Dulloo ME (2008) Conserving plant genetic diversity in protected areas. CABI, Wallingford, 212 pp

IUCN (2010) A new vision for biodiversity conservation. Position paper on strategic plan for the convention on biological diversity 2011–2020. In: Tenth meeting of the conference of parties to the CBD (COP10), Nagoya

Jarvis A, Lane A, Hijmans R (2008) The effect of climate change on crop wild relatives. Agric Ecosyst Environ 126:13–23

Jarvis DI, Hodgkin T, Sthapit BR, Fadda C, Lopez-Noriega I (2011) An heuristic framework for identifying multiple ways of supporting the conservation and use of traditional crop varieties within the agricultural production system. Crit Rev Plant Sci 30:125–176. doi:10.1080/07352689.2011.554358

Kontoleon, A., Pascual, U., and Smale, M. 2009. Introduction: agrobiodiversity for economic development: what do we know? Agrobiodiversity conservation and economic development. In: *Environmental Economics*. pp. 1–24. Kontoleon, A., Pascual, U., and Smale, M., Eds., Routledge Explorations, UK

Laikre L, Larsson LC, Palme A, Charlier J, Josefsson M, Ryman N (2008) Potentials for monitoring gene level biodiversity: using Sweden as an example. Biodivers Conserv 17:893–910

Laikre L, Allendorf FD, Aroner LC, Baker CS, Gregovich DP, Hansen MH, Jackson JA, Kendall KC, Mckelvey K, Neel MC, Olivieri I, Ryman N, Schwartz MK, Bull RS, Jeffrey B, Stetz JB, Tallmon DA, Taylor BL, Vojta CD, Waller DM, Waples RS (2010) Neglect of genetic diversity in implementation of the convention of biological diversity. Conserv Biol 24(1):86–88

Lane A, Jarvis A (2007) Changes in climate will modify the geography of crop suitability: agricultural biodiversity can help with adaptation. J SAT Agric Res 4(1):1–12

Mabberley DJ (1997) The plant book a portable dictionary of vascular plants, 2nd edn. Cambridge University Press, Cambridge

Maxted N, Kell S (2009) Establishment of a global network for the in situ conservation of crop wild relatives: status and needs. FAO Commission on Genetic Resources for Food and Agriculture, Rome, 266 pp

Maxted N, Ford-Lloyd BV, Hawkes JG (1997) Plant genetic conservation: in situ approach. Chapman and Hall, London

Maxted N, Ford-Lloyd BV, Kell SP, Iriondo J, Dulloo E, Turok J (2008) Crop wild relative conservation and use. CABI, Wallingford

Millennium Ecosystem Assessment (2005) Ecosystems and human well-being: biodiversity synthesis. World Resources Institute, Washington, DC

Mittermeier RA, Myers N, Thomsen JB, da Fonseca GAB, Olivieri S (1998) Biodiversity hotspots and major wilderness tropical areas: approaches to setting conservation priorities. Conserv Biol 12(3):516–520

Moore G, Goldberg E (2010) International treaty on plant genetic resources for food and agriculture: learning module. Produced by the CGIAR system-wide genetic resources programme (SGRP). Bioversity international and the CGIAR generation challenge programme. Bioversity International, Rome

Myers N, Mittermeier RA, Mittermeier CG, da Fonseca GAB, Kent J (2000) Biodiversity hotspots for conservation priorities. Nature 403:853–858

Palacios XF (1998) Contribution to the estimation of countries' interdependence in the area of plant genetic resources. Background study paper no 7 rev 1, FAO commission on genetic resources for food and agriculture, Rome. Available at www.fao.org/ag/cgrfa/docs.htm. Accessed 17 July 2011

Plucknett DL, Smith NJH, Williams JT, Anishetty NM (1987) Genebanks and the world's food. Princeton University Press, Princeton, 247 pp

Prance GT, Beentje H, Dransfield J, Johns R (2000) The tropical flora remains undercollected. Ann Mo Bot Gard 87:67–71

Prescott-Allen R, Prescott-Allen C (1990) How many plants feed the world? Conserv Biol 4(4):365–374

Sadiki M, Jarvis D, Rijal D, Bajracharya J, Hue NN, Camacho TC, Burgos-May LA, Sawadogo M, Balma D, Lope D, Arias L, Mar I, Karamura D, Williams D, Chavez-Servia J, Sthapit B, Rao VR (2007) Variety names: an entry point to crop genetic diversity and distribution in agroecosystems? In: Jarvis DI, Padoch C, Cooper D (eds) Managing biodiversity in agricultural ecosystems. Columbia University Press, New York, pp 34–76

SBSTTA (2010) Proposals for a consolidated update of the global strategy for plant conservation. UNEP/CBD/SBSTTA/14/9, – http://www.cbd.int/doc/?meeting=SBSTTA-14. Last accessed 17 Sept 2010

Tester M, Langridge P (2010) Breeding technologies to increase crop production in a changing world. Science 327(5967):818–822

Thormann I, Dulloo ME, Engels JMM (2006) Techniques of ex situ plant conservation. In: Henry R (ed) Plant conservation genetics. Centre for Plant Conservation Genetics, The Haworth Press, Southern Cross University, Lismore, pp 7–36

Van de Wouw M, Kik C, van Hintum T, Van Treuren R, Visser B (2009) Genetic erosion: concepts, research results and challenges. Plant Genet Resour Charact Util 8:1–15. doi:10.1017/S1479262109990062

Vavilov NI (1931) The problem of the origin of world agriculture in the light of the latest investigations. In: Bukharin NI (ed) Science at the crossroads, London. (Reprint New York 1971) pp 1–20

Walpole M, Rosamunde EAA, Besançon C, Butchart SHM, Campbell-Lendrum D, Carr GM, Collen B, Collette L, Davidson NC, Dulloo E, Fazel AM, Galloway JN, Gill M, Goverse T, Hockings M, Leaman DJ, Morgan DHW, Revenga C, Rickwood CJ, Schutyser F, Simons S, Stattersfield AJ, Tyrrell TD, Vié JC, Zimsky M (2009) Tracking progress towards 2010 biodiversity target and beyond. Science 135:1503–1504

Watson JW, Eyzaguirre PB (2002) Home gardens and in situ conservation of plant genetic resources in farming systems. In: Proceedings of the second international home gardens workshop, Witzenhausen, 17–19 Jul 2001

Wilson EO (1989) Threats to biodiversity. Sci Am 261:108–116

Chapter 20
Major Research Challenges and Directions for Future Research

Florent Engelmann and Ramanatha Rao

20.1 Introduction

Plant genetic resources, i.e. the diversity of genetic material which determines the characteristics of plants and hence their ability to adapt and survive, are the biological basis of world food security. These resources are used as food, feed for domesticated animals, fibre, clothing, shelter and energy. The conservation and sustainable use of plant genetic resources is necessary to ensure crop production for a growing world population and meet growing environmental challenges and climate change. Improved conservation and use of these resources are of vital importance in a world which saw an estimated 925 million people suffering from hunger in 2010, and whose population is expected to reach nine billion by 2050 (FAO 2010).

Although conservation includes a spectrum of methods ranging from strictly in situ to completely static ex situ, for pragmatic reasons, two basic conservation strategies, each composed of various techniques, are employed to conserve genetic diversity, i.e., in situ and ex situ conservation. Article 2 of the Convention on Biological Diversity provides the following definitions for these categories (UNCED 1992): "Ex situ conservation means the conservation of components of biological diversity outside their natural habitat. In situ conservation means the conservation of ecosystems and natural habitats and the maintenance and recovery

F. Engelmann (✉)
Director of Research, Honorary Research Fellow, IRD, UMR DIADE, Montpellier, France

Bioversity International, Rome, Italy
e-mail: florent.engelmann@ird.fr

Ramanatha Rao
Honorary Research Fellow, Adjunct Senior Fellow, Bioversity International, Rome, Italy

ATREE, Bangalore, India
e-mail: vramanatharao@gmail.com

M.N. Normah et al. (eds.), *Conservation of Tropical Plant Species*,
DOI 10.1007/978-1-4614-3776-5_20, © Springer Science+Business Media New York 2013

of viable populations of species and, in the case of domesticated or cultivated species, in the surroundings where they have developed their distinctive properties." There is an obvious fundamental difference between these two strategies: ex situ conservation involves the sampling, transfer and storage of target taxa from the collecting area, whereas in situ conservation involves the designation, management and monitoring of target taxa where they are encountered (Maxted et al. 1997a). Another difference lies with the more dynamic nature of in situ conservation and the more static of ex situ conservation (Engelmann and Engels 2002). However, it should be emphasized that ex situ conservation is a continuous process, with new accessions showing interesting characteristics being regularly incorporated in the genebank (under whichever form it is), thereby also providing a dynamic component to this mode of conservation. These two basic conservation strategies are further subdivided into specific techniques including seed storage, in vitro storage, DNA storage, pollen storage, field genebank and botanic garden conservation for ex situ, and protected area, on-farm and home garden conservation for in situ, each technique presenting its own advantages and limitations (Engels and Wood 1999). Ex situ conservation techniques are in particular appropriate for the conservation of crops and their wild relatives, while in situ conservation is especially appropriate for wild species and for landrace material on farm.

Ex situ and in situ conservation strategies and methods should be approached and utilized in a complementary manner, as already advocated for a long time notably by IPGRI/Bioversity, (Arora and Rao 1995; Rao et al. 1998; Engelmann 1999; Engelmann et al. 2007). Indeed, there is no unique, "ideal" conservation strategy and technique, each having its own objectives (e.g. short, medium long-term), advantages and drawbacks. It is necessary to design the best combination of strategies and methods to optimize and secure the conservation of a given genepool. This is a dynamic process, as it is also necessary to revisit the technical framework employed whenever a new technological development becomes available, such as cryopreservation. A good example of such changes in the conservation approach is given by the work performed in IPK (Gatersleben, Germany) with their potato collection, where the establishment of a complete cryopreserved duplicate directly impacts the numbers of samples and/or replicates conserved under other forms, such as tubers or in vitro plantlets by reducing them (Keller et al. 2011). Additional research and, wherever possible, systematic application, are urgently needed in the area of integration of conservation strategies.

Until recently, most conservation efforts, apart from work on forest genetic resources, have concentrated on ex situ conservation, particularly seed genebanks. Today, as a result of this global effort, there are some 1,750 genebanks worldwide, maintaining an estimated 7.4 million accessions and over 2,500 botanical gardens, which maintain samples of some 80,000 species, including numerous crop wild relatives (FAO 2010). It is important to mention that the collecting and conservation activities performed by genebank staff focused largely on the major food crops, including cereals and some legumes, i.e., species that have orthodox seeds and can thus be conserved easily in this form. This has resulted in over-representation of those species (and under-representation of other important but minor useful plant species) in the world's major genebanks, as well as in the fact that conservation strategies and concepts were biased towards such material. It is estimated that over 90% of the accessions conserved in genebanks are seeds showing orthodox storage behaviour (FAO 1996a). It is only more recently that the establishment of field genebanks, allowing to conserve species for which seed conservation is not appropriate or impossible, including non-orthodox seed and vegetatively propagated species, as well as the development of new storage technologies, including in vitro conservation and cryopreservation, was given due attention by the international community (Engelmann and Engels 2002). Similarly, the importance of field genebanks grow more recently with increased work on fruit and other non-orthodox seed producing species (Saad and Ramanatha Rao 2001).

Despite efforts to stop the loss of biodiversity, there are indications that such diversity is being eroded and that the world has not succeeded in halting biodiversity loss, as reported recently by the Global Biodiversity Outlook (Secretariat of the Convention on Biological Diversity 2010). This is a global trend but is particularly strong in tropical countries, which show the highest population growth rate compared with other geographical regions. There has been very little work to quantify the magnitude of any trends in the loss of genetic diversity and there is no routine global scale monitoring of genetic diversity over time (Chap. 19 by Dulloo, this volume). However, in several chapters of this book, authors have highlighted the rapid loss of diversity occurring in their species/crops of interest, including notably fruit trees (Normah et al.), leguminous food crops (Ng), forest trees (Mansor, Marna), spices and tree borne oil seeds (Chaudhury) and medicinal plants (Sharma and Pandey) due to various reasons including human pressure, habitat loss, deforestation, over-exploitation, etc., thus making research for and application of efficient and cost-effective conservation techniques highly relevant.

This book presented the various strategies and methods employed for the conservation of tropical agrobiodiversity, and illustrated its current status for a series of species, including fruits, ornamentals, plantation crops, legumes, root crops, cereals, forestry species, vegetables, spices and medicinal plants. Despite the obvious fact that there are specificities for the conservation of these different groups of plants, due to numerous factors such as the existing knowledge on e.g. their genetic diversity, distribution, conservation or reproductive biology, some commonalities can be found for these species regarding the major research challenges to be faced and directions for future research. In this final section of the book, we will try to briefly highlight those priorities and challenges.

20.2 Collecting

As mentioned by most contributors of case studies presented in this book, there remains a lot of collecting to be performed before valuable germplasm is lost, both for crops and for their wild relatives. Compared with the situation prevailing only 10 years ago, a number of new tools are available to collectors to improve planning and implementation of collecting expeditions (Guarino et al. 2011). Using climate data and specific software, it is possible to predict the shift in areas suitable for localization of a wide range of plant species caused by climate change (Lane and Jarvis 2007). The recent development of new molecular tools in combination with new spatial methods (e.g. Geographic Information Systems [GIS]) and increased computer capacity has created opportunities for new applications of genetic diversity analyses (Manel et al. 2003; Scheldeman and van Zonneveld 2010; Holderegger et al. 2010; van Zonneveld et al. 2012). Such new tools allow notably more detailed prioritization of areas for in situ conservation and more precise definition of areas for collecting. Computer-assisted plant identification systems, which can be used in the field, are also being developed (Ouertani et al. 2010). In addition, improved ability of collectors to document, map and sample plant populations (i.e. taking DNA samples) has added new dimensions to collecting. A wealth of information on the wild relatives of crops can be found from web-based databases that can help guide in planning collecting missions. To the extent possible collectors of plant germplasm need to promote the characterization and evaluation of collected materials.

Collecting of plant germplasm is faced with challenges of different natures. Most importantly, there is, at the global level, a shortage of experts (botanists, taxonomists) that can accurately identify plant germplasm, especially in the case of crop wild relatives (see e.g. Chap. 11 by Ng, this volume) and nowadays, there is less funding available for field studies than in the past. There is a challenge of "time" because of the rapid pace of habitat destruction, particularly the disturbed habitats where close wild relatives of crops are often found. In vitro collecting techniques can significantly improve the efficiency of collecting missions (Pence et al. 2002; Pence and Engelmann 2011). This is particularly relevant in tropical regions of the globe. Finally, especially in these times of budgetary restrictions, another challenge is balancing budget for molecular characterization and non-molecular characterization and evaluation of the plant germplasm collected. Directions for future research and action are also diverse. They include notably a better coordination of genebank information systems, particularly between national and international systems. Efforts are needed to ensure that the original passport number is not "lost" during germplasm transfer, i.e. systems should be established and employed, which trace germplasm from the collecting site right through the international genebank system. Finally, an important task is to fill in the gaps in collecting wild relatives of crops – some of which are large. Greater efforts should notably be made to have in place international transfer systems for crops, including commodity crops, such as sugarcane, coffee, cocoa, which fall outside the remit of the CGIAR centres.

20.3 In Situ/On-Farm Conservation

In situ/on-farm conservation is of high relevance for tropical plant species, especially those from the humid tropics, where the large majority of plants produce recalcitrant seeds and many vegetatively propagated plants are found. As an example, it is estimated that well over 80% of the tree species in tropical rain forests produce recalcitrant seeds (Ouédraogo et al. 1996). For in situ conservation, wild species and crop wild relatives are traditionally conserved in nature reserves and managed areas (Maxted et al. 1997b). However, of central importance is the recognition that if crop genetic resources (including landraces) are to be conserved successfully and sustainably on-farm, such an outcome should be the result of farmers' production activities directed to improve their livelihood ("conservation through use"), as no conservation of diversity can be successful if removed from the people that need it. This means that on-farm conservation efforts must be carried out within the framework of farmers' livelihood needs, and for that reason, the mobilization of support to on-farm conservation needs to be conceived and designed within the broader objective of creating a more enabling environment for agricultural development in its various aspects (Sthapit et al. 2009). Since the time that the Convention on Biological Diversity provided a general framework for ex situ and in situ conservation strategies, most agencies dealing with plant genetic resources conservation have been facing the dilemma of how to implement in practical terms in situ conservation of agricultural biodiversity. A growing number of projects have been implemented and publications produced to provide technical and scientific guidance in the establishment and implementation of in situ/on farm conservation activities (Maxted et al. 1997b; Nguyen et al. 2005; Heywood and Dulloo 2005; Iriondo et al. 2008; Hunter and Heywood 2011; Maxted et al. 2012). A major challenge has been (and still is) to overcome the difficulty in changing the mindset of current PGR institutional set up (Sthapit and Ramanatha Rao 2009). Institutions and researchers need to work closer with farmers and communities. Indeed, the success of on-farm conservation of crop diversity demands not only to provide incentives for conservation but also to empower farmers in self-directed decision making. Any on-farm conservation programme must establish a strong relationship between all the different large multidisciplinary institutions and local farmers (Hunter and Heywood 2011). Too often, local organizations are ignored and the needs of communities are not addressed.

Future developments should work towards understanding the concerns of local farmers and creating awareness on why genetic diversity and gene/allele combinations in the form of landraces should be important to them. We need to better understand how farmers value crop diversity and how they exchange varieties within and between communities. Once these questions are better understood, in situ conservation programs can successfully protect diversity while simultaneously improving the livelihoods of farmers. Even though the genetic basis of on-farm conservation is not very clear (work is underway), serious efforts towards in situ conservation of crop genetic resources would only lead to a win-win situation, i.e. conserving and using crop genetic diversity for the benefit of those who depend on it. In addition, in situ conservation can help us contributing to environmental health through its contribution to ecosystem functions in general. Gradually, over the last decade, many researchers have realized that in situ conservation is important and feasible, but difficult, as it does not fit the scheme of things in formal sector research and development plans. So when presented the option of conserving plant genetic resources, the easier option i.e. ex situ conservation is often chosen, thus neglecting the in situ approach. This needs to be corrected and can only happen when the mindset of researchers is changed and when communities actually get to make decisions which directly affect them. It is also important to note that in situ/on-farm conservation per se is not a panacea on its own. It is neither recommended as a universal practice nor a feasible method in all circumstances; it has a place and time, as in situ/on-farm conservation can be transient and subject to changes over time. It is essential not to see in situ and ex situ conservation as antagonistic, as is still too often the case (Hunter et al. 2012). On the contrary, because of the critical importance of the complementary conservation approach described in a previous section of this Chapter, we would like to emphasize again that they complement each other and that optimal conservation for a given genepool covers a whole array of techniques extending from strictly in situ to completely static ex situ.

20.4 Ex Situ Conservation

Ex situ conservation includes storage of whole plants in the field, of orthodox seeds in seed genebanks, of clonally propagated plants under slow growth in vitro, of non-orthodox seed and vegetatively propagated material, pollen and DNA under cryopreservation. FAO is currently in the process of producing standards for conservation of orthodox seeds, non-orthodox seeds and vegetatively propagated materials. Such documents will be very useful to genebank managers and researchers, as well as to decision makers for the establishment and management of, and continuous financial support to germplasm collections maintained under these various forms.

20.4.1 Conservation in Field Genebanks

Field genebanks are traditionally used for conserving plants, which produce recalcitrant or intermediate seeds, produce very few seeds, are vegetatively propagated or which have a long juvenile period (such as many trees), and are thus of high relevance to tropical species, as they include a large number of plants with one or several of these characteristics. Several scientific and technical documents are available as a guide in the establishment and management of field genebanks (Engelmann 1999; Saad and Rao 2001; Reed et al. 2004; Gotor et al. 2011). No particular research issues or challenges pertain to field genebank conservation, at least in the context of the present publication, except perhaps the development of accurate cost analysis tools, which will not only allow to determine the cost of establishing and maintaining a field genebank but also to compare its cost-effectiveness with other storage methods, as has been performed notably in the case of coffee (Dulloo et al. 2009). Similar studies in other crops will be useful for arguing the case for establishing and managing non-orthodox seed producing species in field genebanks.

20.4.2 Conservation of Orthodox Seeds in Genebanks

Many of the world's major food plants produce seeds that undergo maturation drying, and are thus tolerant to extensive desiccation and low temperature. Storage of such orthodox seeds is the most widely practised method of ex situ conservation of plant genetic resources, since 90% of the 7.4 million accessions stored in genebanks are maintained as seed at low moisture content and temperature. FAO and IPGRI have published genebank standards (FAO and IPGRI 1994), which provide precise instructions on how orthodox seeds should be stored and seed genebanks operated (Ellis et al. 1985a, b; Sackville Hamilton and Chorlton 1997; Engels and Visser 2003; Rao et al. 2006). Based on viability equations (Ellis and Roberts 1980; Roberts and Ellis 1989; Dickie et al. 1990), recovery of plants from orthodox seeds is projected to be possible after centuries to millennia in seed banks (Li and Pritchard 2009). This situation may give the impression that there are no particular challenges to face or research issues to address related to orthodox seed conservation. This is far from being the case, as several topics related to various aspects of orthodox seed conservation are challenges, which need further research. One such topic that requires attention is the longevity of seeds under standard genebank storage conditions. Indeed, over the last 30 years, relatively widespread evidence has emerged of less than expected longevity at conventional seed bank temperatures, as highlighted in a recent review paper by Li and Pritchard (2009). These authors mention that, across nearly 200 species, species from drier (total rainfall) and warmer temperature (mean annual) locations tended to have greater seed P50 (time taken in storage for viability to fall to 50%) under accelerated ageing conditions than species from cool, wet conditions (Probert et al. 2009). Moreover, species P50 values were correlated with the proportion of accessions (not necessarily the same species) in that family that lost a significant amount of viability after 20 years under conditions for long term seed storage, that is, seeds pre-equilibrated with 15% relative humidity air and then stored at –20 °C (FAO and IPGRI 1994). Such relative underperformance at –20 °C was apparent in 26% of the accessions (Probert et al. 2009). Similarly, it has been estimated that half-lives for the seeds of 276 species held for an average of 38 years under cool (–5 °C) and cold (25 years at −18 °C) temperature was >100 years only for 61 (22%) of the species (Walters et al. 2005). Nonetheless, cryogenic storage did prolong the shelf life of lettuce (*Lactuca*) seeds with projected half-lives in the vapour and liquid phases of liquid nitrogen of 500 and 3,400 years, respectively (Walters et al. 2004), up to 20 times greater than that predicted for that species in a conventional seed bank at –20 °C (Roberts and Ellis 1989; Dickie et al. 1990). Although 25 species (from 19 genera) of Cruciferae had high germination (often >90%) after nearly 40 years storage at –5 °C to −10 °C (Perez-Garcia et al. 2007), it is hard not to conclude that, as an extra insurance policy for conservation, cryopreservation should be considered appropriate for all orthodox seeds, and one sub-sample of any accession systematically stored in LN, in addition to the samples stored under classical genebank conditions (Pritchard et al. 2009). As highlighted by Walters et al. (2011), such loss of viability during seed storage under standard conditions reflects molecular mobility within the system, i.e. relaxation of glassy matrixes. Stability of biological glasses is currently not fully understood and research in the thermodynamic principles, which contribute to temperature dependency of glassy relaxation as the context for understanding potential changes in viability of cryogenically stored germplasm is urgently needed.

Another challenge to face is to ensure that the seed genebanks are coordinated at the global level and properly managed in relation to maintenance of viability, characterization of accessions with systems that are user-friendly and ensure easy and fast accessibility for utilization when necessary (Bioversity International 2007; Gotor et al. 2008). The prime factor in the management of a genebank is the presence of complete passport data, which include characterisation and evaluation data, and whenever possible, molecular data (Spooner et al. 2005). The importance of such information for the utilization of plant genetic resources, and the availability of information describing the particular characteristics of genebank accessions are of high importance for users like plant breeders and scientists.

20.4.3 In Vitro Slow Growth Storage

In vitro conservation techniques offer advantages of rapid multiplication, ready availability and secure backup capabilities for germplasm laboratories. Basic laboratory infrastructures already exist in many genebanks and can be utilized to provide backup to field collections. It is also important to develop in vitro protocols that are adaptable to low technology laboratories. Numerous technical publications are available to guide genebank curators and researchers towards the establishment and management of in vitro collections maintained under slow growth (see e.g. Engelmann 1999, 2011; Reed et al. 2004). The main challenges to progress for in vitro collections is improving the introduction of contaminant-free plants into culture and finding the appropriate growth medium for suitable micropropagation. These are challenging projects due to the wide diversity of plants involved in genebanks, but advances are being made in both fields. The use of microbiological media to screen explants for bacteria and fungi early in the culture process will result in healthier collections and can be easily implemented. Within a genepool, there may be large differences in the response to in vitro storage between species/varieties, some responding well while others cannot be conserved using this technology, thus making its application impossible, as observed with an in vitro collection of coffee (Dussert et al. 1997). Developing more customized growth media for unique plant groups should thus result in improved culture response and better storage in vitro.

20.4.4 Cryopreservation of Non-Orthodox Seed and Vegetatively Propagated Materials

Cryopreservation is much more advanced for vegetatively propagated plants, in comparison with non-orthodox plant species, as described by Engelmann and

Dussert (this volume). This is true for tropical plants, particularly those from the humid tropics, as the seeds of many of such species fall within the non-orthodox category. Developing cryopreservation protocols for these species is thus a major challenge, because no other conservation method can offer the long-term storage security and cost-effectiveness equivalent to that provided by cryopreservation and because many of their characteristics (in the first place their extreme sensitivity to desiccation and to exposure to low temperatures) makes such development extremely difficult. Most of these non-orthodox or suspected non-orthodox seed species being wild species, little or nothing is known of their biology and the storage behaviour of their seeds. It is therefore of critical importance to benefit from the assistance and expertise of scientists, particularly botanists and seed physiologists, especially in the countries where such species are found, in order to produce the minimal level of knowledge required to envisage performing cryopreservation research on such species. There is in fact a continuum in seed storage behaviour from high recalcitrance to strong orthodoxy (Berjak and Pammenter 2007), and improved knowledge of seed biology combined with a precise control of desiccation conditions and level may allow succeeding in cryopreserving seeds, which were previously considered impossible to expose to LN temperature, as was notably the case for seeds of a growing number of coffee species (Dussert and Engelmann 2006; Dussert et al. 2011). However, it is compulsory, at the start of any cryopreservation research project, to be able to place the seeds within one of the three storage categories (orthodox, intermediate, recalcitrant) defined, as this will have a direct impact on the technical approach selected to store them. Three such methods are available, which allow categorizing seeds as orthodox, intermediate or recalcitrant (Hong and Ellis 1996; Pritchard et al. 2004; Plant Germplasm Conservation Research 2009). Another, often challenging, prerequisite condition for successful cryopreservation of recalcitrant (and possibly also many intermediate) plant species, is that in vitro culture protocols for the explants selected (embryos, embryonic axes, shoot tips) are operational before cryopreservation trials are initiated (Engelmann 2011). The involvement of tissue culture experts from the beginning of and throughout any cryopreservation project (and, in fact, even well before its initiation) is thus compulsory.

From a technical aspect, various options have been mentioned to improve the response of recalcitrant seeds or axes to cryopreservation (Engelmann and Dussert this volume). Of particular importance is the necessity to desiccate the tissues as rapidly as possible (a process called "flash-drying" [Wesley-Smith et al. 1992]) and to treat them with antioxidants substances, thereby reducing the intensity of and damage caused by the burst of reactive oxygen species (ROS) occurring during dissection and desiccation (Berjak and Pammenter 2007). The positive effect of cathodic water, generated by electrolysis of a solution containing electrolytes, which is a ROS scavenger, on survival of embryonic axes of two recalcitrant-seed species after cryopreservation has been shown recently and may have broader applications (Berjak et al. 2011).

The number of analytical tools at our disposal, which allow better understanding of the biological and physical processes occurring during cryopreservation of plant tissues and organs, is increasing. Measuring thermal events, which take place during cooling and rewarming using differential scanning calorimetry (DSC), has shown that optimal survival is generally achieved when most or all crystallisable water has been extracted from cells (Dumet et al. 1993; Dussert and Engelmann 2006). Recent examples of DSC being instrumental in establishing efficient cryopreservation protocols can be found notably with zygotic embryos of *Parkia speciosa* (Nadarajan et al. 2008), a tropical tree species and several Australian Citrus species (Hamilton et al. 2009). Another important finding, which contradicts the classical rule of thumb that "the lower the storage temperature and the moisture content of a seed, the longer the viability" is that the optimal water content for storage of seeds already in the glassy state increases with decreasing temperature (Vertucci and Roos 1993; Walters et al. 2004). It should be realized that molecular mobility occurs within glasses, even though at a very low rate, and that this phenomenon, which varies in intensity depending on the species, tissue and on numerous additional parameters, may have, over extended periods of several hundred years, an effect on the viability of cryogenically stored germplasm (Walters et al. 2011). Other approaches, such as "omics" studies, also provide original information on mechanisms occurring during desiccation. Notably, it will be particularly interesting to study and compare gene expression, using transcriptomics, during the acquisition of desiccation tolerance between orthodox, intermediate and recalcitrant seeds (Maia et al. 2011; Joet and Dussert 2012; Delahaie et al. 2012) as well as in vegetative tissues during a cryopreservation protocol (Volk et al. 2011). Yet other analytical tools, such as histological or biochemical analyses, can provide additional information. For example, classical histology coupled with image analysis allowed quantifying the changes observed at the cellular level during a cryopreservation protocol (Barraco et al. in preparation; Salma et al. 2011), while real time microscopy (RTM) allowed the dynamic study of such changes throughout the protocol (Gallard 2008; Salma et al. 2011). Using HPLC, it was possible to follow the influx and efflux of cryoprotectants in/out of explants during a cryopreservation protocol (Kim et al. 2004). The next step will be to localize the cryoprotectants within tissues and cells, possibly using marked molecules.

Efficient management of cryopreserved collections will become compulsory, in a situation where the number and size of existing cryobanks is growing steadily. Such management tools specific to cryopreserved collections have been developed, which allow calculate the probability of recovering at least one (or any other fixed number of) plant(s) from a cryobank sample using four given parameters: the percentage of plant recovery observed from a control sample, the number of propagules used for this control, the number of propagules in the cryobank sample and a chosen risk for the calculation of a confidence interval for the observed plant recovery (Dussert et al. 2003). They are routinely used in various cryobanks (Dussert et al. 2011; Keller et al. 2011; Panis, pers. comm.). Finally, as mentioned earlier in this chapter, the development of a new storage technology (in our case cryopreservation) will have an impact on the overall conservation strategy employed for the species of interest. Good case studies should be selected to study this particular aspect, among species for which cryopreservation is already routinely used.

20.4.5 Conservation and Use

Although this topic has not been specifically addressed in this book, it is something that threads through in all its chapters. The major motive for conservation of plant genetic resources is their use for present and future crop/plant improvement. Hence, all plant genetic resources conservation methods/approaches that we use should promote the use of conserved resources.

While conserving landraces on-farm, the major consideration is the benefit that the farmers derive from such an endeavour – no benefit, no conservation. Conservation of crop wild relatives in a fenced off area with 'no go' signboards is of little value. One needs to study the extent and distribution of crop wild relatives in wild areas and should be able to move part of them into ex situ collections when needed for utilization. Conserving large numbers of accessions in genebanks with little characterization and evaluation and no access to information on such data will make those genebanks only museums, which will not be very useful to society and will slowly fade away. The genebank curators should be able to 'sell' the accessions that they conserve to realize the full potential of conservation actions. Cryopreservation should be seen as a method ensuring safe and cost-effective long-term conservation of accessions. In the genebank context, in contrast to accessions conserved using other methods, cryopreserved samples will not be used, i.e. retrieved from liquid nitrogen and regrown into fully developed plants, or only in case of particular necessity. Indeed, the whole process of producing a reproductively active tree/plant from a tiny cryopreserved explant will in most cases be costly, long and time-consuming, especially for species with long juvenility periods. International collaboration and genebank networking have been suggested in most sections of this book as these help to mobilize the resources required for crop/plant improvement from various sources.

20.5 Conclusion

In addition to all scientific and technical issues addressed in this chapter, conservation of tropical (and of any) plant species includes also a fundamental human dimension. Indeed, plant diversity plays an important role in food security and is also a key strategy for farmers around the world to guarantee their sustenance. There are a number of legally binding conventions, treaties, agreements and other non-binding international and regional initiatives that pertain to biodiversity conservation and which provide the global context for the conservation of biodiversity. It is generally agreed that no country in the world will be self sufficient in genetic resources and countries will rely on each other for access to genetic resources for agricultural development. This interdependency calls for increased collaboration among countries to fully support, participate and implement the conventions, treaties and agreements, which aim at the effective conservation and sustainable use of the biological diversity essential for food and agriculture. The Convention on Biological Diversity (UNCED 1992) is a legally binding convention that aims at the conservation of biological diversity, the sustainable use of its components and the fair and equitable sharing of benefits arising form out of the utilization of genetic resources. The tenth Conference of Parties (COP-10 [UNEP/CBD/COP/10/27 2010]) of the CBD in Nagoya, Japan adopted the new *CBD Strategic Plan for the post 2010 period*, including a revised biodiversity target for 2020 and a revised *Global Strategy on Plant Conservation* (UNEP/CBD/COP/10/27 2010). More importantly it adopted a *Protocol on Access and Benefit-sharing*. Another legally binding agreement is the International Treaty on Plant Genetic Resources for Food and Agriculture (PGRFA) adopted by the FAO conference in 2001 (FAO 2009). The Treaty sets out general provision for the ex situ and in situ conservation and use of PGRFA and recognizes Farmers' rights. The core of the Treaty is that it provides a multilateral system which is a virtual, international pool of PGRFA from 35 crops and 29 forage species (commonly referred to as Annex 1 crops to the Treaty) agreed by parties to the Treaty to provide facilitated access to these crop genepools. The non-binding rolling Global Plan of Action for Plant Genetic Resources for Food and Agriculture (GPA [FAO 1996b]) for the conservation and sustainable utilization of plant genetic resources for food and agriculture, first adopted in 1996, and updated in 2011 (GPA2), is one of the main pillars of the FAO Global system for the conservation and sustainable use of PGRFA. These instruments provide an important framework for international cooperation for use and conservation of PGRFA. Lastly, it is important to remember that the most of the plant genetic resources conservation efforts should lead to their utilization in crop/plant improvement and benefit all levels of human society, particularly the poorer ones.

Acknowledgements The authors of this chapter duly acknowledge the valuable inputs provided by the authors of all other chapters of this publication, without whose assistance writing this concluding and prospective section would have been a much more difficult exercise.

References

Arora RK, Ramanatha Rao V (1995) Proceedings of an expert consultation on tropical fruit species of Asia, held at the Malaysian Agricultural Research and Development Institute, Serdang, 17–19 May 1994. IPGRI APO Regional Office, Serdang

Barraco G, Sylvestre I, Collin M, Escoute J, Lartaud M, Verdeil JL, Engelmann F. Histological study of yam shoot tips during their cryopreservation using the encapsulation-dehydration technique (in preparation)

Berjak P, Pammenter N (2007) From *Avicennia* to *Zizania*: seed recalcitrance in perspective. Ann Bot 1–16. doi:10.1093/aob/mcm168

Berjak P, Sershen, Varghese B, Pammenter NW (2011) Cathodic amelioration of the adverse effects of oxidative stress accompanying procedures necessary for cryopreservation of embryonic axes of recalcitrant-seeded species. Seed Sci Res 1–17 doi:10.1017/S0960258511000110

Bioversity International (2007) Guidelines for the development of crop descriptor lists, Bioversity technical bulletin series. Bioversity International, Rome

Delahaie J, Hundertmark M, Deguercy X, Leprince O, Buitink J, Bove J (2012) Characterization of a recalcitrant legume seed towards a better understanding of dessiccation tolerance mechanisms. In: Abstracts of the 6th international workshop: desiccation tolerance and sensitivity of seeds and vegetative plant tissues, Cradle of Humankind, South Africa, 8–13 Jan 2012, p 13

Dickie JB, Ellis RH, Kraak HL, Ryder K, Tompsett PB (1990) Temperature and seed storage longevity. Ann Bot 65:197–204

Dulloo ME, Ebert AW, Dussert S, Gotor E, Astorga C, Vasquez N, Rakotomalala JJ, Rabemiafara A, Eira M, Bellachew B, Omondi C, Engelmann F, Anthony F, Watts J, Qamar Z, Snook L (2009) Cost efficiency of cryopreservation as a long term conservation method for coffee genetic resources. Crop Sci 49:2123–2138

Dumet D, Engelmann F, Chabrillange N, Duval Y (1993) Cryopreservation of oil palm (*Elaeis guineensis* Jacq.) somatic embryos involving a desiccation step. Plant Cell Rep **12**, 352–355.

Dussert S, Engelmann F (2006) New determinants of coffee (*Coffea arabica* L.) seed tolerance to liquid nitrogen exposure. Cryo Letters 27:169–178

Dussert S, Chabrillange N, Anthony F, Engelmann F, Recalt C, Hamon S (1997) Variability in storage response within a coffee (*Coffea* spp.) core collection under slow growth conditions. Plant Cell Rep 16:344–348

Dussert S, Engelmann F, Noirot M (2003) Development of probabilistic tools to assist in the establishment and management of cryopreserved plant germplasm collections. Cryo Letters 24:149–160

Dussert S, Couturon E, Bertrand B, Joët T, Engelmann F (2011) Progress in the implementation of the French coffee cryobank. In: Abstracts of the COST Action 871 "Cryopreservation of crop species in Europe" final meeting, Angers, 8–11 Feb 2011

Ellis RH, Roberts EH (1980) Improved equations for the prediction of seed longevity. Ann Bot 45:13–30

Ellis RH, Hong TD, Roberts EH (1985a) Handbook of seed technology for genebanks volume I: principles and methodology. IBPGR, Rome

Ellis RH, Hong TD, Roberts EH (1985b) Handbook of seed technology for genebanks volume II: compendium of specific germination information and test recommendation. IBPGR, Rome

Engelmann F (1999) (ed) Management of field and in vitro germplasm collections In: Proceedings of a consultation meeting, CIAT/International Plant Genetic Resources Institute, Cali/Rome, 15–20 Jan 1996

Engelmann F (2011) Germplasm collection, storage and preservation. In: Altman A, Hazegawa PM (eds) Plant biotechnology and agriculture – prospects for the 21st century. Academic Press, Oxford, pp 255–268

Engelmann F, Engels JMM (2002) Technologies and strategies for ex situ conservation. In: Engels JHH, Rao VR, Brown AHD, Jackson MT (eds) Managing plant genetic diversity. CABI/IPGRI, Wallingford/Rome, pp 89–104

Engelmann F, Dulloo ME, Astorga C, Dussert S, Anthony F (2007) (eds) Conserving coffee genetic resources. Complementary strategies for ex situ conservation of coffee (*Coffea arabica* L.) genetic resources. A case study in CATIE, Costa Rica. Topical reviews in agricultural biodiversity. Bioversity International, Rome

Engels JM, Visser L (2003) A guide to effective management of germplasm collections, vol 6, IPGRI handbook for genebanks. IPGRI, Rome

Engels JMM, Wood D (1999) Conservation of agrobiodiversity. In: Wood D, Lenné JM (eds) Agrobiodiversity: characterisation, utilisation and management. CAB International, Wallingford, pp 355–386

FAO and IPGRI (1994) Genebank standards. Food and Agriculture Organization of the United Nations, Rome, International Plant Genetic Resources Institute, Rome

FAO (1996a) Report on the state of the world's plant genetic resources for food and agriculture. Food and Agriculture Organization of the United Nations, Rome

FAO (1996b) Global plan of action for the conservation and sustainable utilization of plant genetic resources for food and agriculture. Food and Agriculture Organization of the United Nations, Rome

FAO (2009) International treaty on plant genetic resources for food and agriculture. Food and Agriculture Organization of the United Nations, Rome

FAO (2010) The second report on the state of the world's plant genetic resources for food and agriculture. Food and Agriculture Organization of the United Nations, Rome

Gallard A (2008) Etude de la cryoconservation d'apex en vue dune conservation à long terme de collections de ressources génétiques végétales: compréhension des phénomènes mis en jeu et évaluation de la qualité du matériel régénéré sur le modèle *Pelargonium*. Ph.D. thesis, University of Angers, France

Gotor E, Alercia A, Ramanatha Rao V, Watts J, Caracciolo F (2008) The scientific information activity of bioversity international: the descriptor lists. Genet Resour Crop Evol 55:757–772

Gotor E, Alercia A, Ramanatha Rao V, Watts J, Caracciolo F (2011) Scientific information activity of Bioversity International: the descriptor lists. Impact Assessment Discussion Paper 3, Bioversity International, Rome.

Guarino L, Ramanatha Rao V, Goldberg E (2011) (eds) Collecting plant genetic diversity: technical guidelines – 2011 update. Bioversity International, Rome. Available online: http://cropgenebank.sgrp.cgiar.org/index.php?option=com_content&view=article&id=390&Itemid=557. Accessed 1 Nov 2011

Hamilton KN, Ashmore SE, Pritchard HW (2009) Thermal analysis and cryopreservation of seeds of Australian wild *Citrus* species (Rutaceae): *Citrus australasica, C. inodora and C. garrawayi*. Cryo Letters 30:268–279

Heywood VH, Dulloo ME (2005) In situ conservation of wild plant species: a critical global review of best practices, vol 11, IPGRI technical bulletin. IPGRI, Rome

Holderegger R, Buehler D, Gugerli F, Manel S (2010) Landscape genetics of plants. Trends Plant Sci 15:675–683

Hong TD, Ellis R (1996) A protocol to determine seed storage behaviour, vol 1, IPGRI technical bulletin. IPGRI, Rome

Hunter D, Heywood V (2011) Crop wild relatives: a manual of in situ conservation. Earthscan, London

Hunter D, Guarino L, Khoury C, Dempewolf H (2012) A community divided: lessons from the conservation of crop wild relatives around the world. In: Maxted N, Dulloo EM, Ford-Lloyd BV, Frese L, Iriondo JM, Pinheiro de Carvalho MAA (eds) Agrobiodiversity conservation: securing the diversity of crop wild relatives and landraces. CABI, Wallingford, pp 298–304

Iriondo JM, Maxted N, Dulloo ME (2008) Conserving plant genetic diversity in protected areas. CABI, Wallingford

Joët T, Dussert S (2012) Transcriptome analysis of developing coffee seeds: towards deciphering the intermediate seed storage behaviour. In: Abstracts of the 6th international workshop: desiccation tolerance and sensitivity of seeds and vegetative plant tissues, Cradle of Humankind, South Africa, 8–13 Jan 2012, p 21

Keller ERJ, Senula A, Zanke C, Grübe M, Kaczmarczyk A (2011) Cryopreservation and in vitro culture – state of the art as conservation strategy for genebanks. Acta Hort 919:99–111

Kim HH, Kim JB, Baek HJ, Cho EG, Chae YA, Engelmann F (2004) Evolution of DMSO concentration in garlic shoot tips during a citrification procedure. CryoLetters 25:91–100

Lane A, Jarvis A (2007) Changes in climate will modify the geography of crop suitability: agricultural biodiversity can help with adaptation. J Semi-arid Trop Agri Res 4:(1). Available online at http://www.icrisat.org/Journal/specialproject.htm Accessed 1 Nov 2011

Li DZ, Pritchard HW (2009) The science and economics of ex situ plant conservation. Trends Plant Sci 14:614–621

Maia J, Bas J, Dekkers W, Provart NJ, Ligterink W, Hilhors HWM (2011) The re-establishment of desiccation tolerance in germinated Arabidopsis thaliana seeds and its associated transcriptome. PLoS ONE 6:e29123. doi:10.1371/journal.pone.0029123

Manel S, Schwartz MK, Luikart G, Taberlet P (2003) Landscape genetics: combining landscape ecology and population genetics. Trends Ecol Evol 18:189–197

Maxted N, Ford-Lloyd BV, Hawkes JG (1997a) Complementary conservation strategies. In: Maxted N, Ford-Lloyd BV, Hawkes JG (eds) Plant genetic resources conservation. Chapman and Hall, London, pp 15–39

Maxted N, Ford-Lloyd BV, Hawkes JG (1997b) Plant genetic conservation: the in situ approach. Chapman and Hall, London

Maxted N, Dulloo ME, Ford-Lloyd BV, Frese L, Iriondo JM, Pinheiro de Carvalho MAA (2012) Agrobiodiversity conservation: securing the diversity of crop wild relatives and landraces. CABI, Wallingford

Nadarajan J, Mansor M, Krishnapillay B, Staines HJ, Benson EE, Harding K (2008) Applications of differential scanning calorimetry in developing cryopreservation strategies for *Parkia speciosa*, a tropical tree producing recalcitrant seeds. CryoLetters 29:95–110

Hue NTN, Van Tuyen T, Canh NT, Van Hien P, Van Chuong P (2005) In situ conservation of agricultural biodiversity on-farm: lessons learned and policy implications. In: Sthapit BR, Jarvis D (eds) Proceedings of the Vietnamese national workshop, 30 Mar–1 Apr 2004, Hanoi. International Plant Genetic Resources Institute, Rome

Ouédraogo AS, Engels JMM, Kraak L, Engelmann F (1996) Meeting the challenge of conserving tropical tree species with recalcitrant and intermediate seeds. In: Ouédraogo AS, Poulsen K, Stubsgaard F (eds) Intermediate/recalcitrant tropical forest tree seeds, Proceedings of the workshop on improved methods for handling and storage of intermediate/recalcitrant forest tree seeds, Humlebaek, 8–10 Jun 1995, IPGRI/Rome et DANIDA Forest Seed Centre, Humlebaek, pp 3–8

Ouertani W, Bonnet P, Crucianu M, Boujemaa N, Barthélémy D (2010) Iterative search with local visual features for computer assisted plant identification. In: Proceedings of the bioIdentify 2010. "Tools for identifying biodiversity, progress and problems", Museum of Natural History, Paris, 20–22 Sept 2010

Pence VC, Engelmann F (2011) Collecting in vitro for genetic resources conservation. In: Guarino L, Ramanatha Rao V, Goldberg E (eds) Chap 24, Collecting plant genetic diversity: technical guidelines. 2011 update. Bioversity International, Rome. http://cropgenebank.sgrp.cgiar.org/index.php?option=com_content&view=article&id=661. Accessed 1 Nov 2011

Pence VC, Sandoval J, Villalobos V, Engelmann F (eds) (2002) In vitro collecting techniques for germplasm conservation, vol 7, IPGRI technical bulletin. IPGRI, Rome

Perez-Garcia F, Gonzalez-Benito ME, Gomez-Campo C (2007) High viability recorded in ultra-dry seeds of 37 species of Brassicaceae after almost 40 years of storage. Seed Sci Technol 35:143–155

Plant Germplasm Conservation Research (2009) University of Kwazulu-Natal (Westville Campus), Durban. http://plantgermcons.ukzn.ac.za. Accessed 1 Nov 2011

Pritchard HW, Wood CB, Hodges S, Vautier HJ (2004) 100-seed test for dessiccation tolérance and germination: a case study on eight tropical palm species. Seed Sci Technol 32:393–403

Pritchard HW, Ashmore S, Berjak P, Engelmann F, González-Benito ME, Li DZ, Nadarajan J, Panis B, Pence V, Walters C (2009) Storage stability and the biophysics of preservation. In: Proceedings of the plant conservation for the next decade: a celebration of Kew's 250th anniversary, 12–16 Oct 2009, Royal Botanic Garden Kew, London

Probert RJ, Daws MI, Hay F (2009) Ecological correlates of ex situ seed longevity: a comparative study on 195 species. Ann Bot 104:57–69

Ramanatha Rao V, Riley KW, Engels JMM, Engelmann F, Diekmann M (1998) Towards a coconut conservation strategy. In: Ramanatha Rao V, Batugal P (eds) Proceedings of the COGENT regional coconut genebank planning workshop, Pekanbaru, 26–29 Feb 1996. IPGRI-APO, Serdang, pp 4–20

Rao NK, Hanson J, Dulloo ME, Ghosh K, Nowell D, Larinde M (2006) Manual of seed handling in genebanks, vol 8, Handbooks for genebanks. Bioversity International, Rome

Reed BM, Engelmann F, Dulloo E, Engels JMM (2004) Technical guidelines for the management of field and in vitro germplasm collections. IPGRI/FAO/SGRP, Rome

Roberts EH, Ellis RH (1989) Water and seed survival. Ann Bot 63:39–52

Saad MS, Ramanatha Rao V (2001) Establishment and management of field genebanks – a training manual. IPGRI-APO, Serdang

Sackville Hamilton NR, Chorlton KH (1997) Regeneration of accessions in seed collections: a decision guide, vol 5, Handbook for genebanks. International Plant Genetic Resources Institute, Rome

Salma M, Sylvestre I, Collin M, Escoute J, Lartaud M, Verdeil JL, Engelmann F (2011) Effect of the successive steps of a cryopreservation protocol on structural integrity of *Rubia akane* hairy roots. In: Abstracts.Society for Low Temperature Biology, Autumn meeting, London, 6–7 Oct 2011

Scheldeman X, van Zonneveld M (2010) Training manual on spatial analysis of plant diversity and distribution. Bioversity International, Rome

Secretariat of the Convention on Biological Diversity (2010) Global biodiversity outlook 3. Secretariat of the Convention on Biological Diversity, Montréal

Spooner D, van Treuren R, de Vicente MC (2005) Molecular markers for genebank management, vol 10, IPGRI technical bulletin. International Plant Genetic Resources Institute, Rome

Sthapit BR, Ramanatha Rao V (2009) Consolidating community's role in local crop development by promoting farmer innovation to maximise the use of local crop diversity for the well-being of people. Acta Hort 806:669–676

Sthapit B, Padulosi S, Bhag Mal (2009) Role of on-farm/in situ conservation and underutilized crops in the wake of climate change. Paper presented in the national symposium on recent global developments in the management of plant genetic resources, NBPGR, New Delhi, 17–18 Dec 2009

UNCED (1992) Convention on biological diversity. United Nations Conference on Environment and Development, Geneva

UNEP/CBD/COP/10/27 (2010) Report of the tenth meeting of the conference of the parties to the convention on biological diversity, Nagoya, 18–29 Oct 2010

van Zonneveld M, Scheldeman X, Escribano P, Viruel MA, Van Damme P, Garcia W, Tapia C, Romero J, Siguenas JI (2012) Mapping genetic diversity of cherimoya (*Annona cherimola* Mill.): application of spatial analysis for conservation and use of plant genetic resources. PLoS ONE 7:e29845. doi:doi:10.1371/journal.pone.0029845

Vertucci and Roos (1993) Theoretical basis of protocols for seed storage II. The influence of temperature on optimal moisture levels. Seed Sci Res 3:201–213

Volk G, Henk A, Gross B (2011) Application of functional genomics to plant cryopreservation. Cryobiology 63:319

Walters C, Wheeler LJ, Stanwood PC (2004) Longevity of cryogenically stored seeds. Cryobiology 48:229–244

Walters C, Wheeler LJ, Grotenhuis JM (2005) Longevity of seeds stored in a genebank: species characteristics. Seed Sci Res 15:1–20

Walters C, Volk GM, Stanwood PC, Towill LE, Koster KL, Forsline PL (2011) Long-term survival of cryopreserved germplasm: contributing factors and assessments from thirty year old experiments. Acta Hort 908:113–120

Wesley-Smith J, Vertucci CW, Berjak P, Pammenter NW, Crane J (1992) Cryopreservation of desiccation-sensitive axes of *Camellia sinensis* in relation to dehydration, freezing rate and the thermal properties of tissue water. J Plant Physiol 140:596–606

Index

M.N. Normah et al. (eds.), *Conservation of Tropical Plant Species*,
DOI 10.1007/978-1-4614-3776-5,

D

E

F

J

K

L

M

MIX
Papier aus verantwortungsvollen Quellen
Paper from responsible sources
FSC® C105338

If you have any concerns about our products,
you can contact us on
ProductSafety@springernature.com

In case Publisher is established outside the EU,
the EU authorized representative is:
Springer Nature Customer Service Center GmbH
Europaplatz 3, 69115 Heidelberg, Germany

Printed by Libri Plureos GmbH
in Hamburg, Germany